Isolation and Analysis of Characteristic Compounds from Herbal and Plant Extracts

Isolation and Analysis of Characteristic Compounds from Herbal and Plant Extracts

Editors

Jong Seong Kang
Narendra Singh Yadav

MDPI • Basel • Beijing • Wuhan • Barcelona • Belgrade • Manchester • Tokyo • Cluj • Tianjin

Editors

Jong Seong Kang
College of Pharmacy
Chungnam National University
Yooseong-Ku, Daejeon
Korea

Narendra Singh Yadav
Department of Biological
Sciences
The University of Lethbridge
Lethbridge
Canada

Editorial Office
MDPI
St. Alban-Anlage 66
4052 Basel, Switzerland

This is a reprint of articles from the Special Issue published online in the open access journal *Plants* (ISSN 2223-7747) (available at: www.mdpi.com/journal/plants/special_issues/compounds_herbal_plant).

For citation purposes, cite each article independently as indicated on the article page online and as indicated below:

LastName, A.A.; LastName, B.B.; LastName, C.C. Article Title. *Journal Name* **Year**, *Volume Number*, Page Range.

ISBN 978-3-0365-3279-0 (Hbk)
ISBN 978-3-0365-3278-3 (PDF)

Contents

About the Editors

Jong Seong Kang

Jong Seong Kang is a professor at the College of Pharmacy, Chungnam National University in Korea. He graduated in Pharmaceutical Sciences and completed his Ph.D. in Pharmaceutical Analysis at the Julius Maximilian University of Würzburg in Germany. He has been visiting scientist at the National Institute on Alcohol Abuse and Alcoholism at NIH, USA and exchange professor at the University of Pennsylvania, USA. Dr. Kang has published in several peer-reviewed journals, book chapters, and conference papers. His research interests are mainly focused on the quality standardization in the development of herbal drugs, and the specification and test method in the new drug development.

Narendra Singh Yadav

Narendra Singh Yadav is a Researcher at the University of Lethbridge, Canada. He received his Ph.D. (2013) in Plant Biotechnology from the CSIR-Central Salt and Marine Chemicals Research Institute in Gujarat, India. He has also been a Researcher at Jacob Blaustein Institutes for Desert Research, Ben-Gurion University of the Negev, Israel. His research is mainly focused on the 'genetics and epigenetics plant stress response.' Over his career so far, he has published more than 50 articles, book chapters, and conference abstracts/ papers, reviewed more than 50 manuscripts, and served as handling editor for 25 manuscripts.

Preface to "Isolation and Analysis of Characteristic Compounds from Herbal and Plant Extracts"

Herbal and plant extracts show diverse activities and have been used for centuries as natural medicines for many health problems and diseases. Through the isolation and analysis of the compounds in the extracts, it is possible to understand why the extracts exhibit those activities, as well as the chemical metabolism of compounds that occur in plants and herbs. Recently, there have been increasing attempts to develop herbal and plant extracts into functional foods and drugs, but the legal requirements are becoming stricter. We need sophisticatedly defined extracts through the isolation and analysis of compounds comprising them in order to meet the legal requirements and to pursue quality control strategies in the production of functional foods and drugs. This Special Issue Book compiled the 15 recent research and review articles that highlight the isolation, profiling, and analysis of compounds in herbal and plant extracts, as well as quality control and standardized processing strategies for extracts with characteristic compounds. We would like to thank all authors that contributed to this Special Issue Book and the editorial office, especially Ms. Hinata Fang, Section Managing Editor, for their helpful support during the compilation of this Book.

Jong Seong Kang, Narendra Singh Yadav
Editors

Editorial

Special Issue Editorial: Isolation and Analysis of Characteristic Compounds from Herbal and Plant Extracts

Jong-Seong Kang [1,*] and Narendra Singh Yadav [2,*]

1 College of Pharmacy, Chungnam National University, Daejeon 34134, Korea
2 Department of Biological Sciences, University of Lethbridge, Lethbridge, AB T1K 3M4, Canada
* Correspondence: kangjss@cnu.ac.kr (J.-S.K.); nsyadava2004@gmail.com (N.S.Y.)

Citation: Kang, J.-S.; Yadav, N.S. Special Issue Editorial: Isolation and Analysis of Characteristic Compounds from Herbal and Plant Extracts. *Plants* **2021**, *10*, 2775. https://doi.org/10.3390/plants10122775

Received: 6 December 2021
Accepted: 7 December 2021
Published: 15 December 2021

Publisher's Note: MDPI stays neutral with regard to jurisdictional claims in published maps and institutional affiliations.

Herbal and plant extracts exhibit various types of properties and activities that have been applied in the medicinal field to treat diseases and achieve better health. By isolating and characterizing the compounds in extracts, it helps to elucidate the pharmacological activities from the extract and explains the biochemical metabolism of compounds inside plants and herbs. In this Special Issue, thirteen research articles [1–13] and two reviews [14,15] focusing on isolation and analysis of compounds from plant extracts have been published. This Special Issue of *Plants* covers a broad range of recent findings of isolation, profiling, analysis of compounds from herbal and plant extracts, quality control, and standardization of processing extracts with characterized compounds.

The root from *Polygala tenuifolia* used in Asia is known to exhibit anti-inflammatory effects, which is used to treat asthma, bronchitis, and whooping cough. Vinh et al. isolated 15 compounds from methanolic extracts of *P. tenuifolia* root to discover new anti-inflammatory agents. As a result, they suggested potential candidates for the treatment of inflammation [1].

The dried leaves of *Istia indigotica* is well known for its anti-oxidative and anti-inflammatory effects. With those pharmacological activities, Folium isatidis showed superior anti-wrinkle effects compared to other herb and plants. Gao et al. analyzed antiwrinkle effect of extract and found an enriched fraction showing excellent anti-wrinkle effect. They identified 3,4,5-trimethoxycinnamic acid (TMCA) as a potential anti-wrinkle agent. Additionally, they developed analytical method of TMCA from Folium isatidis, which could be applied for quality control [2].

Cardiovascular diseases are disorders of heart and blood vessels including heart attack, hypertensive heart disease, and heart failure. In order to develop cardioprotective compounds from herb, Adhikari et al. investigated vasorelaxant effects from rhizomes of *Boesenbergia rotunda*. The methanol extracts from *B. rotunda* showed vasorelaxation effects and additional studies suggested that vasorelaxation effects are induced by relaxation in the coronary artery rings. Furthermore, a chromatographic study suggested that vasorelaxant effects are from flavonoids compound [3].

Oxidative stress could induce imbalance of free radicals and antioxidants, which could damage the human body's cell, protein, and DNA. Tanruean et al. investigated antioxidant and biological activities of three Clauseneae plants: *Clausena excavate*, *C. harmandiana*, and *Murray koenigii*. Chemical properties of these plants were also evaluated regarding total phenolic and total flavonoids contents and essential oils. Together, this study suggested the extracts of three Clauseneae plants as potential candidate for natural bioactive agents [4].

Avocado oil has been reported to improve sensorineural hearing loss (SNHL). In order to identify the compound that improves SNHL, Park et al. isolated 20 compounds from avocado oil using column chromatography. Improvement of SNHL from the isolated compounds were evaluated using the ototoxicity zebrafish model. Among 20 isolated compounds, seven compounds showed significant improvement of SNHL. This studied confirmed that compounds from avocado oil have protective effect on damaged otic hair cells [5].

The fruit of *Schisandra chinenesis* (Omija) is well known for several biological activities from dibenzocyclooctadiene lignans. Park et al. compared the content of seven bioactive lignans from stem, leaf, pulp, seed, and flower of *S. chinensis* using HPLC-DAD. Additionally, contents of lignan in fermented beverages of *S. chinensis* were compared with different periods and different sugar composition. As a result, contents of lignan were the highest in seed, followed by fruit, flower, leaf, pup, and stem. Lignan content increased proportionally to fermentation period, and the Omija beverage fermented with oligosaccharide/white sugar had the highest contents of lignan compared to beverages fermented with other sugars [6].

Tepals of saffron (*Crocus sativus*) is known for its abundant bio-residues that could exhibit various bioactivities. Specifically, phenolic compounds and anthocyanins have been detected from saffron with antioxidant properties. Stelluti et al. compared extraction method of dried saffron tepals by comparing phytochemical composition. Collectively, content of phenolic compound from ultrasound assisted extraction method with safer solvent were comparable to the maceration method, which refers to the potential application of green extraction method for obtaining high yields of antioxidant agents [7].

Sarcoma is a malignant tumor originating from mesenchymal cells. Fibrosarcoma is malignant neoplasm that takes place in fibroblast. Proteases are involved in various biological processes, including cell cycle progression, cell adhesion, proliferation, and migration, which facilitates tumor progression. Im et al. evaluated plant-derived protease inhibitors to discover potential agents for cancer prevention. Several protease inhibitors were derived from seeds of *Bauhinia bauhiniodes* and *Enterolobium contortisiliquum*. The effect of these protease inhibitors was investigated on the mouse fibrosarcoma regarding cell viability, cell cycle, and cell adhesion. Collectively, with the positive outcome with effect of protease inhibitor, further studies including cell signaling and in vivo study will be needed for comprehensive understanding [8].

A fruit of Sea buckthorn (*Hippophae rhamnoides*) is well known for its nutrient composition, which is widely used. Lee et al. evaluated the phytochemical composition of the fruits of sea buckthorn to discover bioactive compounds. They isolated one malate derivative, five citrate derivatives, and one quinate derivative. The isolated compounds were analyzed by 1D and 2D nuclear magnetic resonance and high-resolution liquid chromatography-mass spectrometry (LC-MS) for structure identification. These compounds were evaluated for stimulatory effects on osteogenesis. Collectively, five compounds stimulated osteogenic differentiation, which can induce osteogenesis of mesenchymal stem cells [9].

The Dendrobium species has been used as herbal medicine for treatment of various disorders. Nam et al. investigated chemical profiles of Dendrobium in two different species, and hybrid and gamma-irradiated mutant lines were investigated with LC-MS. As a result, 17 compounds were identified from the Dendrobium, and the putative markers that contributed to the discrimination of active and inactive mutant lines were identified. Collectively, this study would be helpful for future investigations of mutation mechanisms and quality evaluation for the mutants [10].

1-Deoxynojirimycin (1-DNJ) is one of most widely used iminosugars from mulberry (*Morus* sp.pl.). This compound has been known to inhibit tumor cell metastasis and reduce fat accumulation. These pharmacological effects resulted in the wide use of mulberry leaves, which had the highest content during harvest season. Marchetti et al. investigated the influence of cultivation area and cultivars toward amount of 1-DNJ in leaves extracts using HPLC-MS. They found a comparable amount of 1-DNJ to Asian cultivations and high amount of 1-DNJ in summer and from apical leaves. Since many aspects affect the level of 1-DNJ, further study is needed for the standardization of 1-DNJ in mulberry leaves [11].

Inotodiol is well known for its wide range of pharmacological activities, which is from chaga mushroom, *Inonotus obliquus*. For further pharmacokinetic studies, development of determination method of inotodiol in biological matrix was necessary. Kim et al. developed novel bioanalytical method for the determination of inotodiol using LC-MS/MS. The method was successfully validated and applied to a pharmacokinetic study using mice model. The developed method would be further applied for the mechanistic study of inotodiol [12].

Flowers from Coreopsis have been used for ethnopharmacological purposes due to its known biological activities including anticancer, antioxidant, anti-inflammatory, and anti-diabetic effects. Kim et al. performed metabolic profiling and their inhibitory effect on dipeptidyl peptidase from five original cultivars and mutant cultivars to evaluate the effect of mutation in each species. The results from this study can help to discriminate various cultivars, which could be further processed for dietary supplements [13].

Plants from Bombacoideae are known for their medicinal properties. Das et al. reported systematic reviews to provide information of taxonomy, phytochemistry, biological activities, and application in food from the subfamily of Bombacoideae [14].

Metabolomics is comprehensive study of metabolites in organisms including plants. This approach can provide useful information related to controlling growth and the development of plants. Patel et al. reported a review article that described recent analytical instrument and techniques applied for studying metabolites in plants. Additionally, bioinformatics tools and metabolomics databases for plants are mentioned in this article. Collectively, this review article provides overall workflow and information for plants metabolomics studies [15].

Overall, the 15 contributions to this Special Issue "Isolation and Analysis of Characteristic Compounds from Herbal and Plant Extracts" discuss the biochemical activities of plant extracts and characterized compounds related to activities. These types of research studies ensure the development of new bioactive compounds that could be used as potential agents for the treatment of diseases in years to come.

Author Contributions: Writing—review and editing, J.-S.K. and N.S.Y. All authors have read and agreed to the published version of the manuscript.

Funding: This research received no external funding.

Institutional Review Board Statement: Not applicable.

Informed Consent Statement: Not applicable.

Data Availability Statement: All data included in the main text.

Acknowledgments: We would like to thank all colleagues that contributed to this Special Issue and the editorial office for their helpful support during the compilation of this Special Issue.

Conflicts of Interest: The authors declare no conflict of interest.

References

1. Vinh, L.B.; Heo, M.; Phong, N.V.; Ali, I.; Koh, Y.S.; Kim, Y.H.; Yang, S.Y. Bioactive compounds from *Polygala tenuifolia* and their inhibitory effects on lipopolysaccharide-stimulated pro-inflammatory cytokine production in bone marrow-derived dendritic cells. *Plants* **2020**, *9*, 1240. [CrossRef] [PubMed]
2. Gao, D.; Cho, C.W.; Kim, C.T.; Jeong, W.S.; Kang, J.S. Evaluation of the antiwrinkle activity of enriched isatidis folium extract and an HPLC–UV method for the quality control of its cream products. *Plants* **2020**, *9*, 1586. [CrossRef] [PubMed]
3. Adhikari, D.; Gong, D.-S.; Oh, S.H.; Sung, E.H.; Lee, S.O.; Kim, D.-W.; Oak, M.-H.; Kim, H.J. Vasorelaxant effect of *Boesenbergia rotunda* and its active ingredients on an isolated coronary artery. *Plants* **2020**, *9*, 1688. [CrossRef] [PubMed]
4. Tanruean, K.; Poolprasert, P.; Suwannarach, N.; Kumla, J.; Lumyong, S. Phytochemical analysis and evaluation of antioxidant and biological activities of extracts from three clauseneae plants in Northern Thailand. *Plants* **2021**, *10*, 117. [CrossRef] [PubMed]
5. Park, S.; Jeong, S.Y.; Nam, Y.H.; Park, J.H.; Rodriguez, I.; Shim, J.H.; Yasmin, T.; Kwak, H.J.; Oh, Y.; Oh, M.; et al. Fatty acid derivatives isolated from the oil of persea americana (Avocado) protects against neomycin-induced hair cell damage. *Plants* **2021**, *10*, 171. [CrossRef] [PubMed]
6. Park, W.S.; Koo, K.A.; Bae, J.-Y.; Kim, H.-J.; Kang, D.-M.; Kwon, J.-M.; Paek, S.-M.; Lee, M.K.; Kim, C.Y.; Ahn, M.-J. Dibenzocyclooctadiene lignans in plant parts and fermented beverages of *Schisandra chinensis*. *Plants* **2021**, *10*, 361. [CrossRef] [PubMed]
7. Stelluti, S.; Caser, M.; Demasi, S.; Scariot, V. Sustainable processing of floral bio-residues of saffron (*Crocus sativus* L.) for valuable biorefinery products. *Plants* **2021**, *10*, 523. [CrossRef] [PubMed]

8.   Yoo Im, S.; Ramalho Bonturi, C.; Miti Nakahata, A.; Ryuichi Nakaie, C.; Pott, A.; Pott, V.J.; Vilela Oliva, M.L. Differences in the inhibitory specificity distinguish the efficacy of plant protease inhibitors on mouse fibrosarcoma. *Plants* **2021**, *10*, 602. [CrossRef] [PubMed]

9.   Lee, Y.H.; Jang, H.J.; Park, K.H.; Kim, S.-H.; Kim, J.K.; Kim, J.-C.; Jang, T.S.; Kim, K.H. Phytochemical analysis of the fruits of sea buckthorn (*Hippophae rhamnoides*): Identification of organic acid derivatives. *Plants* **2021**, *10*, 860. [CrossRef] [PubMed]

10.   Nam, B.; Jang, H.-J.; Han, A.-R.; Kim, Y.-R.; Jin, C.-H.; Jung, C.-H.; Kang, K.-B.; Kim, S.-H.; Hong, M.-J.; Kim, J.-B.; et al. Chemical and biological profiles of dendrobium in two different species, their hybrid, and gamma-irradiated mutant lines of the hybrid based on LC-QToF MS and cytotoxicity analysis. *Plants* **2021**, *10*, 1376. [CrossRef] [PubMed]

11.   Marchetti, L.; Saviane, A.; Montà, A.d.; Paglia, G.; Pellati, F.; Benvenuti, S.; Bertelli, D.; Cappellozza, S. Determination of 1-deoxynojirimycin (1-DNJ) in leaves of Italian or Italy-adapted cultivars of mulberry (*Morus* sp.pl.) by HPLC-MS. *Plants* **2021**, *10*, 1553. [CrossRef] [PubMed]

12.   Kim, J.H.; Gao, D.; Cho, C.W.; Hwang, I.; Kim, H.M.; Kang, J.S. A Novel Bioanalytical Method for determination of inotodiol isolated from inonotus obliquus and its application to pharmacokinetic study. *Plants* **2021**, *10*, 1631. [CrossRef] [PubMed]

13.   Kim, B.-R.; Paudel, S.B.; Han, A.-R.; Park, J.; Kil, Y.-S.; Choi, H.; Jeon, Y.G.; Park, K.Y.; Kang, S.-Y.; Jin, C.H.; et al. Metabolite profiling and dipeptidyl peptidase IV inhibitory activity of coreopsis cultivars in different mutations. *Plants* **2021**, *10*, 1661. [CrossRef] [PubMed]

14.   Das, G.; Shin, H.-S.; Ningthoujam, S.S.; Talukdar, A.D.; Upadhyaya, H.; Tundis, R.; Das, S.K.; Patra, J.K. Systematics, phytochemistry, biological activities and health promoting effects of the plants from the subfamily bombacoideae (family Malvaceae). *Plants* **2021**, *10*, 651. [CrossRef] [PubMed]

15.   Patel, M.K.; Pandey, S.; Kumar, M.; Haque, M.I.; Pal, S.; Yadav, N.S. Plants metabolome study: Emerging tools and techniques. *Plants* **2021**, *10*, 2409. [CrossRef] [PubMed]

Review

Plants Metabolome Study: Emerging Tools and Techniques

Manish Kumar Patel [1,*,†] , Sonika Pandey [2,†] , Manoj Kumar [3] , Md Intesaful Haque [4] , Sikander Pal [5] and Narendra Singh Yadav [6,*]

1 Department of Postharvest Science of Fresh Produce, Agricultural Research Organization, Volcani Center, Rishon LeZion 7505101, Israel
2 Independent Researcher, Civil Line, Fathepur 212601, India; sonikapandey14@gmail.com
3 Institute of Plant Sciences, Agricultural Research Organization, Volcani Center, Rishon LeZion 7505101, Israel; manojbiochem16@gmail.com
4 Fruit Tree Science Department, Newe Ya'ar Research Center, Agriculture Research Organization, Volcani Center, Ramat Yishay 3009500, Israel; intesafulhaque@gmail.com
5 Plant Physiology Laboratory, Department of Botany, University of Jammu, Jammu 180006, India; sikanderpal@jammuuniversity.ac.in
6 Department of Biological Sciences, University of Lethbridge, Lethbridge, AB T1K 3M4, Canada
* Correspondence: patelm1402@gmail.com (M.K.P.); nsyadava2004@gmail.com (N.S.Y.)
† These authors have equally contributed: Manish Kumar Patel and Sonika Pandey.

Abstract: Metabolomics is now considered a wide-ranging, sensitive and practical approach to acquire useful information on the composition of a metabolite pool present in any organism, including plants. Investigating metabolomic regulation in plants is essential to understand their adaptation, acclimation and defense responses to environmental stresses through the production of numerous metabolites. Moreover, metabolomics can be easily applied for the phenotyping of plants; and thus, it has great potential to be used in genome editing programs to develop superior next-generation crops. This review describes the recent analytical tools and techniques available to study plants metabolome, along with their significance of sample preparation using targeted and non-targeted methods. Advanced analytical tools, like gas chromatography-mass spectrometry (GC-MS), liquid chromatography mass-spectroscopy (LC-MS), capillary electrophoresis-mass spectrometry (CE-MS), fourier transform ion cyclotron resonance-mass spectrometry (FTICR-MS) matrix-assisted laser desorption/ionization (MALDI), ion mobility spectrometry (IMS) and nuclear magnetic resonance (NMR) have speed up precise metabolic profiling in plants. Further, we provide a complete overview of bioinformatics tools and plant metabolome database that can be utilized to advance our knowledge to plant biology.

Keywords: analytical tools; data analysis; genetically modified crops; mass spectrometry; metabolomics databases; metabolomics software tools; omics; plant biology

Citation: Patel, M.K.; Pandey, S.; Kumar, M.; Haque, M.I.; Pal, S.; Yadav, N.S. Plants Metabolome Study: Emerging Tools and Techniques. *Plants* **2021**, *10*, 2409. https://doi.org/10.3390/plants10112409

Academic Editor: Atsushi Fukushima

Received: 7 July 2021
Accepted: 1 November 2021
Published: 8 November 2021

Publisher's Note: MDPI stays neutral with regard to jurisdictional claims in published maps and institutional affiliations.

1. Metabolomics: Plant Biology Perspective

Metabolomics is one of the fastest developing and attractive disciplines of the omics field, with huge potential and prospects in crop improvement programs. It is vital to review the abiotic/biotic stress tolerances and metabolomics-assisted breeding of crop plants [1]. Recent metabolomics platforms play a crucial role in exploring unknown regulatory networks that control plant growth and development [1]. Further innovative metabolomics application, called ecological metabolomics, deals with studying the biochemical interactions among plants across different temporal and spatial networks [2]. It describes the biochemical nature of various vital ecological phenomena, such as the effects of parasite load, the incidence of disease, and infection. It also helps to decode the potential impact of biotic and abiotic stresses on any critical biochemical process through the detection of metabolites [1]. Modern metabolomics platforms are being exploited to explain complex biological pathways and explore hidden regulatory networks controlling crop growth and health.

The performance of metabolomics study relies on its methodologies and instruments to comprehensively identify and measure each metabolite [3]. The complexity of the various metabolic characteristics and molecular abundances makes metabolomics a challenging task. Metabolomics or metabolite profiling terms are alternatively used to define three types of approaches, such as untargeted metabolomics, targeted metabolomics, and semi-targeted metabolomics [4,5]. Several integrated technologies and methodologies such as mass spectrometry (MS) based methods, including gas chromatography-mass spectrometry (GC-MS), liquid chromatography mass-spectroscopy (LC-MS), capillary electrophoresis-mass spectrometry (CE-MS), fourier transform ion cyclotron resonance-mass spectrometry (FTICR-MS) matrix-assisted laser desorption/ionization (MALDI), ion mobility spectrometry (IMS) and nuclear magnetic resonance (NMR) are used for large-scale analysis of highly complex mixtures of plant extracts [6]. In fact, these analytical methods have shown their potential in many plant species, including halophytes, medicinal plants, and food crops such as *Salicornia brachiata, Cuminum cyminum, Plantago ovata, Solanum lycopersicum, Oryza sativa, Triticum aestivum,* and *Zea mays* [7–14] (Table 1). In the last decade, a significant rise in the use of integrated metabolomics analysis methods has been reported over individual analytical platforms, as the latter does not provide holistic aspects of a plant metabolome [3].

Since the beginning of the 21st century, major developments in various 'omics' fields, such as genomics, transcriptomics, proteomics, metabolomics, and phenomics, have been seen. The various omics platforms have an endless potential to enhance the current understanding of complex biological pathways, allowing us to develop new approaches for crops improvement [15]. Metabolomics is one of the most complex approaches among other omics approaches and has received attention in agriculture science, especially for plant selections in a molecular breeding program. Therefore, metabolomics is used to acquire a vast amount of useful knowledge by accurate and high throughput peak annotation through the snapshot of the plant metabolome for the novel genes and pathways elucidation [16]. The combination of metabolomic integrated with transcriptomic analysis was successfully used to find out several possible approaches such as breeding and genome editing involved in activating metabolic pathways and gene expression [17]. Nevertheless, plant metabolomics has become an effective tool for exploring different aspects of system biology, greatly expanding our knowledge of the metabolic and signaling pathways in plant growth, development, and response to stress for improving the quality and yield of crops [18]. This review describes the plant metabolome (primary and secondary metabolites), metabolomics in genetically modified (GM) crops, including different analytical techniques, bioinformatics tools, and plant metabolome database.

1.1. Primary Metabolites

Primary metabolites are essential for plant growth and development as they are involved in various physiological and biochemical processes [15]. Primary metabolites include different classes of metabolites such as sugars, fatty acids, and amino acids, serving as vital functions such as osmolytes and osmoprotectants in plants under biotic and abiotic stresses [4,19]. Lipidomics is the comprehensive analysis of lipids in a biological system, including quantification and metabolic pathways. Alteration in lipid metabolism and composition are linked to changes in plant growth, development, and responses to a variety of environmental stressors [20]. Lipidomics can be divided into shotgun and targeted analysis. Shotgun lipidomics identifies all lipid species in a sample without prior knowledge of their composition, whereas targeted lipidomics analyzes a specific group of lipids [21]. LC-MS has been used widely in both global and targeted lipidomics [22–24]. Lipidomics is also utilized to understand better the function of genes involved in lipid metabolism in transgenic plants and manipulate complex lipid metabolism to produce long-chain fatty acids, especially omega-3 species in plants [25]. Yu et al. [26] utilized lipidomics analysis based on high-throughput and high-sensitivity mass spectrometry to characterize membrane lipid responses, which also captures a variety of oxidized lipids.

The nutritional markers α-linolenic acid and linoleic acid were detected in the leaves of *P. ovata* [10]. Linoleic acid predominated in the husk of *P. ovata*, followed by oleic acid, palmitic acid, stearic acid, and cis-11,14-eicosadienoic acid [10]. Seed fatty acid composition analysis of the *Paeonia rockii*, *P. potaninii*, and *P. lutea* revealed that α-linolenic acid was the most abundant, followed by oleic and linoleic acids [27]. According to the fatty acid content, all halophytes (non-succulent, succulent and shrubby halophytes) are high in α-linolenic acid, followed by linolenic and palmitic acid [28]. Oil and oleic acid content increased, while palmitic and linolenic acid content decreased during seed development *Jatropha curcas* [29]. The total lipid and fatty acid levels were strongly linked with the different developmental stages of the *P. ovata* fruit, according to principal component analysis (PCA), and the heat map revealed the differential fatty acid composition [9].

The highest content of threonine followed by glutamic acid, tyrosine, and aspartic acid were quantified in *Amaranthus hypochondriacus* and it is notable that amino acids, glutamic acid, and aspartic acid were among the main contributors [30]. The content of histidine, isoleucine, leucine, threonine, and lysine in leaves was considerably higher than in seeds and husks of *P. ovata* [10]. Glucose-6-phosphate, xylose, 2-piperidine carboxylic acid, monoamidomalonic acid, tryptophan, phenylalanine, histidine and carbodiimide were found to be key metabolites play a vital role in the plant metabolism of *Fritillaria thunbergii* [31]. Furthermore, the amino acid profile of *Cuminum cyminum* plants revealed that the levels of most amino acids (except asparagine) increased in plants subjected to salinity stress when compared to control plants [8]. Under salinity stress, two varieties of *Cicer arietinum* (Genesis 836 and Rupali) showed increased levels of sugar alcohols, including galactitol, erythritol, arabitol, xylitol, mannitol, and inositol, showing the importance of these metabolites in salt tolerance [32]. Nitric oxide-induced accumulation of amino acids, sugars, polyols, organic acids, and but not fatty acids and lipids in *C. arietinum* [33] (Table 1).

Table 1. Identification of key metabolites in various plant species using different analytical methods.

Plant Species	Class	Analytical Tools	Key Metabolites	Reference
Primary metabolites				
Plantago ovata	Fatty acids	GC-MS	α-linolenic acid, linoleic acid and palmitic acid	[10]
P. ovata	Fatty acids	GC-MS	Pentadecanoic acid, palmitic acid, heptadecanoic acid, stearic acid, oleic acid, linoleic acid, γ-linolenic acid and arachidic acid	[9]
Jatropha curcas	Fatty acids	GC	Oleic acid, palmitic acid and linolenic acid	[29]
Paeonia rockii, P. potaninii, and P. lutea	Fatty acids	GC-MS	α-linolenic acid, oleic acid and linoleic acid	[27]
Cicer arietinum	Fatty acids	GC-MS	Pentadecanoic acid, palmitic acid, palmitoleic acid, stearic acid, oleic acid, linoleic acid, α-linolenic acid and arachidic acid	[33]
P. ovata	Amino acids	HPLC	Isoleucine, threonine, leucine, histidine and lysine	[10]
P. ovata	Amino acids	HPLC	Aspartate, glutamine, glycine, alanine, arginine, serine, proline, isoleucine and methionine	[9]
Fritillaria thunbergii	Amino acids	GC-MS	Tryptophan, phenylalanine and histidine	[31]

Table 1. *Cont.*

Plant Species	Class	Analytical Tools	Key Metabolites	Reference
C. arietinum	Amino acids	GC-MS	L-glutamic acid, L-tryptophan, phenylalanine, glycine, serine, L-threonine, L-valine, L-ornithine and L-proline	[33]
C. arietinum	Sugars and Sugar alcohols	GC-MS	Sucrose, cellobiose, galactose, methylgalactoside, *myo*-inositol	[33]
C. arietinum	Sugar alcohols	GC-QqQ-MS	Galactitol, erythritol, arabitol, xylitol, mannitol and inositol	[32]
Secondary metabolites				
Beta vulgaris	Terpenes	HPLC-MS	Oleanolic acid, hederagenin, akebonoic acid and gypsogenin	[34]
Ocimum gratissimum	Terpenes	GC-MS	m-chavicol, t-anethole, germacrene-D, naphthalene, ledene, eucalyptol, azulene and comphore	[35]
Mentha piperita	Terpenes	GC-MS	Menthone, menthol, pulegone and menthofuran	[36]
M. arvensis	Terpenes	GLC	Menthol, isomenthone, L-methone and menthyl acetate	[37]
Achyranthes bidentata	Terpenes	HPLC	Oleanolic acid and ecdysterone	[38]
Arabidopsis thaliana	Phenolics	UHPLC-MS	Scopoletin, umbelliferone and esculetin, scopolin, skimmin and esculin	[39]
P. ovata	Phenolics	LC-MS	Luteolin, quercetagetin, syringetin, kaempferol, limocitrin, helilupolone and catechin	[10]
P. ovata	Phenolics	LC-MS	Kaempferol 3-(2″,3″-diacetylrhamnoside)-7-rhamnoside and apigenin 7-rhamnoside	[9]
P. ovata	Alkaloids	LC-MS	Lunamarine, hordatine B and pinidine	[10]
Dendrobium Snowflake 'Red Star'	Alkaloids	^{1}H and 2D NMR	Dendrobine and nobilonine	[40]

GC, gas chromatography; GC-MS, gas chromatography-mass spectrometry; GC-QqQ-MS, gas chromatography-triple quadrupole-mass spectrometry; GLC, Gas liquid chromatography; HPLC, high-performance liquid chromatography; HPLC-MS, high-performance liquid chromatography-mass spectrometry; LC-MS, liquid chromatography-mass spectrometry; ^{1}H-NMR, nuclear magnetic resonance; UHPLC-MS, ultra-high performance liquid chromatography-mass spectrometry.

1.2. Secondary Metabolites

Secondary metabolites (SMs) play a crucial role in protecting plants against various environmental stresses. It has been estimated that approximately 100,000 SMs have been reported within different plant species and are classified into multiple groups, nitrogen-containing compounds, terpenes, thiols, and phenolic compounds [41]. In *Scutellaria baicalensis*, the major flavonoids are accumulated in the roots before the full-bloom stage [42]. Two flavonoids, kaempferol 3-(2″,3″-diacetylrhamnoside)-7-rhamnoside and apigenin 7-rhamnoside were found in all developmental stages of *P. ovata* [9]. The root of *Achyranthes bidentata*, oleanolic acid and ecdysterone levels are increased during the vegetative growth than in reproductive growth [38]. Nutraceutical flavonoids; luteolin, quercetagetin, syringetin, kaempferol, limocitrin, helilupolone and catechin/epicatechin/pavetannin B2 and were identified in leaf extract, whereas alkaloids, lunamarine and hordatine B were identified in the seed extract and pinidine was detected in the husk extract [10]. The plant growth regulators gibberellic acid (GA), indole -3-acetic acid (IAA) and 6-Benzylaminopurine (BAP) show that the main terpenes (methyl chavicol and trans-anethole) and other terpenes (eucalyptol and azulene) undergo certain changes depending on the type of the treatment of plant growth regulators in *O. gratissimum* [35]. The application of growth regulators enhances the production of essential oils (menthone, menthol, pulegone, and menthofuran) in *Mentha piperita*, which is revealed to be rich in economically important terpenes [36]. The foliar application of triacontanol significantly increased the amount of active terpenes (menthol, L-methone, isomenthone, and menthyl acetate) in *Mentha arvensis* [37]. Lin et al. [43] conducted a phytochemical screening of *Pteris vittata* and identified four flavonoids: quercetin, kaempferol, kaempferol-3-O-D-glucopyranoside and rutin [43]. Scoploletin, umbelliferone and esculetin, as well as their glycosides scopolin, skimmin, and esculin were found in *Arabidopsis thaliana* [39] (Table 1).

2. Involvement of Metabolomics in Genetically Modified (GM) Crops

Metabolomic techniques are rapidly being used to analyze genetically modified organisms (GMOs), allowing for a broader and deeper understanding of composition of GMO than standard analytical methods. Metabolomics studies revealed that malic acid, sorbitol, asparagine, and gluconic acid levels increased in *O. sativa* cultivated at different time points. In addition, mannitol, sucrose, and glutamic acid had a significant increase in transgenic rice grains as compared to non-genetically modified rice [44]. Metabolic profiling was performed in *Solanum tuberosum* DREB1A transgenic lines rd29A::DREB1A (D163 and D164), a 35S::DREB1A (35S-3) line, and non-transgenic [45]. Increased levels of the glutathione metabolite, γ-aminobutyric acid (GABA), as well as accumulation of β-cyanoalanine, a byproduct of ethylene biosynthesis, were observed in the DREB1A transgenic lines [45] (Table 2).

Table 2. Identification of important metabolites in transgenic plants using different analytical tools.

Transgenic Plants	Analytical Techniques	Key Metabolites	References
Artemisia annua	GC-TOF-MS	Borneol, phytol, β-farnesene, germacrene D, artemisinic acid, dihydroartemisinic acid, and artemisinin	[46]
Lactuca sativa	NMR	Asparagine, glutamine, valine, isoleucine, α-chetoglutarate, succinate, fumarate, malate, sucrose, and fructose	[47]
Lycopersicon esculentum	GC-MS	γ-aminobutyric acid, histidine, proline, pyrrol-2-carboxylate, galactitiol/sorbitol, glycerol, maltitol, 3-phosphoglyceric acid, allantoin, homo-cystine, caffeate, gluconate, ribonate, lysine, threonine, homo-serine, tyrosine, tryptophan, leucine, arginine and valine	[48]
Nicotiana tabacum	NMR	Chlorogenic acid, 4-O-caffeoylquinic acid, malic acid, threonine, alanine, glycine, fructose, β-glucose, α-glucose, sucrose, fumaric acid and salicylic acid	[49]
N. tabacum	GC-MS	4-Aminobutanoic acid, asparagine, glutamine, glycine, leucine, phenylalanine, proline, serine, threonine, tryptophan, chlorogenic acid, quininic acid, threonic acid, citric acid, malic acid and ethanolamine	[50]
Oryza sativa	GC-MS	Glycerol-3-phosphate, citric acid, linoleic acid, oleic acid, hexadecanoic acid, 2,3-dihydroxypropyl ester, sucrose, 9-octadecenoic acid, 2,3-dihydroxypropyl ester, sucrose, mannitol and glutamic acid	[44]
O. sativa	LC-MS	Tryptophan, phytosphingosine, palmitic acid, 5-hydroxy-2-octadenoic acid 9,10,13-trihydroxyoctadec-11-enoic acid and ethanolamine	[51]
Populus	GC-MS, HPLC	Caffeoyl and feruloyl conjugates, syringyl-to-guaiacyl ratio, asparagine, glutamine, aspartic acid, γ-amino-butyric acid, 5-oxo-proline, salicylic acid-2-O-glucoside, 2, 5-dihydroxybenzoic acid-5-O-glucoside, 2-methoxyhydroquinone-1-O-glucoside, 2-methoxyhydroquinone-4-O-glucoside, salicin, gallic acid, and dihydroxybenzoic acid	[52]

Table 2. *Cont.*

Transgenic Plants	Analytical Techniques	Key Metabolites	References
Solanum tuberosum	LC-TOF-MS	Glutathione, γ-aminobutyric acid, β-cyanoalanine, 5-oxoproline, sucrose, glucose-1-phosphate, glucose-6-phosphate, fructose-6-phosphate, ethanolamine, adenosine, and guanosine	[45]
Triticum aestivum	GC-MS	Guanine and 4-hydroxycinnamic acid	[53]
T. aestivum	LC-MS	Aminoacyl-tRNA biosynthesis, phenylalanine, tyrosine, tryptophan glyoxylic, tartaric acid, oxalic acids, sucrose, galactose, mannitol, leucine, valine, glutamate, proline, pyridoxamine, glutathione, arginine, citrulline, adenosine, hypoxanthine, allantoin, and adenosine monophosphate	[54]
Zea mays	^{1}H NMR	Lactic acid, citric acid, lysine, arginine, glycine-betaine, raffinose, trehalose, galactose, and adenine	[55]

GC-MS, gas chromatography-mass spectrometry; GC-TOF-MS, gas chromatography-time of flight-mass spectrometry; HPLC, high-performance liquid chromatography; LC-MS, liquid chromatography-mass spectrometry; LC-TOF-MS, liquid chromatography-time of flight-mass spectrometry; ^{1}H-NMR, nuclear magnetic resonance.

Metabolomic profiling also demonstrated that introduction of the *cold and drought regulatory-protein encoding CORA-like gene (SbCDR)* from *S. brachiata* into tobacco could enhance salt and drought tolerance by increasing the stress related metabolites such as proline, threonine, valine, glyceric acid, fructose, 4-aminobutanoic acid, asparagine [50]. Overexpression of a native *UGPase2* gene induced several metabolites related to amino acid, phenolic glycosides such as asparagine, γ-amino-butyric acid, aspartic acid, glutamine, 5-oxo-proline, 2-methoxyhydroquinone-1-*O*-glucoside, 2-methoxyhydroquinone-4-*O*-glucoside, salicylic acid-2-*O*-glucoside, 2,5-dihydroxybenzoic acid-5-*O*-glucoside, salicin in transgenic *Populus* lines [52]. Overexpression of *GmDREB1* in *T. aestivum* substantially impacts numerous metabolic pathways involved in the biosynthesis of amino acids [54]. Tryptophan, leucine phenylalanine, valine, and tyrosine were significantly changed [54]. Some urea cycle-related metabolites, such as adenosine, arginine, allantoin, citrulline, adenosine monophosphate (AMP), hypoxanthine, and guanine, were significantly changed in the transgenic *T. aestivum* line [54]. The combination of modern analytical methodologies and bioinformatics tools in metabolomics provides extensive metabolites data that helps to confirm the significant equivalency and incidence of unanticipated alterations caused by genetic transformation (Table 2).

3. Significance of Sample Preparation in Plant Metabolites

In plant metabolomics study, plant samples are harvested, stored, metabolites extraction and quantification, followed by data interpretation. Sample preparation is a key step in plant metabolomics as it significantly changes the quantity of the metabolites. Thus, considering all the factors, harvesting and storage of plant samples should be quick as to reduce the changes of biochemical reaction in the plant cells [56]. Inappropriate handling during the sample collection is the most likely source of bias in plant metabolomic studies [57]. Sample harvesting, storage, and extract preparation should ideally follow the Metabolomics Standards Initiative (MSI) to justify plant metabolomics studies [58].

3.1. Sample Harvesting and Storage

Commonly, four major steps are involved in plant metabolomics; harvesting, storage, extraction, and sample analysis (Figure 1). Plant sample harvesting must be carried out with caution, as the metabolome of the plant is sensitive to enzymatic reactions that can degrade different metabolites. In addition, metabolites vary with the different development stages, plant age, and time of sample harvesting [6]. Mostly, 10–100 mg of plant samples are required for each biological sample in metabolomics studies. Usually, immediately after harvesting, the plant samples are snap-frozen in liquid nitrogen to prevent metabolic changes. Similarly, various storage techniques, such as freeze-drying, oven-drying, and air-drying, are essential for the processing of metabolomics [57,59].

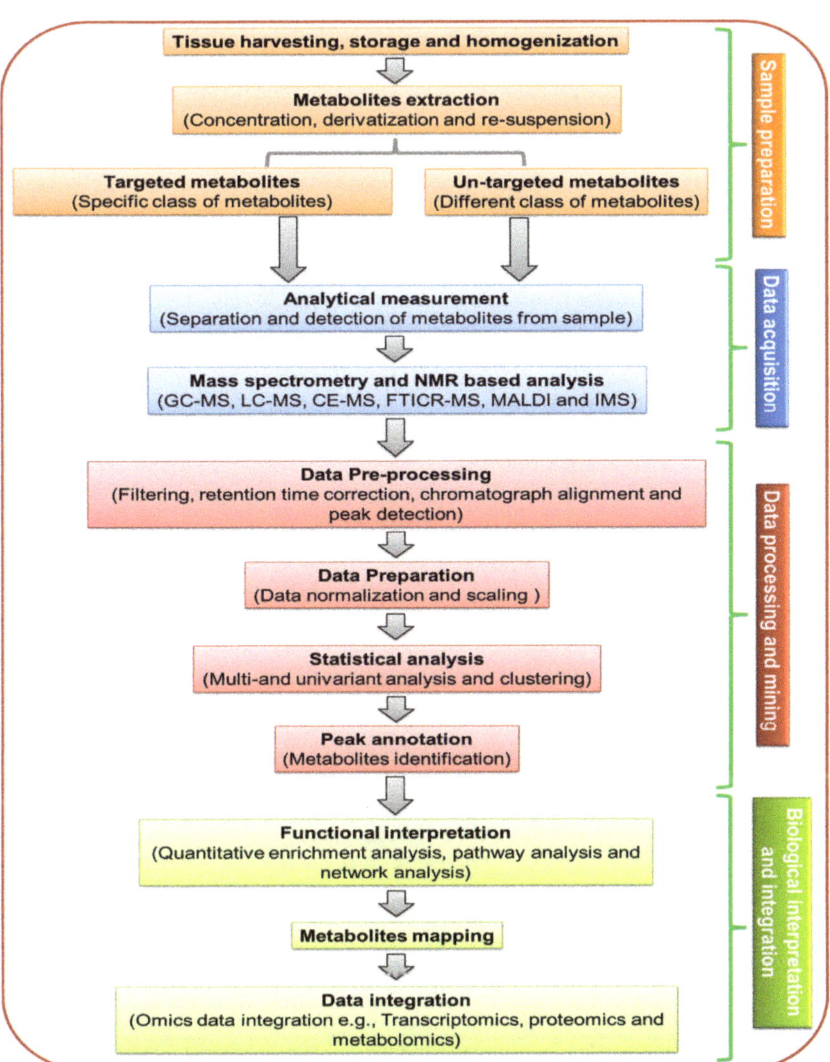

Figure 1. Schematic representation of the multi-step workflow of a plant metabolomics study. Sample preparation, data acquisition, data processing and biological interpretation are key steps in plant metabolomics. Nowadays, for data acquisition, different MS-based analytical tools (GC-MS, LC-MS CE-MS, FTICR-MS, MALDI, and IMS) and NMR are available. The most important step in data processing and mining includes correction of baseline shifts, background noise reduction, chromatograph alignment and peaks detection. Biological interpretation and integration include enrichment analysis, networks, and pathways analysis for a comprehensive scope of the metabolome. GC-MS, gas chromatography-mass spectrometry; IMS, ion mobility spectrometry; LC-MS, liquid chromatography mass-spectroscopy; CE-MS, capillary electrophoresis-mass spectrometry; FTICR-MS, fourier transform ion cyclotron resonance-mass spectrometry; MALDI, matrix-assisted laser desorption/ionization; NMR, nuclear magnetic resonance.

3.2. Sample Preparation

Sample preparation plays a key role in metabolomic study, as it includes the extraction of metabolites using different extraction methods (Figure 1). Among the extraction methods, quenching, mechanical and ultrasound extraction methods are promising in the metabolomic analysis [60]. In addition, high quality, yield and chemical versatility can be obtained by integrating ultrasound extraction method and mechanical grinding [61]. Apart from extraction methods, the choice of solvents is also crucial, as a single solvent cannot extract a variety of metabolites (e.g., polar or nonpolar). A wide variety of metabolites can be isolated using a solvent system composed of chloroform: methanol: water [62,63]. This solvent system is widely used for a wide variety of metabolites such as polar compounds, nonpolar compounds, and hydrophilic metabolites. Diverse solvent systems were reported for the plant metabolomics, such as extraction with pure methanol [64,65], the mixture of methanol: water [66], and methanol: methyl-tert-butyl-ether: water [67]. A specific solvent gradient extraction method was developed to recover almost all types of metabolites in a single protocol [68]. In addition, hot methanol (70% *v/v*) was used to extract phenolic compounds from *Brassica oleracea* using ultra-high-performance liquid chromatography–diode array detector–tandem mass spectrometry [69]. Various methods are used for sample preparation, such as microwave-assisted extraction [70], ultrasound-assisted extraction [71], Swiss rolling technique [72], and enzyme-assisted extraction [73].

Targeted metabolite identification and quantification are the primary approaches for metabolomics investigation [74]. Sample preparation for target metabolites extracted from plant components such as leaves, stems, roots, etc., includes enrichment for metabolites of interest and removal of contaminants such as proteins and salts that hamper the analysis. Targeted metabolomics-based quantification aims for enhanced metabolite coverage by analyzing the selected metabolites [75]. The targeted metabolites extracted using different extraction methods such as different proportion of organic solvents [67], liquid–liquid extraction [75], and solid phase extraction method [76]. To increase analytical reliability, single or multiple internal standards can be spiked into the sample mixture during sample preparation [77]. In the final step of sample preparation for LC-MS, the solvents were evaporated, followed by re-dissolving the sample with a suitable solvent for LC-MS analysis [75]. Targeted metabolite quantification has been considered as the key method because of its reliable quantification accuracy, sensitivity and stability [78]. However, this method is typically confined to measuring a small number of known pre-selected analysts and is incapable of detecting unknown and novel metabolites. LC-multiple reaction monitoring (MRM)-MS approach has been employed for targeted metabolomics quantification analysis due to its rapid scan speed and good analytic stability [79]. New techniques have been developed to broaden the choices for targeted metabolomics research, using high-resolution equipment such as parallel reaction monitoring (PRM) [78]. In plant metabolomics, new extraction methods are also developing day by day in line depending on the nature of the compounds and selection of analytical systems.

4. Analytical Techniques Used for Plant Metabolome

Along with sample preparation, different MS-based analytical systems are available for data acquisition. In plant metabolomics, single analytical tools cannot be used to identify all the metabolites present in a sample; instead, a set of various techniques are needed to provide the largest amount of metabolite coverage [1]. Various metabolomics tools include MS-based techniques, namely GC-MS, LC-MS CE-MS, FTICR-MS MALDI, IMS, and NMR for sensitive and specific qualitative and quantitative analyses of metabolites (Figure 1) [6,80]. All seven mentioned analytical methods identifying metabolites in plant tissue directly or indirectly have advantages and disadvantages (Table 3). Also, the combination of analytical methods can be used to ensure the efficacy of metabolite profiling.

Table 3. Advantages and disadvantages of common analytical techniques used in MS-based and NMR metabolomics.

Analytical Method	Advantage	Disadvantage
GC-MS	• Suitable for the identification of thermally stable and volatile compounds • Large commercial and public libraries • Identification of low molecular weight metabolites (~500 daltons)	• Sample pre-processing process and requires derivatization • Many metabolites are thermally unstable or unsuitable for non-volatile compounds
LC-MS	• Easy sample preparation • No derivatization • Several separation modes are available • Multiple MS detectors • Large number of detectable metabolites	• Few commercial libraries • Adduct ions are needed for metabolites detection
CE-MS	• Evaluating ionic metabolites based on the proportion of charge and size ratio • Fast and high-resolution of charged compounds • No derivatization	• Low sensitivity and reproducibility • Poor migration time and lack of reference libraries
FTICR-MS	• Mass resolving power • Mass accuracy and dynamic range	• Expensive • Lack of detection for non-ionizable compounds • Slow MS/MS
MALDI-MSI	• Quantification by peak intensities • Resolution up to 10 μm • Direct on tissue identification by tandem–MS fragmentation • Mass range up to 20 kDa	• Unsuitable for higher molecular mass compound • Expensive equipment to purchase • Time consuming • Limited by size of the metabolites
IMS	• Ion fragmentation with high versatility • Gold standard CCS values • High resolution; IMS^n (Ion mobility spectrometry)	• Low ion mobility resolution • Resolution depends on the number of passes • Mass range depends on ion mobility resolution
NMR	• Precise quantification and reproducibility • Simple steps of sample preparation • Separation is not required. • Provide detailed information about the structure of known and undiscovered metabolite • Acceptable with liquids and solids samples	• Expensive cost of instrument • Low sensitivity • Inadequate bioinformatics platform • A large amount of sample is required. • Spectral analysis is a tough and time-consuming process.

4.1. Gas Chromatography-Mass Spectrometry (GC-MS)

GC-MS is an ideal technique for the identification and quantification of small metabolites (~500 Daltons). These molecules include amino acids, fatty acids, hydroxyl acids, alcohols, sugars, sterols, and amines, which are identified mostly using chemical derivatization to make them volatile enough for gas chromatography [81]. Moreover, different methods of derivatization, such as alkylation, acylation, methoximation, trimethylsilylation, and silylation, can also be used. Two derivatization steps are required for the extraction and identification of metabolites using GC-MS. The first step requires the conversion of all the carbonyl groups using methoxyamine hydrochloride into corresponding oximes. The seconnhd step is followed by a trimethylsilylation reaction to increasing the volatility of the derivative metabolites using derivatizing reagents such as N-Methyl-N-(trimethylsilyl) trifluoroacetamide (MSTFA) and N,O-bis-(trimethylsilyl)-trifluoroacetamide (BSTFA) [82–84]. In this procedure, the hydrogen is replaced from the -NH, -SH, -OH and -COOH of specific metabolites with [-Si(CH3)3] and are converted into thermally stable, less polar and volatile trimethylsilyl (TMS)-ether, TMS-ester, TMS-amine, or TMS-sulphide groups, respectively [83]. Also, GC-MS is the preferable chromatographic technique for identifying low molecular weight compounds that are either volatile or can be converted into volatile and thermally stable metabolites by chemical derivatization prior to analysis [85]. The technique includes primary metabolites such as sugars, fatty acids, amino acids, long-chain alcohols, amines, organic acids, and sterols.

There are two major forms of ionization used in GC-MS that comprises of electron ionization (EI) and chemical ionization (CI). Till now, the majority of GC-MS methods in metabolomics utilize EI. GC with EI detector equipped with single quadrupole (Q) mass analyzer is the oldest and most advanced analytical tool with robustness, high sensitivity, resolution and reproducibility, but suffers from sluggish scanning speeds and also poor mass accuracy (~50–200 ppm). Therefore, GC with a time-of-flight mass spectrometry (TOF-MS) analyzer is more preferred for metabolic profiling as it provides higher mass accuracy, faster acquisition times, and improved deconvolution for complex mixtures [86]. Among all metabolomics techniques, GC-MS is one of the most standardized, efficient, productive technique in plant metabolomics and it is considered a most versatile platform for metabolites analysis [87]. In addition, GC-MS has the availability of the huge number of well-established libraries of both commercial and in-house metabolite databases [88–90]. Metabolite profiling is utilized as an essential tool for screening of GM crops with regard to quality and health requirements and in categorization to an investigation of potential changes in metabolic contents, e.g., *T. aestivum* [53], *O. sativa* [44], and *Z. mays* [91].

4.2. Liquid Chromatography-Mass Spectrometry (LC-MS)

LC-MS is one of the most comprehensive analytical techniques in plant metabolome research, which is used to measure a wide variety of complex metabolites. The LC-MS approach is appropriate for high molecular weight (>500 kDa) plant metabolites, heat-labile functional groups, chemically unstable functional groups, and high-vapor-point. It does not require volatilization of the metabolites. LC-MS is also quite effective techniques in profiling of SMs (e.g., alkaloids, phenolics, flavonoids and terpenes), lipids (e.g., phospholipids, sphingolipids and glycerolipids) and sterols, and steroids [19,24,92,93].

LC-MS can also be used with various ionization methods and depending on the choice of specific separating columns based on the chemical characteristics of both mobile and stationary phases [94]. Currently, reverse-phase columns such as C18 or C8 are the most widely used columns for LC gradient separation. In reverse-phase separations, organic solvent/aqueous mixed mobile phases are often used, such as water: acetonitrile or water: methanol. Atmospheric pressure ionization (API) and electron spray ionization (ESI) are the most widely used ionization tools for LC-MS [94,95]. ESI and API have provided limited structural information of the compound because they introduce less internal energy and produce only a few fragments [95]. Structural information is typically obtained by number of fragments using collision-induced dissociation (CID) on tandem MS^n. Commonly, two tandem MS^n analytical tool configurations are commonly available with the LC-MS-based metabolite analysis: tandem-in-time and tandem-in-space. The ion trap MS is used by tandem-in-time instruments, such as quadruple ion traps (QIT-MS), FTICR-MS and orbitrap. The tandem-in-space tool facilitates two sequential steps of mass spectrometric analysis (MS2); it includes two mass analyzers separated by a collision cell [96,97]. Although LC-MS requires standard reference compounds to identify and quantify SMs, this restricts the analysis of metabolites that are not commercially available [98,99].

4.3. Capillary Electrophoresis-Mass Spectrometry (CE-MS)

CE-MS is a strong analytical technique for evaluating a large variety of ionic metabolites based on the proportion of charge and size ratio [93]. It provides fast and high-resolution of charged compounds from small injection volumes and enables the metabolites characterization based on mass fragmentation [57]. The coverage of CE-MS metabolites majorly overlaps with GC-MS, but requires no derivatization, thus this technique save time and consumables. CE is performed in a fused silica capillary tube, the ends of which are dipped in buffer solutions and across which high voltages (20–30 kV) are employed [84]. Furthermore, CE has low sensitivity and reproducibility, poor migration time and lack of reference libraries; therefore, it is the least appropriate platform for studying metabolites from complex plant samples [100,101]. However, CE has some distinct rewards over other metabolomics tools; primarily the fact that it uses low volume of separation, which is especially appropriate for the study of plant metabolome [57,102].

4.4. Fourier Transform ion Cyclotron Resonance-Mass Spectrometry (FTICR-MS)

FTICR-MS provides the highest resolving power and mass accuracy among all kinds of mass spectrometry [103]. Its specific analytical features have made FTICR an important technique for proteomics and metabolomics. The ability of FTICR–MS to provide ultimate high resolution and high mass accuracy data is now frequently used as part of metabolomics procedures [84]. It's also well compatible with multi-stage mass spectrometry (MSn) analyzers. However, the instrument associated with a high magnetic field, complex ion-ion interactions and high cost are major barriers to its widespread application and use in plant metabolomics studies [56].

4.5. Matrix-Assisted Laser Desorption/Ionization (MALDI)

Recently, the applications of MALDI-Mass Spectrometry Imaging (MSI) and other MSI tools use a non-target approach for the qualitative or quantitative imaging of a broad variety of metabolites [104]. In plants, many studies have used MALDI-MSI to assess the spatial distribution of lipids, sugars and other classes of metabolites from plant parts such as flowers, leaves and roots [105,106]. In addition, MALDI-MSI has permitted the simultaneous analysis of the distribution of many peptides and proteins actively from a plant tissue section. This method involves coating a thin film of a matrix comprising either sinapinic acid, α-Cyano-4-hydroxycinnamic acid (CHCA) and 2,5-dihdroxybenzoic acid (2,5-DHBA) on the tissue surface. At each stage, a laser beam is inserted across the matrix-coated tissue to obtain a mass spectrum. For protein/metabolites imaging, MALDI is the most used method of ionization, combined with a wide variety of different mass analyzers, namely ToF, ToF-ToF, QqToF (quadrupole time of flight), Fourier ICR transform (FT-ICR), and ion-trap (both linear and spherical). All of these have their own merits and have previously been addressed and reviewed [107]. Other different ionization techniques such as secondary ion mass spectrometry (SIMS), desorption electrospray ionization (DESI) and laser ablation electrospray ionization (LAESI) have been also investigated [108].

4.6. Ion Mobility Spectrometry (IMS)

Ion mobility spectrometry (IMS), which separates gas ions based on their size-to-charge ratio, has become a robust separation method. IMS has been widely employed in a variety of research fields ranging from environmental to pharmaceutical applications [109–114]. The use of ion mobility has gained significance in bioanalysis due to the potential improvement of the sensitivity and the ability of the technique to distinguish highly related molecules based on conformational differences of molecules [115]. The IMS-derived collision cross-section indicates the effective area for the interaction between a particular ion and gas through which it travels [116]. Initially, IMS was utilized largely as a stand-alone technique; however, in recent years, the IMS coupling with MS (IMS-MS) has developed rapidly into a robust and extensively used separation technique with applications in many fields across the biological sciences, including the glycosciences [117]. IMS-MS developed quickly into a ready-to-use technique that became commercially accessible, particularly for glycan analysis [118]. The biological applications of IMS-MS for biomolecules include the analysis of oligonucleotides carbohydrates, steroid, lipids, peptides, and proteins [119–123]. Furthermore, IMS-MS may be hyphenated with front-end liquid chromatography (LC) separation to increase peak capacity and separation capabilities [123]. LC–IMS-MS technique has numerous significant benefits over other technologies in terms of increased peak capacity, isomer separation, and metabolite identification [123,124].

IMS-MS derived collision cross-section (CCS) value is high reproducible characteristic of metabolite ion, allowing for metabolite identification [125]. Therefore, the most essential aspect of metabolite identification in IMS-MS is the curation of the CCS database. Many in silico CCS databases, such as LipidCCS [126], MetCCS [127], and ISiCLE [128], have been curated and include over one million CCS values. Zhou et al. [129] developed the ion mobility new CCS atlas, namely, AllCCS for metabolite annotation using known or unknown chemical structures [129]. The AllCCS atlas included a wide range of chemical structures with >5000 experimental CCS records and ~12 million predicted CCS values for >1.6 million chemical molecules [129]. McCullagh et al. [130] used the ^{TW}CCSN$_2$ library to screen the steviol glycosides in 55 food commodities. Schroeder et al. [131] identified 146 plant natural compounds, 343 CCS values, and 29 isomers annotated (various flavonoids and isoflavonoids) in *Medicago truncatula* based on CCS, retention time, accurate mass, and molecular formula. The combination of a large-scale CCS database and different MS/MS spectra will assist in the discovery of new metabolites.

4.7. Nuclear Magnetic Resonance (NMR)

NMR is another popular analytical tool for investigating the varied metabolome in plants, involving the structure, content, and purity of molecules in the sample. As a result, metabolic profiling can provide qualitative and quantitative data from biological extracts [132]. The basic principle of NMR-based metabolite identification is to capture the radio frequency electromagnetic radiations emitted by atomic nuclei that have an odd atomic number (^{1}H) or an odd mass number (^{13}C) when placed in a strong magnetic field. Because there is no requirement for chromatographic separation or sample derivatization, the use of NMR has grown dramatically in recent years [94,133,134]. Furthermore, easy sample preparation procedures and excellent repeatability, non-destructive nature enables high throughput and quick analysis in NMR metabolomics but has less sensitivity than MS [135,136]. NMR is pH sensitive, buffered solutions are usually needed to keep the pH stable. A combination of methanol and aqueous phosphate buffer (pH 6.0, 1:1 v/v) or ionic liquids such as 1-butyl-3-methylimidazolium chloride has been shown to be the most effective in providing a comprehensive overview of both primary and secondary metabolites [137]. ^{1}H NMR is quick and easy, it has been the leading metabolites profiling technique, but it suffers from signal overlapping in the complex mixture of plant extracts during metabolites profiling. However, other advanced 2D NMR-based techniques includes two-dimensional (2D) ^{1}H J-resolved NMR, heteronuclear single quantum coherence spectroscopy (HSQC), heteronuclearmultiple quantum coherence (HMBC), total correlation spectroscopy (TOCSY) and nuclear overhauser effect spectroscopy (NOESY) [137]. High-resolution magic angle spinning (HRMAS)-NMR is particularly well suited for solid lyophilized tissue without the need for chemical extraction, which is essential for both MS and liquid state NMR practices [86]. The acquisition time for 2D NMR (2D J-resolved spectroscopy) is around 20 min, whereas for one-dimensional (1D) NMR it is approximately 1 min. However, due to the dispersion of the resonance peaks in a second dimension, spectral overlapping can be reduced in 2D NMR J-resolved spectroscopy to detect signals in crowded spectral regions [138]. Using advanced NMR, glycine-betaine, citric acid, trehalose and ethanol levels were higher in *Cry1Ab* gene transformed maize plants than non-transgenic maize plants showed [55]. Transgenic maize plants showed lower levels of pyruvic, isobutyric, succinic, lactic, and fumaric acids than non-transgenics [55]. During seed germination in chickpea, the exogenous uptake of glucose in presence of nitric oxide donor was quantified by using ^{1}H-NMR [33].

5. Metabolomic Data Processing, Annotation, Database and Bioinformatics Tools for Plants METABOLOME Analysis

GC-MS, LC-MS CE-MS, FTICR-MS MALDI, IMS and NMR are perhaps the most important techniques within the context of natural product discovery. Metabolomics generate a huge amount of metabolic data using wide range of analytical instruments. During the last decade, different software tools (web-based programs) have been designed for metabolomics raw data processing, data mining, data assessment, data interpretation, and statistical analysis as well as mathematical modelling of metabolomic networks (Figure 1).

5.1. Data Processing and Annotation

Several software programs are available for in silico data analysis of a large quantity of spectrum data of metabolites generated by various analytical instruments. The web-based programs were used for raw data processing, mining, and integration of metabolites. In general, acquired data is processed for the correction of baseline shifts, background noise reduction, peak detection and alignment, and finally, deconvolution of mass spectra (Figure 1, Table 4). Many bioinformatic tools are designed for the data pre-processing, including XCMS (https://xcmsonline.scripps.edu, accessed on 29 June 2021), METLIN (http://metlin.scripps.edu, accessed on 29 June 2021) AMDIS (Automated Mass Spectral Deconvolution and Identification System), MeltDB, MetaboAnalys, MetAlign, MZmine 2, and AnalyzerPro for different analytical techniques (Table 1). XCMS is an online bioinformatics platform that facilitates the direct uploading of raw data and assists the user in data processing and statistical analysis [139]. For LC-MS experiments, XCMS has been developed for programmed data transfer that has reduced data processing time and improved the effectiveness of an online system [140]. METLIN is another online database, which has been used in various studies related to plant metabolic profiling of stress response. It is useful for plant metabolic profiling of specific metabolites, and it is not time-consuming for data processing, mining, and annotation [141].

MeltDB (https://meltdb.cebitec.uni-bielefeld.de, accessed on 29 June 2021) is an important web-based platform used for data assessment, processing, and statistical analysis in plant metabolomics [142]. In addition, MetaboAnalyst online platform also includes a flexible enrichment analysis tool including some topological and visualization possibilities [143]. Global natural product social molecular networking (GNPS; http://gnps.ucsd.edu, accessed on 29 June 2021) is web-based mass spectrometry (MS/MS) for processing and annotation of metabolites [144]. GNPS assists with the identification and discovery of metabolites throughout the data, from data acquisition/analysis to post-publication [144]. Finally, the MZmine 2 is a publicly accessible data processing module that supports high-resolution spectral analysis. MZmine 2 is suitable for both targeted and non-targeted metabolomic studies, and it is well suited for processing large batches of data [145]. Various computational web-based, statistical and online bioinformatics tools are commonly used for data analysis in plant metabolomics (Table 4).

Table 4. Available/accessible bioinformatics and statistical tools for metabolite identification.

Database Name	Website (URL, Accessed on 29 June 2021)	Data Input	Major Function	Reference
ADAP	http://www.du-lab.org/software.htm/	GC/TOF-MS	Data processing	[146]
AllCSS	http://allccs.zhulab.cn/	DTIM-MS TWIM-MS	Metabolite prediction and annotation	[129]
AMDIS	http://www.amdis.net/	GC-MS	Data processing	[147]
BinBase	http://fiehnlab.ucdavis.edu/db or https://fiehnlab.ucdavis.edu/projects/binbase-setup	GC-MS	Metabolite annotation	[148]
FiehnLib	http://fiehnlab.ucdavis.edu/db or https://fiehnlab.ucdavis.edu/projects/fiehnlib	GC-qTOF-MS	Metabolic profiling	[149]
GMDB	https://jcggdb.jp/rcmg/glycodb/Ms_ResultSearch	MALDI-TOF	Metabolite annotation	[150]

Table 4. *Cont.*

Database Name	Website (URL, Accessed on 29 June 2021)	Data Input	Major Function	Reference
GNPS	https://gnps.ucsd.edu/ProteoSAFe/static/gnps-splash.jsp	GC-MS-EI LC-MS	Data processing, visualization and metabolite annotation	[144]
KEGG	http://www.genome.jp/kegg/	–	Metabolic models	[151]
KNApSAcK	http://kanaya.naist.jp/KNApSAcK/	FT/ICR-MS	Metabolite database	[152]
MarVis	http://marvis.gobics.de/	LC-MS	Metabolite annotation	[153]
MassBase	http://webs2.kazusa.or.jp/massbase/	MS	Metabolite annotation	[154]
MAVEN	https://maven.apache.org/	LC-MS	Data processing	[155]
MeltDB 2.0	https://meltdb.cebitec.uni-bielefeld.de	GC-MS & LC-MS	Data processing	[142]
MetaboAnalyst	www.metaboanalyst.ca/	GC-MS & LC-MS	Statistical analysis	[156]
Metabolome Express	https://www.metabolome-express.org	GC-MS	Data processing, visualization and statistical analysis	[157]
MetaboSearch	http://omics.georgetown.edu/metabosearch.html	MS	Data annotation	[158]
Metabox	https://github.com/kwanjeeraw/metabox	MS	Analysis workflow	[159]
MetAlign	www.metalign.nl	GC-MS & LC-MS	Data processing & Statistical analysis	[160]
metaP-server	http://metabolomics.helmholtz-muenchen.de/metap2/	LC-MS/MS	Data analysis	[161]
MetAssign	http://mzmatch.sourceforge.net/	LC-MS	Data annotation	[162]
MetFrag	https://ipb-halle.github.io/MetFrag/	MS	Metabolite annotation	[163]
MET-IDEA	http://bioinfo.noble.org/gateway/index.php?option=com_wrapper&Itemid=57	GC-MS & LC-MS	Data processing	[164]
MetiTree	http://www.metitree.nl/	MS	Data annotation	[165]
METLIN	https://metlin.scripps.edu/	LC-MS & MS/MS	Metabolite annotation	[141]
MMCD	http://mmcd.nmrfam.wisc.edu/ or https://www.g6g-softwaredirectory.com/bio/metabolomics/dbs-kbs/20670-Univ-Madison-WI-MMCD.php	MS	Metabolite annotation	[166]

Table 4. *Cont.*

Database Name	Website (URL, Accessed on 29 June 2021)	Data Input	Major Function	Reference
Molfind	http://metabolomics.pharm.uconn.edu/Software.html	HPLC/MS	Metabolite annotation	[167]
Mzcloud	https://www.mzcloud.org/	MS/MS & MSn	Metabolite annotation	[168]
MZedDB	http://maltese.dbs.aber.ac.uk:8888/hrmet/index.html	MS	Data annotation	[169]
MZmine2	http://mzmine.github.io/	LC-MS	Data processing	[145]
NIST	http://www.nist.gov/srd/nist1a.cfm or https://www.nist.gov/srd/nist-standard-reference-database-1a	GC-MS, LC-MS & MS/MS	Metabolite annotation	[170]
PRIMe	http://prime.psc.riken.jp/	GC-MS, LC-MS & CE-MS	Metabolite annotation	[171]
XCMS	https://xcmsonline.scripps.edu	GC-MS, LC-MS & MS2	Data processing	[139]

CE-MS, capillary electrophoresis-mass spectrometry; DTIM-MS, drift tube ion mobility–mass spectrometry; EI, electrospray ionization; FTICR-MS, fourier transform ion cyclotron resonance-mass spectrometry; GC-TOF-MS, gas chromatography-time of flight-mass spectrometry; GC-MS, gas chromatography-mass spectrometry; HPLC, high-performance liquid chromatography; LC-MS, liquid chromatography-mass spectrometry; MALDI-TOF, matrix-assisted laser desorption/ionization- time of flight; TWIM-MS, traveling wave ion mobility–mass spectrometry.

5.2. Network Analysis

The basic goal of pathway analysis is to combine biochemical information with collected metabolomics data to recognize metabolite patterns that match with metabolic pathways [172]. It is possible to consider metabolic pathways as groups of metabolites that share a common biological process and are related by one or more enzymatic reactions. A broad set of metabolic pathways are covered by comprehensive metabolic pathway databases, such as the KEGG database [173], MetaCyc [174], AraCyc [175] and the small molecule pathway database (SMPDB) [176] (Table 5). A number of software, such as, metabolite set enrichment analysis (MSEA), MPEA, IMPaLA, MBRole, VANTED, Metabo-Analyst, Paintomics, ProMeTra, Metscape2, and MetaMapRR can perform statistical and other metabolite enrichment analyses (Table 5). MSEA methods can be methodically distinguished into over-representation (ORA), single-sample profiling (SSP) and quantitative enrichment (QEA) analysis [177]. Metscape2 [178], which is an add-on to the common Cytoscape software [179] that allows data on metabolites, genes, and pathways to be displayed in the scope of metabolic networks. In addition, platform-independent online resources such as Paintomics [180], ProMeTra [181] and MetaMapRR [182] are also accessible.

Table 5. Database for metabolite enrichment analysis and pathway visualization.

Database	Website (URL, Accessed on 29 June 2021)	References
AraCyc	https://www.plantcyc.org/typeofpublication/aracyc	[175]
Cytoscape	http://www.cytoscape.org/	[183]
IMPaLA	http://impala.molgen.mpg.de	[184]
iPath	http://pathways.embl.de/	[185]

Table 5. *Cont.*

Database	Website (URL, Accessed on 29 June 2021)	References
KEGG	http://www.genome.jp/kegg/	[173]
MapMan	http://mapman.gabipd.org/web/guest/mapman	[186]
MBRole	http://csbg.cnb.csic.es/mbrole/	[187]
Metabolonote	http://metabolonote.kazusa.or.jp/	[188]
MetaCrop	http://metacrop.ipk-gatersleben.de	[189]
MetaCyc	http://www.metacyc.org	[174]
MetPA	http://metpa.metabolomics.ca/MetPA/	[190]
MPEA	http://ekhidna.biocenter.helsinki.fi/poxo/mpea/	[191]
MSEA	http://www.msea.ca. or http://www.metaboanalyst.ca	[177]
Pathcase	http://nashua.case.edu/PathwaysMAW/Web/	[192]
PathwayExplorer	http://genome.tugraz.at/pathwayexplorer/ pathwayexplorer_description.shtml	[193]
SMPDB	http://www.smpdb.ca	[176]
VANTED	https://immersive-nalytics.infotech.monash.edu/vanted/	[194]
WikiPathways	http://wikipathways.org	[195]

6. Conclusions

Metabolomics has achieved a prominent role in plant science research. It has wide applications ranging from investigating the stress-specific metabolites for different climatic stresses, evaluating candidate metabolic gene functions to analyzing the biological mechanism in plant cells, and dissecting the genotype-phenotype relationship in response to the various biotic and abiotic stresses. This review provides an overview of different sample collection, harvesting methods, storage, and sample preparation in the plant metabolomics experiments. Furthermore, the most widely used analytical tools in metabolomics for agriculture research viz. GC-MS, LC-MS, CE-MS, FTICR-MS, MALDI, IMS, and NMR with new development in their applications. In addition, we discussed computational software and database employed for metabolomics data processing in plant science. The integration of comprehensive bioinformatics tools with omics strategies professionally dissects novel metabolic networks for crop improvement. Metabolomics has excelled classical approach for novel metabolites discovery and simultaneously explores the complexity and enormous chemical diversity of metabolites in any crop plant. The integration of metabolomics with other "omics" technologies, e.g., genomics, transcriptomics, proteomics, can deliver novel insights into crop plants' genetic regulations in the context of their cellular function and metabolic network. The complete elucidation of physio-biochemical and molecular mechanisms underlying plant developmental and stress-responsive biology primarily depends on the comprehensive investigations using omics techniques that make metabolomics more applicable in agriculture sciences. Metabolomics has tremendous potential in plant research, as metabolites are more appropriate to the plant phenotype than DNAs, RNAs, or proteins. Therefore, studies in this field will effort on both ways, one is the systematic study of the biochemical and genetic mechanisms of metabolic variations in crop plants using both targeted and non-targeted methods; other is metabolomic platform can be used for metabolic profiling of genome-edited plants using CRISPR/Cas9 system for risk evaluation and regulatory affairs related with genetically modified crops [196]. Thus, we can say metabolomics will be able to contribute a lot to agriculture science, such as crop breeding and genome editing for crop improvement, better grain yield, and elucidating their unknown and novel metabolic pathways.

Author Contributions: Conceptualization, M.K.P. and N.S.Y.; writing—original draft preparation, M.K.P. and S.P. (Sonika Pandey); writing—review and editing, M.K.P., S.P. (Sonika Pandey), M.K., M.I.H., S.P. (Sikander Pal) and N.S.Y.; supervision, M.K.P. and N.S.Y.; project administration, M.K.P.; funding acquisition, N.S.Y. All authors have read and agreed to the published version of the manuscript.

Funding: This research received no external funding and the APC was funded by MDPI to N.S.Y.

Institutional Review Board Statement: Not applicable.

Informed Consent Statement: Not applicable.

Data Availability Statement: All data included in the main text.

Conflicts of Interest: The authors declare no conflict of interest.

References

1. Razzaq, A.; Sadia, B.; Raza, A.; Khalid Hameed, M.; Saleem, F. Metabolomics: A way forward for crop improvement. *Metabolites* **2019**, *9*, 303. [CrossRef]
2. Peters, K.; Worrich, A.; Weinhold, A.; Alka, O.; Balcke, G.; Birkemeyer, C.; Bruelheide, H.; Calf, O.W.; Dietz, S.; Dührkop, K.; et al. Current challenges in plant eco-metabolomics. *Int. J. Mol. Sci.* **2018**, *19*, 1385. [CrossRef]
3. Hong, J.; Yang, L.; Zhang, D.; Shi, J. Plant Metabolomics: An indispensable system biology tool for plant science. *Int. J. Mol. Sci.* **2016**, *17*, 767. [CrossRef] [PubMed]
4. Shulaev, V.; Cortes, D.; Miller, G.; Mittler, R. Metabolomics for plant stress response. *Physiol Plant.* **2008**, *132*, 199–208. [CrossRef] [PubMed]
5. Piasecka, A.; Kachlicki, P.; Stobiecki, M. Analytical methods for detection of plant metabolomes changes in response to biotic and abiotic stresses. *Int. J. Mol. Sci.* **2019**, *20*, 379. [CrossRef] [PubMed]
6. Patel, M.K.; Mishra, A.; Jha, B. Untargeted metabolomics of halophytes. In *Marine Omics: Principles and Applications*; Kim, S., Ed.; CRC Press: Boca Raton, FL, USA, 2016; pp. 309–325.
7. Mishra, A.; Patel, M.K.; Jha, B. Non–targeted metabolomics and scavenging activity of reactive oxygen species reveal the potential of *Salicornia brachiata* as a functional food. *J. Funct. Foods* **2015**, *13*, 21–31. [CrossRef]
8. Pandey, S.; Patel, M.K.; Mishra, A.; Jha, B. Physio-biochemical composition and untargeted metabolomics of cumin (*Cuminum cyminum* L.) make it promising functional food and help in mitigating salinity stress. *PLoS ONE* **2015**, *10*, e0144469. [CrossRef]
9. Patel, M.K.; Mishra, A.; Jaiswar, S.; Jha, B. Metabolic profiling and scavenging activities of developing circumscissile fruit of psyllium (*Plantago ovata* Forssk.) reveal variation in primary and secondary metabolites. *BMC Plant Biol.* **2020**, *20*, 116. [CrossRef]
10. Patel, M.K.; Mishra, A.; Jha, B. Non-targeted metabolite profiling and scavenging activity unveil the nutraceutical potential of psyllium (*Plantago ovata* Forsk). *Front. Plant Sci.* **2016**, *7*, 431. [CrossRef]
11. Bénard, C.; Bernillon, S.; Biais, B.; Osorio, S.; Maucourt, M.; Ballias, P.; Deborde, C.; Colombié, S.; Cabasson, C.; Jacob, D. Metabolomic profiling in tomato reveals diel compositional changes in fruit affected by source–sink relationships. *J. Exp. Bot.* **2015**, *66*, 3391–3404. [CrossRef] [PubMed]
12. Xie, Z.; Wang, C.; Zhu, S.; Wang, W.; Xu, J.; Zhao, X. Characterizing the metabolites related to rice salt tolerance with introgression lines exhibiting contrasting performances in response to saline conditions. *Plant Growth Regul.* **2020**, *92*, 157–167. [CrossRef]
13. Francki, M.G.; Hayton, S.; Gummer, J.; Rawlinson, C.; Trengove, R.D. Metabolomic profiling and genomic analysis of wheat aneuploid lines to identify genes controlling biochemical pathways in mature grain. *Plant Biotechnol. J.* **2016**, *14*, 649–660. [CrossRef] [PubMed]
14. Rao, J.; Cheng, F.; Hu, C.; Quan, S.; Lin, H.; Wang, J.; Chen, G.; Zhao, X.; Alexander, D.; Guo, L. Metabolic map of mature maize kernels. *Metabolomics* **2014**, *10*, 775–787. [CrossRef]
15. Kumar, R.; Bohra, A.; Pandey, A.K.; Pandey, M.K.; Kumar, A. Metabolomics for plant improvement: Status and prospects. *Front. Plant Sci.* **2017**, *8*, 1302. [CrossRef] [PubMed]
16. Tohge, T.; De Souza, L.P.; Fernie, A.R. Genome-enabled plant metabolomics. *J. Chromatogr. B* **2014**, *966*, 7–20. [CrossRef] [PubMed]
17. Xu, J.; Chen, Z.; Wang, F.; Jia, W.; Xu, Z. Combined transcriptomic and metabolomic analyses uncover rearranged gene expression and metabolite metabolism in tobacco during cold acclimation. *Sci. Rep.* **2020**, *10*, 1–13.
18. Hamany Djande, C.Y.; Pretorius, C.; Tugizimana, F.; Piater, L.A.; Dubery, I.A. Metabolomics: A tool for cultivar phenotyping and investigation of grain crops. *Agronomy* **2020**, *10*, 831. [CrossRef]
19. Patel, M.K.; Kumar, M.; Li, W.; Luo, Y.; Burritt, D.J.; Alkan, N.; Tran, L.-S.P. Enhancing salt tolerance of plants: From metabolic reprogramming to exogenous chemical treatments and molecular approaches. *Cells* **2020**, *9*, 2492. [CrossRef]
20. Rupasinghe, T.W.; Roessner, U. Extraction of plant lipids for LC-MS-based untargeted plant lipidomics. *Plant Metab.* **2018**, *1778*, 125–135.
21. Shulaev, V.; Chapman, K.D. Plant lipidomics at the crossroads: From technology to biology driven science. *BBA—Mol. Cell. Biol. Lipids* **2017**, *1862*, 786–791. [CrossRef]

22. Kofeler, H.C.; Fauland, A.; Rechberger, G.N.; Trötzmüller, M. Mass spectrometry based lipidomics: An overview of technological platforms. *Metabolites* **2012**, *2*, 19–38. [CrossRef] [PubMed]
23. Ni, Z.; Milic, I.; Fedorova, M. Identification of carbonylated lipids from different phospholipid classes by shotgun and LC-MS lipidomics. *Anal. Bioanal. Chem.* **2015**, *407*, 5161–5173. [CrossRef]
24. Okazaki, Y.; Kamide, Y.; Hirai, M.Y.; Saito, K. Plant lipidomics based on hydrophilic interaction chromatography coupled to ion trap time-of-flight mass spectrometry. *Metabolomics* **2013**, *9*, 121–131. [CrossRef]
25. Abbadi, A.; Domergue, F.; Bauer, J.; Napier, J.A.; Welti, R.; Zähringer, U.; Cirpus, P.; Heinz, E. Biosynthesis of very-long-chain polyunsaturated fatty acids in transgenic oilseeds: Constraints on their accumulation. *Plant Cell* **2004**, *16*, 2734–2748. [CrossRef]
26. Yu, D.; Boughton, B.A.; Hill, C.B.; Feussner, I.; Roessner, U.; Rupasinghe, T.W. Insights into oxidized lipid modification in barley roots as an adaptation mechanism to salinity stress. *Front. Plant Sci.* **2020**, *11*, 1. [CrossRef] [PubMed]
27. Zhang, Q.Y.; Yu, R.; Xie, L.H.; Rahman, M.M.; Kilaru, A.; Niu, L.X.; Zhang, Y.L. Fatty acid and associated gene expression analyses of three tree peony species reveal key genes for α-linolenic acid synthesis in seeds. *Front. Plant Sci.* **2018**, *9*, 106. [CrossRef]
28. Patel, M.K.; Pandey, S.; Brahmbhatt, H.R.; Mishra, A.; Jha, B. Lipid content and fatty acid profile of selected halophytic plants reveal a promising source of renewable energy. *Biomass Bioenergy* **2019**, *124*, 25–32. [CrossRef]
29. Sinha, P.; Islam, M.A.; Negi, M.S.; Tripathi, S.B. Changes in oil content and fatty acid composition in *Jatropha curcas* during seed development. *Ind. Crops. Prod.* **2015**, *77*, 508–510. [CrossRef]
30. Nimbalkar, M.S.; Pai, S.R.; Pawar, N.V.; Oulkar, D.; Dixit, G.B. Free amino acid profiling in grain Amaranth using LC–MS/MS. *Food Chem.* **2012**, *134*, 2565–2569. [CrossRef]
31. Cui, M.C.; Chen, S.J.; Wang, H.H.; Li, Z.H.; Chen, H.J.; Chen, Y.; Zhou, H.B.; Li, X.; Chen, J.W. Metabolic profiling investigation of *Fritillaria thunbergii* Miq. by gas chromatography–mass spectrometry. *J. Food Drug Anal.* **2018**, *26*, 337–347. [CrossRef]
32. Dias, D.A.; Hill, C.B.; Jayasinghe, N.S.; Atieno, J.; Sutton, T.; Roessner, U. Quantitative profiling of polar primary metabolites of two chickpea cultivars with contrasting responses to salinity. *J. Chromatogr. B.* **2015**, *1000*, 1–13. [CrossRef]
33. Pandey, S.; Kumari, A.; Shree, M.; Kumar, V.; Singh, P.; Bharadwaj, C.; Loake, G.J.; Parida, S.K.; Masakapalli, S.K.; Gupta, K.J. Nitric oxide accelerates germination via the regulation of respiration in chickpea. *J. Exp. Bot.* **2019**, *70*, 4539–4555. [CrossRef]
34. Mikołajczyk-Bator, K.; Błaszczyk, A.; Czyżniejewski, M.; Kachlicki, P. Characterization and identification of triterpene saponins in the roots of red beets (*Beta vulgaris* L.) using two HPLC–MS systems. *Food Chem.* **2016**, *192*, 979–990. [CrossRef] [PubMed]
35. Hazzoumi, Z.; Moustakime, Y.; Joutei, K.A. Effect of gibberellic acid (GA), indole acetic acid (IAA) and benzylaminopurine (BAP) on the synthesis of essential oils and the isomerization of methyl chavicol and trans-anethole in *Ocimum gratissimum* L. *SpringerPlus* **2014**, *3*, 321–327. [CrossRef] [PubMed]
36. Santoro, M.V.; Nievas, F.; Zygadlo, J.; Giordano, W.; Banchio, E. Effects of growth regulators on biomass and the production of secondary metabolites in peppermint (*Mentha piperita*) micropropagated in vitro. *Am. J. Plant Sci.* **2013**, *4*, 49. [CrossRef]
37. Naeem, M.; Khan, M.M.A.; Idrees, M.; Aftab, T. Triacontanol-mediated regulation of growth yield, physiological activities and active constituents of *Mentha arvensis* L. *Plant Growth Regul.* **2011**, *65*, 195–206. [CrossRef]
38. Li, J.T.; Hu, Z.H. Accumulation and dynamic trends of triterpenoid saponin in vegetative organ of *Achyranthus bidentata*. *J. Integr. Plant Biol.* **2009**, *51*, 122–129. [CrossRef]
39. Perkowska, I.; Siwinska, J.; Olry, A.; Grosjean, J.; Hehn, A.; Bourgaud, F.; Lojkowska, E.; Ihnatowicz, A. Identification and quantification of coumarins by UHPLC-MS in *Arabidopsis thaliana* natural populations. *Molecules* **2021**, *26*, 1804. [CrossRef]
40. Morita, H.; Fujiwara, M.; Yoshida, N.; Kobayashi, J. New picrotoxin-type and dendrobine-type sesquiterpenoids from *Dendrobium snowflake* 'Red Star'. *Tetrahedron* **2000**, *56*, 5801–5805. [CrossRef]
41. Zagorchev, L.; Seal, C.E.; Kranner, I.; Odjakova, M. A central role for thiols in plant tolerance to abiotic stress. *Int. J. Mol. Sci.* **2013**, *14*, 7405–7432. [CrossRef] [PubMed]
42. Xu, J.; Yu, Y.; Shi, R.; Xie, G.; Zhu, Y.; Wu, G.; Qin, M. Organ-specific metabolic shifts of flavonoids in *Scutellaria baicalensis* at different growth and development stages. *Molecules* **2018**, *23*, 428. [CrossRef]
43. Lin, L.; Huang, X.; Lv, Z. Isolation and identification of flavonoids components from *Pteris vittata* L. *SpringerPlus* **2016**, *5*, 1649. [CrossRef] [PubMed]
44. Zhou, J.; Ma, C.; Xu, H.; Yuan, K.; Lu, X.; Zhu, Z.; Wu, Y.; Xu, G. Metabolic profiling of transgenic rice with *cryIAc* and *sck* genes: An evaluation of unintended effects at metabolic level by using GC-FID and GC–MS. *J. Chromatogr. B.* **2009**, *877*, 725–732. [CrossRef]
45. Iwaki, T.; Guo, L.; Ryals, J.A.; Yasuda, S.; Shimazaki, T.; Kikuchi, A.; Watanabe, K.N.; Kasuga, M.; Yamaguchi-Shinozaki, K.; Ogawa, T.; et al. Metabolic profiling of transgenic potato tubers expressing *Arabidopsis* dehydration response element-binding protein 1A (DREB1A). *J. Agric. Food Chem.* **2013**, *61*, 893–900. [CrossRef]
46. Ma, C.; Wang, H.; Lu, X.; Wang, H.; Xu, G.; Liu, B. Terpenoid metabolic profiling analysis of transgenic *Artemisia annua* L. by comprehensive two-dimensional gas chromatography time-of-flight mass spectrometry. *Metabolomics* **2009**, *5*, 497–506. [CrossRef]
47. Sobolev, A.P.; Testone, G.; Santoro, F.; Nicolodi, C.; Iannelli, M.A.; Amato, M.E.; Ianniello, A.; Brosio, E.; Giannino, D.; Mannina, L. Quality traits of conventional and transgenic lettuce (*Lactuca sativa* L.) at harvesting by NMR metabolic profiling. *J. Agric. Food Chem.* **2010**, *58*, 6928–6936. [CrossRef]
48. Roessner-Tunali, U.; Hegemann, B.; Lytovchenko, A.; Carrari, F.; Bruedigam, C.; Granot, D.; Fernie, A.R. Metabolic profiling of transgenic tomato plants overexpressing hexokinase reveals that the influence of hexose phosphorylation diminishes during fruit development. *Plant Physiol.* **2003**, *133*, 84–99. [CrossRef] [PubMed]

49. Choi, H.K.; Choi, Y.H.; Verberne, M.; Lefeber, A.W.; Erkelens, C.; Verpoorte, R. Metabolic fingerprinting of wild type and transgenic tobacco plants by ^{1}H NMR and multivariate analysis technique. *Phytochemistry* **2004**, *65*, 857–864. [CrossRef]
50. Jha, R.K.; Patel, J.; Patel, M.K.; Mishra, A.; Jha, B. Introgression of a novel cold and drought regulatory-protein encoding CORA-like gene, *SbCDR*, induced osmotic tolerance in transgenic tobacco. *Physiol. Plant* **2021**, *172*, 1170–1188. [CrossRef]
51. Chang, Y.; Zhao, C.; Zhu, Z.; Wu, Z.; Zhou, J.; Zhao, Y.; Lu, X.; Xu, G. Metabolic profiling based on LC/MS to evaluate unintended effects of transgenic rice with *cry1Ac* and *sck* genes. *Plant Mol. Biol.* **2012**, *78*, 477–487. [CrossRef] [PubMed]
52. Payyavula, R.S.; Tschaplinski, T.J.; Jawdy, S.S.; Sykes, R.W.; Tuskan, G.A.; Kalluri, U.C. Metabolic profiling reveals altered sugar and secondary metabolism in response to UGPase overexpression in *Populus. BMC Plant Boil.* **2014**, *14*, 1–14. [CrossRef] [PubMed]
53. Stamova, B.S.; Roessner, U.; Suren, S.; Laudencia-Chingcuanco, D.; Bacic, A.; Beckles, D.M. Metabolic profiling of transgenic wheat over-expressing the high-molecular-weight Dx5 glutenin subunit. *Metabolomics* **2009**, *5*, 239–252. [CrossRef]
54. Niu, F.; Jiang, Q.; Sun, X.; Hu, Z.; Wang, L.; Zhang, H. Metabolic profiling of DREB-overexpressing transgenic wheat seeds by liquid chromatography–mass spectrometry. *Crop J.* **2020**, *8*, 1025–1036. [CrossRef]
55. Piccioni, F.; Capitani, D.; Zolla, L.; Mannina, L. NMR metabolic profiling of transgenic maize with the *Cry1A(b)* gene. *J. Agric. Food Chem.* **2020**, *57*, 6041–6049. [CrossRef]
56. Tanna, B.; Mishra, A. Metabolomics of seaweeds: Tools and techniques. In *Plant Metabolites and Regulation Under Environmental Stress*; Ahmad, P., Ahanger, M.A., Singh, V.P., Tripathi, D.K., Alam, P., Alyemeni, M.N., Eds.; Academic Press: Cambridge, MA, USA, 2018; pp. 37–52.
57. Salem, M.A.; Perez de Souza, L.; Serag, A.; Fernie, A.R.; Farag, M.A.; Ezzat, S.M.; Alseekh, S. Metabolomics in the context of plant natural products research: From sample preparation to metabolite analysis. *Metabolites* **2020**, *10*, 37. [CrossRef]
58. Sumner, L.W.; Amberg, A.; Barrett, D.; Beale, M.H.; Beger, R.; Daykin, C.A.; Fan, T.W.M.; Fiehn, O.; Goodacre, R.; Griffin, J.L.; et al. Proposed minimum reporting standards for chemical analysis Chemical Analysis Working Group (CAWG) Metabolomics Standards Initiative (MSI). *Metabolomics* **2007**, *3*, 211–221. [CrossRef] [PubMed]
59. Harbourne, N.; Marete, E.; Jacquier, J.C.; O'Riordan, D. Effect of drying methods on the phenolic constituents of meadowsweet (*Filipendula ulmaria*) and willow (*Salix alba*). *LWT—Food Sci. Technol.* **2009**, *42*, 1468–1473. [CrossRef]
60. Parida, A.K.; Panda, A.; Rangani, J. Metabolomics-guided elucidation of abiotic stress tolerance mechanisms in plants. In *Plant Metabolites and Regulation Under Environmental Stress*; Ahmad, P., Ahanger, M.A., Singh, V.P., Tripathi, D.K., Alam, P., Alyemeni, M.N., Eds.; Academic Press: Cambridge, MA, USA, 2018; pp. 89–131.
61. Gong, Z.G.; Hu, J.; Wu, X.; Xu, Y.J. The recent developments in sample preparation for mass spectrometry-based metabolomics. *Crit. Rev. Anal. Chem.* **2017**, *8347*, 1–7. [CrossRef]
62. Silva-Navas, J.; Moreno-Risueno, M.A.; Manzano, C. Flavonols mediate root phototropism and growth through regulation of proliferation-to differentiation transition. *Plant Cell* **2016**, *28*, 1372–1387. [CrossRef]
63. Corrales, A.R.; Carrillo, L.; Lasierra, P. Multifaceted role of cycling DOF factor 3 (CDF3) in the regulation of flowering time and abiotic stress responses in *Arabidopsis. Plant Cell Environ.* **2017**, *40*, 748–764. [CrossRef]
64. Sánchez-Parra, B.; Frerigmann, H.; Pérez Alonso, M.-M. Characterization of four bifunctional plant IAM/PAM-amidohydrolases capable of contributing to auxin biosynthesis. *Plants* **2014**, *3*, 324–347. [CrossRef] [PubMed]
65. Lehmann, T.; Janowitz, T.; Sánchez-Parra, B. *Arabidopsis* NITRILASE 1 contributes to the regulation of root growth and development through modulation of auxin biosynthesis in seedlings. *Front. Plant Sci.* **2017**, *8*, 36. [CrossRef] [PubMed]
66. T'Kindt, R.; Morreel, K.; Deforce, D. Joint GC–MS and LC–MS platforms for comprehensive plant metabolomics: Repeatability and sample pre-treatment. *J. Chromatogr. B* **2009**, *877*, 3572–3580. [CrossRef]
67. Giavalisco, P.; Li, Y.; Matthes, A. Elemental formula annotation of polar and lipophilic metabolites using (13) C, (15) N and (34) S isotope labelling, in combination with high-resolution mass spectrometry. *Plant J.* **2011**, *68*, 364–376. [CrossRef]
68. Yuliana, N.D.; Khatib, A.; Verpoorte, R.; Choi, Y.H. Comprehensive extraction method integrated with NMR metabolomics: A new bioactivity screening method for plants, adenosine a1 receptor binding compounds in *Orthosiphon stamineus*, Benth. *Anal. Chem.* **2011**, *83*, 6902–6906. [CrossRef]
69. Gratacós-Cubarsí, M.; Ribas-Agustí, A.; García-Regueiro, J.A.; Castellari, M. Simultaneous evaluation of intact glucosinolates and phenolic compounds by UPLC-DAD-MS/MS in *Brassica oleracea* L. var. botrytis. *Food Chem.* **2010**, *121*, 257–263. [CrossRef]
70. Teo, C.C.; Chong, W.P.K.; Ho, Y.S. Development and application of microwave-assisted extraction technique in biological sample preparation for small molecule analysis. *Metabolomics* **2013**, *9*, 1109–1128. [CrossRef]
71. Altemimi, A.; Watson, D.G.; Choudhary, R.; Dasari, M.R.; Lightfoot, D.A. Ultrasound assisted extraction of phenolic compounds from peaches and pumpkins. *PLoS ONE* **2016**, *11*, e0148758. [CrossRef]
72. Velickovic, D.; Chu, R.K.; Myers, G.L.; Ahkami, A.H.; Anderton, C.R. An approach for visualizing the spatial metabolome of an entire plant root system inspired by the swiss-rolling technique. *J. Mass Spectrom.* **2020**, *55*, 4363. [CrossRef]
73. Zuorro, A.; Lavecchia, R.; Medici, F.; Piga, L. Enzyme-assisted production of tomato seed oil enriched with lycopene from tomato pomace. *Food Bioprocess Tech.* **2013**, *6*, 3499–3509. [CrossRef]
74. Zhou, J.; Yin, Y. Strategies for large-scale targeted metabolomics quantification by liquid chromatography-mass spectrometry. *Analyst* **2016**, *141*, 6362–6373. [CrossRef]
75. Cajka, T.; Fiehn, O. Toward merging untargeted and targeted methods in mass spectrometry-based metabolomics and lipidomics. *Anal. Chem.* **2016**, *88*, 524–545. [CrossRef] [PubMed]

76. Bojko, B.; Reyes-Garcés, N.; Bessonneau, V.; Goryński, K.; Mousavi, F.; Silva, E.A.S.; Pawliszyn, J. Solid-phase microextraction in metabolomics. *Trends Analyt. Chem.* **2014**, *61*, 168–180. [CrossRef]

77. Ciccimaro, E.; Blair, I.A. Stable-isotope dilution LC–MS for quantitative biomarker analysis. *Bioanalysis* **2010**, *2*, 311–341. [CrossRef] [PubMed]

78. Zhou, J.; Liu, H.; Liu, Y.; Liu, J.; Zhao, X.; Yin, Y. Development and evaluation of a parallel reaction monitoring strategy for large-scale targeted metabolomics quantification. *Anal. Chem.* **2016**, *88*, 4478. [CrossRef] [PubMed]

79. Guo, B.; Chen, B.; Liu, A.; Zhu, W.; Yao, S. Liquid chromatography-mass spectrometric multiple reaction monitoring-based strategies for expanding targeted profiling towards quantitative metabolomics. *Curr. Drug Metab.* **2012**, *13*, 1226–1243. [CrossRef]

80. Bianchi, F.; Ilag, L.; Termopoli, V.; Mendez, L. Advances in MS-based analytical methods: Innovations and future trends. *J. Anal. Methods Chem.* **2018**, *2018*, 1–2. [CrossRef] [PubMed]

81. Fiehn, O. Metabolomics by gas chromatography–mass spectrometry: Combined targeted and untargeted profiling. *Curr. Protoc. Mol. Biol.* **2016**, *114*, 1–32. [CrossRef]

82. Kopka, J. Current challenges and developments in GC–MS based metabolite profiling technology. *J. Biotechnol.* **2006**, *124*, 312–322. [CrossRef]

83. Harvey, D.J.; Vouros, P. Mass spectrometric fragmentation of trimethylsilyl and related alkylsilyl derivatives. *Mass Spectrom. Rev.* **2020**, *39*, 105–211. [CrossRef]

84. Jorge, T.F.; Rodrigues, J.A.; Caldana, C.; Schmidt, R.; Van Dongen, J.T.; Thomas-Oates, J.; António, C. Mass spectrometry-based plant metabolomics: Metabolite responses to abiotic stress. *Mass Spectrom. Rev.* **2016**, *35*, 620–649. [CrossRef]

85. Hall, R.D. Plant metabolomics: From holistic hope, to hype, to hot topic. *New Phytol.* **2006**, *169*, 453–468. [CrossRef]

86. Kumar, M.; Kuzhiumparambil, U.; Pernice, M.; Jiang, Z.; Ralph, P.J. Metabolomics: An emerging frontier of systems biology in marine macrophytes. *Algal Res.* **2016**, *16*, 76–92. [CrossRef]

87. Tsugawa, H.; Tsujimoto, Y.; Arita, M.; Bamba, T.; Fukusaki, E.; Fiehn, O. GC/MS based metabolomics: Development of a data mining system for metabolite identification by using soft independent modeling of class analogy (simca). *BMC Bioinform.* **2011**, *12*, 131. [CrossRef] [PubMed]

88. Koek, M.M.; Jellema, R.H.; Van der Greef, J.; Tas, A.C.; Hankemeier, T. Quantitative metabolomics based on gas chromatography mass spectrometry: Status and perspectives. *Metabolomics* **2011**, *7*, 307–328. [CrossRef] [PubMed]

89. Mastrangelo, A.; Ferrarini, A.; Rey-Stolle, F.; García, A.; Barbas, C. From sample treatment to biomarker discovery: A tutorial for untargeted metabolomics based on GC-(EI)-Q-MS. *Anal. Chim. Acta* **2015**, *900*, 21–35. [CrossRef] [PubMed]

90. Beale, D.J.; Pinu, F.R.; Kouremenos, K.A.; Poojary, M.M.; Narayana, V.K.; Boughton, B.A.; Dias, D.A. Review of recent developments in GC–MS approaches to metabolomics-based research. *Metabolomics* **2018**, *14*, 1–31. [CrossRef] [PubMed]

91. Sissener, N.H.; Hemre, G.-I.; Lall, S.P.; Sagstad, A.; Petersen, K.; Williams, J.; Rohloff, J.; Sanden, M. Are apparent negative effects of feeding genetically modified MON810 maize to Atlantic salmon, Salmo salar caused by confounding factors? *Br. J. Nutr.* **2011**, *106*, 42–56. [CrossRef]

92. Matsuda, F.; Hirai, M.Y.; Sasaki, E.; Akiyama, K.; Yonekura-Sakakibara, K.; Provart, N.J.; Sakurai, T.; Shimada, Y.; Saito, K. AtMetExpress development: A phytochemical atlas of *Arabidopsis* development. *Plant Physiol.* **2010**, *152*, 566–578. [CrossRef]

93. Obata, T.; Fernie, A.R. The use of metabolomics to dissect plant responses to abiotic stresses. *Cell Mol. Life Sci.* **2012**, *69*, 3225–3243. [CrossRef]

94. Okazaki, Y.; Saito, K. Recent advances of metabolomics in plant biotechnology. *Plant Biotechnol. Rep.* **2012**, *6*, 1–15. [CrossRef] [PubMed]

95. Holčapek, M.; Jirásko, R.; Lísa, M. Recent developments in liquid chromatography–mass spectrometry and related techniques. *J. Chromatogr.* **2012**, *1259*, 3–15. [CrossRef] [PubMed]

96. Allwood, J.W.; Goodacre, R. An introduction to liquid chromatography–mass spectrometry instrumentation applied in plant metabolomic analyses. *Phytochem. Anal.* **2010**, *21*, 33–47. [CrossRef]

97. Schiffmann, C.; Hansen, R.; Baumann, S.; Kublik, A.; Nielsen, P.H.; Adrian, L.; Von Bergen, M.; Jehmlich, N.; Seifert, J. Comparison of targeted peptide quantification assays for reductive dehalogenases by selective reaction monitoring (SRM) and precursor reaction monitoring (PRM). *Anal. Bioanal. Chem.* **2014**, *406*, 283–291. [CrossRef]

98. Last, R.L.; Jones, A.D.; Shachar-Hill, Y. Towards the plant metabolome and beyond. *Nat. Rev. Mol. Cell Biol.* **2007**, *8*, 167–174. [CrossRef] [PubMed]

99. Alseekh, S.; Fernie, A.R. Metabolomics 20 years on: What have we learned and what hurdles remain? *Plant J.* **2018**, *94*, 933–942. [CrossRef] [PubMed]

100. Soga, T.; Imaizumi, M. Capillary electrophoresis method for the analysis of inorganic anions, organic acids, amino acids, nucleotides, carbohydrates and other anionic compounds. *Electrophoresis* **2001**, *22*, 3418–3425. [CrossRef]

101. Williams, B.J.; Cameron, C.J.; Workman, R.; Broeckling, C.D.; Sumner, L.W.; Smith, J.T. Amino acid profiling in plant cell cultures: An inter-laboratory comparison of CE-MS and GC-MS. *Electrophoresis* **2007**, *28*, 1371–1379. [CrossRef]

102. Ren, J.L.; Zhang, A.H.; Kong, L.; Wang, X.J. Advances in mass spectrometry-based metabolomics for investigation of metabolites. *RSC Adv.* **2018**, *8*, 22335–22350. [CrossRef]

103. Nikolaev, E.N.; Kostyukevich, Y.I.; Vladimirov, G.N. Fourier transform ion cyclotron resonance (FTICR) mass spectrometry: Theory and simulations. *Mass Spectrom. Rev.* **2016**, *35*, 219–258. [CrossRef] [PubMed]

104. Baker, T.C.; Han, J.; Borchers, C.H. Recent advancements in matrix-assisted laser desorption/ionization mass spectrometry imaging. *Curr. Opin. Biotechnol.* **2017**, *43*, 62–69. [CrossRef]
105. Cha, S.; Zhang, H.; Ilarslan, H.I.; Wurtele, E.S.; Brachova, L.; Nikolau, B.J.; Yeung, E.S. Direct profiling and imaging of plant metabolites in intact tissues by using colloidal graphite-assisted laser desorption ionization mass spectrometry. *Plant J.* **2008**, *55*, 348–360. [CrossRef] [PubMed]
106. Jun, J.H.; Song, Z.; Liu, Z.; Nikolau, B.J.; Yeung, E.S.; Lee, Y.J. High-spatial and high-mass resolution imaging of surface metabolites of *Arabidopsis thaliana* by laser desorption-ionization mass spectrometry using colloidal silver. *Anal. Chem.* **2010**, *82*, 3255–3265. [CrossRef]
107. Goodwin, R.J.; Pennington, S.R.; Pitt, A.R. Protein and peptides in pictures: Imaging with MALDI mass spectrometry. *Proteomics* **2008**, *8*, 3785–3800. [CrossRef] [PubMed]
108. Lee, Y.J.; Perdian, D.C.; Song, Z.; Yeung, E.S.; Nikolau, B.J. Use of mass spectrometry for imaging metabolites in plants. *Plant J.* **2012**, *70*, 81–95. [CrossRef]
109. Kathirvel, S.; Gayatri Ramya, M.; Rajesh, A. An overview on the benefits and applications of high performance ion mobility spectrometer in pharmaceutical arena-focus on current research. *World J. Pharm. Pharm. Sci.* **2017**, *6*, 402–406.
110. Hernandez-Mesa, M.; Escourrou, A.; Monteau, F.; Le Bizec, B.; Dervilly-Pinel, G. Current applications and perspectives of ion mobility spectrometry to answer chemical food safety issues. *Trends Anal. Chem.* **2017**, *94*, 39–53. [CrossRef]
111. Campuzano, I.D.G.; Lippens, J.L. Ion mobility in the pharmaceutical industry: An established biophysical technique or still niche? *Curr. Opin. Chem. Biol.* **2018**, *42*, 147–159. [CrossRef]
112. Burnum-Johnson, K.E.; Zheng, X.; Dodds, J.N.; Ash, J.; Fourches, D.; Nicora, C.D.; Wendler, J.P.; Metz, T.O.; Waters, K.M.; Jansson, J.K.; et al. Ion mobility spectrometry and the omics: Distinguishing isomers, molecular classes and contaminant ions in complex samples. *Trends Anal. Chem.* **2019**, *116*, 292–299. [CrossRef]
113. Odenkirk, M.T.; Baker, E.S. Utilizing drift tube ion mobility spectrometry for the evaluation of metabolites and xenobiotics. *Methods Mol. Biol.* **2020**, *2084*, 35–54.
114. Armenta, S.; Esteve-Turrillas, F.A.; Alcala, M. Analysis of hazardous chemicals by "stand alone" drift tube ion mobility spectrometry: A review. *Anal. Methods* **2020**, *12*, 1163–1181. [CrossRef]
115. Garcia, X.; Sabaté, M.D.M.; Aubets, J.; Jansat, J.M.; Sentellas, S. Ion mobility–mass spectrometry for bioanalysis. *Separations* **2021**, *8*, 33. [CrossRef]
116. May, J.C.; Morris, C.B.; McLean, J.A. Ion mobility collision cross section compendium. *Anal. Chem.* **2017**, *89*, 1032–1044. [CrossRef]
117. Mu, Y.; Schulz, B.L.; Ferro, V. Applications of ion mobility-mass spectrometry in carbohydrate chemistry and glycobiology. *Molecules* **2018**, *23*, 2557. [CrossRef] [PubMed]
118. Hofmann, J.; Pagel, K. Glycan analysis by ion mobility-mass spectrometry. *Angew. Chem. Int. Ed.* **2017**, *56*, 8342–8349. [CrossRef]
119. Li, H.; Bendiak, B.; Kaplan, K.; Davis, E.; Siems, W.F.; Hill, H.H. Evaluation of ion mobility-mass spectrometry for determining the isomeric heterogeneity of oligosaccharide-alditols derived from bovine submaxillary mucin. *Int. J. Mass Spectrom.* **2013**, *352*, 9–18. [CrossRef]
120. Ahonen, L.; Fasciotti, M.; Gennäs, G.B.A.; Kotiaho, T.; Daroda, R.J.; Eberlin, M.; Kostiainen, R. Separation of steroid isomers by ion mobility mass spectrometry. *J. Chromatogr. A* **2013**, *1310*, 133–137. [CrossRef] [PubMed]
121. Clowers, B.H.; Dwivedi, P.; Steiner, W.E.; Hill, H.H.; Bendiak, B. Separation of sodiated isobaric disaccharides and trisaccharides using electrospray ionization-atmospheric pressure ion mobility-time of flight mass spectrometry. *J. Am. Soc. Mass Spectrom.* **2005**, *16*, 660–669. [CrossRef]
122. Struwe, W.B.; Benesch, J.L.; Harvey, D.J.; Pagel, K. Collision cross sections of high-mannose N-glycans in commonly observed adduct states–identification of gas-phase conformers unique to [M-H]⁻ ions. *Analyst* **2015**, *140*, 6799–6803. [CrossRef] [PubMed]
123. Lapthorn, C.; Pullen, F.; Chowdhry, B.Z. Ion mobility spectrometry-mass spectrometry (IMS-MS) of small molecules: Separating and assigning structures to ions. *Mass Spectrom Rev.* **2013**, *32*, 43–71. [CrossRef]
124. Lanucara, F.; Holman, S.W.; Gray, C.J.; Eyers, C.E. The power of ion mobility-mass spectrometry for structural characterization and the study of conformational dynamics. *Nat. Chem.* **2014**, *6*, 281. [CrossRef]
125. Luo, M.D.; Zhou, Z.W.; Zhu, Z.J. The application of ion mobility-mass spectrometry in untargeted metabolomics: From separation to identification. *J. Anal. Test.* **2020**, *4*, 163–174. [CrossRef]
126. Zhou, Z.; Tu, J.; Xiong, X.; Shen, X.; Zhu, Z.J. LipidCCS: Prediction of collision cross-section values for lipids with high precision to support ion mobility–mass spectrometry-based lipidomics. *Anal. Chem.* **2017**, *89*, 9559–9566. [CrossRef]
127. Zhou, Z.; Xiong, X.; Zhu, Z.J. MetCCS Predictor: A web server for predicting collision cross-section values of metabolite in metabolomics. *Bioinformatics* **2017**, *33*, 2235–2237. [CrossRef]
128. Colby, S.M.; Thomas, D.G.; Nuñez, J.R.; Baxter, D.J.; Glaesemann, K.R.; Brown, J.M.; Pirrung, M.A.; Govind, N.; Teeguarden, J.G.; Metz, T.O.; et al. ISiCLE: A quantum chemistry pipeline for establishing in silico collision cross section libraries. *Anal. Chem.* **2019**, *91*, 4346–4356. [CrossRef]
129. Zhou, Z.; Luo, M.; Chen, X.; Yin, Y.; Xiong, X.; Wang, R.; Zhu, Z.J. Ion mobility collision cross-section atlas for known and unknown metabolite annotation in untargeted metabolomics. *Nat. Commun.* **2020**, *11*, 1–13. [CrossRef] [PubMed]
130. McCullagh, M.; Douce, D.; Van Hoeck, E.; Goscinny, S. Exploring the complexity of steviol glycosides analysis using ion mobility mass spectrometry. *Anal. Chem.* **2018**, *90*, 4585–4595. [CrossRef] [PubMed]

131. Schroeder, M.; Meyer, S.W.; Heyman, H.M.; Barsch, A.; Sumner, L.W. Generation of a collision cross section library for multi-dimensional plant metabolomics using UHPLC-trapped ion mobility-MS/MS. *Metabolites* **2019**, *10*, 13. [CrossRef] [PubMed]
132. Kim, H.K.; Choi, Y.H.; Verpoorte, R. NMR-based metabolomic analysis of plants. *Nat. Proto.* **2010**, *5*, 536–549. [CrossRef]
133. Foroutan, A.; Goldansaz, S.A.; Lipfert, M.; Wishart, D.S. Protocols for NMR analysis in livestock metabolomics. In *Metabolomics*; Humana: New York, NY, USA, 2019; pp. 311–324.
134. Ward, J.L.; Baker, J.M.; Beale, M.H. Recent applications of NMR spectroscopy in plant metabolomics. *FEBS J.* **2007**, *274*, 1126–1131. [CrossRef]
135. Emwas, A.H.M. The strengths and weaknesses of NMR spectroscopy and mass spectrometry with particular focus on metabolomics research. *Methods Mol. Biol.* **2015**, *1277*, 161–193. [PubMed]
136. Deborde, C.; Moing, A.; Roch, L.; Jacob, D.; Rolin, D.; Giraudeau, P. Plant metabolism as studied by NMR spectroscopy. *Prog. Nucl. Magn. Reson. Spectrosc.* **2017**, *102*, 61–97. [CrossRef]
137. Kim, H.K.; Choi, Y.H.; Verpoorte, R. NMR-based plant metabolomics: Where do we stand, where do we go? *Trends Biotechnol.* **2011**, *29*, 267–275. [CrossRef]
138. Viant, M.R. Improved methods for the acquisition and interpretation of NMR metabolomic data. *Biochem. Biophys. Res. Commun.* **2003**, *310*, 943–948. [CrossRef]
139. Tautenhahn, R.; Patti, G.J.; Rinehart, D.; Siuzdak, G. XCMS Online: A web-based platform to process untargeted metabolomic data. *Anal. Chem.* **2012**, *84*, 5035–5039. [CrossRef] [PubMed]
140. Montenegro-Burke, J.R.; Aisporna, A.E.; Benton, H.P.; Rinehart, D.; Fang, M.; Huan, T.; Warth, B.; Forsberg, E.; Abe, B.T.; Ivanisevic, J. Data streaming for metabolomics: Accelerating data processing and analysis from days to minutes. *Anal. Chem.* **2017**, *89*, 1254–1259. [CrossRef]
141. Smith, C.A.; O'Maille, G.; Want, E.J.; Qin, C.; Trauger, S.A.; Brandon, T.R.; Custodio, D.E.; Abagyan, R.; Siuzdak, G. METLIN: A metabolite mass spectral database. *Ther. Drug Monit.* **2005**, *27*, 747–751. [CrossRef] [PubMed]
142. Kessler, N.; Neuweger, H.; Bonte, A.; Langenkämper, G.; Niehaus, K.; Nattkemper, T.W.; Goesmann, A. MeltDB 2.0–advances of the metabolomics software system. *Bioinformatics* **2013**, *29*, 2452–2459. [CrossRef] [PubMed]
143. Xia, J.; Wishart, D.S. Using MetaboAnalyst 3.0 for comprehensive metabolomics data analysis. *Curr. Protoc. Bioinform.* **2016**, *55*, 14.10.1–14.10.91. [CrossRef]
144. Wang, M.; Carver, J.J.; Phelan, V.V.; Sanchez, L.M.; Garg, N.; Peng, Y.; Nguyen, D.D.; Watrous, J.; Kapono, C.A.; Luzzatto-Knaan, T.; et al. Sharing and community curation of mass spectrometry data with Global Natural Products Social Molecular Networking. *Nat. Biotechnol.* **2016**, *34*, 828–837. [CrossRef]
145. Pluskal, T.; Castillo, S.; Villar-Briones, A.; Orešič, M. MZmine 2: Modular framework for processing, visualizing, and analyzing mass spectrometry-based molecular profile data. *BMC Bioinform.* **2010**, *11*, 1–11. [CrossRef]
146. Jiang, W.; Qiu, Y.; Ni, Y.; Su, M.; Jia, W.; Du, X. An automated data analysis pipeline for GC−TOF−MS metabonomics studies. *J. Proteome Res.* **2010**, *9*, 5974–5981. [CrossRef] [PubMed]
147. Behrends, V.; Tredwell, G.D.; Bundy, J.G. A software complement to AMDIS for processing GC-MS metabolomic data. *Anal. Biochem.* **2011**, *415*, 206–208. [CrossRef] [PubMed]
148. Skogerson, K.; Wohlgemuth, G.; Barupal, D.K.; Fiehn, O. The volatile compound BinBase mass spectral database. *BMC Bioinform.* **2011**, *12*, 1–15. [CrossRef]
149. Kind, T.; Wohlgemuth, G.; Lee, D.Y.; Lu, Y.; Palazoglu, M.; Shahbaz, S.; Fiehn, O. FiehnLib: Mass spectral and retention index libraries for metabolomics based on quadrupole and time-of-flight gas chromatography/mass spectrometry. *Anal. Chem.* **2009**, *81*, 10038–10048. [CrossRef] [PubMed]
150. Kameyama, A.; Kikuchi, N.; Nakaya, S.; Ito, H.; Sato, T.; Shikanai, T.; Takahashi, Y.; Takahashi, K.; Narimatsu, H. A strategy for identification of oligosaccharide structures using observational multistage mass spectral library. *Anal. Chem.* **2005**, *77*, 4719–4725. [CrossRef]
151. Aoki, K.F.; Kanehisa, M. Using the KEGG database resource. *Curr. Protoc. Bioinform.* **2005**, *11*, 1–12. [CrossRef]
152. Shinbo, Y.; Nakamura, Y.; Altaf-Ul-Amin, M.; Asahi, H.; Kurokawa, K.; Arita, M.; Saito, K.; Ohta, D.; Shibata, D.; Kanaya, S. KNApSAcK: A comprehensive species-metabolite relationship database. In *Plant Metabolomics*; Saito, K., Richard, A.D., Willmitzer, L., Eds.; Springer: Berlin/Heidelberg, Germany, 2006; pp. 165–181.
153. Kaever, A.; Landesfeind, M.; Feussner, K.; Mosblech, A.; Heilmann, I.; Morgenstern, B.; Feussner, I.; Meinicke, P. MarVis-Pathway: Integrative and exploratory pathway analysis of non-targeted metabolomics data. *Metabolomics* **2015**, *11*, 764–777. [CrossRef]
154. Ara, T.; Sakurai, N.; Suzuki, H.; Aoki, K.; Saito, K.; Shibata, D. MassBase: A large-scaled depository of mass spectrometry datasets for metabolome analysis. *Plant Biotechnol.* **2021**, *38*, 167–171. [CrossRef]
155. Clasquin, M.F.; Melamud, E.; Rabinowitz, J.D. LC-MS data processing with MAVEN: A metabolomic analysis and visualization engine. *Curr. Protoc. Bioinform.* **2012**, *37*, 1–23.
156. Xia, J.; Psychogios, N.; Young, N.; Wishart, D.S. MetaboAnalyst: A web server for metabolomic data analysis and interpretation. *Nucleic Acids Res.* **2009**, *37*, 652–660. [CrossRef] [PubMed]
157. Carroll, A.J.; Badger, M.R.; Millar, A.H. The Metabolome Express Project: Enabling web-based processing, analysis and transparent dissemination of GC/MS metabolomics datasets. *BMC Bioinform.* **2010**, *11*, 1–13. [CrossRef] [PubMed]
158. Zhou, B.; Wang, J.; Ressom, H.W. MetaboSearch: Tool for mass-based metabolite identification using multiple databases. *PLoS ONE* **2012**, *7*, e40096. [CrossRef]

159. Wanichthanarak, K.; Fan, S.; Grapov, D.; Barupal, D.K.; Fiehn, O. Metabox: A toolbox for metabolomic data analysis, interpretation and integrative exploration. *PLoS ONE* **2017**, *12*, e0171046.
160. Lommen, A.; Kools, H.J. MetAlign 3.0: Performance enhancement by efficient use of advances in computer hardware. *Metabolomics* **2012**, *8*, 719–726. [CrossRef] [PubMed]
161. Kastenmüller, G.; Römisch-Margl, W.; Wägele, B.; Altmaier, E.; Suhre, K. MetaP-server: A web-based metabolomics data analysis tool. *BioMed Res. Int.* **2010**, *2011*, 1–7. [CrossRef] [PubMed]
162. Daly, R.; Rogers, S.; Wandy, J.; Jankevics, A.; Burgess, K.E.; Breitling, R. MetAssign: Probabilistic annotation of metabolites from LC–MS data using a Bayesian clustering approach. *Bioinformatics* **2014**, *30*, 2764–2771. [CrossRef]
163. Ruttkies, C.; Schymanski, E.L.; Wolf, S.; Hollender, J.; Neumann, S. MetFrag relaunched: Incorporating strategies beyond in silico fragmentation. *J. Cheminform.* **2016**, *8*, 3. [CrossRef]
164. Lei, Z.; Li, H.; Chang, J.; Zhao, P.X.; Sumner, L.W. MET-IDEA version 2.06; Improved efficiency and additional functions for mass spectrometry-based metabolomics data processing. *Metabolomics* **2012**, *8*, 105–110. [CrossRef]
165. Rojas-Chertó, M.; Van Vliet, M.; Peironcely, J.E.; Van Doorn, R.; Kooyman, M.; Te Beek, T.; Van Driel, M.A.; Hankemeier, T.; Reijmers, T. MetiTree: A web application to organize and process high-resolution multi-stage mass spectrometry metabolomics data. *Bioinformatics* **2012**, *28*, 2707–2709. [CrossRef] [PubMed]
166. Cui, Q.; Lewis, I.A.; Hegeman, A.D.; Anderson, M.E.; Li, J.; Schulte, C.F.; Westler, W.M.; Eghbalnia, H.R.; Sussman, M.R.; Markley, J.L. Metabolite identification via the madison metabolomics consortium database. *Nat. Biotechnol.* **2008**, *26*, 162. [CrossRef]
167. Menikarachchi, L.C.; Cawley, S.; Hill, D.W.; Hall, L.M.; Hall, L.; Lai, S.; Wilder, J.; Grant, D.F. MolFind: A software package enabling HPLC/MS-based identification of unknown chemical structures. *Anal. Chem.* **2012**, *84*, 9388–9394. [CrossRef]
168. Mistrik, R.; Lutisan, J.; Huang, Y.; Suchy, M.; Wang, J.; Raab, M. mzCloud: A key conceptual shift to understand 'Who's Who' in untargeted metabolomics. In Proceedings of the Metabolomics Society 2013 Conference, Glasgow, UK, 1–13 July 2013; pp. 1–4.
169. Draper, J.; Enot, D.P.; Parker, D.; Beckmann, M.; Snowdon, S.; Lin, W.; Zubair, H. Metabolite signal identification in accurate mass metabolomics data with MZedDB, an interactive m/z annotation tool utilising predicted ionisation behaviour 'rules'. *BMC Bioinform.* **2009**, *10*, 1–16. [CrossRef]
170. Rumble, J.R., Jr.; Bickham, D.M.; Powell, C.J. The NIST x-ray photoelectron spectroscopy database. *Surf. Interface Anal.* **1992**, *19*, 241–246. [CrossRef]
171. Sakurai, T.; Yamada, Y.; Sawada, Y.; Matsuda, F.; Akiyama, K.; Shinozaki, K.; Hirai, M.Y.; Saito, K. PRIMe update: Innovative content for plant metabolomics and integration of gene expression and metabolite accumulation. *Plant Cell Physiol.* **2013**, *54*, e5. [CrossRef] [PubMed]
172. Pérez-Alonso, M.M.; Carrasco-Loba, V.; Pollmann, S. Advances in plant metabolomics. *Annu. Plant. Rev. Online* **2018**, *1*, 557–588.
173. Kanehisa, M.; Furumichi, M.; Tanabe, M. KEGG: New perspectives on genomes, pathways, diseases and drugs. *Nucleic Acids Res.* **2017**, *45*, 353–361. [CrossRef] [PubMed]
174. Caspi, R.; Foerster, H.; Fulcher, C.A. The MetaCyc database of metabolic pathways and enzymes and the BioCyc collection of pathway/genome databases. *Nucleic Acids Res.* **2008**, *36*, 623–631. [CrossRef]
175. Mueller, L.A.; Zhang, P.; Rhee, S.Y. AraCyc: A biochemical pathway database for Arabidopsis. *Plant Physiol.* **2003**, *132*, 453–460. [CrossRef]
176. Jewison, T.; Su, Y.; Disfany, F.M. SMPDB 2.0: Big improvements to the small molecule pathway database. *Nucleic Acids Res.* **2014**, *42*, 478–484. [CrossRef]
177. Xia, J.; Wishart, D.S. MSEA: A web-based tool to identify biologically meaningful patterns in quantitative metabolomic data. *Nucleic Acids Res.* **2010**, *38*, 71–77. [CrossRef] [PubMed]
178. Karnovsky, A.; Weymouth, T.; Hull, T. Metscape 2 bioinformatics tool for the analysis and visualization of metabolomics and gene expression data. *Bioinformatics* **2012**, *28*, 373–380. [CrossRef] [PubMed]
179. Kohl, M.; Wiese, S.; Warscheid, B. Cytoscape: Software for visualization and analysis of biological networks. *Methods Mol. Biol.* **2011**, *696*, 291–303. [PubMed]
180. Garcia-Alcalde, F.; Garcia-Lopez, F.; Dopazo, J. Paintomics: A web based tool for the joint visualization of transcriptomics and metabolomics data. *Bioinformatics* **2011**, *27*, 137–139. [CrossRef]
181. Neuweger, H.; Persicke, M.; Albaum, S.P. Visualizing post genomics data-sets on customized pathway maps by ProMeTra-aeration-dependent gene expression and metabolism of *Corynebacterium glutamicum* as an example. *BMC Syst. Biol.* **2009**, *3*, 82. [CrossRef] [PubMed]
182. Grapov, D.; Wanichthanarak, K.; Fiehn, O. MetaMapR: Pathway independent metabolomic network analysis incorporating unknowns. *Bioinformatics* **2015**, *31*, 2757–2760. [CrossRef]
183. Smoot, M.E.; Ono, K.; Ruscheinski, J.; Wang, P.L.; Ideker, T. Cytoscape 2.8: New features for data integration and network visualization. *Bioinformatics* **2011**, *27*, 431–432. [CrossRef]
184. Kamburov, A.; Cavill, R.; Ebbels, T.M.; Herwig, R.; Keun, H.C. Integrated pathway-level analysis of transcriptomics and metabolomics data with IMPaLA. *Bioinformatics* **2011**, *27*, 2917–2918. [CrossRef] [PubMed]
185. Letunic, I.; Yamada, T.; Kanehisa, M.; Bork, P. iPath: Interactive exploration of biochemical pathways and networks. *Trends Biochem. Sci.* **2008**, *33*, 101–103. [CrossRef]

186. Thimm, O.; Bläsing, O.; Gibon, Y.; Nagel, A.; Meyer, S.; Krüger, P.; Selbig, J.; Müller, L.A.; Rhee, S.Y.; Stitt, M. MAPMAN: A user-driven tool to display genomics data sets onto diagrams of metabolic pathways and other biological processes. *Plant J.* **2004**, *37*, 914–939. [CrossRef]
187. Chagoyen, M.; Pazos, F. MBRole: Enrichment analysis of metabolomic data. *Bioinformatics* **2011**, *27*, 730–731. [CrossRef]
188. Ara, T.; Enomoto, M.; Arita, M.; Ikeda, C.; Kera, K.; Yamada, M.; Nishioka, T.; Ikeda, T.; Nihei, Y.; Shibata, D.; et al. Metabolonote: A wiki-based database for managing hierarchical metadata of metabolome analyses. *Front. Bioeng. Biotechnol.* **2015**, *3*, 38. [CrossRef] [PubMed]
189. Schreiber, F.; Colmsee, C.; Czauderna, T.; Grafahrend-Belau, E.; Hartmann, A.; Junker, A.; Junker, B.H.; Klapperstück, M.; Scholz, U.; Weise, S. MetaCrop 2.0: Managing and exploring information about crop plant metabolism. *Nucleic Acids Res.* **2012**, *40*, 1173–1177. [CrossRef]
190. Xia, J.; Wishart, D.S. MetPA: A web-based metabolomics tool for pathway analysis and visualization. *Bioinformatics* **2010**, *26*, 2342–2344. [CrossRef] [PubMed]
191. Kankainen, M.; Gopalacharyulu, P.; Holm, L.; Orešič, M. MPEA—Metabolite pathway enrichment analysis. *Bioinformatics* **2011**, *27*, 1878–1879. [CrossRef] [PubMed]
192. Elliott, B.; Kirac, M.; Cakmak, A.; Yavas, G.; Mayes, S.; Cheng, E.; Wang, Y.; Gupta, C.; Ozsoyoglu, G.; Meral Ozsoyoglu, Z. PathCase: Pathways database system. *Bioinformatics* **2008**, *24*, 2526–2533. [CrossRef]
193. Mlecnik, B.; Scheideler, M.; Hackl, H.; Hartler, J.; Sanchez-Cabo, F.; Trajanoski, Z. PathwayExplorer: Web service for visualizing high-throughput expression data on biological pathways. *Nucleic Acids Res.* **2005**, *33*, 633–637. [CrossRef] [PubMed]
194. Junker, B.H.; Klukas, C.; Schreiber, F. VANTED: A system for advanced data analysis and visualization in the context of biological networks. *BMC Bioinform.* **2006**, *7*, 1–13. [CrossRef]
195. Kelder, T.; Van Iersel, M.P.; Hanspers, K. WikiPathways: Building research communities on biological pathways. *Nucleic Acids Res.* **2012**, *40*, 1301–1307. [CrossRef]
196. Razzaq, A.; Saleem, F.; Kanwal, M.; Mustafa, G.; Yousaf, S.; Imran Arshad, H.M.; Hameed, M.K.; Khan, M.S.; Joyia, F.A. Modern trends in plant genome editing: An inclusive review of the CRISPR/Cas9 toolbox. *Int. J. Mol. Sci.* **2019**, *20*, 4045. [CrossRef]

Article

Metabolite Profiling and Dipeptidyl Peptidase IV Inhibitory Activity of *Coreopsis* Cultivars in Different Mutations

Bo-Ram Kim [1,2,†], Sunil Babu Paudel [3,†], Ah-Reum Han [1,†], Jisu Park [1], Yun-Seo Kil [3], Hyukjae Choi [3,4], Yeo Gyeong Jeon [5], Kong Young Park [5], Si-Yong Kang [6], Chang Hyun Jin [1], Jin-Baek Kim [1] and Joo-Won Nam [3,*]

[1] Advanced Radiation Technology Institute, Korea Atomic Energy Research Institute, Jeongeup-si 56212, Jeollabuk-do, Korea; boram0307@hnibr.re.kr (B.-R.K.); arhan@kaeri.re.kr (A.-R.H.); parksj94@kaeri.re.kr (J.P.); chjin@kaeri.re.kr (C.H.J.); jbkim74@kaeri.re.kr (J.-B.K.)

[2] Natural Product Research Division, Honam National Institute of Biological Resources, Mokpo-si 58762, Jeollanam-do, Korea

[3] College of Pharmacy, Yeungnam University, Gyeongsan-si 38541, Gyeongsangbuk-do, Korea; phrsunil@gmail.com (S.B.P.); yskil@yu.ac.kr (Y.-S.K.); h5choi@yu.ac.kr (H.C.)

[4] Research Institute of Cell Culture, Yeungnam University, Gyeongsan 38541, Gyeongbuk, Korea

[5] Uriseed Group, Icheon-si 17408, Gyeonggi-do, Korea; ygjeon@uriseed.com (Y.G.J.); uriseeds@naver.com (K.Y.P.)

[6] Department of Horticulture, College of Industrial Sciences, Kongju National University, Yesan-gun 32439, Chungcheongnam-do, Korea; sykang@kongju.ac.kr

* Correspondence: jwnam@yu.ac.kr; Tel.: +82-53-810-2818

† These authors contributed equally to this work.

check for **updates**

Citation: Kim, B.-R.; Paudel, S.B.; Han, A.-R.; Park, J.; Kil, Y.-S.; Choi, H.; Jeon, Y.G.; Park, K.Y.; Kang, S.-Y.; Jin, C.H.; et al. Metabolite Profiling and Dipeptidyl Peptidase IV Inhibitory Activity of *Coreopsis* Cultivars in Different Mutations. *Plants* **2021**, *10*, 1661. https://doi.org/10.3390/plants10081661

Academic Editors: Jong Seong Kang and Narendra Singh Yadav

Received: 20 July 2021
Accepted: 9 August 2021
Published: 12 August 2021

Publisher's Note: MDPI stays neutral with regard to jurisdictional claims in published maps and institutional affiliations.

Abstract: *Coreopsis* species have been developed to produce cultivars of various floral colors and sizes and are also used in traditional medicine. To identify and evaluate mutant cultivars of *C. rosea* and *C. verticillata*, their phytochemical profiles were systematically characterized using ultra-performance liquid chromatography time-of-flight mass spectrometry, and their anti-diabetic effects were evaluated using the dipeptidyl peptidase (DPP)-IV inhibitor screening assay. Forty compounds were tentatively identified. This study is the first to provide comprehensive chemical information on the anti-diabetic effect of *C. rosea* and *C. verticillata*. All 32 methanol extracts of *Coreopsis* cultivars inhibited DPP-IV activity in a concentration-dependent manner (IC_{50} values: 34.01–158.83 µg/mL). Thirteen compounds presented as potential markers for distinction among the 32 *Coreopsis* cultivars via principal component analysis and orthogonal partial least squares discriminant analysis. Therefore, these bio-chemometric models can be useful in distinguishing cultivars as potential dietary supplements for functional plants.

Keywords: *Coreopsis rosea*; *Coreopsis verticillata*; mutant cultivar; metabolomics; dipeptidyl peptidase-IV

1. Introduction

Coreopsis species are annual or perennial plants belonging to the Asteraceae (Compositae) family [1]. Approximately 80 species of *Coreopsis* are native to North America and are currently widespread in America, Asia, and Oceania regions [2–4]. They are usually cultivated for ornamental purposes in gardens or on roadsides. The plants are in the range of 46–120 cm in height and the petals of the flowers are primarily yellow in color and are serrated [5,6]. The color and size of *Coreopsis* flowers have commercially important value and are the reason for *Coreopsis* breeding. In addition, the *Coreopsis* flower has been ethnopharmacologically used for the treatment of diarrhea, vomiting, and hemorrhage in North America, where the *Coreopsis* species originates [7,8]. It has also been used as a drink to control diabetes in China and Portugal, and as an herbal tea to eliminate toxins and fever from the body in China [8–10]. Nowadays, owing to scientific proof of its traditional use, several studies have been conducted on the phytochemical and biological activities of *C. lanceolata* and *C. tinctoria*, in particular [2–19]. Diverse types of flavonoids, such as aurone, chalcone, flavanone, and flavanol have been identified from *C. lanceolata* and *C. tinctoria*. In addition, unique polyacetylene compounds have also been found in these plants [10,12,13]. Various pharmacological activities such as anticancer [2,5], antioxidant [6,14–18], anti-inflammatory [6,10,19], and anti-diabetic effects [8,12] have been reported of compounds isolated from *C. lanceolata* and *C. tinctoria*. Limited reports on the chemical composition and biological activities of *Coreopsis* species exist.

In this study, several new cultivars of *C. rosea* and *C. verticillata* throughout γ-irradiated mutation or herbicide-induced artificial mutation were developed and registered in the Korea Seed and Variety Service (Table 1 and Figure 1) [20]. For horticultural purposes and the improvement of quality and functionality, numerous cultivars have been developed by hybridization, and mutations were induced by chemical mutagens and ionizing radiation in plant breeding programs [21]. Previous studies on *C. rosea* and *C. verticillata* have primarily focused on plant growth impact assessment, new variety development, and horticulture [22–25]. However, there has been no report on the phytochemical and biological activity of *C. rosea* and *C. verticillata*, except for our previous study on the volatiles' composition and antioxidant activity of *C. rosea* cultivars [18].

Table 1. The list of the original and mutant cultivars of *Coreopsis rosea* and *Coreopsis verticillata* used in this study.

Group (Plant Name)	No.	Cultivar Names	Registration No.	Application No.	Breeding Process
	1	Heaven's gate	-	-	Original cultivar
	2	Luckyten 6	3869	-	Herbicide-induced artificial mutation
	3	Redfin	4408	-	γ-Irradiated mutation
	4	Lemon candy	4418	-	γ-Irradiated mutation
	5	Shiny pink	4420	-	γ-Irradiated mutation
I	6	Uri-dream 01	3993	-	Herbicide-induced artificial mutation
(*C. rosea*)	7	Luckyten5	4411	-	Herbicide-induced artificial mutation
	8	Luckyten9	4413	-	Herbicide-induced artificial mutation
	9	Uri-dream red	6001	-	γ-Irradiated mutation
	10	Uri-dream 07	3998	-	Herbicide-induced artificial mutation
	11	Uri-dream 06	3997	-	Herbicide-induced artificial mutation
	12	Pink sherbet	4415	-	γ-Irradiated mutation
	13	Citrine	-	-	Original cultivar
	14	Golden ball No.18	6421	-	γ-Irradiated mutation
II	15	Golden ball No.21	6422	-	γ-Irradiated mutation
(*C. verticillata*)	16	Golden ball No.26	5995	-	γ-Irradiated mutation
	17	Golden ball No.42	5997	-	γ-Irradiated mutation
	18	Golden ball No.48	5999	-	γ-Irradiated mutation
	19	Pumpkin Pie	-	-	Original cultivar
	20	Gold ring	7523	-	γ-Irradiated mutation
III	21	Golden ring	5994	-	γ-Irradiated mutation
(*C. rosea*)	22	Mini ball yellow	6453	-	γ-Irradiated mutation
	23	Box tree	6462	-	γ-Irradiated mutation
	24	Orange ball	6005	-	γ-Irradiated mutation
	25	Route 66	-	-	Original cultivar
	26	Golden sunlight	-	2018-406	γ-Irradiated mutation
IV	27	Red sunlight	-	2018-410	γ-Irradiated mutation
(*C. verticillata*)	28	Bright sunlight	-	2018-408	γ-Irradiated mutation
	29	Yellow sunlight	-	2018-411	γ-Irradiated mutation
	30	Orange sunlight	-	2018-399	γ-Irradiated mutation
V	31	Moonbeam	-	-	Original cultivar
(*C. verticillata*)	32	Moonlight sonata	-	2018-401	Selection of phenotypic variant

Group I (*Coreopsis rosea*)

Group II (*Coreopsis verticillata*)

Group III (*Coreopsis rosea*)

Figure 1. *Cont.*

Figure 1. Photos of original and mutant cultivars of *Coreopsis rosea* and *Coreopsis verticillata* used in this study.

As part of our investigation of the effects of mutation on metabolic changes between the original mutant cultivars and their biological functions, we analyzed metabolite profiling of the five original cultivars and each mutant cultivar. Given that *Coreopsis* species have been known to be effective for diabetes in folk medicine, 70% methanol extracts of 32 *Coreopsis* samples were evaluated for their inhibitory effect against dipeptidyl peptidase (DPP)-IV, a target of incretin-based therapies for the treatment of type 2 diabetes mellitus.

2. Results

2.1. Subsection Identification of Metabolites in Coreopsis Cultivars Using UPLC-QTof-MS

Metabolites in *Coreopsis* cultivars were tentatively identified using UPLC-QTof-MS. The Metabolites were separated with high resolution within 10 min in the base peak ion (BPI) chromatogram. BPI chromatograms of the original *Coreopsis* cultivars are shown in Figure S1. The mass spectrum of each peak was carefully interpreted by analyzing its experimental and theoretical high resolution MS (the deprotonated molecular ion, $[M - H]^-$), error ppm, molecular formula, and MS/MS fragmentation. Additionally, these were compared with data from the literature of plants belonging to the same genus, such as *C. tinctoria* (known as snow chrysanthemum) and *C. lanceolata* [3,5,6,12,26–29]. Moreover, its mass spectrum was compared to that in Waters Traditional Medicine Library that is built in UNIFI software (Waters, Milford, MA, USA) and MassBank available online (a public database for sharing mass spectral data) [30,31]. Forty compounds, including phenolic acids, flavonoids, and a polyacetylene were identified in methanol extracts of original and mutant cultivars of *C. rosea* and *C. verticillata* (Table 2). However, a peak observed in total ion chromatograms of all, or some *Coreopsis* cultivars could not be identified in this study.

2.1.1. Phenolic Acids

Chlorogenic acid (peak 2, t_R 4.50 min) produced a major molecular ion at m/z 353.0864 $[M - H]^-$ (calculated for $C_{16}H_{17}O_9{}^-$, 353.0878). At high energy scan, a fragment ion for a quinic acid was observed at m/z 191.0556 [3,26,27]. Peak 4 (t_R 5.21 min) produced a molecular ion at m/z 329.0865 $[M - H]^-$ corresponding to $C_{14}H_{17}O_9{}^-$ and produced two fragment ions at m/z 167.0338 and 151.0026, assigned to the loss of glucose and $[M - H - glucose - O]^-$, respectively. This peak was tentatively identified as vanillic acid-4-glucoside, which was confirmed by the UNIFI local library and first detected in *Coreopsis* species. Peak 27 (t_R 7.68 min) produced a major ion at m/z 515.1183 $[M - H]^-$ (calculated for $C_{25}H_{23}O_{12}{}^-$, 515.1195) with two stable fragment ions at m/z 353.0869 $[M - H - C_9H_6O_3]^-$ and 191.0557 $[M - H - 2C_9H_6O_3]^-$ assigned to dicaffeoylquinic acid. Peak 31 (t_R 8.04 min) had the same precursor and fragment ions with peak 27. Peaks 27 and 31 were tentatively identified as 3,5-dicaffeoylquinic acid and 4,5-dicaffeoylquinic acid, respectively, by comparing with values in the literature of clearly identified constituents in *C. tinctoria* [26].

Table 2. Characterization and tentative identification of metabolites found in original and mutant cultivars of *Coreopsis rosea* and *Coreopsis verticillata* using ultra-performance liquid chromatography time-of-flight mass spectrometry (UPLC-QTof MS).

Peak No.	ESI-MS t_R (min)	Observed Mass (m/z)	Caculated Mass (m/z)	Error (ppm)	Molecular Formula	Key MS^E Fragment Ions (m/z)	Identification
1	4.48	465.1030	465.1039	−0.8	$C_{21}H_{22}O_{12}$	303.0503, 285.0397, 151.0034, 125.0239	Taxifolin-7-*O*-glucoside
2	4.50	353.0864	353.0878	−1.4	$C_{16}H_{18}O_9$	191.0556, 133.0290	Chlorogenic acid
3	4.96	465.1030	465.1039	−0.8	$C_{21}H_{22}O_{12}$	303. 0503, 287.0550, 285.0397, 151.0034, 125.0234	Taxifolin-3-*O*-glucoside
4	5.21	329.0865	329.0878	−1.3	$C_{14}H_{18}O_9$	167.0338, 151.0026	Vanillic acid-4-glucoside
5	5.74	449.1085	449.1089	−0.4	$C_{21}H_{22}O_{11}$	287.0554, 269.0446, 151.0034, 135.0449	Flavanomarein
6	5.85	595.1649	595.1668	−1.9	$C_{27}H_{32}O_{15}$	449.1069, 287.0548, 269.0428, 151.0028, 135.0447	Isookanin-7-*O*-rutinoside
7	5.93	609.1454	609.1461	−0.7	$C_{27}H_{30}O_{16}$	447.0932, 285.0392, 151.0033	Luteolin-7-*O*-sophoroside
8	5.98	433.1135	433.1140	−0.5	$C_{21}H_{22}O_{10}$	271.0605, 253.0499, 135.0449	Butin-7-*O*-glucoside
9	6.17	479.0825	479.0831	−0.6	$C_{21}H_{20}O_{13}$	317.0291, 166.9963	8-Methoxyeriodictyol-7-*O*-glucoside

Table 2. *Cont.*

Peak No.	ESI-MS t_R (min)	Observed Mass (m/z)	Caculated Mass (m/z)	Error (ppm)	Molecular Formula	Key MS^E Fragment Ions (m/z)	Identification
10	6.23	463.1239	463.1246	−0.7	$C_{22}H_{24}O_{13}$	301.0708, 165.0188, 135.0449	Coreolanceoline B
11	6.34	463.1251	463.1246	−0.5	$C_{22}H_{24}O_{11}$	301.0708, 165.0188, 135.0449	Lanceolin
12	6.48	433.1134	433.1140	−0.6	$C_{21}H_{22}O_{10}$	271.0602, 151.0029, 119.0488	Naringenin-7-*O*-glucoside
13	6.51	611.1612	611.1618	−0.6	$C_{27}H_{32}O_{16}$	449.1080, 287.0551, 269.0393, 135.0447	Okanin-3′,4′-*O*-diglucoside
14	6.52	595.1664	595.1668	−0.4	$C_{27}H_{32}O_{15}$	433.1121, 271.0604, 135.0447	4′,7,8-Trihydroxyflavone-*O*-diglucoside
15	6.58	609.1454	609.1461	−0.7	$C_{27}H_{30}O_{16}$	447.0932, 285.0394, 135.0082	Fisetin-3,7-*O*-diglucoside
16	6.64	287.0555	287.0561	−0.6	$C_{15}H_{12}O_6$	151.0031, 135.0449	Isookanin
17	6.87	303.0502	303.0510	−0.8	$C_{15}H_{12}O_7$	285.0399, 151.0084, 135.0447, 125.0240	Taxifolin
18	6.91	581.1501	581.1512	−1.1	$C_{26}H_{30}O_{15}$	287.0552, 167.0342, 151.0029	4′,5,7,8-Tetrahydroxyflavanone-7-*O*-(6-*O*-arabinosyl-glucoside)
19	6.94	431.0977	431.0984	−0.7	$C_{21}H_{20}O_{10}$	269.0447, 135.0447, 133.0290	Sulfuretin-6-*O*-glucoside
20	6.97	463.0885	463.0882	0.3	$C_{21}H_{20}O_{12}$	301.0346, 151.0031	Quercetin-7-*O*-glucoside
21	7.01	447.0929	447.0927	0.2	$C_{21}H_{20}O_{11}$	285.0397, 135.0447, 133.0291	Maritimein
22	7.03	447.0929	447.0927	0.2	$C_{21}H_{20}O_{11}$	285.0397, 151.0033	Luteolin-7-*O*-glucoside
23	7.15	449.1081	449.1089	−0.8	$C_{21}H_{22}O_{11}$	287.0551, 269.0445, 151.0033, 135.0448	Marein
24	7.26	493.0984	493.0988	−0.4	$C_{22}H_{22}O_{13}$	331.0447, 316.0200, 164.9830	Taxifolin 3′,7-dimethyl ether 3-*O*-glucoside
25	7.33	461.1085	461.1089	−0.4	$C_{22}H_{22}O_{11}$	299.0547, 283.0242, 165.0188, 133.0291	3,3′,4′-Trihydroxy-7-methoxyflavone 3-*O*-glucoside
26	7.58	641.1141	641.1148	−0.7	$C_{30}H_{26}O_{16}$	317.0294, 301.0342, 285.0381, 179.0343, 161.0224, 135.0447, 133.0289	Qurcetagetin-7-*O*-(6″-caffeoylglucoside)
27	7.68	515.1183	515.1195	−1.2	$C_{25}H_{24}O_{12}$	353.0869, 191.0557, 179.0346, 135.0447	3,5-Dicaffeoylquinic acid
28	7.77	269.0449	269.0450	−0.1	$C_{15}H_{10}O_5$	135.0447, 133.0287	Sulfuretin
29	7.91	431.0978	431.0984	−0.6	$C_{21}H_{20}O_{10}$	285.0398, 151.0031, 133.0289	Luteolin-6-*O*-rhamnoside
30	8.02	433.1135	433.1140	−0.5	$C_{21}H_{22}O_{10}$	271.0606, 135.0448	Coreopsin
31	8.04	515.1187	515.1195	−0.8	$C_{25}H_{24}O_{12}$	353.0862, 191.0556, 179.0340	4,5-Dicaffeoylquinic acid
32	8.26	287.0556	287.0561	−0.5	$C_{15}H_{12}O_6$	151.0032, 134.0368, 123.0083	Okanin
33	8.46	611.1398	611.1406	−0.8	$C_{30}H_{28}O_{14}$	449.1109, 287.0559, 269.0441, 151.0024	Eriodictyol chalcone-*O*-diglucoside
34	8.73	287.0553	287.0561	−0.8	$C_{15}H_{12}O_6$	151.0032, 135.0047	Eriodictyol chalcone
35	8.74	299.0555	299.0561	−0.6	$C_{16}H_{12}O_6$	284.0319, 151.0032, 133.0291	Kaempferide
36	8.80	285.0398	285.0405	−0.7	$C_{15}H_{10}O_6$	151.0032, 133.0291	Luteolin
37	8.84	477.1396	477.1402	−0.6	$C_{23}H_{26}O_{11}$	315.0864, 300.0624, 282.0527, 148.0524, 135.0435	4-Methoxylanceoletin-4′-*O*-glucoside
38	9.20	271.0605	271.0612	−0.7	$C_{15}H_{12}O_5$	253.0496, 135.0448, 227.0351, 117.0341	Butein
39	9.28	269.0447	269.0450	−0.3	$C_{15}H_{10}O_5$	227.0351, 117.0341	Apigenin
40	9.39	831.3595	831.3597	−0.2	$C_{46}H_{56}O_{14}$	785.3536, 666.2998, 545.2401, 145.0291	Unknown
41	9.45	557.2244	557.2240	−2.1	$C_{26}H_{38}O_{13}$	233.0650, 191.0554, 149.0441	Lobetyolinin

2.1.2. Flavanones and Flavanonols

It has been reported that a retro Diels–Alder reaction, as well as the loss of H_2O, sugar (usually glucose), and carbonyl groups were observed in the ion fragmentation pathways of flavonoids [3]. Flavanones, chalcones, and their glycosides have been known as the major types of flavonoids found in *Coreopsis* species [28]. These compounds usually showed the loss of H_2O caused by the disposition of hydroxyls at C-3′ and C-4′ in the flavanone structure or at C-3 and C-4 in the chalcone structure, and the loss of a glucose at C-7 in the flavone structure or at C-4′ in the chalcone structure [3]. These phenomena were observed in mass spectra of flavanones and chalcones identified in this study. Abundant fragment ions, $[M - H - glucose]^-$ and $[M - H - glucose - H_2O]^-$ for their glycosides and $[M - H - H_2O]^-$ for aglycones were produced by the loss of glucose and H_2O, respectively. The loss of C_8H_6O (118 Da) or $C_8H_6O_2$ (134 Da) were also characteristic fragment ions for flavanones and chalcones [3].

Peaks 1 (t_R 4.48 min) and 3 (t_R 4.96 min) produced a major molecular ion at m/z 465.1030 $[M - H]^-$ (calculated for $C_{21}H_{21}O_{12}{}^-$, 465.1038) and yielded fragment ions at m/z 303.0503 $[M - H - glucose]^-$ and 285.0397 $[M - H - glucose - H_2O]^-$ by the loss of glucose and H_2O, indicating the presence of a 3′,4′-dihydroxyphenyl group. The fragment ion at m/z 151.0034 $[M - H - glucose - H_2O - C_8H_6O_2]^-$ showed the presence of a 3-hydroxy group. Peak 3 had the fragment ion at 287.0550 $[M - H - glucose - O]^-$, assuming that the sugar was attached at C-3 in the C ring. Therefore, peaks 1 and 3 were tentatively identified as taxifolin-7-*O*-glucoside [3] and taxifolin-3-*O*-glucoside [27], respectively. The aglycone of these compounds, taxifolin (peak 17, t_R 6.87 min), showed the same fragment ions with those of its glycosides (peaks 1 and 3) [3,26]. Peak 5 (t_R 5.74 min) produced a molecular ion at m/z 449.1085 $[M - H]^-$, corresponding to the molecular formula $C_{21}H_{21}O_{11}{}^-$. At high energy scan, fragment ions at m/z 287.0554 $[M - H - glucose]^-$ and 269.0446 $[M - H - glucose - H_2O]^-$ were detected by the loss of glucose and H_2O and indicated the presence of a 3′,4′-dihydroxyphenyl group for the B ring. Fragment ions at m/z 151.0034 $[M - H - glucose - H_2O - C_8H_6O]^-$ suggested no hydroxyl group at C-3. Thus, it was tentatively identified as flavanomarein [26]. The fragmentation pattern of peak 6 (t_R 5.85 min) was identical to that of peak 5, except for the major molecular ion at m/z 595.1649 $[M - H]^-$ (calculated for $C_{27}H_{31}O_{15}{}^-$, 595.1668) and the fragment ion at m/z 449.1069 $[M - H - rhamnose]$. Therefore, peak 6 was predicted as isookanin-7-*O*-rutinoside, which has been described previously [27]. Given that peak 16 (t_R 6.64 min) also exhibited the same fragment ions with those of peaks 5 and 6, it was tentatively identified as their aglycone, isookanin [3,26]. Peak 8 (t_R 5.98 min) produced a molecular ion at m/z 433.1135 $[M - H]^-$ (calculated for $C_{21}H_{21}O_{10}{}^-$, 433.1140), which is 16 Da less than that of peak 5, and the fragment ion at m/z 135.0449 $[M - H - glucose - H_2O - C_8H_6O]^-$ indicated the presence of a 3′,4′-dihydroxyphenyl group and no hydroxyl group at C-3 in the C ring, and consequently confirmed the presence of a hydroxyl group in the A ring. Thus, peak 8 was tentatively identified as butin-7-*O*-glucoside [3]. Peak 9 (t_R 6.17 min) produced a molecular ion at m/z 479.0825 $[M - H]^-$, corresponding to the molecular formula $C_{21}H_{19}O_{13}{}^-$. The fragment ions at m/z 317.0291 $[M - H - glucose]^-$ and 166.9963 $[M - H - glucose - CH_3 - H_2O - C_8H_6O]^-$ represented the presence of two hydroxy groups at C-3′ and C-4′ in the B ring and the absence of a hydroxy group at C-3 in the C ring, and consequently predicted the presence of 5,7-dihydroxy-8-methoxyphenyl group for the A ring. Therefore, peak 9 was tentatively identified as 8-methoxyeriodictyol-7-*O*-glucoside. Its aglycone, 8-methoxyeriodictyol, has been isolated from several plants [32–34]; however, its glycoside form has not been described previously. Peaks 10 (t_R 6.23 min) and 11 (t_R 6.34 min) produced the same molecular ion at m/z 463.1239 $[M - H]^-$ (calculated for $C_{22}H_{23}O_{11}{}^-$, 463.1246) and the same fragment ion at m/z 165.0188 $[M - H - glucose - H_2O - C_8H_6O]^-$, indicating the presence of a 3′,4′-dihydroxyphenyl group for the B ring and no hydroxyl group at C-3 in the C ring, and consequently predicted the presence of the 7-hydroxy-8-methoxyphenyl group for the A ring. Therefore, peaks 10 and 11 were tentatively identified as coreolanceoline B [12] and lanceolin [6], respectively. Peak

12 (t_R 6.48 min) produced a molecular ion at m/z 433.1134 [M – H]$^-$ (calculated for $C_{21}H_{21}O_{10}^-$, 433.1140), exhibiting the same molecular ions as peak 8; however, an [M – H – glucose – C_8H_8O]$^-$ ion, instead of a [M – H – glucose – H_2O]$^-$ ion of peak 8, was observed in the fragmentation pattern of peak 12, indicating the presence of one hydroxy group in the B ring and two hydroxy groups in the A ring. Accordingly, peak 12 was tentatively identified as naringenin-7-*O*-glucoside [26]. Peak 14 (t_R 6.52 min) produced a molecular ion at m/z 595.1664 [M – H]$^-$ (calculated for $C_{27}H_{31}O_{15}^-$, 595.1668) and the fragment ions at m/z 433.1121 [M – H – glucose]$^-$ and 271.0604 [M – H – 2glucose]$^-$, indicating the presence of two glucose groups. The fragment ion at m/z 135.0447 [M – H – 2glucose – $C_8H_8O_2$]$^-$ without the loss of H_2O resulted in the presence of two hydroxy groups at C-3$'$ and C-5$'$ in the B ring. Thus, peak 14 was tentatively identified as 7,3$'$,5$'$-trihydroxyflavanone-*O*-diglucoside. Given that 7,3$'$,5$'$-trihydroxyflavanone-7-*O*-glucoside and its aglycone have been found in *Coreopsis* species [3,28], three types of this compound, 7,3$'$,5$'$-trihydroxyflavanone-7-*O*-(glucosyl glucoside), 7,3$'$,5$'$-trihydroxyflavanone-7,3$'$-*O*-di glucoside, and 7,3$'$,5$'$-trihydroxyflavanone-7,5$'$-*O*-diglucoside were predicted as possible structures; however, the three compounds have not been described previously. Peak 18 (t_R 6.91 min) produced a molecular ion at m/z 581.1501 [M – H]$^-$ (calculated for $C_{26}H_{29}O_{15}^-$, 581.1512) and the fragment ion at m/z 287.0552 [M – H – arabinose – glucose] by a loss of the arabinosyl-glucose. The fragment ion at m/z 167.0342 [M – H – arabinosyl-glucose – C_8H_6O] without the loss of H_2O indicated the presence of a hydroxy group in the B ring and three hydroxyl groups in the A ring. Thus, peak 18 was tentatively identified as 4$'$,5,7,8-tetrahydroxyflavanone-7-*O*-(6-*O*-arabinosyl-glucoside), which has not been described previously. Its aglycone, isocarthamidin (4$'$,5,7,8-tetrahydroxyflavanone), has not been reported in *Coreopsis* species; however, it has been isolated from the Asteraceae plant [35]. Peak 24 (t_R 7.26 min) produced a molecular ion at m/z 493.0984 [M – H]$^-$ (calculated for $C_{22}H_{21}O_{13}^-$, 493.0988) and fragment ions at 331.0447 [M – H – glucose]$^-$ and 316.0200 [M – H – glucose – CH_3]$^-$ by loss of a glucose and a methyl of the methoxy group, respectively. The fragment ion at m/z 164.9830 [M – H – glucose – CH_3 – H_2O – $C_8H_6O_2$]$^-$ produced a 3$'$-methoxy-4$'$-hydroxyphenyl group (or 3$'$-hydroxy-4$'$-methoxy phenyl group) for the B ring and a 3-hydroxy group in the C ring. Accordingly, peak 24 was tentatively identified as taxifolin 3$'$,7-dimethyl ether 3-*O*-glucoside [36]. Another candidate, taxifolin 4$'$,7-dimethyl ether 3-*O*-glucoside, has not been described previously; however, aglycones, taxifolin 4$'$,7-dimethyl ether, and taxifolin 3$'$,7-dimethyl ether have been reported in Asteraceae plants [37–39].

2.1.3. Chalcones

Peak 23 has the same fragment rules as flavanomarein; however, it has been known that flavanones have shorter retention times than chalcones in chromatographic elution [29]. Therefore, peak 23 (t_R 7.15 min) was identified as marein [3,26,27]. The aglycone of this compound, okanin (peak 32, t_R 8.26 min), produced identical fragment ions with peak 23 [3,26,27]. Similarly, peak 30 (t_R 8.02 min) showed the same molecular ion and fragment ions as peak 8, thus identified as coreopsin [3,26]. Butein (peak 38, t_R 9.20 min), the aglycone of peak 30, showed a molecular ion at m/z 271.0605 [M – H]$^-$ (calculated for $C_{15}H_{11}O_5^-$, 271.0612) and identical fragment ions with peak 30 [3,27]. Peak 13 (t_R 6.51 min) produced a molecular ion at m/z 611.1612 [M – H]$^-$, corresponding to the molecular formula $C_{27}H_{31}O_{16}^-$. At high energy scan, the fragment ions at m/z 449.1080, 287.0551, and 269.0393 were formed by the loss of one glucose, two glucoses, and H_2O, respectively, indicating the presence of two glucose groups and a 3,4-dihydroxyphenyl group for the B ring. Hence, peak 13 was tentatively identified as okanin-3$'$,4$'$-*O*-diglucoside, which has been isolated from *Bidens pilosa* [40]. Peak 33 (t_R 8.46 min) produced a molecular ion at m/z 611.1398 [M – H]$^-$, corresponding to the molecular formula $C_{30}H_{27}O_{14}^-$. At high energy scan, fragment ions at m/z 449.1109 [M – H – glucose]$^-$ and 287.0559 [M – H – 2glucose]$^-$ were produced by the loss of two glucoses and 269.0441 [M – H – 2glucose – H_2O]$^-$ by the loss of H_2O, and the fragment ion at m/z 151.0024 [M – H – 2glucose – H_2O

$- C_8H_6O]^-$ indicated the presence of a $3',4'$-dihydroxyphenyl group for the B ring and the absence of a 3-hydroxy group in the C ring. Therefore, peak 33 was tentatively identified as eriodictyol chalcone-7-O-(glucosyl glucoside) or eriodictyol chalcone-O-diglucoside, which have not been described previously. Eriodictyol chalcone-7-O glucoside, which has one glucose, has been found in *Antirrhinum majus* [41]. Peak 34 (t_R 8.73 min) produced a molecular ion at m/z 287.0553 [M – H]$^-$ and exhibited the same fragment pathway with that of peak 33, suggesting that it was an aglycone of peak 33, eriodictyol chalcone, which has been identified in *Coreopsis* species [42]. Peak 37 (t_R 8.84 min) produced a molecular ion at m/z 477.1396 [M – H]$^-$, corresponding to the molecular formula $C_{23}H_{25}O_{11}^-$. At high energy scan, fragment ions were produced at m/z 315.0864 [M – H –glucose]$^-$, 300.0624 [M – H –glucose – CH_3]$^-$, 297.0754 [M – H –glucose – H_2O]$^-$, 282.0527 [M – H – glucose – H_2O – CH_3]$^-$, 163.0747 [M – H – glucose – H_2O – C_8H_6O]$^-$, and 148.00524 [M – H – glucose – H_2O – CH_3 – C_8H_6O]$^-$. Therefore, peak 37 was tentatively identified as 4-methoxylanceoletin-$4'$-O-glucoside, which has been isolated from *C. lanceolata* [12], or lanceolein $2'$-methyl ether, which has not been described previously.

2.1.4. Flavones and Flavanols

Flavone having a double bond between C-2 and C-3 exhibits a molecular ion that is 2 Da less than that of flavanone or chalcone and characteristic fragment ions by the loss of C_8H_4O (116 Da) or $C_8H_4O_2$ (132 Da) [3]. Peak 7 (t_R 5.93 min) produced a molecular ion at m/z 609.1454 [M – H]$^-$ (calculated for $C_{27}H_{29}O_{16}^-$, 609.1461) and fragment ions at m/z 447.0932 [M – H – glucose]$^-$ and 285.0392 [M – H – glucose – glucose]$^-$ by the loss of two glucoses. Another fragment ion at m/z 151.0033 [M – H – 2glucose – H_2O – C_8H_4O]$^-$ indicated the presence of a $3',4'$-dihydroxyphenyl group for the B ring without a 3-hydroxy group in the C ring. Thus, peak 7 was tentatively identified as luteolin-7-O-sophoroside [3]. Peaks 22 (t_R 7.03 min) and 36 (t_R 8.80 min) produced molecular ions at m/z 447.0929 [M – H]$^-$ (calculated for $C_{21}H_{19}O_{11}^-$, 447.0927) and m/z 285.0398 [M – H]$^-$ (calculated for $C_{15}H_9O_6^-$, 285.0405) that were 162 Da and 324 Da less than that of peak 8, respectively, indicating that the sugar moiety was removed from C-$2''$ and C-7 in luteolin-7-O-sophoroside (peak 7). Accordingly, peaks 22 and 36 were tentatively identified as luteolin-7-O-glucoside and luteolin, respectively [3,26,27]. In addition, a molecular ion at m/z 431.0978 [M – H]$^-$ (calculated for $C_{21}H_{19}O_{10}^-$, 431.0984) for peak 29 (t_R 7.91 min) was 146 Da more than that of peak 36, indicating the addition of a rhamnose. Moreover, its fragment ions were similar to those of peaks 22 and 36. Thus, it was identified as luteolin-7-O-rhamnoside, which was first detected in *Coreopsis* species; however, it has been found in other plants, such as *Glechoma grandis* Kuprianova var. longituba, *Rumex algeriensis*, and *Cornulaca monacantha* [43–45]. Peak 15 (t_R 6.58 min) produced a molecular ion at m/z 609.1454 [M – H]$^-$ (calculated for $C_{27}H_{29}O_{16}^-$, 609.1461) and fragment ions at m/z 447.0932 [M – H – glucose]$^-$ and 285.0394 [M – H – 2glucose]$^-$ by the loss of two glucoses. The fragment ion at m/z 151.0033 [M – H – 2glucose – H_2O – $C_8H_4O_2$]$^-$ suggested the presence of a $3',4'$-dihydroxyphenyl group for the B ring and a 3-hydroxy group in the C ring. As a result, peak 15 was tentatively identified as fisetin-3,7-O-diglucoside, which was first detected in *Coreopsis* species; however, it has been found in other *Sophora* species [46]. Other glycosides of fisetin have not been previously described. Peak 20 (t_R 6.97 min) produced a molecular ion at 463.0885 [M – H]$^-$, corresponding to the molecular formula $C_{21}H_{19}O_{12}^-$. At a high energy scan, the fragment ion at m/z 301.0346 [M – H – glucose]$^-$ was observed by the loss of a glucose, and fragment ions at m/z 151.0034 [M – H – glucose – H_2O – $C_8H_4O_2$]$^-$ indicated the presence of a $3',4'$-dihydroxyphenyl group by the loss of H_2O and the presence of a 3-hydroxy group. Therefore, peak 20 was tentatively identified as quercetin-7-O-glucoside [3,26]. Peak 25 (t_R 7.33 min) showed a molecular ion at m/z 461.1085 [M – H]$^-$ (calculated for $C_{22}H_{21}O_{11}^-$, 461.1089). At high energy scan, fragment ions at m/z 299.0547 [M – H – glucose]$^-$, 283.0242 [M – H – glucose – O]$^-$, 165.0188 [M – H – glucose – O – H_2O – C_8H_4O]$^-$, and 133.0291 [M – H – glucose – O – $C_8H_6O_3$]$^-$ were observed, indicating the loss of a glucoside at a 3-hydroxy group in the C ring and the presence of 3′,4′-dihydroxyphenyl group for the B ring. Therefore, peak 25 was tentatively identified as 3,3′,4′-trihydroxy-7-methoxyflavone 3-*O*-glucoside, which has been reported in *Aptenia cordifolia* [47]. Peak 26 (t_R 7.58 min) produced a molecular ion at m/z 641.1141 [M – H]$^-$, corresponding to the molecular formula $C_{30}H_{25}O_{16}{}^-$. At high energy scan, fragment ions were produced at m/z 317.0294 [M – H – caffeoylglucose]$^-$, 301.0342 [M – H – caffeoylglucose – O]$^-$, 285.0381 [M – H – caffeoylglucose – O – O]$^-$, 179.0343 [M – H – $C_{15}H_9O_8$ – $C_6H_{10}O_4$]$^-$, 161.0224 [M – H – $C_{15}H_9O_8$ – $C_6H_{10}O_4$ – O]$^-$, 135.0447 [M – H – $C_{15}H_9O_8$ – $C_6H_{10}O_4$ – CO_2]$^-$, and 133.0289 M – H – caffeoylglucose – O – O – $C_7H_3O_4$]$^-$. This peak was tentatively identified as qurcetagetin-7-*O*-(6″-caffeoylglucoside), which was confirmed using the UNIFI local library and was first detected in *Coreopsis* species. However, this compound has been found in Asteraceae plants, such as *Gnaphalium uliginosum* and *Tagetes maxima* [48,49]. Peak 35 (t_R 8.74 min) produced a molecular ion at m/z 299.0555 [M – H]$^-$ (calculated for $C_{16}H_{11}O_6{}^-$, 299.0561). The fragment ions produced at m/z 284.0319 [M – H – CH_3]$^-$ by the loss of a methyl of a methoxy group were observed. Moreover, 151.0032 [M – H – CH_3 – $C_8H_5O_2$]$^-$ indicated the presence of a hydroxyl group at the B ring and a 3-hydroxy group in the C-ring. Other fragment ions at m/z 151.0032 [M – H – CH_3 – $C_8H_6O_2$]$^-$ and 133.0447 [M – H – CH_3 – $C_7H_3O_4$]$^-$ indicated the presence of a 4′-hydroxyphenyl group for the B ring with a 3-hydroxy group and two hydroxyl groups in the A ring, respectively. Thus, peak 35 was tentatively identified as kaempferide, by comparison of its mass spectrum with that in the MassBank database [31]. This compound was first detected in *Coreopsis* species; however, it has been found in Asteraceae plants, such as *Chrysanthemum morifolium*, *C. coronarium*, *Artemisia annua*, *Chromolaena odorata*, and *Filago germanica* [50–52]. Peak 39 (t_R 9.28 min) produced a molecular ion at m/z 269.0447 [M – H]$^-$ (calculated for $C_{15}H_9O_5{}^-$, 269.0450) and fragment ions at m/z 227.0351 [M – H– C_2H_2O]$^-$ and 117.0341 [M – H– $C_7H_4O_4$]$^-$. Thus, it was tentatively identified as apigenin [26].

2.1.5. Aurones

Aurones are also one of the characteristic flavonoids found in *Coreopsis* species [3,12,29]. Peak 19 (t_R 6.94 min) produced a molecular ion at m/z 431.0977 [M – H]$^-$ (calculated for $C_{21}H_{19}O_{10}{}^-$, 431.0984) and a fragment ion at m/z 269.0447 [M – H – glucose]$^-$ by the loss of a glucose. In addition, fragment ions at m/z 135.0447 [M – H – glucose – H_2O – C_8H_4O]$^-$ and 133.0447 [M – H – glucose – $C_7H_4O_3$]$^-$ were also observed, indicating the presence of a 3′,4′-dihydroxyphenyl group for the B ring. Thus, peak 19 was tentatively identified as sulfurein (sulfuretin-6-*O*-glucoside) [35,53]. Sulfuretin (peak 28, t_R 7.77 min), the aglycone of peak 19, showed a molecular ion at m/z 269.0449 [M – H]$^-$ (calculated for $C_{15}H_9O_5{}^-$, 269.0450) and identical fragment ions with peak 19 [27]. Peak 21 (t_R 7.01 min) showed a molecular ion at m/z 447.0929 [M – H]$^-$ (calculated for $C_{21}H_{19}O_{11}{}^-$, 447.0927), which is 16 Da more than that of peak 20, and fragment ions at m/z 285.0397 [M – H – glucose]$^-$, 135.0447 [M – H – glucose – H_2O – C_8H_4O – O]$^-$, and 133.0447 [M – H – glucose – $C_7H_4O_3$]$^-$, thus, identified as maritimein [3]. This identification was also confirmed by comparison with ESI-QTof-MS (negative ion mode) of maritimein in the MassBank database [31].

2.1.6. Polyacetylene

Polyacetylenes of various structures have been isolated from the genus *Coreopsis* [12,28]. In this study, peak 41 (t_R 9.45 min) produced a molecular ion at m/z 557.2219 [M – H]$^-$, a molecular formula of $C_{26}H_{37}O_{13}{}^-$. At high energy scan, fragment ions were produced at m/z 233.0650 [M – H – 2glucose]$^-$, 191.0554 [M – H – 2glucose – C_3H_6]$^-$, and 149.0441 [M – H – caffeoylglucose – C_5H_8O]$^-$. This peak was tentatively identified as lobetyolinin, which was confirmed by the UNIFI local library and first detected in *Coreopsis* species; however, it has primarily been found in *Lobelia* species [54,55].

2.2. DPP-IV Inhibitory Effects of the 70% Ethanol Extract Obtained from Coreopsis cultivars

Type 2 diabetes mellitus is determined by several factors, including pancreas β-cell dysfunction, insulin resistance, increased hepatic and intestinal glucose production, or deficient insulin secretion [56]. Recently, the incretin effect has been observed to be reduced in patients with type 2 diabetes mellitus, which is a symptom of increased insulin secretion induced by oral administration, such as eating a meal, compared to intravenous administration of glucose [56]. This effect is mediated by incretin hormones, glucagon-like peptide-1 (GLP-1), and glucose-dependent insulinotropic polypeptide (GIP), which stimulate insulin secretion from pancreatic β-cells and consequently increase the blood glucose level [57,58]. In the incretin system, an increase of the elimination of GLP-1 and GIP occurs primarily through enzymatic degradation of DPP-IV [59]. Thus, DPP-IV inhibition enhances the function of insulinotropic hormones. It improves glucose tolerance in patients with type 2 diabetes mellitus [58]. Hence, DPP-IV inhibitors have emerged as a new class of oral anti-diabetic agents, and synthetic compounds have mainly been used in current treatments with these inhibitors [59]. However, there have also been studies that show that DPP-inhibitors are derived from natural sources as promising candidates of functional foods or pharmaceuticals [12,60–63].

In this study, the 70% ethanol extract of original and mutant cultivars of *C. rosea* and *C. verticillata* confirmed their anti-diabetic effect using an in vitro DPP-IV inhibitor screening assay. All extracts inhibited DPP-IV activity in a concentration-dependent manner with IC_{50} values from 34.01 to 134.28 µg/mL (Table 3). The positive control, sitagliptin, exhibited an IC_{50} of 0.095 µM. In two different species, the cultivars of *C. rosea* (Groups I and III) showed less inhibition of DPP-IV than the cultivars of *C. verticillata* (Groups II, IV, and V). Of the 32 samples, 'Orange sunlight (No. 30)', which belongs to Group IV (*C. verticillata*), showed the greatest DPP-IV inhibitory effects. Thereafter, the most active cultivars with IC_{50} values less than 65 µg/mL were in the order of 'Golden sunlight (No. 26)', 'Golden ball No.48 (No. 18)', 'Golden ball No.42 (No. 17)', 'Red sunlight (No. 27)', 'Bright sunlight (No. 28)', and 'Golden ball No.21 (No. 15)', all of which belonged to *C. verticillata*. The DPP-IV inhibitory effects of six mutant cultivars, 'Lemon candy (No. 4)', 'Shiny pink (No. 5)', 'Uri-dream 01 (No. 6)', 'Luckyten5 (No. 7)', 'Luckyten9 (No. 8)', and 'Uri-dream red (No. 9)' were greater by 24–47 % than that of the original cultivar, 'Heaven's gate (No. 1)' in Group I, while other mutant cultivars, 'Luckyten 6 (No. 2)', 'Redfin (No. 3)', 'Uri-dream 07 (No. 10)', 'Uri-dream 06 (No. 11)', and 'Pink sherbet (No. 12)' had similar or lower efficacy. In Group II, except for 'Golden ball No.26 (N. 16)', four mutant cultivars, 'Golden ball No.18 (No. 14)', 'Golden ball No.21 (No. 15)', 'Golden ball No.42 (No. 17)', and 'Golden ball No.48 (No. 18)' showed a 5–26 % increase in the inhibitory effect of DPP-IV compared to the original cultivar, 'Citrine (No. 9)'. In Groups III and V, mutant cultivars exhibited similar or lower DPP-IV inhibitory effects than original cultivars. In Group IV, compared to the original cultivar, 'Route 66 (No. 25)', all mutant cultivars presented 7–37 % higher inhibitory effects of DPP-IV.

Among all Coreopsis cultivars samples, 'Orange sunlight (No. 30)' showed the best efficacy with an IC_{50} value of 34.01 µg/mL; however, 'Uri-dream red (No. 9)' (IC_{50}, 66.46 µg/mL) had the highest increase with 47% DPP-IV inhibitory activity compared to the original cultivar (IC_{50}, 125.29 µg/mL). Therefore, 'Orange sunlight (No. 30)' had the potential to develop as a functional food, such as a tea ingredient or a food additive for the prevention or treatment of type 2 diabetes. The mutant cultivars with a greater increase in activity compared to the original cultivar, such as 'Uri-dream red (No. 9)' may be used for studies to identify metabolites changed by mutation using multivariate analysis, and for further research on genomic mutation mechanism.

Table 3. Effects of the 70% ethanol extract of *Coreopsis* cultivars on dipeptidyl peptidase (DPP)-IV activity.

Group (Plant Name)	No.	Cultivar Names	DPP-IV Inhibitory Effects (IC$_{50}$, μg/mL) [1]
I (*C. rosea*)	1	Heaven's gate	125.29
	2	Luckyten 6	158.83
	3	Redfin	117.55
	4	Lemon candy	95.39
	5	Shiny pink	76.92
	6	Uri-dream 01	95.53
	7	Luckyten5	78.06
	8	Luckyten9	78.60
	9	Uri-dream red	66.46
	10	Uri-dream 07	118.13
	11	Uri-dream 06	134.28
	12	Pink sherbet	117.70
II (*C. verticillata*)	13	Citrine	56.86
	14	Golden ball No.18	53.55
	15	Golden ball No.21	49.64
	16	Golden ball No.26	63.84
	17	Golden ball No.42	45.01
	18	Golden ball No.48	41.44
III (*C. rosea*)	19	Pumpkin Pie	76.40
	20	Gold ring	87.62
	21	Golden ring	89.22
	22	Mini ball yellow	74.57
	23	Box tree	76.83
	24	Orange ball	124.88
IV (*C. verticillata*)	25	Route 66	54.87
	26	Golden sunlight	40.37
	27	Red sunlight	45.42
	28	Bright sunlight	47.58
	29	Yellow sunlight	50.45
	30	Orange sunlight	34.01
V (*C. verticillata*)	31	Moonbeam	60.61
	32	Moonlight sonata	61.15
		Sitagliptin [2]	0.095 (μM)

[1] Values are presented as the mean ± SD of three independent experiments. [2] Sitagliptin was used as the positive control.

2.3. Multivariate Analysis

Metabolite differences among original and mutant cultivars, *C. rosea* and *C. verticillate*, were examined based on the metabolite profiles analyzed by UPLC-QTof-MS. However, it was difficult to find differences among the samples in chromatograms. Therefore, PCA and OPLS-DA were used to provide an effective visualization for the classification and differentiation of a metabolome system.

To compare metabolites from different cultivars of *C. rosea* and *C. verticillata*, we performed PCA analysis on negative ion mode data obtained from UPLC-QTof-MS analysis. PCA analysis was performed with three principal components (PC1–PC3) describing variation explained, 0.66 of R^2X and predictive capability, 0.366 of Q^2. Eigenvalues for PC1 and PC2 were found to be 9.94 and 8.05, respectively, indicating these first two principal components explain a large amount of the variance in the data. PC3 showed a comparatively smaller eigenvalue of 3.13, which led us to choose only PC1 and PC2 for further analysis. As shown in Figure 2A, the first two principal components described 56.2% of the total variation (31.1% and 25.1% by PC1 and PC2, respectively), and 32 *Coreopsis* samples were clearly clustered into four groups. Group I and Group III were clustered together, indicating similar chemical profiles among samples, and these two groups were of the same species, *C. rosea*. This cluster also suggested that there is no distinct difference between original cultivars and other mutation cultivars induced from each original one. However, an exception was found in 'Luckyten 6 (No. 2)', which is one of the mutant cultivars artificially induced using herbicide from the original cultivar of Group I, 'Heaven's gate (No. 1)'. Alternatively, a distinct separation was observed from cultivars in Group II, Group IV, and Group V, although they were all included in *C. verticallata*. Group II demonstrated different chemical profiles compared with Group IV and Group V. However, Group II showed one clustering with no substantial deviations between the γ-irradiated mutant cultivars (No.14–No. 18) and the original cultivar 'Citrine (No. 13)'. In Group IV, the γ-irradiated mutant cultivar, 'Orange sunlight (No. 30)' was shown as the outlier, indicating that it had a different chemical profile than samples within the same group. Figure 2B, shows the derivation of markers primarily distributed among the four groups. However, this resulted in whole variability directions, with no distinction of variabilities among groups. Accordingly, we performed OPLS-DA analysis on the metabolite profiles between *C. rosea* cultivars (Group Cr) and *C. verticallata* cultivars (Group Cv) to find the differentiation and significant variances in these two species. Two clusters were clearly differentiated from each other according to species in the OPLS-DA model, with a cumulative R^2Y value of 1.00 and a cumulative Q^2 value of 0.94 (Figure 2C). However, 'Luckyten 6 (No. 2)' and 'Orange sunlight (No. 30)' were marginally out of each grouped sample area. The internal validation of OPLS-DA model was performed by a permutation test ($n = 200$). In permutation test, the intercept values of R^2 and Q^2 were 0.425 and -1.09 respectively. All permutations of the R^2 and Q^2 values to the left were lower than the original points to the right and the intersection of regression lines of the R^2 and Q^2 points on vertical axis was below 0.4 and -1.1, respectively (Figure S43). These values indicated OPLS-DA model of this analysis was strongly validated without overfitting of the original model. As shown in Figure 2D, the corresponding OPLS-DA S-plot enabled the derivation of 13 potential marker compounds responsible for separating two groups by being far from the center. Eight marker metabolites which were shifted in the same direction as Group Cv from the OPLS-DA score plot were peaks 2, 8, 19, 21, 27, 30, 36, and 38, indicating the most abundant markers in Group Cv. Five marker metabolites, peaks 1, 5, 20, 23, and 34, were at the highest level in Group Cr. The variable importance plot (VIP) (Figure 2E) confirms these 13 selected marker compounds are primarily responsible for the discrimination between Group Cr and Group Cv with high VIP values (VIP $\geq$ 1). Moreover, the variable average by group clearly shows differences of selected marker compounds (Figure 2F) in these groups.

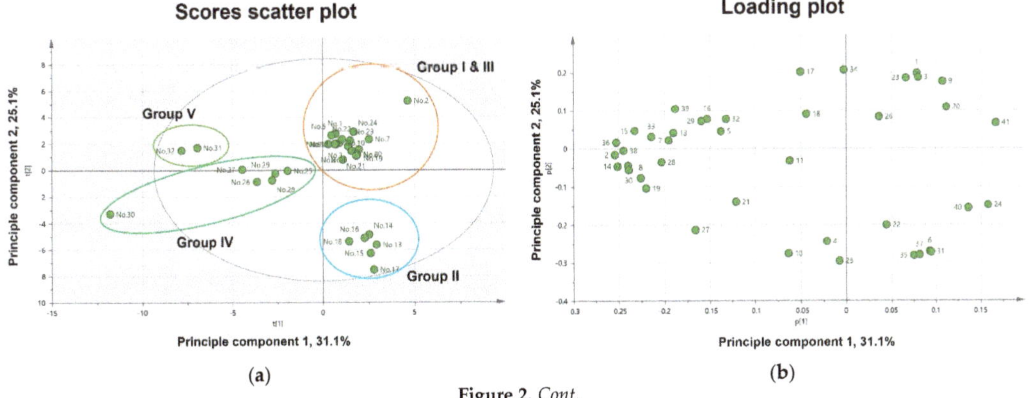

Figure 2. *Cont.*

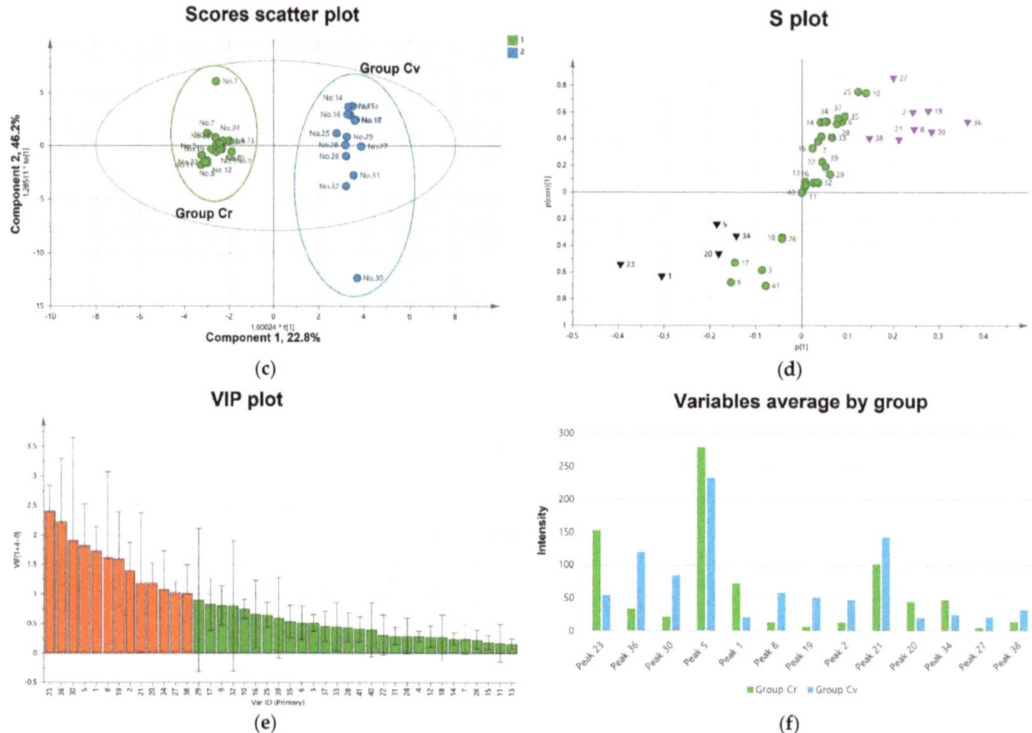

Figure 2. Principal component analysis (PCA) (**a**) score plot and (**b**) loading plot of metabolome analysis of the 32 *Coreopsis* cultivars; orthogonal partial least-squares discriminant analysis (OPLS-DA) (**c**) score plot and (**d**) S-plot show selected markers for differentiating *Coreopsis rosea* and *Coreopsis verticillata*; (**e**) Variable importance plot (VIP) scores of selected markers; (**f**) variables averages by group of selected potential marker compounds.

The similarities in chemical composition and relative quantitative differences among different cultivars of *C. rosea* and *C. verticallata* were clearly visualized on a heatmap with a dendrogram, while a hierarchical cluster analysis exhibited the same pattern of clustering as observed in PCA analysis (Figure 3). Heatmap is considered as one of the best tools for converting qualitative data into quantitative. Group I (No. 1–No. 12) and Group III (No. 19–No. 24) were clustered as one big cluster with similar distribution of areas of peaks 1, 3, 4, 5, 9, 11, 12, 13, 17, 20, 21, 22, 23, 24, 26, 27, 34, 40, and 41. 'Luckyten 6' (No. 2) was observed to have comparatively higher area values for peaks 1, 3, 20, and 34 than other cultivars in Group I, indicating the relatively high contents of these four peaks when compared to other samples in Group I. These four peaks could be responsible for making 'Luckyten 6' (No. 2) an outlier. Peaks 6, 25, 31, 35, and 37 appear with intense color in heatmap representing high quantity in comparison with other samples, which was responsible for the clustering of group II (No. 13–No. 18). Group IV has peaks 2, 8, 10, 14, 19, 28, 30, and 36 in abundance, while 'Orange sunlight' (No. 30) is rich in peaks 8, 10, 14, 19, 28, and 30 among groups. These six peaks' composition and relatively higher content could turn 'Orange sunlight' (No. 30) into an outlier in this statistical study. The contents of eight peaks 7, 14, 15, 29, 32, 33, 36, and 39 determine the clustering of group V (No. 31–No. 32), adjacent to group IV, sharing some similarities between them.

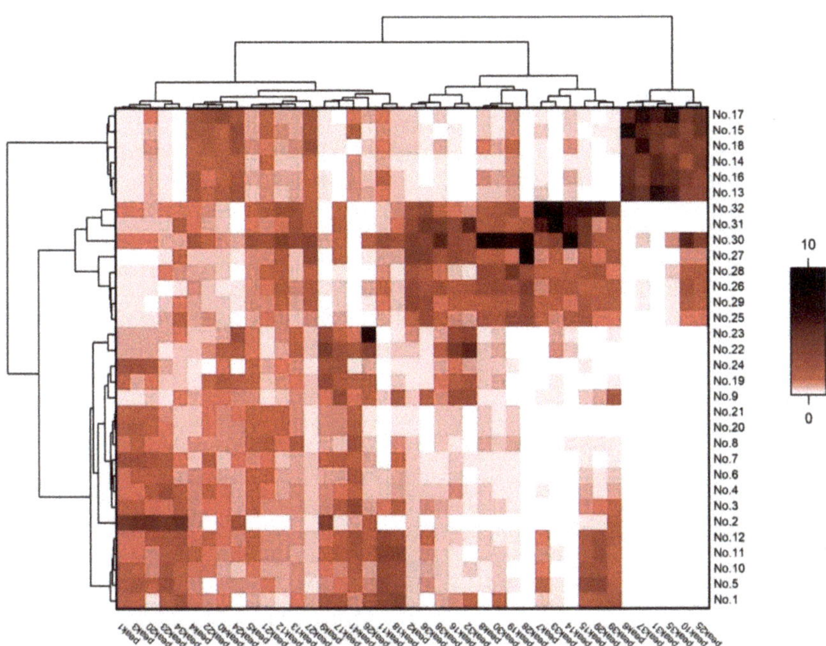

Figure 3. Hierarchical clustering analysis (HCA) with a heatmap from original and mutant cultivars of *Coreopsis* species.

The results of multivariate analyses to verify the correlation between metabolites and DPP-IV activities of the 32 *Coreopsis* samples were similar to the chemometric patterns between the two species. Given that DPP-IV inhibitory activities of *C. verticallata* cultivars appeared greater than that of *C. rosea* cultivars, distinguished metabolites between the active and inactive groups were almost identical to metabolites that showed differences between the two *Coreopsis* species presented in Table 3 (*C. verticallata*: $IC_{50} < 65$ μg/mL and *C. rosea*: $IC_{50} > 65$ μg/mL). Notably, 'Orange sunlight (No. 30)' and 'Luckyten 6 (No. 2)', which are outliers of Group Cv and Group Cr, respectively, were found to have the greatest and lowest DPP-IV inhibitory activity, respectively. These results suggested that the composition and relative content of distinguishable markers between *C. rosea* and *C. verticallata* cultivars were evaluated as key markers for the classification of species and contribution to the correlation of active and inactive cultivars.

3. Materials and Methods

3.1. Plant Material

Coreopsis cultivars were grown and collected from a wild cultivation field at Uriseed Group, Icheon-si, Gyeonggi-do, Republic of Korea and authenticated by Yeo Gyeong Jeon and Kong Young Park. These cultivars were selected according to their diverse phenotypic variants and exhibited a stable inheritance of these phenotypes for 4 years. Among them, five γ-irradiated mutants (Redfin, Lemone candy, Shiny pink, Uri-dream red, pink sherbet) of the original cultivar (Heaven's gate) and the series of γ-irradiated mutants of original cultivars (Citirne, Pumpkin pie, Route 66) were generated using γ (^{60}Co) irradiation (150 TBq capacity; AECL, Ottawa, ON, Canada). Six other mutant cultivars of 'Heaven's gate' (Luckyten 6, Uri-dream 01, Luckyten5, Luckyten9, Uri-dream 07, Uri-dream 06) were artificially mutated using an herbicide. 'Moonlight sonata' was selected as the phenotypic variation of the original cultivar 'Moonbeam'. Flowers used in this study were handpicked at the flowering stage in August 2018. These flowers were freeze-dried and stored at

−20 °C for further analysis. Voucher specimens were deposited at the Uriseed Group Corporation.

3.2. Sample Preparation

Freeze-dried flowers of *Coreopsis* cultivars were ground into powder using a mixer. Extractions were performed with 200 mg of this powder in 20 mL of 70% methanol using an ultrasonic bath for 60 min, and subsequently evaporated to achieve a dry product. Thereafter, these dried extracts (1 mg each) were dissolved in 1 mL of 70% methanol and filtered through a 0.20 μm polyvinylidene fluoride filter. Samples (1000 ppm) were diluted with 70% methanol to a concentration of 200 ppm for further liquid chromatography–mass spectrometry (LC-MS) analysis. For the evaluation of bioactivity, methanol extracts were initially dissolved in dimethyl sulfoxide (DMSO) at a concentration of 10 mg/mL stock solution. All extraction and chromatographic solvents used in this study were of analytical grade (J. T. Baker, Phillipsburg, NJ, USA).

3.3. Ultra-Performance Liquid Chromatography Time-of-Flight Mass Spectrometry (UPLC-QTof-MS) Analysis

A Waters ACQUITY UPLC I-Class system equipped with a binary solvent pump and an autosampler combined with a Xevo G2-XS QTof-MS (Waters, Milford, MA, USA) was used. Each sample (1 μL) was injected into a ACQUITY UPLC BEH C_{18} column (2.1 mm i.d. × 100 mm, 1.7 μm) at a flow rate of 0.5 mL/min. The temperature of the column oven was maintained at 15 °C. The mobile phase was composed of 0.1% formic acid in water (*v/v*; solvent A) and 0.1% formic acid in acetonitrile (*v/v*; solvent B). Gradient elution was carried out as follows: 0–1.0 min, 1% B; 1.0–7.0 min, 1–20% B; 7.0–11.0 min, 20–40% B; 11.0–14.0 min, 60–100% B; 14.0–14.5 min, 100% B; 14.5–15.0 min, 100–1% B and 15.0–17.0 min, 1% B. The mass spectrometer was operated in negative ion mode with the following parameters: source temperature, 120 °C; desolvation temperature, 400 °C; capillary voltage, 3.0 kV; cone voltage, 40 V; cone gas flow: 50 L/h; flow rate of desolvation gas (N_2), 1000 L/h; mass scan range, 550–1500 Da; scan time, 0.1 s. Leucine-enkephalin was used for the lock mass ([M - H]$^-$ m/z 554.2615). Full scan data, MS/MS spectra, accurate mass, and elemental composition were calculated using UNIFI software (Waters, Milford, MA, USA).

3.4. DPP-IV Inhibitor Screening Assay

DPP-IV activity of *Coreopsis* cultivars was analyzed using a DPP-IV inhibitor screening assay kit (Cayman Chemical, Ann Arbor, MI, USA) which provides a fluorescence-based method for screening DPP-IV inhibitors. The assay uses the fluorogenic substrate, Gly-Pro-Aminomethylcoumarine (AMC), to measure DPP-IV activity. Cleavage of the peptide bond by DPP releases the free AMC group, resulting in fluorescence that can be analyzed using an excitation wavelength of 350–360 nm and an emission wavelength of 450–465 nm. The tested samples dissolved in DMSO at a concentration of 10 mg/mL were subsequently diluted to a final concentration of 20 to 200 μg/mL using DMSO and were added to a 96-well plate to a final volume of 10 μL and a final concentration of 50 μM. The assay procedure is described in our previous studies [12,62,63]. Briefly, diluted assay buffer (30 μL) and diluted enzyme solution (10 μL) were added to the 96-well plate containing 10 μL of solvent (blank) or solvent-dissolved test samples. The reaction was initiated by adding 50 μL of a diluted substrate solution, and the plate was incubated for 30 min at 37 °C. Following incubation, fluorescence with an excitation wavelength of 350 nm and an emission wavelength of 450 nm was monitored using a plate reader (TECAN, Männedorf, Switzerland). The percent inhibition was expressed as ([DPP-IV level of vehicle-treated control − DPP-IV level of test samples]/DPP-IV level of vehicle-treated control) × 100. Subsequently, the 50% inhibitory concentration (IC_{50}) was determined using GraphPad Prism software (GaraphPad Software, La Jolla, CA, USA) via dose–response analysis.

3.5. Chemometric Data Analysis

Data management for the UPLC-QTof-MS analysis was performed using UNIFI software (Waters, Milford, MA, USA). MS data were processed using UNIFI to obtain a data matrix containing retention times, accurate masses, and normalized peak intensities. Parameters included retention time (t_R, range of 0.0–15.0 min), mass-to-charge ratio (m/z, range of 100–1500 Da), and a mass tolerance of 0.04 Da. The resulting data were evaluated using SIMCA 15.0.2 (Umetrics, Umeå, Sweden) for multivariate statistical analysis. Unsupervised principal component analysis (PCA) was performed using UV (univariate)-scaled and supervised orthogonal partial least-squares discriminant analysis (OPLS-DA) was used to identify and compare different metabolite sizes of the 32 samples. The quality of the OPLS-DA model was evaluated with R^2Y value and cumulative Q^2 value. The model was further validated with a permutation test (n = 200). Markers for the difference between groups were identified by analyzing the S-plot with pareto scaling, which were generated with covariance (p) and correlation (pcorr) data. These data sets were normalized by dividing with mean value to get a value between 0 and 10 and a heatmap with dendrograms was generated using OriginPro 2021 (OriginLab Corporation, Northampton, MA, USA) selecting ward for cluster method. Marker compounds were tentatively identified by comparison to published MS data in literature and databases such as Waters Local Library in UNIFI and Massbank [3,5,6,26–31].

4. Conclusions

To the best of our knowledge, a comparative metabolomics approach to identify metabolite composition and DPP-IV inhibitory activities in various cultivars of *C. rosea* and *C. verticillata*, were demonstrated for the first time in this study. UPLC-QTof-MS techniques were used to identify several phenolic acids, flavonoids and a polyacetylene in mutant cultivars compared to original cultivars. PCA and OPLS-DA results showed that metabolites discriminate between the mutant and original cultivars and between the two species. In addition, significant changes in metabolite content were observed under different DPP-IV inhibitory activities of cultivars, and chlorogenic acid, butin-7-*O*-glucoside, sulfuretin-6-*O*-glucoside, maritimein, 3,5-dicaffeoylquinic acid, coreopsin, luteolin, and butein were abundant in the active extracts. Therefore, the DPP-IV inhibitory cultivars and the metabolites influencing their activities would be favorable for the development of functional foods and the information of the metabolites accumulated differently for each mutant cultivar would be useful as a scientific reference for further studies on plant mutation mechanisms.

Supplementary Materials: The following are available online at https://www.mdpi.com/article/10.3390/plants10081661/s1, Figure S1: title, Table S1: title, Video S1: title. Figure S1: Representative UPLC-QTof-MS chromatograms of the 70% methanol extracts of the original cultivars at low CE scan (6 eV) for precursor (up) and high CE scan (20-50 eV) for fragment ions (down). (a) 'Heaven's gate' (No. 1), (b) 'Citrine' (No. 13), (c) 'Pumpkin Pie' (No. 19), (d) 'Route 66' (No. 25), and (e) 'Moonbeam' (No. 31), Figure S2: ESI-QTof-MS spectrum of taxifolin-7-*O*-glucoside (peak 1), Figure S3: ESI-QTof-MS spectrum of chlorogenic acid (peak 2), Figure S4: ESI-QTof-MS spectrum of taxifolin-3-*O*-glucoside (peak 3), Figure S5: ESI-QTof-MS spectrum of vanillic acid-4-glucoside (peak 4), Figure S6: ESI-QTof-MS spectrum of flavanomarein (peak 5), Figure S7: ESI-QTof-MS spectrum of isookanin-7-*O*-rutinoside (peak 6), Figure S8: ESI-QTof-MS spectrum of luteolin-7-*O*-sophoroside (peak 7), Figure S9: ESI-QTof-MS spectrum of butin-7-*O*-glucoside (peak 8), Figure S10: ESI-QTof-MS spectrum of 8-methoxyeriodictyol-7-*O*-glucoside (peak 9), Figure S11: ESI-QTof-MS spectrum of coreolanceoline B (peak 10), Figure S12: ESI-QTof-MS spectrum of lanceolin (peak 11), Figure S13: ESI-QTof-MS spectrum of naringenin-7-*O*-glucoside (peak 12), Figure S14: ESI-QTof-MS spectrum of okanin-3′,4′-*O*-diglucoside (peak 13), Figure S15: ESI-QTof-MS spectrum of 4′,7,8-trihydroxyflavone-*O*-diglucoside (peak 14), Figure S16: ESI-QTof-MS spectrum of fisetin-3,7-*O*-diglucoside (peak 15), Figure S17: ESI-QTof-MS spectrum of isookanin (peak 16), Figure S18: ESI-QTof-MS spectrum of taxifolin (peak 17), Figure S19: ESI-QTof-MS spectrum of 4′,5,7,8-tetrahydroxyflavanone-7-*O*-(6-*O*-arabinosyl-glucoside) (peak 18), Figure S20: ESI-QTof-MS spectrum of sulfuretin-6-*O*-glucoside (peak 19), Figure S21: ESI-QTof-MS spectrum of quercetin-7-*O*-glucoside (peak 20), Figure S22: ESI-QTof-MS spectrum of maritimein (peak 21), Figure S23: ESI-QTof-MS spectrum of luteolin-7-*O*-glucoside (peak 22), Figure S24: ESI-QTof-MS spectrum of marein (peak 23), Figure S25: ESI-QTof-MS spectrum of taxifolin 3′,7-dimethyl ether 3-*O*-glucoside (peak 24), Figure S26: ESI-QTof-MS spectrum of 3,3′,4′-trihydroxy-7-methoxyflavone 3-*O*-glucoside (peak 25), Figure S27: ESI-QTof-MS spectrum of qurcetagetin-7-*O*-(6″-caffeoylglucoside) (peak 26), Figure S28: ESI-QTof-MS spectrum of 3,5-dicaffeoylquinic acid (peak 27), Figure S29: ESI-QTof-MS spectrum of sulfuretin (peak 28), Figure S30: ESI-QTof-MS spectrum of luteolin-6-*O*-rhamnoside (peak 29), Figure S31: ESI-QTof-MS spectrum of coreopsin (peak 30), Figure S32: ESI-QTof-MS spectrum of 4,5-dicaffeoylquinic acid (peak 31), Figure S33: ESI-QTof-MS spectrum of okanin (peak 32), Figure S34: ESI-QTof-MS spectrum of eriodictyol chalcone-*O*-diglucoside (peak 33), Figure S35: ESI-QTof-MS spectrum of eriodictyol chalcone (peak 34), Figure S36: ESI-QTof-MS spectrum of kaempferide (peak 35), Figure S37: ESI-QTof-MS spectrum of luteolin (peak 36), Figure S38: ESI-QTof-MS spectrum of 4-methoxylanceoletin-4′-*O*-glucoside (peak 37), Figure S39: ESI-QTof-MS spectrum of butein (peak 38), Figure S40: ESI-QTof-MS spectrum of apigenin (peak 39), Figure S41: ESI-QTof-MS spectrum of unknown (peak 40), Figure S42: ESI-QTof-MS spectrum of lobetyolinin (peak 41), Figure S43: Validation plot of the OPLS-DA obtained from 200 permutation test.

Author Contributions: Conceptualization, A.-R.H., and S.-Y.K.; methodology, A.-R.H., C.H.J., and J.-W.N.; software, B.-R.K., S.B.P., J.P., and Y.-S.K.; validation, A.-R.H., H.C., and J.-W.N.; formal analysis, B.-R.K., S.B.P., J.P., and Y.-S.K.; investigation, B.-R.K., and S.B.P.; resources, Y.G.J., K.Y.P., S.-Y.K., and J.-B.K.; data curation, B.-R.K., S.B.P., J.P., and Y.-S.K.; writing—original draft preparation, B.-R.K., S.B.P., and A.-R.H.; writing—review and editing, A.-R.H., H.C., and J.-W.N.; visualization, B.-R.K., S.B.P., J.P., and Y.-S.K.; supervision, A.-R.H., and J.-W.N.; project administration, A.-R.H., C.H.J., and J.-B.K.; funding acquisition, A.-R.H., C.H.J., and J.-B.K. All authors have read and agreed to the published version of the manuscript.

Funding: This research was supported by Radiation Technology R&D program (No. 2017M2A2A6A05 018541) through the National Research Foundation of Korea (NRF) funded by the Ministry of Science, ICT & Future Planning.

Acknowledgments: The authors thank Waters Korea (Seoul, Republic of Korea) for technical support for UPLC-QTof-MS experiments.

Conflicts of Interest: The authors declare no conflict of interest.

References

1. Kim, S.-C.; Crawford, D.J.; Tadesse, M.; Berbee, M.; Ganders, F.R.; Pirseyedi, M.; Esselman, E.J. ITS sequences and phylogenetic relationships in *Bidens* and *Coreopsis* (Asteraceae). *Syst. Bot.* **1999**, *24*, 480–493. [CrossRef]
2. Pardede, A.; Mashita, K.; Ninomiya, M.; Tanaka, K.; Koketsu, M. Flavonoid profile and antileukemic activity of *Coreopsis lanceolata* flowers. *Bioorg. Med. Chem. Lett.* **2016**, *26*, 2784–2787. [CrossRef] [PubMed]
3. Yang, Y.; Sun, X.; Liu, J.; Kang, L.; Chen, S.; Ma, B.; Guo, B. Quantitative and qualitative analysis of flavonoids and phenolic acids in snow chrysanthemum (*Coreopsis tinctoria* Nutt.) by HPLC-DAD and UPLC-ESI-QTOF-MS. *Molecules* **2016**, *21*, 1307. [CrossRef]
4. Nakabo, D.; Okano, Y.; Kandori, N.; Satahira, T.; Kataoka, N.; Akamatsu, J.; Okada, Y. Convenient synthesis and physiological activities of flavonoids in *Coreopsis lanceolata* L. Petals and their related compounds. *Molecules* **2018**, *23*, 1671. [CrossRef]
5. Kim, H.-G.; Oh, H.-J.; Ko, J.-H.; Song, H.S.; Lee, Y.-G.; Kang, S.C.; Lee, D.Y.; Baek, N.-I. Lanceolein A–G, hydroxychalcones, from the flowers of *Coreopsis lanceolate* and their chemopreventive effects against human colon cancer cells. *Bioor. Chem.* **2019**, *85*, 274–281. [CrossRef]
6. Kim, H.-G.; Jung, Y.S.; Oh, S.M.; Oh, H.-J.; Ko, J.-H.; Kim, D.-O.; Kang, S.C.; Lee, Y.-G.; Baek, N.-I. Coreolanceolins A–E, new flavanones from the flowers of *Coreopsis lanceolata*, and their antioxidant and anti-inflammatory effects. *Antioxidants* **2020**, *9*, 539. [CrossRef] [PubMed]
7. Gaspar, L.; Oliveira, A.P.; Silva, L.R.; Andrade, P.B.; Pinho, P.G.D.; Botelho, J.; Valentão, P. Metabolic and biological prospecting of *Coreopsis tinctoria*. *Rev. Bras. Farmacogn.* **2012**, *22*, 350–358. [CrossRef]
8. Dias, T.; Bronze, M.R.; Houghton, P.J.; Mota-Filipe, H.; Paulo, A. The flavonoid-rich fraction of *Coreopsis tinctoria* promotes glucose tolerance regain through pancreatic function recovery in streptozotocin-induced glucose-intolerant rats. *J. Ethnopharmacol.* **2010**, *132*, 483–490. [CrossRef]
9. Wang, T.; Xi, M.; Guo, Q.; Shen, Z. Chemical components and antioxidant activity of volatile oil of a Compositae tea (*Coreopsis tinctoria* Nutt.) from Mt. Kunlun. *Ind. Crop. Prod.* **2015**, *67*, 318–323. [CrossRef]
10. Zhang, Y.; Shi, S.; Zhao, M.; Chai, X.; Tu, P. Coreosides A–D, C_{14}-polyacetylene glycosides from the capitula of *Coreopsis tinctoria* and its anti-inflammatory activity against COX-2. *Fitoterapia* **2013**, *87*, 93–97. [CrossRef] [PubMed]

11. Shang, Y.F.; Oidovsambuu, S.; Jeon, J.-S.; Nho, C.W.; Um, B.-H. Chalcones from flowers of *Coreopsis lanceolata* and thier in vitro antioxidative activity. *Planta Med.* **2013**, *79*, 295–300.
12. Kim, B.-R.; Paudel, S.B.; Nam, J.-W.; Jin, C.H.; Lee, I.-S.; Han, A.-R. Constituents of *Coreopsis lanceolate* flower and their dipeptidyl peptidase IV inhibitory effects. *Molecules* **2020**, *25*, 4370. [CrossRef]
13. Kimura, Y.; Hiraoka, K.; Kawano, T.; Fujioka, S.; Shimada, A. Nematicidal activities of acetylene compounds from *Coreopsis lanceolata* L. *J. Biosci.* **2008**, *63*, 843–847. [CrossRef]
14. Okada, Y.; Okita, M.; Murai, Y.; Okano, Y.; Nomura, M. Isolation and identification of flavonoids from *Coreopsis lanceolata* L. pentals. *Nat. Prod. Res.* **2014**, *28*, 201–204. [CrossRef]
15. Tanimoto, S.; Miyazawa, M.; Inoue, T.; Okada, Y.; Nomura, M. Chemical constituents of *Coreopsis lanceolata* L. and their physiological activities. *J. Oleo Sci.* **2009**, *58*, 141–146. [CrossRef] [PubMed]
16. Ma, Z.; Zheng, S.; Han, H.; Meng, J.; Yang, X.; Zeng, S.; Zhou, H.; Jiang, H. The bioactive components of *Coreopsis tinctoria* (Asteraceae) capitula: Antioxidant activity in vitro and profile in rat plasma. *J. Funct. Foods* **2015**, *20*, 575–586. [CrossRef]
17. Kim, H.-G.; Oh, H.-J.; Ko, J.-H.; Jung, Y.S.; Oh, S.-M.; Lee, Y.-G.; Kim, D.-O.; Baek, N.-I. Phenolic compounds from the flowers of *Coreopsis lanceolata*. *J. Appl. Biol. Chem.* **2019**, *62*, 323–326. [CrossRef]
18. Kim, B.-R.; Kim, H.M.; Jin, C.H.; Kang, S.-Y.; Kim, J.-B.; Jeon, Y.G.; Park, K.Y.; Lee, I.-S.; Han, A.-R. Composition and antioxidant activities of volatile organic compounds in radiation-bred *Coreopsis* cultivars. *Plants* **2020**, *9*, 717. [CrossRef] [PubMed]
19. Hou, Y.; Li, G.; Wang, J.; Pan, Y.; Jiao, K.; Du, J.; Chen, R.; Wang, B.; Li, N. Okanin, effective constituent of the flower tea *Coreopsis tinctoria*, attenuates LPS-induced microglial activation through inhibition of the TLR4/NF-κB signaling pathways. *Sci. Rep.* **2017**, *7*, 45705. [CrossRef]
20. Korea Seed & Variety Service. Available online: http://www.seed.go.kr/sites/seed_eng/index..do (accessed on 19 July 2021).
21. Ali, H.; Ghori, Z.; Sheikh, S.; Gul, A.E. Effects of gamma radiation on crop production. In *Crop Production and Global Environmental Issues*; Hakeem, K., Ed.; Springer: Cham, Switzerland, 2016; pp. 27–78.
22. Burnett, S.E.; Keever, G.J.; Kessler, J.R.; Cilliam, C.H. Foliar application of plant growth retardants to *Coreopsis rosea* 'American dream'. *J. Environ. Hort.* **2000**, *18*, 39–62. [CrossRef]
23. Park, K.-Y.; Hwang, H.-J.; Chae, W.-B.; Choi, G.-W. Development of a new *Coreopsis* variety 'Uridream Pink' by gamma-ray irradiation. *Kor. J. Hort. Sci. Technol.* **2014**, *32*, 906–911.
24. Sorrie, B.A.; LeBlond, R.J.; Weakley, A.S. Identification, distribution, and habitat of *Coreopsis* section *Eublepharis* (Asteraceae) and description of a new species. *J. Bot. Res. Inst. Tex.* **2013**, *7*, 299–310.
25. Kessler, J.R., Jr.; Keever, G.J. Plant growth retardants affect growth and flowering of *Coreopsis verticillata* 'Moonbeam'. *J. Environ. Hortic.* **2007**, *25*, 229–233. [CrossRef]
26. Peng, A.; Lin, L.; Zhao, M.; Sun, B. Classification of edible chrysanthemums based on phenolic profiles and mechanisms underlying the protective effects of characteristic phenolics on oxidatively damaged erythrocyte. *Food Res. Int.* **2019**, *123*, 64–74. [CrossRef] [PubMed]
27. Li, Y.; Yang, P.; Luo, Y.; Gao, B.; Sun, J.; Lu, W.; Liu, J.; Chen, P.; Zhang, Y.; Yu, L. Chemical compositions of chrysanthemum teas and their anti-inflammatory and antioxidant properties. *Food Chem.* **2019**, *286*, 8–16. [CrossRef]
28. Shen, J.; Hu, M.; Tan, W.; Ding, J.; Jiang, B.; Xu, L.; Hamulati, H.; He, C.; Sun, Y.; Xiao, P. Traditional uses, phytochemistry, pharmacology, and toxicology of *Coreopsis tinctoria* Nutt.: A review. *J. Ethnopharmacol.* **2021**, *269*, 113690. [CrossRef]
29. Zălaru, C.; Crişan, C.; Călinescu, I.; Moldovan, Z.; Ţârcomnicu, I.; Litescu, S.; Tatia, R.; Moldovan, L.; Boda, D.; Iovu, M. Polyphenols in *Coreopsis tinctoria* Nutt. fruits and the plant extracts antioxidant capacity evaluation. *Cent. Eur. J. Chem.* **2014**, *12*, 858–867. [CrossRef]
30. *UNIFI Scientific Information System, Driver Pack 2020 Release 1*; Waters: Milford, MA, USA, 2020.
31. MassBank. Available online: www.massbank.jp/Search (accessed on 19 July 2021).
32. Saleem, M.; Hareem, S.; Khan, A.; Naheed, S.; Raza, M.; Hussain, R.; Imran, M.; Choudhary, M.I. Dual inhibitors of urease and carbonic anhydrase-II from *Iris* species. *Pure Appl. Chem.* **2019**, *91*, 1695–1707. [CrossRef]
33. Wollenweber, E.; Valant-Vetschera, K.M.; Fernandes, G.W. Chemodiversity of exudate flavonoids in *Baccharis concinna* and three further south-american *Baccharis* species. *Nat. Prod. Commun.* **2006**, *1*, 627–632. [CrossRef]
34. Wollenweber, E.; Mann, K.; Doerr, M.; Fritz, H.; Roitman, J.N.; Yatskievych, G. Exudate flavonoids in three Ambrosia species. *Nat. Prod. Lett.* **1995**, *7*, 109–116. [CrossRef]
35. Abraham, J.; Thomas, T.D. Isolation, characterization and evaluation of antibacterial activity of a flavanone derivative 8-hydroxyl naringenin from *Elephantopus scaber* Linn. *World J. Pharm. Res.* **2015**, *4*, 2232–2240.
36. Rani, G.; Yadav, L.; Kalidhar, S.B. Phytochemical investigation of *Citrus sinensis* flavedo variety Blood Red. *J. Indian Chem. Soc.* **2011**, *88*, 887–888.
37. Wang, Y.-M.; Zhao, J.-Q.; Yang, J.-L.; Tao, Y.-D.; Mei, L.-J.; Shi, Y.-P. Separation of antioxidant and α-glucosidase inhibitory flavonoids from the aerial parts of *Asterothamnus centrali-asiaticus*. *Nat. Prod. Res.* **2017**, *31*, 1365–1369. [CrossRef]
38. Elshamy, A.I.; Mohamed, T.A.; Marzouk, M.M.; Hussien, T.A.; Umeyama, A.; Hegazy, M.E.F.; Efferth, T. Phytochemical constituents and chemosystematic significance of *Pulicaria jaubertii* E.Gamal-Eldin (Asteraceae). *Phytochem. Lett.* **2018**, *24*, 105–109. [CrossRef]
39. Shimokoriyama, M.; Hattori, S. Anthoclor pigments of *Cosmos sulphureus*, *Coreopsis lanceolata*, and C saxicola. *J. Am. Chem. Soc.* **1953**, *75*, 1900–1904. [CrossRef]

40. Hoffmann, B.; Hoelzl, J. Chalcone glucosides from *Bidens pilosa*. *Phytochemistry* **1988**, *28*, 247–249. [CrossRef]
41. Sato, T.; Nakayama, T.; Kikuchi, S.; Fukui, Y.; Yonekura-Sakakibara, K.; Ueda, T.; Nishino, T.; Tanaka, Y.; Kusumi, T. Enzymatic formation of aurones in the extracts of yellow snapdragon flowers. *Plant Sci.* **2001**, *160*, 229–236. [CrossRef]
42. Kaintz, C.; Molitor, C.; Thill, J.; Kampatsikas, I.; Michael, C.; Halbwirth, H.; Rompel, A. Cloning and functional expression in *E. coli* of a polyphenol oxidase transcript from Coreopsis grandiflora involved in aurone formation. *FEBS Lett.* **2014**, *588*, 3417–3426.
43. Li, J.; Wen, Q.; Feng, Y.; Zhang, J.; Luo, Y.; Tan, T. Characterization of the multiple chemical components of Glechomae Herba using ultra high performance liquid chromatography coupled to quadrupole-time-of-flight tandem mass spectrometry with diagnostic ion filtering strategy. *J. Sep. Sci.* **2019**, *42*, 1312–1322. [CrossRef] [PubMed]
44. Ammar, S.; Abidi, J.; Vlad Luca, S.; Boumendjel, M.; Skalicka-Wozniak, K.; Bouaziz, M. Untargeted metabolite profiling and phytochemical analysis based on RP-HPLC-DAD-QTOF-MS and MS/MS for discovering new bioactive compounds in *Rumex algeriensis* flowers and stems. *Phytochem. Anal.* **2020**, *31*, 616–635. [CrossRef]
45. Kandil, F.E.; Grace, M.H. Polyphenols from *Cornulaca monacantha*. *Phytochemistry* **2001**, *58*, 611–613. [CrossRef]
46. Ruiz, E.; Donoso, C.; Gonzalez, F.; Becerra, J.; Marticorena, C.; Silva, M. Phenetic relationships between Juan Fernandez and continental Chilean species of *Sophora* (Fabaceae) based on flavonoid patterns. *Bol. Soc. Chil. Quím.* **1999**, *44*, 351–356. [CrossRef]
47. Elgindi, M.R.; Elgindi, O.D.; Mabry, T.J. Flavonoids of *Aptenia cordifolia*. *Asian J. Chem.* **1999**, *11*, 1525–1527.
48. Olennikov, D.N.; Chirikova, N.K.; Kashchenko, N.I. Spinacetin, a New Caffeoylglycoside, and Other Phenolic Compounds from *Gnaphalium uliginosum*. *Chem. Nat. Compd.* **2015**, *51*, 1085–1090. [CrossRef]
49. Parejo, I.; Bastida, J.; Viladomat, F.; Codina, C. Acylated quercetagetin glycosides with antioxidant activity from *Tagetes maxima*. *Phytochemistry* **2005**, *66*, 2356–2362. [CrossRef]
50. Lai, J.-P.; Lim, Y.H.; Su, J.; Shen, H.-M.; Ong, C.N. Identification and characterization of major flavonoids and caffeoylquinic acids in three *Compositae* plants by LC/DAD-APCI/MS. *J. Chromatogr. B* **2007**, *848*, 215–225. [CrossRef]
51. Hung, T.M.; Cuong, T.D.; Nguyen, H.D.; Zhu, S.; Long, P.Q.; Komatsu, K.; Min, B.S. Flavonoid glycosides from *Chromolaena odorata* leaves and their *in vitro* cytotoxic activity. *Chem. Pharm. Bull.* **2011**, *59*, 129–131. [CrossRef]
52. Saleem, H.; Zengin, G.; Locatelli, M.; Tartaglia, A.; Ferrone, V.; Htar, T.T.; Naidu, R.; Mahomoodally, M.F.; Ahemad, N. *Filago germanica* (L.) Huds. bioactive constituents: Secondary metabolites fingerprinting and *in vitro* biological assays. *Ind. Crop. Prod.* **2020**, *152*, 112505. [CrossRef]
53. Nicholls, K.W.; Bohm, B.A. Flavonoids and affinities of *Coreopsis bigelovii*. *Phytochemistry* **1979**, *186*, 1076. [CrossRef]
54. Ishimaru, K.; Sadoshima, S.; Neera, S.; Koyama, K.; Takahashi, K.; Shimomura, K. A polyacetylene gentiobioside from hairy roots of *Lobelia inflate*. *Phytochemistry* **1992**, *31*, 1577–1579. [CrossRef]
55. Zhou, Y.; Wang, Y.; Wang, R.; Guo, F.; Yan, C. Two-dimensional liquid chromatography coupled with mass spectrometry for the analysis of *Lobelia chinensis* Lour. using an ESI/APCI multimode ion source. *J. Sep. Sci.* **2008**, *31*, 2388–2394. [CrossRef] [PubMed]
56. Nauck, M. Incretin therapies: Highlighting common features and differences in the modes of action of glucagon-like peptide-1 receptor agonists and dipeptidyl peptidase-4 inhibitors. *Diabetes Obes. Metab.* **2016**, *18*, 203–216. [CrossRef]
57. Mentlein, R. Dipeptidyl-peptidase IV (CD26)-role in the inactivation of regulatory peptides. *Regul. Pept.* **1999**, *85*, 9–24. [CrossRef]
58. Langley, A.K.; Suffoletta, T.J.; Jennings, H.R. Dipeptidyl peptidase IV inhibitors and the incretin system in type 2 Diabets Mellitus. *Pharmacotherapy* **2007**, *27*, 1163–1180. [CrossRef]
59. Gao, Y.; Zhang, Y.; Zhu, J.; Li, B.; Li, Z.; Zhu, W.; Shi, J.; Jia, Q.; Li, Y. Recent progress in natural products as DPP-4 inhibitors. *Future Med. Chem.* **2015**, *7*, 1079–1089. [CrossRef]
60. Kalhotra, P.; Chittepu, V.; Osorio-Revilla, G.; Gallardo-Velázquez, T. Structure-activity relationship and molecular docking of natural product library reveal chrysin as a novel dipeptidyl peptidase-4 (DPP-4) inhibitors: An integrated in silico and In Vitro study. *Molecules* **2018**, *23*, 1368. [CrossRef] [PubMed]
61. Parmar, H.S.; Jain, P.; Chauhan, D.S.; Bhinchar, M.K.; Munjal, V.; Yusuf, M.; Choube, K.; Tawani, A.; Tiwari, V.; Manivannan, E.; et al. DPP-IV inhibitory potential of naringin: An *in silico, in vitro* and *in vivo* study. *Diabetes Res. Clin. Pract.* **2012**, *97*, 105–111. [CrossRef] [PubMed]
62. Kim, B.-R.; Kim, H.Y.; Choi, I.H.; Kim, J.-B.; Jin, C.H.; Han, A.-R. DPP-IV Inhibitory Potentials of Flavonol Glycosides Isolated from the Seeds of *Lens culinaris*: In Vitro and *Molecular Docking* Analyses. *Molecules* **2018**, *23*, 1998. [CrossRef]
63. Kim, B.-R.; Thapa, P.; Kim, H.M.; Jin, C.H.; Kim, S.H.; Kim, J.-B.; Choi, H.J.; Han, A.-R.; Nam, J.-W. Purification of phenylpropanoids from the scaly bulbs of *Lilium longiflorum* by CPC and determination of their DPP-IV inhibitory potentials. *ACS Omega* **2020**, *5*, 4050–4057. [CrossRef]

Article

A Novel Bioanalytical Method for Determination of Inotodiol Isolated from Inonotus Obliquus and Its Application to Pharmacokinetic Study

Jin Hyeok Kim †, Dan Gao †, Chong Woon Cho, Inkyu Hwang, Hyung Min Kim * and Jong Seong Kang *

College of Pharmacy, Chungnam National University, Daejeon 34134, Korea; oojh52@cnu.ac.kr (J.H.K.);
gaodan521361@hotmail.com (D.G.); chongw113@naver.com (C.W.C.); hwanginkyu@cnu.ac.kr (I.H.)
* Correspondence: kimhm@cnu.ac.kr (H.M.K.); kangjss@cnu.ac.kr (J.S.K.)
† These authors contributed equally to this work.

Abstract: In this study, we developed a bioanalytical method using liquid chromatography coupled to triple quadrupole tandem mass spectrometry (LC-MS/MS) to apply to a pharmacokinetic study of inotodiol, which is known for its anti-cancer activity. Plasma samples were prepared with alkaline hydrolysis, liquid–liquid extraction, and solid-phase extraction. Inotodiol was detected in positive mode with atmospheric pressure chemical ionization by multiple-reaction monitoring mode using LC-MS/MS. The developed method was validated with linearity, accuracy, and precision. Accuracy ranged from 97.8% to 111.9%, and the coefficient of variation for precision was 1.8% to 4.4%. The developed method was applied for pharmacokinetic study, and the mean pharmacokinetic parameters administration were calculated as follows: λ_z 0.016 min^{-1}; $T_{1/2}$ 49.35 min; C_{max} 2582 ng/mL; Cl 0.004 ng/min; AUC_{0-t} 109,500 ng×min/mL; MRT_{0-t} 32.30 min; Vd 0.281 mL after intravenous administration at dose of 2 mg/kg and λ_z 0.005 min^{-1}; $T_{1/2}$ 138.6 min; T_{max} 40 min; C_{max} 49.56 ng/mL; AUC_{0-t} 6176 ng×in/mL; MRT_{0-t} 103.7 min after oral administration. The absolute oral bioavailability of inotodiol was 0.45%, similar to nonpolar phytosterols. Collectively, this is the first bioanalytical method and pharmacokinetic study for inotodiol.

Keywords: *Inonotus obliquus*; inotodiol; noncompartment analysis; pharmacokinetic study

check for updates

Citation: Kim, J.H.; Gao, D.; Cho, C.W.; Hwang, I.; Kim, H.M.; Kang, J.S. A Novel Bioanalytical Method for Determination of Inotodiol Isolated from Inonotus Obliquus and Its Application to Pharmacokinetic Study. *Plants* **2021**, *10*, 1631. https://doi.org/ 10.3390/ plants10081631

Academic Editors: Antonella Smeriglio and Sebastian Granica

Received: 25 June 2021
Accepted: 6 August 2021
Published: 9 August 2021

Publisher's Note: MDPI stays neutral with regard to jurisdictional claims in published maps and institutional affiliations.

1. Introduction

Belonging to the family *Hymenochaetaceae*, *Inonotus obliquus*, known as chaga mushroom, is a polypore fungus collected from birch trees that have been used to treat cancer in Russia [1]. Many bioactive components have been isolated, and their pharmacological and biological studies have been also reported. For example, lanosterol, inotodiol, 3β-hydroxylanosta-8,24-dien-21-al, 3β,22R-dihydroxylanosta-8,24-dien-11-one, and ergosterol peroxide act on various cancer cells such as those of the Michigan cancer foundation-7 breast adenocarcinoma cell line [2] and the P388 and L1210 leukemia cell lines [3,4], and other studies also have reported antitumor activity from various component inhibiting the growth of COLO 205 gastric adenocarcinoma cells, HeLa cervical adenocarcinoma cells, A-549 lung carcinoma cells, and PC3 prostate carcinoma cells [5]. Among those bioactive components, inotodiol has been studied for various types of biological or pharmacological effects: not only just antitumor activity but also stabilizing mast cells to alleviate food allergy and inhibiting apoptosis of PC12 cells induced by oxygen/glucose deprivation/reoxygenation, which is effective in reversing the effects of ischemic stroke [6].

Previous studies have already demonstrated various bioactivities of inotodiol, but the mechanism has not been well elucidated. This is a common problem in the development of nutraceuticals or functional food [7]. Therefore, a bioanalytical study for a pharmacokinetic and pharmacodynamics study is necessary to better understand the pharmacological mechanism. There are few previous studies that developed analytical methods for inotodiol determination [8,9]. However, these methods were developed for the detection of inotodiol in the chaga mushroom, which is not suitable for the analysis of inotodiol in the biological matrix because the content and criteria of bioanalytical method validation are distinct from common analytical method validation. Therefore, developing a determination method of inotodiol in the biological matrix is needed.

To develop an analytical method for inotodiol determination, applying an analytical method for sterols should be considered since the structure of inotodiol is similar to sterols (Figure S1). Analysis of sterols, especially sterol lipids, is challenging because their solubilities are poor in blood and lipoproteins, phospholipids and triglycerides interrupt to analyze sterols. Moreover, the majority of sterols circulate as steryl esters, which are esterified with fatty acids, and a small portion of sterols circulate as free sterol form in animals [10]. These two forms of sterols make it more difficult to quantify in an animal sample.

The Bligh and Dyer method and the Folch method have been used to extract sterols [11,12]. In both methods, alkaline hydrolysis is applied to degrade abundant lipids, such as triglycerides and phospholipids, and to change esterified sterol into free sterol. Then, solid-phase extraction (SPE) is applied for the sample cleanup step to isolate sterols from the biological matrix. Although the extraction method for the analysis of sterols is well studied, a method for the analysis of inotodiol is lacking.

In this study, we developed a bioanalytical method to quantify inotodiol in plasma using high-performance liquid chromatography with a triple quadrupole mass spectrometer. Validation was performed to evaluate the validity of the method, and a pharmacokinetic study was conducted to confirm the applicability. This is the first bioanalytical and pharmacokinetic study of inotodiol, and this study can be applied for a better understanding of absorption, distribution, metabolism, excretion, and toxicity.

2. Results and Discussion

2.1. Sample Preparation

Sample preparation steps consist of alkaline hydrolysis, liquid-liquid extraction and SPE. Suitable temperature and time are essential for alkaline hydrolysis since high temperature can cause degradation of sterol compounds as well as inotodiol. Samples prepared with alkaline hydrolysis showed more precise results with regard to the coefficient of variation (Table 1). The alkaline hydrolysis step significantly improved the coefficient of variation, which made the method suitable in accordance with the guideline on bioanalytical method validation of the European Medicines Agency (EMA), which recommends maintaining lower than 5% for the coefficient of variation in precision. Therefore, the alkaline hydrolysis step was included in the sample preparation steps.

Table 1. Plasma concentration of inotodiol (ng/mL) at 30 min after oral bolus administration with or without alkaline hydrolysis step.

	Dose of 40 mg/kg		Dose of 20 mg/kg	
	Without Hydrolysis	With Hydrolysis	Without Hydrolysis	With Hydrolysis
1	108	123	64	77
2	115	118	58	74
3	130	121	67	80
Mean	118	121	63	77
SD [1]	9.0	2.0	4.0	2.0
CV [2]	8.0	2.0	6.0	3.0

[1] SD: standard deviation, [2] CV: coefficient of variation (%).

After hydrolysis, dichloromethane (DCM) was used in this study to extract inotodiol. Hexane, DCM, and chloroform were commonly used for the extraction of sterols and oxysterols because of their low polarity. Chloroform and n-hexane are more toxic than

DCM. In addition, phosgene can be formed in chloroform, which could react with KOH to form dichlorocarbene [13].

2.2. Instrumental Analysis

Acetonitrile (ACN) and methanol with water are commonly used as a mobile phase with water for reverse-phase chromatography [14]. ACN was selected as a mobile phase in this study because it has stronger elution strength than methanol. Inotodiol has low polarity and should be analyzed with ACN to reduce the elution time.

Liquid chromatography coupled to triple quadrupole tandem mass spectrometry (LC–MS/MS) has been widely used to develop bioanalytical methods since it has good sensitivity and selectivity toward a complex sample matrix. As mass spectrometry is able to analyze the ionized compounds, the ionization method of inotodiol should be considered to acquire optimized results. Sterols are nonpolar compound which is not susceptible to ionization with electrospray ionization. Since the structure of inotodiol resembles sterols, ionization with atmospheric-pressure chemical ionization (APCI) interface was applied with neutral mobile phase without acidic additives [13,15].

Inotodiol showed high intensity at positive modes because it has two hydroxyl groups that normally are positively charged. The most common ionization form of hydroxyl groups is the protonated form $[M+H]^+$, and the second most common form is the loss of a hydroxyl group $[M-OH]^+$ [16]. Inotodiol, which has a molecular weight of 442, is ionized as the later form, and so the mass to charge ratio (m/z) value of the selected precursor ion was 425 at positive mode Figure 1a. The internal standard, triamcinolone acetonide (Figure S2), was detected in $[M+H]^+$ with an m/z value of 435. The selected-reaction monitoring program was set to monitor precursor-to-product-ion transitions for inotodiol (m/z 425→247) and triamcinolone (m/z 435→339) at -15 eV of the collision energy. According to the literature about the fragmentation pathway of sterols [17], the fragmentation process is illustrated in Figure 1c with MS/MS spectra. The precursor ion showed m/z 425, which means that carbocation occurred after dehydration $[M-H_2O+H]^+$. Inotodiol has two hydroxyl groups, which can be lost. Indeed, the signal observed at m/z 407 with similar intensity as one at m/z 425 corresponds to the removal of two hydrogen groups, $[M-2H_2O+H]^+$. A double bond after dehydration subsequently led to the loss of the side chain, which produced m/z 327 by a retro-ene reaction. In the second MS spectra (Figure 1b), signals with similar intensities were observed at m/z 247 and m/z 229. The mass difference between two ions (-18) indicated dehydration. Therefore, the fragment m/z 247 containing one hydroxyl group could be formed by ring cleavage of cyclopentane, which would lead to ring opening, loss of neutral ethylene, and the formation of a double bond [18].

2.3. Bioanalytical Method Validation

Specificity was confirmed by analysis of blank plasma and the lower limit of quantification (LLOQ) sample. The peaks of inotodiol and the internal standard were not detected with the developed the LC–MS/MS method in blank plasma samples from seven different mice (Figure 2).

The formula of calibration curve was y = 0.0053x − 0.0015 ($n = 6$). The calibration curve of inotodiol showed good linearity in the range of 4–300 ng/mL; the coefficient of determination was 0.9993. Additionally, the back-calculated concentrations of calibration standards were within 15% except that of LLOQ; the back-calculated concentration of LLOQ was less than 20% (Table S1).

The inotodiol peak was not observed in the blank sample (Figure S3) after analysis of the upper limit of the quantification (ULOQ) sample. This result confirmed that the previous run did not affect the later run for detecting inotodiol, and carryover was not observed.

Precision and accuracy were suitable for LLOQ and quality control samples (Table 2). The coefficient of variation for within-run precision ranged from 1.8% to 3.5%, and for between-run precision ranged from 3.6% to 4.4%, which accepted range was within 5%. Within-run accuracy ranged from 97.8% to 111.9%, and between-run accuracy ranged from 104.8% to 112.2%. The accepted range was from 85% to 115% except for the LLOQ sample.

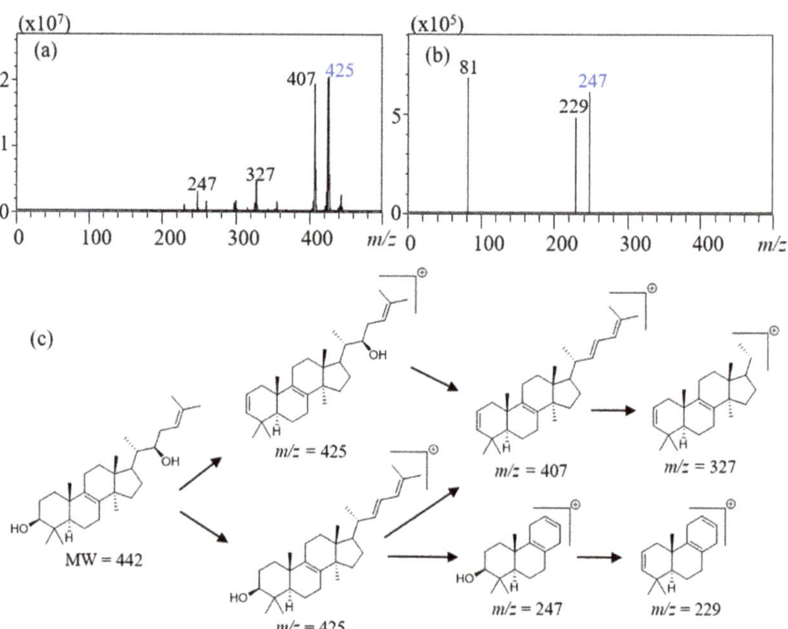

Figure 1. Atmospheric pressure chemical ionization coupled to tandem mass spectrometry and proposed fragmentation pathway; (**a**) precursor ion spectrum, (**b**) product ion spectrum, and (**c**) proposed fragmentation pathway. Mass spectrometry conditions: Interface; APCI, nebulizing gas flow; 3 L/min, interface temperature; 350 °C, desolvation line temperature; 200 °C, heat block temperature; 200 °C, drying gas flow; 5 L/min.

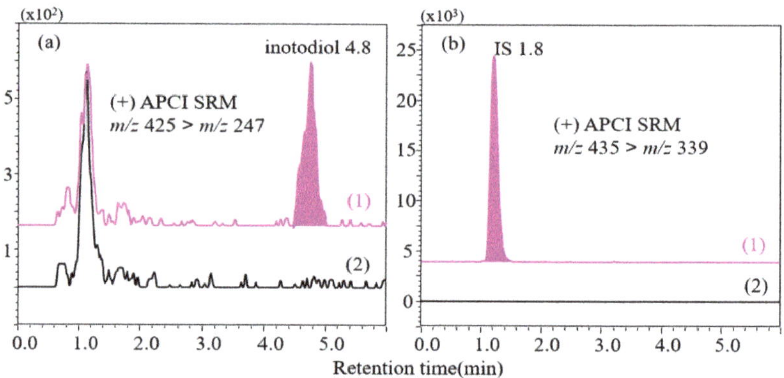

Figure 2. Selected reaction monitoring chromatograms of (**a**) inotodiol and (**b**) triamcinolone acetonide used as internal standard (IS); (**a1**) 4 ng/mL inotodiol in plasma and (**a2**) blank plasma; (**b1**) 150 ng/mL IS in plasma and (**b2**) blank plasma. Liquid chromatographic conditions: column; HECTOR-M C8 (75 × 2.1 mm, 3 μm), flow rate; 0.2 mL/min, eluent; 85% acetonitrile (isocratic).

Table 2. Within-run and between-run precision and accuracy of inotodiol (*n* = 3).

	Within Run		
Conc (ng/mL) [1]	Determined (ng/mL)	Accuracy (%)	CV (%) [2]
4	4.5 ± 0.1	111.9	1.8
10	9.8 ± 0.2	97.8	2.2
120	124.3 ± 4.4	103.6	3.5
230	233.0 ± 4.7	101.3	2.0
	Between Run		
Conc (ng/mL)	Determined (ng/mL)	Accuracy (%)	CV (%)
4	4.5 ± 0.2	112.2	3.6
10	10.7 ± 0.4	107.3	3.6
120	125.7 ± 4.2	104.8	3.4
230	247.0 ± 11.0	107.5	4.4

[1] Conc: concentration; [2] CV: coefficient of variation.

The internal standard–normalized matrix factors and their coefficients of variation for quality control samples are listed in Table S2. These matrix factors and low value of coefficients of variation indicate that the developed method reduced matrix effects, and the matrix did not significantly affect the analysis of inotodiol.

Table 3 showed the analysis results for stability of quality control samples. The ratios of the calculated concentration to the nominal concentration ranged from 96.2% to 99.8% for freeze-thaw stability and from 101.8% to 108.7% for short-term stability. Lastly, the percentages of calculated concentration divided by nominal concentration ranged from 100.2% to 112.5% for long-term stability.

The coefficient of variation for diluted samples ranged from 1.8% to 3.7%, and accuracy ranged from 94.8% to 107.5% (Table S3). Therefore, samples with concentrations higher than the ULOQ can be quantified by dilution.

Table 3. The stability of inotodiol in quality control samples (*n* = 3).

Freeze-Thaw Stability (Two Freeze/Thaw Cycles at −80 °C)			
[1] Conc (ng/mL)	Calculated (ng/mL)	Accuracy (%)	[1] CV (%)
10	10.0 ± 0.4	99.8	3.6
120	115.5 ± 1.9	96.2	1.6
230	228.1 ± 4.3	99.2	1.9
Short-Term Stability (Storage at Room Temperature for 2 Days)			
[1] Conc (ng/mL)	Calculated (ng/mL)	Accuracy (%)	[2] CV (%)
10	10.2 ± 0.4	101.8	3.7
120	125.2 ± 2.0	104.3	1.7
230	250.0 ± 6.4	108.7	2.6
Long-Term Stability (Storage for 30 Days at −80 °C)			
[1] Conc (ng/mL)	Calculated (ng/mL)	Accuracy (%)	[2] CV (%)
10	10.0 ± 0.62	100.2	6.2
120	131.1 ± 3.0	109.3	2.3
230	258.8 ± 2.7	112.5	1.0

[1] Conc: concentration; [2] CV: coefficient of variation.

2.4. Application of Pharmacokinetic Study

The developed method was applied to quantify inotodiol in plasma after oral and intravenous bolus administration. The mean plasma concentration-time curve is illustrated in Figure 3. In total, nine pharmacokinetic parameters were estimated: first-order rate constant associated with terminal portion of the log-linear curve (λ_z), half-life ($T_{1/2}$), peak time (T_{max}), peak plasma concentration (C_{max}), clearance (Cl), area under the plasma drug concentration-time curve from time zero to the latest measurable time (AUC_{0-t}), mean residence time (MRT_{0-t}), volume of distribution (Vd), and bioavailability (F, fraction of drug absorbed), and they are listed in Table 4.

After oral administration, inotodiol showed poor absorption with a mean C_{max} of 49.56 ng/mL and T_{max} of 40 min. AUC after oral administration was found to be 6176 ng×min/mL. After intravenous administration, Vd was estimated as 0.281 mL, which refers to poor distribution of inotodiol in organs and tissues. The absolute bioavailability is a very important parameter because bioavailability indicates delivery to the systemic circulation. The mean F was 0.45%, and such low F means that inotodiol was almost wasted. This poor F is known to result from poor intestinal absorption. According to a previous study, intestinal absorption of beta-sitosterol and campesterol were 0.42% and 0.63% after oral and intravenous administration, which were similar to our data [19].

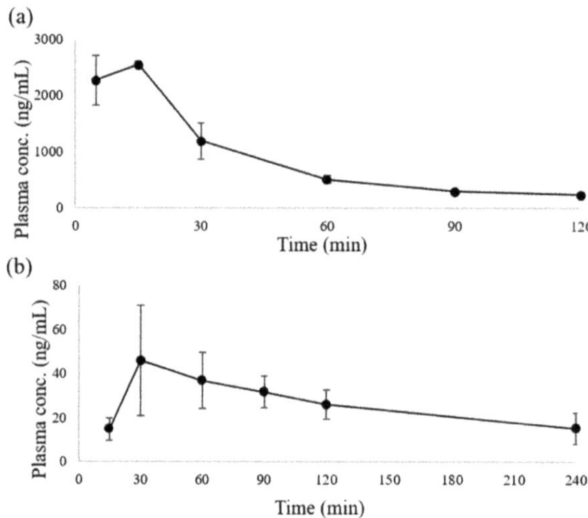

Figure 3. The mean plasma inotodiol concentration–time curve after (**a**) intravenous bolus administration at a dose of 2 mg/kg and (**b**) oral bolus administration at a dose of 20 mg/kg.

Table 4. Pharmacokinetic parameters of inotodiol after intravenous bolus administration and oral bolus administration (*n* = 3 for each group).

Parameters	[1] Intravenous	[1] Oral
λ_z (1/min)	0.016 ± 0.01	0.005 ± 0.00
$T_{1/2}$ (min)	49.35 ± 14.1	138.6 ± 30.3
T_{max} (min)		40 ± 14.1
C_{max} (ng/mL)	2582 ± 74.2	49.56 ± 14.2
Cl (ng/min)	0.004 ± 0.00	
AUC_{0-t} (ng×min/mL)	109500 ± 5670	6176 ± 455
MRT_{0-t} (min)	32.30 ± 0.55	103.7 ± 3.54
Vd (mL)	0.281 ± 0.08	
F (%)		0.45 ± 0.0

[1] data was represented as mean ± standard deviation.

Further study to improve absorption of inotodiol, such as investigation of pharmaceutical dosages and routes of administration, seems necessary to determine functional foods or health supplements. Additionally, a pharmacokinetic study with compartmental analysis should be conducted to better understand absorption, distribution, metabolism, and excretion of inotodiol.

2.5. Computational Prediction of Pharmacokinetics

The bioavailability radar (Figure 4a) provides information on the drug-likeness of a molecule, which describes lipophilicity, size, polarity, solubility, and saturation flexibility [20]. A pin area in the radar represents a suitable physicochemical range for oral bioavailability. According to the radar, inotodiol is not orally available because it is too lipophilic and insoluble. This finding corresponds to the absolute bioavailability described in Section 2.4.

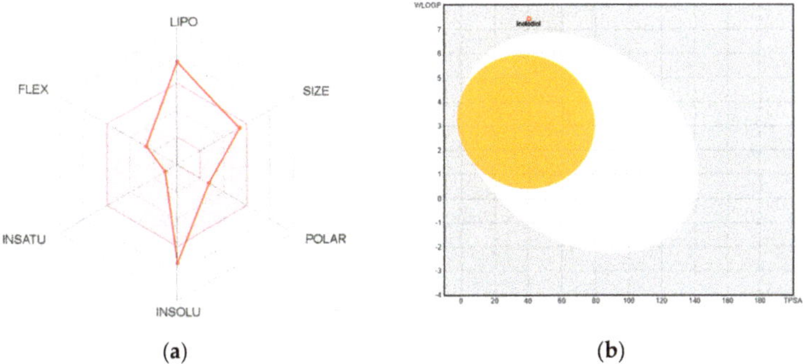

(a)   (b)

Figure 4. Computational prediction of bioavailability, gastrointestinal absorption, and brain penetration by SwissADME. (**a**) The bioavailability radar; LIPO: lipophilicity; POLAR: polarity; INSOLU: solubility; INSATU: saturation; FLEX: flexibility. (**b**) Boiled egg model for intuitive evaluation of passive gastrointestinal absorption and brain penetration.

Lipophilicity was presented by the partition coefficient between n-octanol and water (log Po/w), which is calculated by several computational methods. The log Po/w was 6.62, which indicates that inotodiol is poorly soluble in water.

The white region of the boiled egg model (Figure 4b) represents a high probability of passive absorption by the gastrointestinal tract, and the yellow region (yolk) represents a high probability of brain penetration. Since inotodiol is positioned in the outer area of the boiled egg region, we could speculate that inotodiol has poor gastrointestinal absorption and rarely permeate the blood–brain barrier.

3. Materials and Methods

3.1. Materials

Potassium hydroxide and polysorbate 80 (Tween 80) were purchased from Samchun Chemicals (Pyeongtaek, Republic of Korea). Triamcinolone acetonide, purchased from Acros Organics (Incheon, Republic of Korea), was used as the internal standard. Guaranteed-grade reagents such as n-hexane, chloroform, acetone, methyl alcohol, and ethyl alcohol for isolation were purchased from SK Chemicals (Seongnam, Republic of Korea). Water was purified using the Milli-Q system by MilliporeSigma (Burlington, MA, USA). High-performance liquid chromatography-grade ACN and DCM were purchased from Honeywell–Burdick & Jackson (Muskegon, MI, USA) for instrumental analysis.

3.2. Animals

Eight-week-old female BALB/c mice were purchased from Samtako BioKorea Co. (Osan, Republic of Korea) and accommodated in the Core Animal Facility in Chungnam National University, Daejeon, Republic of Korea. The animal study protocol used in this study was approbated by the Animal Ethics Committee of Chungnam National University (Approval Number: CNU-00570), and animal experiments followed the approbated protocol. The animals were housed for 1 week before the experiment.

3.3. Isolation and Purification of Inotodiol

The spectra (hydrogen-1 and carbon-13) were obtained on a 400-MHz Fourier transform nuclear magnetic resonance spectrometer (Bruker, Billerica, MA, USA), the standard for which was tetramethylsilane. Medium-pressure liquid chromatography was used on an Isolera One system (Biotage, Uppsala, Sweden) and silica gel mesh (70–230 mesh and 230–400 mesh; Merck, Whitehouse Station, NJ, USA) and YMC reversed-phase (RP)–18 resins (75 μm; Fuji Silysia Chemical Ltd., Kasugai, Japan) were used as absorbents in the column chromatography. Thin-layer chromatography with the YMC RP-18 resins was performed with the use of precoated silica gel 60 F254 and RP-18 F254S plates (both 0.25 mm; Merck, Darmstadt, Germany), and the spots were detected under ultraviolet light at 254 and 365 nm wavelengths and with 10% H2SO4, followed by heating for 3 to 5 min.

Pulverized *I. obliquus* (2.0 kg) was freeze-dried and extracted with 6 L ethanol in ultrasonic conditions for 6 h. Then, ethyl alcohol residue (600 g) was obtained after the evaporation of the solvent under reduced pressure. This residue was suspended in water and partitioned with DCM to produce DCM residue (200 g) and a water layer, respectively. The DCM residue was separated by elution with silica gel column chromatography with combinations of n-hexane and acetone (100:1, 50:1, 25:1, 10:1, and 5:1) to produce five subfractions (C-1 to C-5). Inotodiol (400 mg) was separated from subfraction C-2 (10.0 g) by medium-pressure liquid chromatography using acetone and water (3:1, v/v) as mobile phase.

Inotodiol: White crystals, $C_{30}H_{50}O_2$, 1H-NMR (δH ppm, 400 MHz); (0.67, 3H, H-18), (0.93, 3H, H-19), (0.88, 3H, H-21), (1.60, 3H, H-26), (1.69, 3H, H-27), (0.90, 3H, H-28), (0.76, 3H, H-29), and (0.82, 3H, H-30); 13C-NMR δC ppm (100 MHz): 35.4 (C-1), 27.6 (C-2), 78.7 (C-3), 38.7 (C-4), 50.2 (C-5), 18.9 (C-6), 26.3 (C-7), 134.4 (C-8), 134.0 (C-9), 36.8 (C-10), 20.8 (C-11), 28.9 (C-12), 44.5 (C-13), 49.2 (C-14), 30.7 (C-15), 30.7 (C-16), 47.0 (C-17), 15.5 (C-18), 18.0 (C-19), 41.5 (C-20), 12.4 (C-21), 73.2 (C-22), 27.0 (C-23), 121.2 (C-24), 134.8 (C-25), 25.7 (C-26), 17.9 (C-7), 27.7 (C-28), 15.2 (C-29), and 24.1 (C-30).

3.4. Sample Preparation

A stock solution of inotodiol was prepared at a concentration of 0.25 mg/mL of ACN. Working solutions were prepared by diluting the stock solution with ACN. The final concentrations of working solutions were 900, 690, 600, 360, 300, 150, 120, 75, and 12 ng/mL for calibration curve and quality control samples. A stock solution of the internal standard was prepared at a concentration of 0.65 mg/mL in ACN and then diluted to 450 ng/mL with ACN.

A total of 30 μL of the working solutions of inotodiol was added to 20 μL of plasma in 2-mL Axygen microcentrifuge tubes (Corning Incorporated, Corning, NY, USA). Then, 300 μL of 6 M KOH and methanol were also added. The tube was incubated at 55 °C for 90 min in a JSOF-Series forced convection oven (JS Research, Inc., Tokyo, Japan) to achieve alkaline hydrolysis. After incubation, 1 mL of DCM was added to the tube, and the sample was centrifuged at 325,953× g relative centrifugal field (RCF) for 6 min in a Smart R17 centrifuge (Hanil Scientific, Daejeon, Republic of Korea). The supernatant was decanted into another tube, and the residual DCM layer was evaporated by nitrogen gas. Then, 1 mL of DCM was added to the tube containing supernatant, and the tube was centrifuged as the earlier method. The upper layer was removed, and the lower layer was decanted into the tube. Lastly, the transferred solvent was evaporated by nitrogen gas, and 0.2 mL of 70% ACN was added to dissolve the extracts.

The next step was SPE. Inotodiol was extracted by Sep-Pak C18 3 cc Vac silica SPE column (Waters Corporation, Milford, MA, USA). The column was conditioned with 3 mL of ACN. The sample was loaded into the SPE column, and then the column was flushed with 5 mL of 70% ACN and 0.4 mL of ACN. Finally, inotodiol was eluted by 7 mL of ACN, and the eluent was placed in a 10-mL glass test tube. The sample was dried completely with nitrogen at 45 °C. Lastly, 60 μL of ACN and 30 μL of the internal standard working solution were added to the test tube, and it was sonicated for 30 s. The sample was filtered through a 0.2 μm polyvinylidene fluoride (PVDF) syringe filter (Whatman plc, Maidstone, UK) for instrumental analysis.

3.5. Instrumental Conditions

To perform liquid chromatography coupled to triple quadrupole tandem mass spectrometry (LC-MS/MS) analysis, a Prominence UFLC system (Shimadzu, Kyoto, Japan) was connected to an LCMS-8040 system (Shimadzu). The Prominence UFLC system was equipped with a degasser, a column oven, a pump, a module and an autosampler. After a 10 μL injection of the sample, analytes were identified through a HECTOR-M C8 high-performance liquid chromatography column (75 × 2.1 mm, 3 μm; RStech Corporation, Daejeon, Republic of Korea). The elution started from an 85 % ACN for 6 min, increased to 100% ACN immediately, held for 11 min, returned to 85 % ACN and equilibrated for 6 min at a flow rate of 0.2 mL/min. Detection was carried out by the APCI by multiple-reaction monitoring mode. Inotodiol and the internal standard were ionized mainly as $[M-OH]^+$ and $[M+H]^+$. Inotodiol was fragmented to m/z values of 229 and 247, and the internal standard was transmitted at m/z values of 339, 397, and 415.

3.6. Bioanalytical Method Validation

The developed method was validated in accordance with the guideline by EMA for selectivity, carryover, LLOQ, calibration curve, accuracy, precision, matrix effect, and stability [21]. Blank plasma for validation was obtained from seven different mice and was used for selectivity. Selectivity was proven by analysis of the blank plasma and spiked plasma samples. Carryover was evaluated in blank samples after analysis of the ULOQ sample. LLOQ was calculated from the signal-to-noise ratio. To draw the calibration curve in the range of 4 to 300 ng/mL, the blank matrix was spiked with six concentrations. The back-calculation was the calculated concentration by calibration divided by nominal concentration to prove the suitability of calibration.

To validate within-run precision and accuracy, quality control samples and LLOQ samples were analyzed six times. To determine between-run precision and accuracy, quality control and LLOQ samples were analyzed in different batches for 3 consecutive days. All samples for analysis of precision and accuracy were freshly made. Precision was calculated as the coefficient of variation, and accuracy was calculated as percentages of the nominal value divided by the value calculated from the calibration curve. The matrix effect (also known as recovery) was determined by the internal standard-normalized matrix factor calculated as the ratio of peak area in the presence of matrix to the peak area in the absence of matrix. To assess stability, quality control samples were spiked with three different conditions in triplicates for short-term stability (2 days at room temperature), long-term stability (30 days at −80 °C) and freeze-and-thaw stability (two freeze/thaw cycles at −80 °C). The dilution integrity was assessed by mean accuracy, and to obtain the coefficient of variation, five replicates of one-third of the sample that was diluted (from 330, 500, and 700 ng/mL) were analyzed to quantify samples with concentrations higher than the ULOQ.

3.7. Pharmacokinetic Study

Inotodiol was suspended in 0.5% polysorbate 80 and administered to two groups of mice for the pharmacokinetic study. In one group, a dose of 2 mg/kg was administered by intravenous bolus injection, and blood samples were collected in tubes containing heparin 5, 15, 30 60, 90, 120 min after intravenous administration. In the other group, a dose of 20 mg/kg was orally administered as a bolus, and blood samples were collected into heparinized tubes 15, 30, 60, 90, 120, 240 min after oral administration. Obtained blood samples were centrifuged at 1124205 RCF for 5 min to isolate plasma, and all samples were stored at $-80\ ^\circ$C before sample preparation. The noncompartment package of RStudio (freeware version 3.5.1) was used to estimate pharmacokinetic parameters.

The C_{max} and T_{max} were obtained from experimental data. λ_z was calculated as a slope of the linear regression, which was drawn as a semi-log plot. The $T_{1/2}$ was calculated by $0.693/\lambda_z$. The AUC_{0-t} was calculated from the plasma drug concentration-time curve in accordance with the trapezoidal rule. Cl was calculated as $\lambda_z \times Vd$. MRT_{0-t} was calculated as $AUMC_{0-t}/AUC_{0-t}$, which $AUMC_{0-t}$ means the area under the first moment curve. F was calculated as the dose-corrected AUC after oral administration divided by AUC after intravenous administration.

3.8. Computational Pharmacokinetic Study

Pharmacokinetic properties were predicted using SwissADME (http://www.swis sadme.ch/; retrived 7 November 2020) to compare the results of this pharmacokinetic study with the computational result and to predict additional physicochemical properties, absorption, distribution, metabolism, elimination, and toxicity properties. The isomeric simplified molecular input line entry system was found on Pubchem (https://pubche m.ncbi.nlm.nih.gov/; retrived 7 November 2020) and was used as input data for the computational study.

The bioavailability radar included six axes with six major properties for oral bioavailability. The pink area illustrated optimal values of six properties. Six parameters were estimated as follows: Saturation, the ratio of sp3 hybridized carbons over the total carbon count of the molecule should be at least 0.25; Size, the molecular weight calculated should be between 150 and 500 g/mol for size; Polarity, the topological polar surface area should be between 20 and 130 Å^2. Solubility, log S calculated with the estimating aqueous solubility directly from molecular structure mode should not exceed 6; Lipophilicity, octanol/water partition coefficients by guiding an additive model should be in the range from -0.7 to $+6.0$; Flexibility, the molecule should not have more than 9 rotatable bonds.

Passive gastro-intestinal absorption and blood-brain barrier permeation are predicted with the BOILED-Egg model, which defines favorable and unfavorable zones in the log Po/w versus polar surface area physicochemical space for passive diffusion through both physiological barriers.

4. Conclusions

To date, many studies have evaluated the bioactivities of inotodiol. However, pharmacokinetic and bioanalytical studies have not been reported. In this study, a reliable LC–MS/MS method focusing on precise and accurate quantification of inotodiol in mouse plasma was developed. This method was applied to pharmacokinetic profiles after intravenous and oral bolus administration of 2 mg/kg and 20 mg/kg of inotodiol, respectively. The developed method was validated in accordance with the guidelines of the EMA. Inotodiol was successfully detected in plasma, and pharmacokinetic profiles were well defined. The pharmacokinetic results revealed poor bioavailability. Therefore, oral administration of inotodiol is not an efficient way of administration. As such, other routes of administration or formulation should be further studied. Additionally, computational prediction results supported and explained the pharmacokinetic results. This is the first bioanalytical and pharmacokinetic study of inotodiol and can be employed in assessing metabolic profiles and other investigations.

Supplementary Materials: The following are available online at https://www.mdpi.com/artic le/10.3390/plants10081631/s1, Figure S1: Chemical structure of (a) inotodiol and (b) cholesterol, Figure S2: Chemical structure of Triamcinolone acetonide used for internal standard (IS), Figure S3: SRM of inotodiol for carry over (injection order: (1) ULOQ sample, (2) blank sample), Figure S4: The boiled egg model for intuitive evaluation of passive gastrointestinal absorption and brain penetration by Swiss ADME, Table S1: Percentages of the back-calculated concentration from the equation of calibration curve for inotodiol ($n = 3$), Table S2: IS-normalised matrix factor and coefficient of variation of quality control samples, Table S3: Dilution integrity of inotodiol after 3 times dilution ($n = 5$).

Author Contributions: Conceptualization, J.H.K., C.W.C., J.S.K. and H.M.K.; animal experiment, I.H.; analysis, J.H.K., D.G. and C.W.C.; data processing, J.H.K. and D.G.; writing, J.H.K., H.M.K. and J.S.K. All authors have read and agreed to the published version of the manuscript.

Funding: This work was funded by research fund of Chungnam National University.

Institutional Review Board Statement: The study accommodated in the Core Animal Facility in Chungnam National University, Daejeon, Republic of Korea. The animal study protocol used in this study was approbated by the Animal Ethics Committee of Chungnam National University (Approval Number: CNU-00570), and animal experiments followed the approbated protocol.

Informed Consent Statement: Not applicable.

Data Availability Statement: Not applicable.

Acknowledgments: This work was supported by research fund of Chungnam National University.

Conflicts of Interest: The authors declare no conflict of interest.

References

1. Ma, L.; Chen, H.; Dong, P.; Lu, X. Anti-inflammatory and anticancer activities of extracts and compounds from the mushroom *Inonotus obliquus*. *Food Chem.* **2013**, *139*, 503–508. [CrossRef] [PubMed]
2. Kahlos, K.; Kangas, L.; Hiltunen, R. Antitumor activity of triterpenes in *Inonotus obliquus*. *Planta Med.* **1986**, *52*, 495–554. [CrossRef] [PubMed]
3. Zheng, W.; Miao, K.; Liu, Y.; Zhao, Y.; Zhang, M.; Pan, S.; Dai, Y. Chemical diversity of biologically active metabolites in the sclerotia of *Inonotus obliquus* and submerged culture strategies for up-regulating their production. *Appl. Microbiol. Biotechnol.* **2010**, *87*, 1237–1254. [CrossRef] [PubMed]
4. Kim, Y.J. Chemical constituents from the sclerotia of *Inonotus obliquus*. *J. Korean Soc. Appl. Biol. Chem.* **2011**, *54*, 287–294. [CrossRef]
5. Nikitina, S.A.; Habibrakhmanova, V.R.; Sysoeva, M.A. Chemical composition and biological activity of triterpenes and steroids of chaga mushroom. *Biochem. Mosc. Suppl. Ser. B Biomed. Chem.* **2016**, *10*, 63–69. [CrossRef]
6. Yan, L.; Wenting, Z.; Chun, C.; Chunping, Z.; Jingyu, D.; Huankai, Y.; Qunli, W.; Aiguo, M.; Jun, S. Inotodiol protects PC12 cells against injury induced by oxygen and glucose deprivation/restoration through inhibiting oxidative stress and apoptosis. *J. Appl. Biomed.* **2018**, *16*, 126–132.
7. Le, V.N.H.; Zhao, Y.; Cho, C.W.; Na, M.; Quan, K.T.; Kim, J.H.; Hwang, S.Y.; Kim, S.W.; Kim, K.T.; Kang, J.S. Pharmacokinetic study comparing pure desoxo-narchinol A and nardosinonediol with extracts from *Nardostachys jatamansi*. *J. Chromatogr. B Analyt. Technol. Biomed. Life Sci.* **2018**, *1102–1103*, 152–158. [CrossRef] [PubMed]
8. Baek, J.; Roh, H.S.; Baek, K.H.; Lee, S.; Lee, S.; Song, S.S.; Kim, K.H. Bioactivity-based analysis and chemical characterization of cytotoxic constituents from Chaga mushroom (*Inonotus obliquus*) that induce apoptosis in human lung adenocarcinoma cells. *J. Ethnopharmacol.* **2018**, *224*, 63–75. [CrossRef] [PubMed]
9. Du, D.; Zhu, F.; Chen, X.; Ju, X.; Feng, Y.; Qi, L.W.; Jiang, J. Rapid isolation and purification of inotodiol and trametenolic acid from *Inonotus obliquus* by high-speed counter-current chromatography with evaporative light scatting detection. *Phytochem. Anal.* **2011**, *22*, 419–423. [CrossRef] [PubMed]
10. Quehenberger, O.; Armando, A.M.; Brown, A.H.; Milne, S.B.; Myers, D.S.; Merrill, A.H.; Bandyopadhyay, S.; Jones, K.N.; Kelly, S.; Shaner, R.L.; et al. Lipidomics reveals a remarkable diversity of lipids in human plasma. *J. Lipid Res.* **2010**, *51*, 3299–3305. [CrossRef] [PubMed]
11. Bligh, E.G.; Dyer, W.J. A rapid method of total lipid extraction and purification. *Can. J. Biochem. Physiol.* **1959**, *37*, 911–917. [CrossRef] [PubMed]
12. Folch, J.; Lees, M.; Stanley, G.H.S. A simple method for the isolation and purification of total lipides from animal tissues. *J. Biol. Chem.* **1957**, *226*, 497–509. [CrossRef]
13. McDonald, J.G.; Smith, D.D.; Stiles, A.R.; Russell, D.W. A comprehensive method for extraction and quantitative analysis of sterols and secosteroids from human plasma. *J. Lip. Res.* **2012**, *53*, 1399–1409. [CrossRef] [PubMed]
14. Jandera, P.; Hajek, T. Mobile phase effects on the retention on polar columns with special attention to the dual hydrophilic interaction-reversed-phase liquid chromatography mechanism, a review. *J. Sep. Sci.* **2017**, *41*, 145–162. [CrossRef] [PubMed]
15. Honda, A.; Yamashita, K.; Miyazaki, H.; Shirai, M.; Ikegami, T.; Xu, G.; Numazawa, M.; Hara, T.; Matsuzaki, Y. Highly sensitive analysis of sterol profiles in human serum by LC-ESI-MS/MS. *J. Lipid Res.* **2008**, *49*, 2063–2073. [CrossRef] [PubMed]
16. Le, V.N.H.; Lee, W.J.; Kim, Y.H.; Chae, G.H.; Chin, Y.W.; Kim, K.T.; Kang, J.S. High-performance liquid chromatography method development for the quality control of Ginkgonis Semen. *Arab. J. Chem.* **2017**, *10*, 792–800. [CrossRef]
17. Munger, L.H.; Boulos, S.; Nystrom, L. UPLC-MS/MS based identification of dietary steryl glucosides by investigation of corresponding free sterols. *Front. Chem.* **2018**, *6*, 1–19. [CrossRef] [PubMed]
18. Banerjee, S.; Mazumdar, S. Electrospray ionization mass spectrometry: A technique to access the information beyond the molecular weight of the analyte. *Int. J. Anal. Chem.* **2012**, *2012*, 1–40. [CrossRef] [PubMed]

19. Ostlund, R.E.; Mcgill, J.B.; Zeng, C.M.; Covey, D.F.; Stearns, J.; Stenson, W.F.; Spilburg, C.A. Gastrointestinal absorption and plasma kinetics of soy 5-phytosterols and phytostanols in humans. *Am. J. Physiol. Endocrinol. Metab.* **2002**, *282*, E911–E916. [CrossRef] [PubMed]
20. Daina, A.; Michielin, O.; Zoete, V. SwissADME: A free web tool to evaluate pharmacokinetics, drug-likeness and medicinal chemistry friendliness of small molecules. *Sci. Rep.* **2017**, *7*, 1–13. [CrossRef] [PubMed]
21. European Medicines Agency. Guideline on Bioanalytical Method Validation. 2011. Available online: https://www.ema.europa.e u/documents/scientific-guideline/guideline-bioanalytical-method-validation_en.pdf (accessed on 2 August 2019).

plants

Article

Determination of 1-Deoxynojirimycin (1-DNJ) in Leaves of Italian or Italy-Adapted Cultivars of Mulberry (*Morus* sp.pl.) by HPLC-MS

Lucia Marchetti [1,2], Alessio Saviane [3], Antonella dalla Montà [3], Graziella Paglia [3], Federica Pellati [1], Stefania Benvenuti [1], Davide Bertelli [1,*] and Silvia Cappellozza [3]

1   Department of Life Sciences, University of Modena and Reggio Emilia, Via G. Campi 103, 41125 Modena, Italy; lucia.marchetti@unimore.it (L.M.); federica.pellati@unimore.it (F.P.); stefania.benvenuti@unimore.it (S.B.)
2   Doctorate School in Clinical and Experimental Medicine (CEM), University of Modena and Reggio Emilia, 41125 Modena, Italy
3   Consiglio per la Ricerca in Agricoltura e L'Analisi Dell'Economia Agraria (CREA)-Centro per la Ricerca Agricoltura e Ambiente, Laboratorio di Gelsibachicoltura, Via Eulero, 6a, 35143 Padova, Italy; alessio.saviane@crea.gov.it (A.S.); antonella.dallamonta@crea.gov.it (A.d.M.); graziella.paglia@crea.gov.it (G.P.); silvia.cappellozza@crea.gov.it (S.C.)
*   Correspondence: davide.bertelli@unimore.it; Tel.: +39-059-205-8561

check for updates

Citation: Marchetti, L.; Saviane, A.; Montà, A.d.; Paglia, G.; Pellati, F.; Benvenuti, S.; Bertelli, D.; Cappellozza, S. Determination of 1-Deoxynojirimycin (1-DNJ) in Leaves of Italian or Italy-Adapted Cultivars of Mulberry (*Morus* sp.pl.) by HPLC-MS. *Plants* **2021**, *10*, 1553. https://doi.org/10.3390/plants10081553

Academic Editors: Jong-Seong Kang and Narendra Singh Yadav

Received: 25 June 2021
Accepted: 26 July 2021
Published: 28 July 2021

Abstract: Recently, 1-DNJ has been widely studied by scientists for its capacity to inhibit α-glucosidase and reduce postprandial blood glucose and fat accumulation. To the best of our knowledge, this is the first analytical determination of 1-DNJ in *Morus* sp.pl. leaves carried out on Italian crops, and it could be used as a reference to assess the quality of the plant material in comparison to Far Eastern Asia cultivations. The effects of two thermal treatments were compared to test the incidence of the drying process on the 1-DNJ extractability. In addition, two harvesting seasons in the same year (2017) and two subsequent harvesting years (2017–2018) were considered. The amount of 1-DNJ herein found was comparable to that reported in the scientific literature for Asian cultivations. The increase in 1-DNJ along the summer and the higher level of this compound in the apical leaves also complies with previous findings. However, a strong implication for the climatic conditions in the different years and a significant interaction between climate and genotypes suggest exploring very carefully the agronomic practices and selecting cultivars according to different environmental conditions with a view to standardize the 1-DNJ amount in leaves.

Keywords: *Morus* sp. pl.; cultivar; mulberry; 1-DNJ; HPLC-ESI-MS; HILIC

1. Introduction

Some peculiar characteristics of the third millennium in the Old Continent can be summarized in a few words: increased life expectancy, ageing population, and wide media coverage of health-related topics. Consequently, Europe is a promising market for natural ingredients and companies are continuously launching new natural health products; on the other hand, European consumers are very sensitive to any advertising regarding complementary and alternative medicine. These trends are not expected to change soon so that demand for pharmaceuticals, supplements, and alternative medicine products are forecast to increase in the European market as well as the demand for nutraceuticals. These substances are capable of giving physiological benefits or providing protection against chronic diseases [1]. Among natural products extracted from plants and other sources, iminosugars are analogs of sugars in which an atom of nitrogen replaces the ring oxygen atom. This substitution results in the inhibition of glycosidases and glycosyl-transferases, therefore preventing normal carbohydrate metabolism [2]. The compound 1-Deoxynojirimycin (1-DNJ) is one of the most popular iminosugars and was first extracted from mulberry (*Morus* sp.pl.) roots in 1976 [3]. This active ingredient, along with others, interferes with sugar metabolism, and was found at surprisingly high concentrations in the mulberry leaf latex. Indeed, it is involved in the mechanism of defense against insects, as the mulberry leaves are highly toxic for most species of Lepidopterans others than *Bombyx mori* [4]. Even though active ingredients contained in the mulberry leaf are harmful to herbivorous insects, man learnt to use their therapeutic effect from ancient times [5,6]. Furthermore, from the end of the last century 1-DNJ has been widely studied by scientists for its capacity to inhibit α-glucosidase and tumor cell metastasis, reducing postprandial blood glucose and fat accumulation, contrasting aging-related behaviors and HIV progression [7]. Due to its powerful and wide range of possible applications in therapeutics, various efforts have also been carried out to identify the highest-yielding mulberry varieties [8–12], the richest branch portions in 1-DNJ leaf content [8,9,11,13], the optimal harvesting season [14], the influence of the cultivation area and its interaction with the mulberry varieties [14], in addition to the best leaf processing and extraction method [7,9,15]. Furthermore, as mulberry leaves are the only feed of the silkworm *B. Mori*, which is efficient at concentrating the active ingredient in its body tissues, researchers have also been investigating this insect as a precious bio-accumulator [7,16,17]. However, most current research on 1-DNJ has been carried out in Far Eastern Asia (China, India, Korea, Thailand, and Japan), while few data are available on the content of 1-DNJ in the mulberry cultivars used in Europe. Therefore, in the current paper, we tried to fill this gap of knowledge with regard to the European genetic material, by also taking already-tested Japanese mulberry cultivars, which were grown under European conditions, as a reference. Furthermore, since the correlation among the cultivation area, the genetic material, and the environmental conditions (season of harvesting) has already been proven [14], we repeated the test on different cultivars in two subsequent years. The aim was to ascertain whether the difference in the 1-DNJ level, although variable in absolute value along the years, would respect the same ranking, being affected by the plant genetic constitution more than environmental conditions. Another aspect we took into consideration was the treatment of the collected material (leaf) before extraction, which was freezing of fresh samples or drying. In fact, regarding this aspect information is not very exhaustive because it is generally accepted that 1-DNJ is quite resistant to high temperatures [9,18]; but this topic has not been definitely tested.

Different methods for the determination of 1-DNJ are available in the literature, and other quantitative techniques are constantly being developed. Regardless, some issues have to be considered from an analytical point of view. First, due to the high hydrophilicity and small molecular weight of this compound, the interaction with the stationary phase of conventional reversed phase columns is so weak that 1-DNJ is not retained in the column. Moreover, since the molecule lacks a chromophore, direct UV detection is not suitable like for many other aminoglycosides. Some authors reported pre-column derivatization methods, which, however, involve sample manipulation and ultimately may result in the introduction of impurities or degradation products [19]. On the other hand, derivatization is unnecessary for an HPLC-evaporative light scattering detector (HPLC–ELSD) and HPLC-tandem mass spectrometry (HPLC–MS/MS). Kimura et al. successfully used an evaporative light scattering detector (ELSD), which is a universal detector, for relatively non-volatile analytes [20]. The main drawbacks are the poor sensitivity and the long time required to complete the analysis. Hydrophilic interaction chromatography (HILIC), coupled with mass spectrometry, has proven to be a promising technique for 1-DNJ quantitation. In our opinion, this is the most convenient method since it showed the lowest LOD and LOQ and the shortest retention time.

2. Results and Discussion

The herein used analytical method employs hydrophilic interaction chromatography (HILIC). The hybrid quadrupole-orbitrap mass parameters were optimized under positive ion electrospray ionization. The compound 1-Deoxynojirimicin was detected by MS/MS with parallel reaction monitoring (PRM) for transition of the parent ion to the product ions. In the reported experimental conditions, retention time (t_R) of 1-DNJ was 8.57 min, with minimal shifts between samples (Figure 1). The chromatogram of 1-DNJ standard solution (6.67 ppm) is shown in Figure 2.

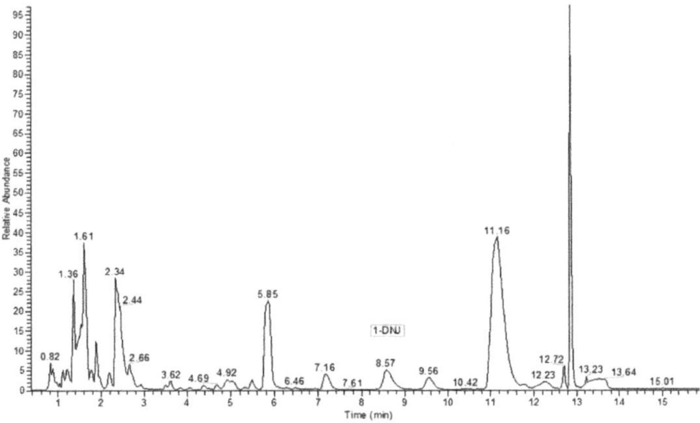

Figure 1. Typical base peak chromatogram of mulberry leaf hydroalcoholic extract.

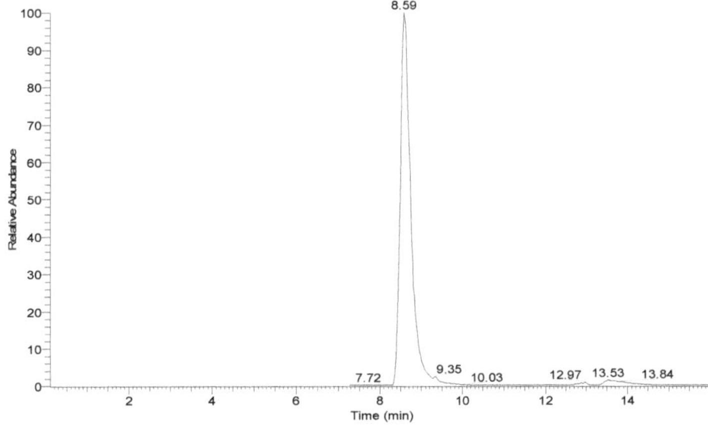

Figure 2. Base peak chromatogram of 1-DNJ standard solution (6.67 ppm).

Figure 3 shows 1-DNJ mass spectrum; an intense molecular ion [M+H]$^+$ m/z 164.0919 and other intense fragment ions generated by the consecutive loss of water molecules are present: m/z 60.0453 (8), 69.0343 (8), 80.0502 (8), 110.0604 (14) [M+H$-$3H$_2$O]$^+$, 128.0708 (11) [M+H$-$2H$_2$O]$^+$, 146.0812 (33) [M+H$-$H$_2$O]$^+$, 164.0919 (100). The integral of the product ion peak (m/z 146.0811) was used for the quantification.

The residual precursor ion and other fragment peaks were used as qualifier ions (Figure 4).

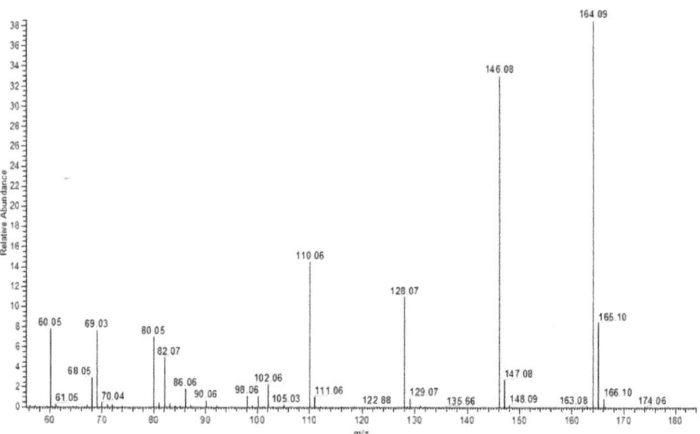

Figure 3. ESI mass spectrum of 1-DNJ.

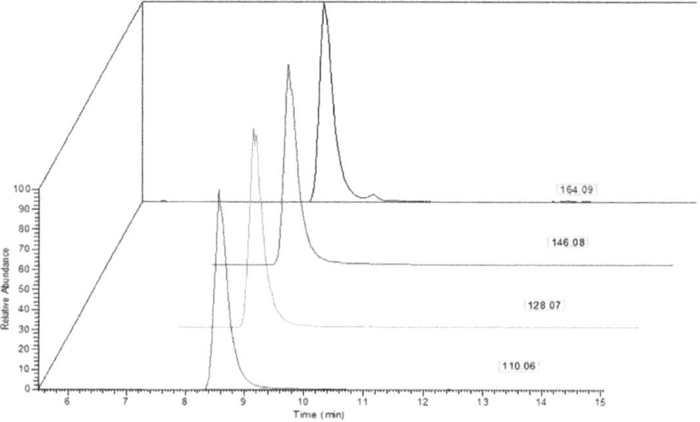

Figure 4. Chromatograms of extracted ion peaks used for quantification of 1-DNJ and as qualifier ions.

Nakagawa et al. previously applied HILIC–MS/MS for the simultaneous determination of DNJ and its derivatives. In this case, the selective detection of DNJ was achieved by monitoring the transition m/z 164 to 69 $[M+H-95]^+$ [21], thus indicating that MRM was adaptable for the HILIC–MS/MS analysis of iminosugars in general. During recent decades, HILIC has proven to be an efficient strategy for the analysis of highly polar compounds, including sugars and peptides. Moreover, the presence of amino groups is prone to the elongation of retention time [22]. Under the herein optimized conditions, DNJ was clearly detected at a reasonable retention time.

Data in Tables 1–3 are expressed as the overall mean of two sample replicates, each analyzed twice, ±SD (standard deviation). Data were obtained from the analysis of dried leaves and are expressed as 1-DNJ mg/g dry weight (DW). The content of 1-DNJ was determined in both fresh frozen and dried leaves in order to evaluate the influence of the drying process (50 °C) on the active compound extractability, and to assess the potential loss due to thermal degradation. Indeed, we tried to assess if the preservation of the mulberry leaf (freezing or drying) is influent on the extractable DNJ content. With this aim, a representative portion of frozen leaves was ground in an automatic mill and extracted with ethanol 50%. The same procedure was repeated on dried leaves.

Table 1. 1-DNJ content of 31 cultivars from five different *Morus* species harvested at the end of May 2017. Data are expressed as mg of 1-DNJ/g of dry leaf ± SD. Dry samples data are shown in comparison with fresh samples corrected by the relative humidity.

Cultivar	End-May 2017		
	Humidity (%)	Fresh Samples (mg/g of Dry Matter) [a]	Dry Samples (mg/g of Dry Matter) [a]
Aobanezumi (*M. alba* L.)	69.4	0.55 ± 0.02 m,n	0.44 ± 0.06 z
Cattaneo female (*M. alba* L.)	66.3	0.66 ± 0.02 m,n	0.59 ± 0.03 z,v
Cattaneo male (*M. alba* L.)	70.5	1.87 ± 0.02 g,h,i	1.50 ± 0.03 o,p,q
Egyptienne (*M. alba* L.)	52.9	1.78 ± 0.04 g,h,i	1.66 ± 0.08 o,n
Florio (*M. alba* L.)	70.7	0.74 ± 0.02 n,m	0.55 ± 0.02 v,z
Giazzola (*M. alba* L.)	67.7	1.33 ± 0.05 i,l	1.20 ± 0.05 s,t
Kayriounezumigaeshi (*M. alba* L.)	68.1	1.99 ± 0.04 g,h	1.86 ± 0.08 l,m
Limoncina (*M. alba* L.)	64.2	1.01 ± 0.02 l,m	0.93 ± 0.05 u
Morettiana (*M. alba* L.)	72.5	1.36 ± 0.02 i,l	1.26 ± 0.04 r,s
Nervosa (*M. alba* L.)	74.4	1.70 ± 0.04 h,i	1.62 ± 0.06 n,o,p
Pendula (*M. alba* L.)	72.6	1.49 ± 0.05 h,i,l	1.43 ± 0.04 q,r
Pyramidalis (*M. alba* L.)	70.2	2.19 ± 0.06 g	2.07 ± 0.08 i
Spain Black fruit (*M. alba* L.)	70.3	1.81 ± 0.02 g,h,i	1.76 ± 0.07 m,n
Ukraina (*M. alba* L.)	70.5	2.99 ± 0.03 f	2.94 ± 0.06 f
Filippine (*M. alba* L.)	65.9	2.26 ± 0.11 n	2.22 ± 0.11 h
Kokosou 21 (*M. alba* L.)	70.4	3.17 ± 0.11 f	2.50 ± 0.04 g
Restelli (*M. alba* L.)	72.3	4.56 ± 0.11 d	3.45 ± 0.05 e
Gorgeous (*M. alba* L.)	70.6	5.65 ± 0.11 c	4.24 ± 0.01 d
Queensland black (*M. alba* L.)	64.4	3.56 ± 0.11 f,e	2.96 ± 0.02 f
Romana lhou (*M. alba* L.)	70.7	8.55 ± 0.11 b	6.86 ± 0.22 b
Wildtype (*M. nigra* L.)	71.9	5.99 ± 0.11 c	4.46 ± 0.04 c
Chirtut (*M. nigra* L.)	55.2	0.56 ± 0.11 m,n	0.66 ± 0.03 v
Date-Akagi (*M. alba* L.)	68.7	1.61 ± 0.11 h,i	1.45 ± 0.21 p,q
Akagi (*M. alba* L.)	61.2	1.57 ± 0.11 h,i	1.22 ± 0.03 s
Illinois everbearing (*M. rubra* L.)	59.8	1.74 ± 0.11 g,h,i	1.96 ± 0.02 i,l
Muki (*M. alba* L.)	69.5	5.66 ± 0.11 c	4.49 ± 0.03 c
Korin (*M. alba* L.)	72.2	9.55 ± 0.11 a	7.07 ± 0.14 a
Miura (*M. alba* L.)	70.7	4.98 ± 0.1 d	4.32 ± 0.03 c.d
Platanoide (*M. alba*.)	69.0	3.14 ± 0.11 f	2.57 ± 0.02 g
Sinuense (*M. alba* L.)	67.5	1.44 ± 0.11 h,i,l	1.04 ± 0.01 u,t
Nagazaki (*M. alba* L.)	70.2	3.75 ± 0.11 e	2.94 ± 0.02 f

[a] Means followed by distinct letters in the same column statistically differ according to Tukey's post hoc test, ($p < 0.05$). Homogeneous subsets are indicated by the same letter. Two-way factorial ANOVA: Cultivar $F_{30, 186} = 1392$, $p < 0.05$. Treatment $F_{1, 186} = 597$, $p < 0.05$. Cultivar*Treatment $F_{30, 186} = 34$, $p < 0.05$.

Therefore, data reported in Table 1 are shown in comparison with the 1-DNJ content obtained from fresh leaves and then corrected by the value of humidity related to the specific cultivar.

Table 1 shows the 1-DNJ contents of 31 cultivars from different *Morus* species, harvested at the end of May 2017. The object of the study was not to identify the compositional characteristics of the different varieties in terms of secondary metabolites and iminosugars. The purpose of the current research was to screen among the most spread varieties largely represented in Italy (country taken as an example of the Mediterranean area) and used for sericulture or fruit production, to enlarge the chance of exploiting the already existing mulberry fields for the pharmaceutical industry. The relative percentages of humidity ranged from 52.9 (Egyptienne, *M. alba* L.) to 74.4% (Nervosa, *M. alba* L.). Two-way ANOVA was carried out by taking into consideration two main factors (thermal treatment and cultivars) and their interaction. The statistical analysis confirmed a significant difference in the 1-DNJ content among the different cultivars at $p < 0.05$, and a better 1-DNJ yield in the fresh-frozen samples in comparison to the dried ones at $p < 0.05$. The average 1-DNJ loss in dried samples was 12.8%. Only in two cases (Chirtut and Illinois everbearing) an increase in the iminosugar content was observed. This slight variation in the DNJ content might be attributable to random sampling or the analysis error; however, it is worthy of noting that it occurred in the only two varieties belonging to species other than *M. alba*. Therefore, it might be due to a different leaf morphology (hairs, waxes, other leaf structures, thickness …) or physiology (enzymatic reactions). The interaction between the two factors (cultivar and thermal treatment) was also significant. The post hoc Tukey's test following ANOVA analysis of the different cultivars demonstrated that a great variation was present in the genetic material. Table 1 reports cultivars and their species; the genus *Morus* shows a very complex and uncertain taxonomy because of its broad geographical distribution, long domestication history, morphological plasticity, and documented inter-species hybridization [23]; according to Nepal and Ferguson [24] it comprises 10–13 species, but Zeng et al. recently identified eight species only: *M. alba*, *M. nigra*, *M. notabilis*, *M. serrata*, *M. celtidifolia*, *M. insignis*, *M. rubra*, and *M. mesozygia*, based on the technique of internal transcribed spacer-based phylogeny [25]. Therefore, according to this classification, cultivars reported in Table 1 all belong to *M. alba*, except for Hicks fancy, which is ascribable to *M. rubra*, for the sample of *M. nigra* wild accession, which does not belong to any selected cultivar, and for the *M. nigra* cultivar Chirtut. It is quite clear that the variation in the 1-DNJ content is as broad among cultivars belonging to the same species as it is among different species.

Table 2. 1-DNJ content of 14 *M. alba* cultivars harvested in 2018. Data are expressed as mg of 1-DNJ/g of dry leaf ± SD.

M. alba Cultivars	End-May 2018		End-July 2018	
	Humidity (%)	Dry Samples (mg/g of Dry Matter) [a]	Humidity (%)	Dry Samples (mg/g of Dry Matter) [a]
Aobanezumi	67.0	0.38 ± 0.02 l	65.3	0.77 ± 0.02 h,i
Cattaneo female	65.2	0.84 ± 0.02 b	64.0	3.88 ± 0.05 c
Cattaneo male	69.4	0.74 ± 0.03 c	69.0	2.60 ± 0.03 f
Egyptienne	62.2	0.42 ± 0.02 l	59.2	0.74 ± 0.02 h,i
Florio	52.0	0.36 ± 0.01 l	50.3	0.88 ± 0.02 h
Giazzola	68.2	0.53 ± 0.02 i	-	-
Kayriounezumigaeshi	66.1	0.97 ± 0.03 a	65.3	3.48 ± 0.05 d
Limoncina	59.2	0.42 ± 0.02 l	58.8	0.44 ± 0.02 l
Morettiana	74.1	0.74 ± 0.04 c	73.3	2.75 ± 0.09 e
Nervosa	66.0	0.92 ± 0.03 a	65.2	5.01 ± 0.08 b
Pendula	70.2	0.99 ± 0.05 a	68.9	5.48 ± 0.10 a
Pyramidalis	69.3	0.63 ± 0.05 d	68.4	2.20 ± 0.05 g
Spain Black Fruit	74.4	0.97 ± 0.03 a	74.0	2.13 ± 0.05 g
Ukraina	63.8	0.53 ± 0.02 i	62.1	0.65 ± 0.02 i

[a] Means followed by distinct letters in the same column statistically differ according to Tukey's post hoc test ($p < 0.05$), homogeneous subsets are indicated by the same letter. The sample Giazzola relative to the end-July 2018 harvest is missing due to the inability to collect leaves (poor harvest due to bad weather conditions). Two-way factorial ANOVA: Cultivar $F_{13, 123} = 2294$, $p < 0.05$. Season $F_{2, 123} = 16512$, $p < 0.05$. Cultivar*Season $F_{25, 123} = 1663$, $p < 0.05$.

Since the drying procedure was undertaken at 50 °C, far below the degradation limit temperature, the hypothetical heat-due deterioration of the compound determined in dried leaves seems unlikely. The explanation for the phenomenon is likely simple: in fact, we should take into account the worsening of the compound extractability according to the reduction of the water content, a condition that occurs with the loss of moisture during desiccation [26]. However, it should also be considered that low temperatures (below 50 °C) allow time for enzymatic conversions and respiratory losses, whereas high temperatures (above 80 °C) can cause thermochemical degradation [27]. Therefore, it is worth exploring the difference between short thermal treatments at higher temperatures and longer thermal treatments at lower temperatures to analyze how the duration of the desiccation process affects the 1-DNJ stability in the leaf, especially in the first drying phase, when the water content is still considerable. Furthermore, an interaction between the two factors (thermal treatment and CVs) was demonstrated ($p < 0.05$). This interaction was found to be particularly significant for Restelli (*M. alba* L.), Gorgeous (*M. alba* L.),

Romana lhou (*M. alba* L.) Wildtype (*M. nigra* L.), and Korin (*M. alba* L.). In these cases, the 1-DNJ decrease in dry leaves was remarkable. The morphology of the leaf surface, the wax content, the thickness of the leaf blade and leaf hairs can otherwise affect the drying process, determining different efficiencies in the preservation of the compound.

Table 3. 1-DNJ content of *Morus alba* cultivars in 2017 and 2018 samplings.

	End-May 2017		End-May 2018	
M. alba Cultivars	Humidity (%)	Dry Samples (mg/g of Dry Matter) [a]	Humidity (%)	Dry Samples (mg/g of Dry Matter) [a]
Aobanezumi	69.4	0.44 ± 0.06 i	67.0	0.38 ± 0.02 f
Cattaneo Female	66.3	0.59 ± 0.03 h	65.2	0.84 ± 0.02 b
Cattaneo Male	70.5	1.50 ± 0.03 e	69.4	0.74 ± 0.03 c
Egyptienne	52.9	1.66 ± 0.08 d	62.2	0.42 ± 0.02 f
Florio	70.7	0.55 ± 0.02 h,i	52.0	0.36 ± 0.01 f
Giazzola	67.7	1.20 ± 0.05 f	68.2	0.53 ± 0.02 e
Kayriounezumigaeshi	68.1	1.86 ± 0.08 c	66.1	0.97 ± 0.03 a
Limoncina	64.2	0.93 ± 0.05 g	59.2	0.42 ± 0.02 f
Morettiana	72.5	1.26 ± 0.04 f	74.1	0.74 ± 0.04 c
Nervosa	74.4	1.62 ± 0.06 d,e	66.0	0.92 ± 0.03 a
Pendula	72.6	1.43 ± 0.04 e	70.2	0.99 ± 0.05 a
Pyramidalis	70.2	2.07 ± 0.08 b	69.3	0.63 ± 0.05 d
Spain Black Fruit	70.3	1.76 ± 0.07 c,d	74.4	0.97 ± 0.03 a
Ukraina	70.5	2.94 ± 0.06 a	63.8	0.53 ± 0.02 e

[a] Means followed by distinct letters in the same column statistically differ according to Tukey's post hoc test ($p < 0.05$), homogeneous subsets are indicated by the same letter. Two-way factorial ANOVA: Cultivar $F_{13, 84} = 593$, $p < 0.05$. Year $F_{1, 84} = 7615$, $p < 0.05$. Cultivar*Year $F_{13, 84} = 428$, $p < 0.05$.

Hu et al. previously investigated the content of 1-DNJ in mature leaves from 132 *Morus* cultivars belonging to nine Chinese species; among them, *M. alba* was the most represented. The mean content found by the authors ranged from 0.13 to 1.47 mg/g DW [28]; these data are slightly lower with respect to the results herein discovered, in which the concentration of 1-DNJ ranged from 0.44 mg/g (Aobanezumi) to 7.07 mg/g (Korin). In addition, comparing the values for Kayriounezumigaeshi and Aobanezumi, recovered under our conditions with the ones shown by Kimura et al. [9], we can observe a higher 1-DNJ level for Kayriounezumigaeshi, but a lower level for Aobanezumi, in May harvests of both years. This fact highlights that the cultivar adaptation to different climatic situations can greatly affect the 1-DNJ content in leaf samples.

Table 2 reports the 1-DNJ content found in 14 *M. alba* cultivars, relative to the two harvests of 2018. The mean humidity percentages recovered were 66.2 ± 5.8 and 64.9 ± 6.4 in May 2018 and July 2018, respectively. The slight diminution in water in the leaf is a marker of the advancement of leaf maturation; young leaves are richer in water, but their content in nutrients is still low. Pendula was the richest in 1-DNJ both at the end-May and the end-July 2018 harvests. As is well known and extensively outlined in the literature, weather conditions and harvest time play a key role in the abundance of active substances in *Morus* sp. pl. [28,29]. Typically, iminosugar levels start increasing in April, and then the concentration gradually decreases until September, with the maximum around June and July [29]. This study was a first test aimed at investigating the seasonal fluctuation of the DNJ content in leaves even in the Mediterranean area and on local (Italian) varieties, which were compared to Japanese varieties, already considered by other authors.

This trend was also observed in data reported in Table 2. In fact, the mean content of 1-DNJ was 0.68 ± 0.23 mg/g and 2.39 ± 1.69 mg/g in the end-May 2018 and the end-July 2018 harvest, respectively. Hence, the mean 1-DNJ content increased more than threefold, starting from May to July. The two-way ANOVA performed considering simultaneously all the CVs and the two seasons assessed the significance of differences between the two seasons and among CVs ($p < 0.05$). The post hoc test showed various homogeneous subsets of CVs were identifiable (lowercase letters in columns). It must be observed that the correlation between seasons and CVs was also significant ($p < 0.05$). Different factors can contribute to this mulberry plants response to the different seasons: first of all, the amount of the average monthly radiation, which, under the conditions of the North-Eastern part of Italy, increases from May to July; the total radiation received by the canopy of mulberry plants might have a direct effect on photosynthesis and, therefore, on the iminosugars synthesis. The same trend is recorded for the average daily temperature, which is also important for plant metabolism. Furthermore, DNJ synthesis by plants might be the defense reaction to the attack of insect populations (increasing their density with subsequent generations in summer), as this iminosugar shows remarkable biological activity against herbivorous predators, by inhibiting intestinal α-glucosidase [30].

Our findings are not conclusive: in fact, the difference in the DNJ-1 content in the leaves of the Italian Cultivars (Cattaneo female and male, Florio, Giazzola, Limoncina, Morettiana, Nervosa, Pendula, Pyramidalis, Spain Black Fruit) should have been tested in the two different seasons both in 2017 and 2018, while data are available for 2018 only. However, our one-year data collection agrees with those of other authors [9,31,32].

In addition, in the end-July 2018 harvest, Morettiana was selected and analyzed both in the form of apical leaves and branch mature leaves to study the distribution of the constituent in the different parts of the plant. The cultivar Morettiana was specifically chosen because this is one of the most representative and widespread Italian cultivars. The 1-DNJ content of the younger leaves (data not shown in Table 2) was equal to 7.34 ± 0.42 mg/g of dry matter (Figure 5).

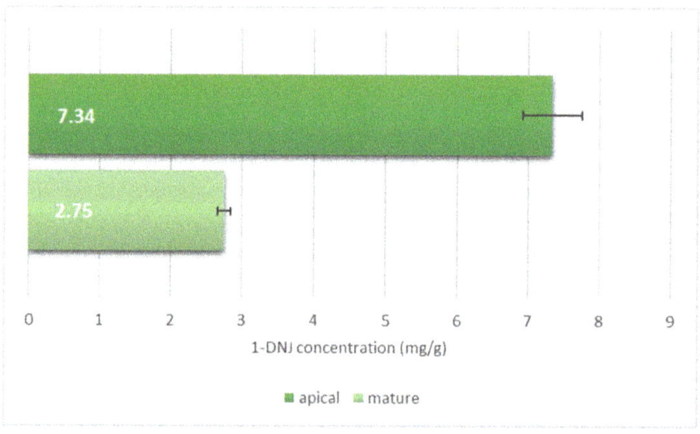

Figure 5. 1-DNJ content by leaf-maturity type in cultivar Morettiana.

These data are consistent with what was previously observed by Hu et al. [28], establishing that apical younger leaves usually contain higher 1-DNJ than mature ones. This fact is strengthened by the major enzymatic activity discovered in young leaves that is responsible for the imination of glucose, the initial step in 1-DNJ biosynthesis [12,33].

In Table 3, data from 2017 and 2018 spring harvests are compared. The two-way ANOVA performed, considering CVs and harvesting year simultaneously, showed a significant effect for the two main factors CVs ($p < 0.05$) and year ($p < 0.05$), and a significant interaction among factors, demonstrating the importance of annual variation in 1-DNJ content, which could be influenced by annual climatic fluctuations. In particular, the amount of 1-DNJ, which was recorded in the mulberry leaf samples at the end of May 2017 was almost double in comparison to the amount in those collected at the end of May 2018. The climatic conditions in the two years were very different: in the Padua province, North-Eastern Italy, where the experiment was carried out, May 2018 was rainier (81.8 mm) than May 2017 (79 mm) [34].

Furthermore, as it is possible to observe in Table 4, the average daily accumulated energy (calculated at the ground level on the horizontal surface) was 5.16 kWh/m^2/day (18.57 MJ/m^2/day) in May 2018 while it was 5.70 kW/m^2/day (20.52 MJ/m^2/day) in May 2017 [34]. It can be assumed that higher solar radiation (recorded in End-May 2017) produced increased photosynthetic activity and 1-DNJ biosynthesis. In addition, due to the humid conditions and high average temperature (May 2018 average monthly temperatures was 19.4, while May 2017 average monthly temperature was 18.6), fungal attacks spread on the mulberry plants, affecting the leaf surface, and by extension the photosynthetic capacity. The methodology used to produce these meteorology and solar radiation data are available online [35].

Table 4. Different climatic conditions recorded in May 2017 and 2018 in Padova province (45°24′57″96 N lat., 11°52′58″08 E long.), Italy [34].

Year	Average Radiation (kWh/m^2/day)	Monthly Total Rainfall (mm)	Average Temperature (°C)
2017	5.70 (20.52 MJ/m^2/day)	79 mm	18.6
2018	5.16 (18.57 MJ/m^2/day)	81.8 mm	19.4

In conclusion, this work aims at the investigation of the 1-DNJ amount in leaves of different CVs of mulberry grown in Italy, and at understanding its seasonal fluctuations, with a view to determining a rational harvesting time. To the best of our knowledge, this is the first analytical determination of 1-DNJ carried out on Italian mulberry crops. This study could be used as a reference to assess the quality of Italian plant material in comparison to cultivations carried out in Far Eastern Asia. Based on our data, we can conclude that the best harvesting time is in the middle of summer rather than in springtime, and that apical young leaves are much richer in 1-DNJ than older ones at the base of the shoot. However, with regard to the cultivar selection for the 1-DNJ production, agronomical tests in subsequent years and under different cultivation conditions are recommended. Indeed, the interactions with the weather conditions are determinant for the 1-DNJ leaf content, which, in turn, depends on the physiological status of the tree and its photosynthetic capacity. Furthermore, the drying procedure should be optimized for industrial use before upscaling the crop for pharmaceutical exploitation to avoid the loss of important compounds as well as thermal inactivation during leaf desiccation. The results of the recovery and precision studies are shown in Table 5.

Table 5. Recovery percentage and precision data obtained with spiked samples.

μg/mL	Intra—Day $n = 3$						
	1	2	3	Mean	SD	RSD (%)	Recovery (%)
0.06	0.056	0.058	0.064	0.059	0.004	7.02	98.89
0.60	0.630	0.550	0.570	0.583	0.042	7.14	97.22
6.00	5.980	6.720	6.400	6.367	0.371	5.83	106.11
	Inter—Day $n = 3$						
	1	2	3	mean	SD	RSD (%)	Recovery (%)
0.06	0.062	0.058	0.054	0.058	0.004	6.90	96.67
0.60	0.570	0.620	0.580	0.590	0.026	4.48	98.33
6.00	6.120	5.940	6.200	6.087	0.133	2.19	101.44

3. Materials and Methods

3.1. Chemicals and Solvents

All solvents were analytical grade and purchased from Sigma-Aldrich (Milan, Italy). The reference standard of 1-DNJ (purity ≥ 95.0%) was purchased from Sigma-Aldrich

(Milan, Italy). Water was purified by using a Milli-Q Plus185 system from Millipore (Milford, MA, USA).

3.2. Plant Material and Extraction Procedure

Morus sp.pl. (L.) leaves were harvested at different times, (end-May 2017, end-May 2018, and end-July 2018), and were provided by CREA—Research Centre for Agriculture and Environment, laboratory of sericulture (Padova, Italy). This experimental mulberry field preserves a germplasm bank of around 60 *Morus* cultivars. The first screening was carried out by considering the more representative cultivars of the collection (the varieties of different geographic origin and the same age and pruning type, which were therefore comparable in this study). Within this group, for the subsequent analyses, a more limited number of varieties was explored; these varieties are the most used for sericulture or fruit production in the country, and therefore easily exploitable for pharmaceutical uses too.

To ensure the uniformity of the sampling, leaves were collected randomly from at least five different trees per cultivar. A representative sample of about 250 g was labelled in plastic bags and stored according to the chosen treatment method. The content of 1-DNJ was determined in both frozen and dried leaves in order to evaluate the influence of the drying process on the active compound extractability and to assess the potential loss due to thermal degradation. *Morus* leaves were weighed as fresh samples; afterwards, a mild drying process was carried out in an oven at 50 °C until reaching constant weight; then the percentage of humidity was calculated per each sample. Samples were stored at -18 °C and protected from light by storage inside amber glass sealed containers. Then, to prepare the fresh samples, with the aim to prevent any oxidation or degradation process, a representative portion of frozen leaves was ground in an automatic mill with dry ice. Immediately after, two grams of powdered frozen leaves were extracted at room temperature with 20 mL of a water:ethanol (50:50) solution under magnetic stirring. The whole procedure required three consecutive extraction steps of 2 h each. The first extract was centrifuged (RCF 7200 g) for 5 min, the supernatant was collected, paper filtered, and the further two extractions were performed on the residue. The three extracts were combined, and the final volume was adjusted to 100 mL with the same hydroalcoholic solution; the extract was then stored at 4 °C in flasks covered with silver paper until analysis. The extraction procedure was repeated twice for each sample, and it was performed in the same way for dried samples too. Before injection in the UHPLC system, the extracts were filtered by using a 0.22 μm cellulose acetate filter (GVS, Bologna, Italy) into an HPLC vial and properly diluted with acetonitrile.

3.3. HPLC-ESI-MS Analysis

The chromatographic method previously developed by Vichasilp et al. [36], was herein properly modified and optimized. A Thermo Scientific Dionex Ultimate 3000 UHPLC system equipped with an autosampler, a quaternary pump, and a thermostated column compartment controlled by Chromeleon 7.2 Software (Thermo Scientific, Waltham, MA, USA) was used. The UHPLC system was coupled with a high-resolution Q Exactive mass spectrometer (Thermo Scientific, Bremen, Germany). The mass spectrometer was calibrated before the analyses. Nitrogen (N_2) (purity > 99.99%), obtained from a Zefiro zero 60 LC-MS nitrogen generator (CINEL, Vigonza, Italy), was employed both as the source gas and the collision gas. The capillary temperature was set at 270 °C and the following N_2 flows (arbitrary units) were used: sheath gas 40, auxiliary gas 30, and sweep gas 3. The MS/MS parameters were optimized with standard 1-DNJ under positive ion electrospray ionization. A Hydrophilic Interaction Liquid Chromatography column (Cortecs UPLC HILIC, 1.6 μm, 2.1 × 100 mm) was used (Waters, Milford, MA, USA). This kind of column provides strong retention of very polar molecules that are typically unretained under conventional reversed phase conditions. 1-DNJ was eluted with a binary gradient consisting of ammonium formate 20 mM in water (solvent A) and acetonitrile (solvent B). The gradient profile was as follows: 0–10 min, 10% A; 11–16 min, 50% A; 16–27 min, 10% A. The flow rate was adjusted to 0.3 mL/min, and the column temperature was maintained at 30 °C. 1-Deoxynojirimycin was detected by MS/MS with Parallel Reaction Monitoring (PRM) for transition of the parent ion to the product ions. The concentration of 1-DNJ in samples was obtained through the calibration curve built with the pure standard compound. The selected range was 0.31 to 40.9 mM, corresponding to 0.050–6.67 ppm.

3.4. Method Validation

The method showed good linearity ($r^2 > 0.9995$) within the selected range of concentrations. The method was validated in terms of precision and accuracy. Precision was evaluated with the relative standard deviation determined from repeated extract injections and triplicate extractions. Method accuracy was evaluated by recoveries, which were carried out by spiking samples with three different concentrations of standard solutions. Spiking recovery (%) was calculated as: $[(C_2 - C_0)/C_1] \times 100$, where C_2 is the analyte concentration in the final solution after spiking with a known concentration of standard, C_0 is the original analyte concentration in the initial solution, and C_1 is the added known concentration of standard. Intra-day precision was determined by analyzing three replicates of spiked samples, and inter-day precision was determined by running the three replicates of spiked samples on three different days.

3.5. Statistical Analysis

Statistical analysis was carried out by using Statistica 6.0 (StatSoft Italia, Vigonza, Italy) and SPSS 13.0 (SPSS Inc., IBM, Segrate, Italy), the significance was established at $p < 0.05$. One-way ANOVA or factorial ANOVA was carried out according to the experimental situation. In addition, post hoc analysis based on Tuckey's HSD test was performed to evaluate differences among means.

Author Contributions: Conceptualization, D.B. and S.C.; methodology, L.M. and D.B.; software and statistical analysis, D.B. and A.S.; validation, L.M. and D.B.; investigation, L.M. and D.B.; resources, D.B. and S.C.; data curation, L.M., A.d.M., G.P., F.P., D.B. and S.C.; writing—original draft preparation, L.M., D.B. and S.C.; writing—review and editing, S.B., D.B. and S.C.; supervision, D.B. and S.C.; funding acquisition, D.B. and S.C. All authors have read and agreed to the published version of the manuscript.

Funding: This research was funded by the Italian Ministry of Agricultural, Food and Forestry Policies, (D.D. no. 2380, 17 July 2017) and the Veneto Region (project "Serinnovation", Programma di sviluppo rurale per il Veneto 2014–2020–DGR N. 2175 del 23/12/2016 Misura 16.1).

Institutional Review Board Statement: Not applicable.

Informed Consent Statement: Not applicable.

Acknowledgments: The authors acknowledge Fondazione Cassa di Risparmio di Modena for funding the UHPLC-ESI-QExactive system located at Centro Interdipartimentale Grandi Strumenti (CIGS) of the University of Modena and Reggio Emilia, Italy.

Conflicts of Interest: The authors declare no conflict of interest. The funders had no role in the design of the study; in the collection, analyses, or interpretation of data; in the writing of the manuscript, or in the decision to publish the results.

References

1. Nasri, H.; Baradaran, A.; Shirzad, H.; Rafieian-Kopaei, M. New Concepts in Nutraceuticals as Alternative for Pharmaceuticals. *Int. J. Prev. Med.* **2014**, *5*, 1487–1499.
2. Zamoner, L.O.B.; Aragão-Leoneti, V.; Carvalho, I. Iminosugars: Effects of Stereochemistry, Ring Size, and N-Substituents on Glucosidase Activities. *Pharmaceuticals* **2019**, *12*, 108. [CrossRef]
3. Yagi, Y.M. The Structure of Moranoline, a Piperidine Alkaloid from Morus Species. *Nippon Nogei Kagaku Kaishi* **1976**, *50*, 571–572. [CrossRef]
4. Konno, K.; Ono, H.; Nakamura, M.; Tateishi, K.; Hirayama, C.; Tamura, Y.; Hattori, M.; Koyama, A.; Kohno, K. Mulberry Latex Rich in Antidiabetic Sugar-Mimic Alkaloids Forces Dieting on Caterpillars. *Proc. Natl. Acad. Sci. USA* **2006**, *103*, 1337–1341. [CrossRef]

5. Lemus, I.; García, R.; Delvillar, E.; Knop, G. Hypoglycaemic Activity of Four Plants Used in Chilean Popular Medicine. *Phytother. Res.* **1999**, *13*, 91–94. [CrossRef]
6. Wang, Z.; Wang, J.; Chan, P. Treating Type 2 Diabetes Mellitus with Traditional Chinese and Indian Medicinal Herbs. *Evid. Based Complement. Alternat. Med.* **2013**, *2013*, 1–17. [CrossRef] [PubMed]
7. Wang, N.; Zhu, F.; Chen, K. 1-Deoxynojirimycin: Sources, Extraction, Analysis and Biological Functions. *Nat. Prod. Commun.* **2017**, *12*, 1934578X1701200934. [CrossRef]
8. Bajpai, S.; Rao, A.V.B. Quantitative Determination of 1-Deoxynojirimycin in Different Mulberry Varieties of India. *J. Pharmacogn. Phytochem.* **2014**, *3*, 17–22.
9. Kimura, T.; Nakagawa, K.; Kubota, H.; Kojima, Y.; Goto, Y.; Yamagishi, K.; Oita, S.; Oikawa, S.; Miyazawa, T. Food-Grade Mulberry Powder Enriched with 1-Deoxynojirimycin Suppresses the Elevation of Postprandial Blood Glucose in Humans. *J. Agric. Food Chem.* **2007**, *55*, 5869–5874. [CrossRef] [PubMed]
10. Yatsunami, K.; Ichida, M.; Onodera, S. The Relationship between 1-Deoxynojirimycin Content and α-Glucosidase Inhibitory Activity in Leaves of 276 Mulberry Cultivars (*Morus* Spp.) in Kyoto, Japan. *J. Nat. Med.* **2008**, *62*, 63–66. [CrossRef] [PubMed]
11. Song, W.; Wang, H.-J.; Bucheli, P.; Zhang, P.-F.; Wei, D.-Z.; Lu, Y.-H. Phytochemical Profiles of Different Mulberry (*Morus* sp.) Species from China. *J. Agric. Food Chem.* **2009**, *57*, 9133–9140. [CrossRef]
12. Vichasilp, C.; Nakagawa, K.; Sookwong, P.; Higuchi, O.; Luemunkong, S.; Miyazawa, T. Development of High 1-Deoxynojirimycin (DNJ) Content Mulberry Tea and Use of Response Surface Methodology to Optimize Tea-Making Conditions for Highest DNJ Extraction. *LWT—Food Sci. Technol.* **2012**, *45*, 226–232. [CrossRef]
13. Esti Wulandari, Y.R.; Prasasty, V.D.; Rio, A.; Geniola, C. Determination of 1-Deoxynojirimycin Content and Phytochemical Profiles from Young and Mature Mulberry Leaves of *Morus* Spp. *OnLine J. Biol. Sci.* **2019**, *19*, 124–131. [CrossRef]
14. Hao, J.-Y.; Wan, Y.; Yao, X.-H.; Zhao, W.-G.; Hu, R.-Z.; Chen, C.; Li, L.; Zhang, D.-Y.; Wu, G.-H. Effect of Different Planting Areas on the Chemical Compositions and Hypoglycemic and Antioxidant Activities of Mulberry Leaf Extracts in Southern China. *PLoS ONE* **2018**, *13*, e0198072. [CrossRef]
15. Przygoński, K.; Wojtowicz, E. The Optimization of Extraction Process of White Mulberry Leaves and the Characteristic Bioactive Properties Its Powder Extract. *Herba Pol.* **2019**, *65*, 12–19. [CrossRef]
16. Tomotake, H.; Katagiri, M.; Yamato, M. Silkworm Pupae (*Bombyx Mori*) Are New Sources of High Quality Protein and Lipid. *J. Nutr. Sci. Vitaminol.* **2010**, *56*, 446–448. [CrossRef]
17. Tong, T.-T.; Zhao, E.-H.; Gao, H.-L.; Xu, Y.-H.; Zhao, Y.-J.; Fu, G.; Cui, H. Recent Research Advances of 1-Deoxynojirimycin and Its Derivatives. *China J. Chin. Mater. Med.* **2018**, *43*, 1990–1997. [CrossRef]
18. Ryu, K.S.; Lee, H.S.; Kim, K.Y.; Kim, M.J.; Kang, P.D. Heat Stability and Glucose-Lowering Effect of 1-Deoxynojirimycin from Silkworm (*Bombyx Mori*) Extract Powder. *Int. J. Ind. Entomol.* **2013**, *27*, 277–281. [CrossRef]
19. Rodríguez-Sánchez, S.; Hernández-Hernández, O.; Ruiz-Matute, A.I.; Sanz, M.L. A Derivatization Procedure for the Simultaneous Analysis of Iminosugars and Other Low Molecular Weight Carbohydrates by GC–MS in Mulberry (*Morus* Sp.). *Food Chem.* **2011**, *126*, 353–359. [CrossRef]
20. Kimura, T.; Nakagawa, K.; Saito, Y.; Yamagishi, K.; Suzuki, M.; Yamaki, K.; Shinmoto, H.; Miyazawa, T. Determination of 1-Deoxynojirimycin in Mulberry Leaves Using Hydrophilic Interaction Chromatography with Evaporative Light Scattering Detection. *J. Agric. Food Chem.* **2004**, *52*, 1415–1418. [CrossRef]
21. Nakagawa, K.; Ogawa, K.; Higuchi, O.; Kimura, T.; Miyazawa, T.; Hori, M. Determination of Iminosugars in Mulberry Leaves and Silkworms Using Hydrophilic Interaction Chromatography—Tandem Mass Spectrometry. *Anal. Biochem.* **2010**, *404*, 217–222. [CrossRef] [PubMed]
22. Tolstikov, V.V.; Fiehn, O. Analysis of Highly Polar Compounds of Plant Origin: Combination of Hydrophilic Interaction Chromatography and Electrospray Ion Trap Mass Spectrometry. *Anal. Biochem.* **2002**, *301*, 298–307. [CrossRef] [PubMed]
23. Burgess, K.S.; Morgan, M.; Deverno, L.; Husband, B.C. Asymmetrical Introgression between Two *Morus* Species (*M. Alba, M. Rubra*) That Differ in Abundance. *Mol. Ecol.* **2005**, *14*, 3471–3483. [CrossRef]
24. Nepal, M.P.; Ferguson, C.J. Phylogenetics of *Morus* (Moraceae) Inferred from ITS and TrnL-TrnF Sequence Data. *Syst. Bot.* **2012**, *37*, 442–450. [CrossRef]
25. Zeng, Q.; Chen, H.; Zhang, C.; Han, M.; Li, T.; Qi, X.; Xiang, Z.; He, N. Definition of Eight Mulberry Species in the Genus *Morus* by Internal Transcribed Spacer-Based Phylogeny. *PLoS ONE* **2015**, *10*, e0135411. [CrossRef] [PubMed]
26. Gąsecka, M.; Siwulski, M.; Magdziak, Z.; Budzyńska, S.; Stuper-Szablewska, K.; Niedzielski, P.; Mleczek, M. The Effect of Drying Temperature on Bioactive Compounds and Antioxidant Activity of Leccinum Scabrum (Bull.) Gray and Hericium Erinaceus (Bull.) Pers. *J. Food Sci. Technol.* **2020**, *57*, 513–525. [CrossRef]
27. Chlumská, Z.; Janeček, S.; Doležal, J. How to Preserve Plant Samples for Carbohydrate Analysis? Test of Suitable Methods Applicable in Remote Areas. *Folia Geobot.* **2013**, *49*, 1–15. [CrossRef]
28. Hu, X.-Q.; Jiang, L.; Zhang, J.-G.; Deng, W.; Wang, H.-L.; Wei, Z.-J. Quantitative Determination of 1-Deoxynojirimycin in Mulberry Leaves from 132 Varieties. *Ind. Crops Prod.* **2013**, *49*, 782–784. [CrossRef]
29. Han, W.; Chen, X.; Yu, H.; Chen, L.; Shen, M. Seasonal Variations of Iminosugars in Mulberry Leaves Detected by Hydrophilic Interaction Chromatography Coupled with Tandem Mass Spectrometry. *Food Chem.* **2018**, *251*, 110–114. [CrossRef]
30. Asano, N.; Oseki, K.; Tomioka, E.; Kizu, H.; Matsui, K. N-Containing Sugars from *Morus Alba* and Their Glycosidase Inhibitory Activities. *Carbohydr. Res.* **1994**, *259*, 243–255. [CrossRef]

31. Sugiyama, M.; Katsube, T.; Koyama, A.; Itamura, H. Seasonal Changes in Functional Component Contents in Mulberry (*Morus Alba* L.) Leaves. *Hortic. J.* **2017**, *86*, 534–542. [CrossRef]
32. Nakanishi, H.; Onose, S.; Kitahara, E.; Chumchuen, S.; Takasaki, M.; Konishi, H.; Kanekatsu, R. Effect of Environmental Conditions on the α-Glucosidase Inhibitory Activity of Mulberry Leaves. *Biosci. Biotechnol. Biochem.* **2011**, *75*, 2293–2296. [CrossRef] [PubMed]
33. Shibano, M.; Fujimoto, Y.; Kushino, K.; Kusano, G.; Baba, K. Biosynthesis of 1-Deoxynojirimycin in Commelina Communis: A Difference between the Microorganisms and Plants. *Phytochemistry* **2004**, *65*, 2661–2665. [CrossRef]
34. Prediction of Worldwide Energy Resource. Available online: https://power.larc.nasa.gov/data-access-viewer/ (accessed on 24 May 2021).
35. Power Data Methodology. Available online: https://power.larc.nasa.gov/docs/methodology/ (accessed on 24 May 2021).
36. Vichasilp, C.; Nakagawa, K.; Sookwong, P.; Suzuki, Y.; Kimura, F.; Higuchi, O.; Miyazawa, T. Optimization of 1-Deoxynojirimycin Extraction from Mulberry Leaves by Using Response Surface Methodology. *Biosci. Biotechnol. Biochem.* **2009**, *73*, 2684–2689. [CrossRef] [PubMed]

plants

Article

Chemical and Biological Profiles of *Dendrobium* in Two Different Species, Their Hybrid, and Gamma-Irradiated Mutant Lines of the Hybrid Based on LC-QToF MS and Cytotoxicity Analysis

Bomi Nam [1,2,†], Hyun-Jae Jang [3,†], Ah-Reum Han [1,†], Ye-Ram Kim [1], Chang-Hyun Jin [1], Chan-Hun Jung [4], Kyo-Bin Kang [5], Sang-Hoon Kim [1], Min-Jeong Hong [1], Jin-Baek Kim [1] and Hyung-Won Ryu [3,*]

1   Advanced Radiation Technology Institute, Korea Atomic Energy Research Institute (KAERI), Jeongeup-si 56212, Jeollabuk-do, Korea; bomi1201@ncn.re.kr (B.N.); arhan@kaeri.re.kr (A.-R.H.); yrkim327@kaeri.re.kr (Y.-R.K.); chjin@kaeri.re.kr (C.-H.J.); shkim80@kaeri.re.kr (S.-H.K.); hongmj@kaeri.re.kr (M.-J.H.); jbkim74@kaeri.re.kr (J.-B.K.)
2   Institute of Natural Cosmetic Industry for Namwon, Namwon-si 55801, Jeollabuk-do, Korea
3   Natural Medicine Research Center, Korea Research Institute of Bioscience & Biotechnology (KRIBB), Cheongju-si 28116, Chungbuk-do, Korea; water815@kribb.re.kr
4   Jeonju AgroBio-Materials Institute, Jeonju-si 54810, Jeollabuk-do, Korea; chjung@jami.re.kr
5   Research Institute of Pharmaceutical Sciences, College of Pharmacy, Sookmyung Women's University, Seoul 04310, Korea; kbkang@sookmyung.ac.kr
*   Correspondence: ryuhw@kribb.re.kr; Tel.: +82-43-240-6117
†   These authors contributed equally to this work.

Citation: Nam, B.; Jang, H.-J.; Han, A.-R.; Kim, Y.-R.; Jin, C.-H.; Jung, C.-H.; Kang, K.-B.; Kim, S.-H.; Hong, M.-J.; Kim, J.-B.; et al. Chemical and Biological Profiles of *Dendrobium* in Two Different Species, Their Hybrid, and Gamma-Irradiated Mutant Lines of the Hybrid Based on LC-QToF MS and Cytotoxicity Analysis. *Plants* **2021**, *10*, 1376. https://doi.org/10.3390/plants10071376

Academic Editors: Igor Jerkovic, Jong Seong Kang and Narendra Singh Yadav

Received: 9 June 2021
Accepted: 1 July 2021
Published: 5 July 2021

Publisher's Note: MDPI stays neutral with regard to jurisdictional claims in published maps and institutional affiliations.

Abstract: The *Dendrobium* species (Orchidaceae) has been cultivated as an ornamental plant as well as used in traditional medicines. In this study, the chemical profiles of Dendrobii Herba, used as herbal medicine, *Dendrobium* in two different species, their hybrid, and the gamma-irradiated mutant lines of the hybrid, were systematically investigated via ultra-performance liquid chromatography coupled with quadrupole time-of-flight mass spectrometry (UPLC-QToF MS). Among the numerous peaks detected, 17 peaks were unambiguously identified. Gigantol (**1**), (1*R*,2*R*)-1,7-hydroxy-2,8-methoxy-2,3-dihydrophenanthrene-4(1*H*)-one (**2**), tristin (**3**), (-)-syringaresinol (**4**), lusianthridin (**5**), 2,7-dihydroxy-phenanthrene-1,4-dione (**6**), densiflorol B (**7**), denthyrsinin (**8**), moscatilin (**9**), lusianthridin dimer (**10**), batatasin III (**11**), ephemeranthol A (**12**), thunalbene (**13**), dehydroorchinol (**14**), dendrobine (**15**), shihunine (**16**), and 1,5,7-trimethoxy-2-phenanthrenol (**17**), were detected in Dendrobii Herba, while **1**, **2**, and **16** were detected in *D. candidum*, **1**, **11**, and **16** in *D. nobile*, and **1**, **2**, and **16** in the hybrid, *D. nobile* × *candidum*. The methanol extract taken of them was also examined for cytotoxicity against FaDu human hypopharynx squamous carcinoma cells, where Dendrobii Herba showed the greatest cytotoxicity. In the untargeted metabolite analysis of 436 mutant lines of the hybrid, using UPLC-QToF MS and cytotoxicity measurements combined with multivariate analysis, two tentative flavonoids (M1 and M2) were evaluated as key markers among the analyzed metabolites, contributing to the distinction between active and inactive mutant lines.

Keywords: *Dendrobium*; Orchidaceae; *D. nobile*; *D. candidum*; *D. nobile* × *candidum*; gamma-irradiated mutant; metabolomics; cytotoxicity

1. Introduction

Dendrobium is a genus of mostly epiphytic and lithophytic orchids in the family *Orchidaceae* [1]. It is a very large genus, containing more than 1800 species that are found in diverse habitats throughout much of South, East, and Southeast Asia, and in many of the islands of the Pacific [1]. The genera with medicinal importance include *D. nobile*, *D. chrysanthum*, *D. officinale*, *D. loddigessi*, *D. fimbriatum* var. *oculatum*, *D. moniliforme*, and *D. candidum* [1,2]. These have been used as traditional folk remedies for the treatment of various diseases, such as chronic atrophic gastritis, diabetes, and cardiovascular disease [3]. Previous phytochemical studies on the *Dendrobium* species have resulted in the isolation of diverse types of compounds, such as alkaloids [4–10], bibenzyls [9,11–19], phenanthrenes [19–31], fluorenones [18,31], sesquiterpenoids [6,32,33], lignans [21,34], flavonoids [35,36], and polysaccharides [37–39]. These compounds have been reported as exhibiting neuroprotective [4,5], anticancer [6,10–17,25–27,34], antioxidant [18,21,32,36–38], anti-inflammatory [22,23,38], and immunomodulatory activities [39]. In particular, there have been many reports on the mechanisms of their anticancer actions, including apoptosis and cell migration [11–17].

In our previous study, nine phenanthrenes, three bibenzyls, and a lignan were isolated from the ethyl acetate fraction of Dendrobii Herba, which were examined for their cytotoxicity against FaDu human hypopharynx squamous carcinoma cells [40]. Among them, densiflorol B, 6,7-dimethoxyphenanthrene-2,5-diol, dehydroorchinol, 1,5,7-trimethoxy-2-phenanthrenol, ephemeranthol A, and 3-[(1*E*)-2-(3-hydroxyphenyl)ethenyl]-5-methoxyphenol exhibited cytotoxicity, and together with moscatilin, exhibited remarkable activity (IC$_{50}$, 2.55 µg/mL). In this study, electrospray ionization and the quadrupole time-of-flight mass spectrometry (ESI-QToF MS) data of these isolated compounds and five standards were obtained, as well as the composition of the methanol extracts of Dendrobii Herba, *D. nobile*, *D. candidum*, and the hybrid *D. nobile* × *candidum* (Figure 1). In addition, the cytotoxicity of these extracts against FaDu cells was evaluated.

Figure 1. Pictures of the *Dendrobium* samples, (**a**) the medicinal herb Dendrobii Herba; (**b**) *D. nobile*; (**c**) *D. candidum*; (**d**) the hybrid *D. nobile* × *candidum*; (**e**) the gamma-irradiated mutant lines of *D. nobile* × *candidum*.

Hybridization, genetic modification, or mutation using chemical or physical mutagens to improve the productivity and quality of plants have been widely used in plant breeding [41,42]. Mutation breeding following gamma irradiation has induced novel mutational traits while preserving the unique properties of the plant [43,44]. Our research group developed mutant lines of the hybrid *D. nobile* × *candidum* generated through the gamma irradiation (20 Gy) of its stems to produce mutants with enhanced biological activities and the improved yield of phytochemicals. Metabolomic studies of the *Dendrobium* species have previously been reported [45–48], however, there have been no studies on the chemical profile of *D. candidum*, the hybrid, *D. nobile* × *candidum*, and its gamma-irradiated mutant lines using LC-MS.

As part of our research project for the development of improved varieties, the chemical compositions of four *Dendrobium* extracts and their cytotoxicities against FaDu cells were compared, and mutant lines of the hybrid developed by radiation breeding were also evaluated for cytotoxicities with untargeted metabolite analysis. In addition, ultra-performance liquid chromatography coupled with quadrupole time-of-flight mass spectrometry (UPLC-QToF MS) and multivariate analysis, including principal component analysis (PCA) and orthogonal partial least squares analysis (OPLS-DA), were applied to characterize the metabolomic differences between the active and inactive mutant lines in FaDu cells.

2. Results and Discussion

2.1. Identification of the Compounds in Dendrobium Samples and Their Cytotoxic Activities

The compounds **1–17** isolated from Dendrobii Herba [40] were ionized in both negative and positive ion modes using electrospray ionization (ESI). The QToF MS data are listed in Table 1. In addition, LC–MS base peak ion (BPI) chromatograms of all of the compounds and individual high-dimension mass spectra are provided in Figures S1–S17. The UPLC-QToF-MS analysis of the methanol extracts of the stems of four *Dendrobium* samples (Dendrobii Herba, *D. nobile*, *D. candidum*, and the hybrid *D. nobile* × *candidum*) was performed in the negative and positive ion modes (6 eV, ESI). Compared to their ESI⁻ and ESI⁺ MS chromatograms, the ESI⁻ of Dendrobii Herba, *D. candidum*, *D. nobile*, and *D. nobile* × *candidum* provided better sensitivity and lower detection limits compared to the ESI⁺ for the detection of the compounds, as shown in Figure 2. Among the numerous peaks detected, peaks 1-9 and 11-17 were unambiguously identified in Dendrobii Herba by comparing their retention times and masses with the previously acquired data of these compounds. Peak 10 gave a molecular ion at *m/z* 481.1639 [M − H]⁻, which is twice the molecular weight of lusianthridin (peak 5) and also exhibited the same fragment pathway as that of peak 5, suggesting that it was tentatively identified as a dimer of lusianthridin. Peak 10 was assumed to be phochinenin G or phochinenin D, as reported in *D. nobile*, because it had the same molecular weight and molecular formula as the lusianthridin dimer [19]. Most of the compounds were hardly detected in *D. nobile*, *D. candidum*, and *D. nobile* × *candidum* (Table 2). *D. nobile* contained three peaks: 1, 11, and 16. Specifically, peaks 1, 2, and 16 were detected in *D. candidum* and *D. nobile* × *candidum*. In previous studies, gigantol (peak 1), moscatilin (peak 9), and dendrobine (peak 15) have been identified in both *D. nobile* [6,18] and *D. candidum* [49,50], whereas densiflorol B (peak 7), denthyrsinin (peak 8), thunalbene (peak 13), (1*R*,2*R*)-1,7-hydroxy-2,8-methoxy-2,3-dihydrophenanthrene-4(1*H*)-one (peak 2), and 2,7-dihydroxy-phenanthrene-1,4-dione (peak 6), which were isolated as new compounds in our previous study [40], have never been reported in either plant. Tristin (peak 3), lusianthridin (peak 5), batatasin III (peak 11), ephemeranthol A (peak 12), dehydroorchinol (peak 14), shihunine (peak 16), and 1,5,7-trimethoxy-2-phenanthrenol (peak 17) have been found in *D. nobile* [19,22,51], but have not been reported in *D. candidum*. The metabolite identification of the hybrid, *D. nobile* × *candidum*, was reported for the first time in this study.

Table 1. ESI QToF MS data of compounds isolated from Dendrobii Herba and four standard compounds.

Peak No.	Identification	t_R (min)	Observed m/z (Da)	Calculated m/z (Da)	Error (ppm)	Molecular Formula	Fragments
1	Gigantol	1.73	275.12688 [M + H]$^+$	275.1278	−2.14	$C_{16}H_{18}O_4$	198
2	(1*R*,2*R*)-1,7-Hydroxy-2,8-methoxy-2,3-dihydrophenanthrene-4(1*H*)-one	1.98	287.09214 [M − H]$^-$	287.0914	4.44	$C_{16}H_{14}O_5$	272, 239
3	Tristin	2.98	259.09822 [M − H]$^-$	259.0965	5.17	$C_{15}H_{16}O_4$	243
4	(-)-Syringaresinol	3.32	417.15652 [M − H]$^-$	417.1555	5.21	$C_{22}H_{26}O_8$	387, 190
5	Lusianthridin	4.69	241.08710 [M − H]$^-$	241.087	−4.67	$C_{15}H_{14}O_3$	116
6	2,7-Dihydroxy-phenanthrene-1,4-dione	5.02	239.03168 [M − H]$^-$	239.0339	−9.26	$C_{14}H_8O_4$	256
7	Densiflorol B	5.07	253.05075 [M − H]$^-$	253.0495	4.11	$C_{15}H_{10}O_4$	238
8	Denthyrsinin	5.49	299.09974 [M − H]$^-$	299.0925	22.73	$C_{17}H_{16}O_5$	284, 254
9	Moscatilin	5.63	305.13779 [M + H]$^+$	305.1384	−3.52	$C_{17}H_{20}O_5$	181
10	Lusianthridin dimer	5.84	481.16393 [M − H]$^-$	481.1657	−4.67	$C_{30}H_{26}O_6$	116
11	Batatasin III	6.05	243.11115 [M − H]$^-$	243.1027	31.18	$C_{15}H_{16}O_3$	-
12	Ephemeranthol A	6.31	273.11093 [M + H]$^+$	273.1121	5.21	$C_{16}H_{16}O_4$	272, 241, 213
13	Thunalbene	6.43	241.08710 [M − H]$^-$	241.087	1.33	$C_{15}H_{14}O_3$	-
14	Dehydroorchinol	7.7	255.10002 [M + H]$^+$	255.1016	−2.21	$C_{16}H_{14}O_3$	240
15	Dendrobine	7.7	264.19496 [M + H]$^+$	264.1958	−2.03	$C_{16}H_{25}NO_2$	-
16	Shihunine	7.7	203.09751 [M]$^+$	203.0952	11.49	$C_{12}H_{13}NO_2$	405, 203
17	1,5,7-Trimethoxy-2-phenanthrenol	7.97	285.11172 [M + H]$^+$	285.1121	−5.40	$C_{17}H_{16}O_4$	253, 225

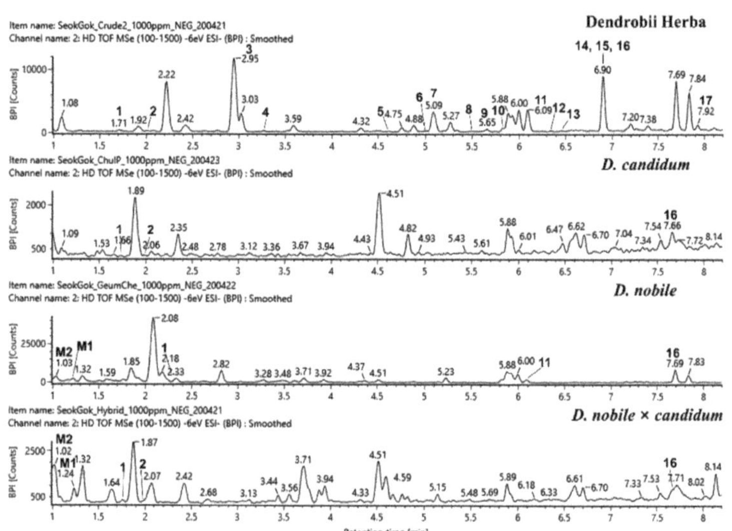

Figure 2. LC–MS base peak ion (BPI) chromatograms of four *Dendrobium* extracts at negative ion mode (6 eV, ESI$^-$). The selected chromatographic peaks are annotated with peak numbers referred to in Table 2.

In addition, the cytotoxicity of the methanol extracts from four *Dendrobium* stems against the human pharynx squamous carcinoma (FaDu) cell line was also determined (Table 2 and Figure S20). The methanol extract of Dendrobii Herba showed the greatest activity, with an IC_{50} value of 63.03 μg/mL. The activities of the other *Dendrobium* samples exhibited IC_{50} values ranging from 84.47 to 91.93 μg/mL, with the hybrid *D. nobile* × *candidum* showing the lowest activity. In addition, the methanol extracts of the stems of 436 mutant lines of *D. nobile* × *candidum* were prepared and tested for cytotoxic ability against FaDu cells at a concentration of 50 µg/mL. The samples that showed less than 50% cell viability were determined as active samples (Table S1).

Table 2. Identification of individual compounds in extracts prepared from Dendrobii Herba, *D. nobile*, *D. candidum*, the hybrid, *D. nobile* × *candidum*, and their cytotoxicities against FaDu cells.

Samples	Cytotoxicity [1] (IC_{50}, µg/mL)	Peak No.																
		1	2	3	4	5	6	7	8	9	10	11	12	13	14	15	16	17
Dendrobii Herba	63.03	+	+	+	+	+	+	+	+	+	+	+	+	+	+	+	+	+
D. candidum	84.47	+	+	−	−	−	−	−	−	−	−	−	−	−	−	−	+	−
D. nobile	89.22	+	−	−	−	−	−	−	−	−	−	+	−	−	−	−	+	−
D. nobile × *candidum*	91.93	+	+	−	−	−	−	−	−	−	−	−	−	−	−	−	+	−

[1] Cisplatin as a positive control showed an IC_{50} value of 1.52 µM.

In our previous study of Dendrobii Herba [40], phytochemical investigation on the ethyl acetate-soluble fraction that showed cytotoxicity against FaDu cells resulted in the isolation of the active compounds, densiflorol B (peak 7), moscatilin (peak 9), ephemeranthol A (peak 12), dehydroorchinol (peak 14), and 1,5,7-trimethoxy-2-phenanthrenol (peak 17). These peaks were only detected in Dendrobii Herba and not in *D. nobile*, *D. candidum*, and their hybrid. Therefore, this result suggests that the distribution of the various constituents of Dendrobii Herba contributed more to its potent biological activity than the chemical composition of the other *Dendrobium* samples. In our further study on the mechanism of action of the most active compound, moscatilin, we have confirmed that moscatilin induced apoptosis in FaDu cells through the extrinsic and intrinsic apoptotic signaling pathways and the c-Jun N-terminal kinase (JNK) signaling pathway [17]. *D. nobile* constituents have been shown to exert anticancer activities: dendrobine has been shown to enhance the anticancer effect of cisplatin on non-small cell lung cancer cells via JNK stress signaling [10] and to induce apoptosis and inhibit cancer cell invasion in human gastric cells [52]. Nudol has been shown to inhibit the cell proliferation of U2OS osteosacorma cells by inducing cell cycle arrest at the G2/M phase and the migration of U2OS cells, and inducing cell apoptosis through the caspase-dependent pathway [53]. The extract of *D. candidum* has been reported to decrease the cell viability of human breast cancer cells by inducing cell cycle arrest at the G2/M phase and regulating the biomarkers (ERα, PGR, and GATA3) and oncogenes (p53, Ki67, and ELF5) [54]. To the best of our knowledge, the cytotoxicities of the methanol extracts from *D. nobile*, *D. candidum*, and *D. nobile* × *candidum* against FaDu cells were reported for the first time.

2.2. Untargeted UPLC-QToF MS Analysis of Gamma-Irradiated Mutant Lines of D. Nobile × Candidum

Recently, multivariate statistical analysis has been shown to be an effective approach in distinguishing differences between experimental setups, in untargeted and targeted metabolomic studies [55]. First, the metabolite profiles of 436 mutant lines grown under the same environmental conditions after γ-irradiation with the same dose as the stems of *D. nobile* × *candidum* were analyzed via UPLC-QToF MS. Second, we performed principal component analysis (PCA), orthogonal partial least squares-discriminant analysis (OPLS-DA), and S-plots, which have been widely used in the field in recent years for the metabolomic analysis of extremely complex samples. There were no significant differences among the mutant lines in the chromatograms (data not shown), however, PCA and OPLS-DA score plots showed a clear separation of the clusters when applying multivariate analysis to the normalized dataset (Figure 3). In particular, supervised OPLS-DA has been widely used to study the differences between two similar groups. The OPLS-DA model quality can be estimated using the cross-validation parameters Q^2 (model predictability) and $R^2(y)$ (total explained variation for the X matrix). OPLS-DA of the samples produced one predictive and one orthogonal (1 + 3) component and showed that the cross-validated predictive ability Q^2 was 0.684, and the variance related to the differences between the two origins $R^2(y)$ was 0.693. In most cases, a Q^2 value greater than 0.5 is adequate, and the difference between R^2 and Q^2 values should be less than 0.3. Third, the *S*-plot (Point, t_R-*m/z* pair) from the OPLS-DA model, which is a useful tool for comparing the magnitude and reliability of a variable, was also analyzed. The markers associated with the mutant lines of *D. nobile* × *candidum* were based on the threshold of variable importance in the projection (VIP) value (VIP > 1.0) from the *S*-plot. The two identified metabolites, M1 and M2, had VIP values greater than 1.0, making them the primary markers of the differences between the two clusters for the mutant lines of *D. nobile* × *candidum* on the OPLS-DA score plot (Figure 3).

(a)

(b)

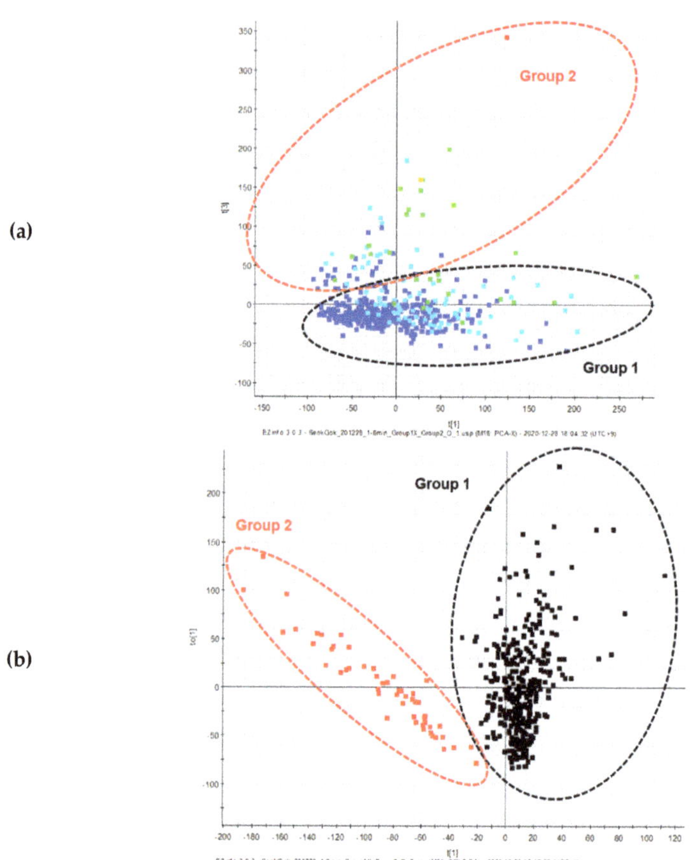

Figure 3. *Cont.*

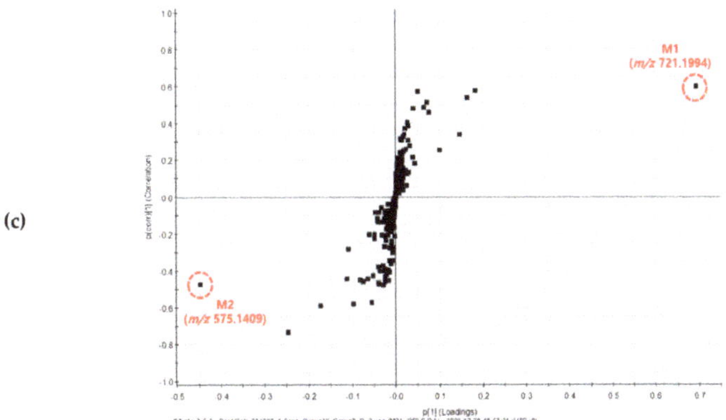

Figure 3. Principal component analysis (PCA) (**a**) and orthogonal partial least squares-discriminant analysis (OPLS-DA) (**b**) score plots and (**c**) *S*-plots of the inactive and active samples analyzed via UPLC-QToF MS.

Marker ions ([M − H]$^-$) at *m/z* 721.1994 (M1) were located far from the center and were in the same direction as group 1, indicating distinguishable markers for group 1. Differential markers in the same direction as group 2 demonstrated molecular ions ([M − H]$^-$) at *m/z* 575.1409 (M2). These markers were identified as unknown metabolites that did not match with the compounds reported here (**1-17**) and any compounds previously reported in the *Dendrobium* species. M1, which belongs to the inactive group, gave a molecular ion at *m/z* 721.1994 [M – H]$^-$ corresponding to the molecular formula $C_{33}H_{38}O_{18}$ and −0.56 ppm of *m/z* error (Figure S18). The marker M1 (t_R = 1.03 min) was classified as the 3-hydroxy-3-methylglutaryl (HMG) moiety of *C*-glucoside and *C*-rhamnoside with MS/MS fragmentation patterns (*m/z* 577, 487, 457, 383, and 353) corresponding to the loss of HMG (*m/z* 144), *C*-glucoside (*m/z* 90 and 120), and *C*-rhamnoside (*m/z* 74 and 104) fragments. Thus, the M1 marker was tentatively identified as apigenin 6-*C*-α-L-rhamnopyranoside-8-*C*-[6″-*O*-(3-hydroxy-3-methylglutaryl)]-β-D-glucopyranoside [56]. Another marker, M2, in the active group, demonstrated a molecular ion at *m/z* 575.1409 [M – H]$^-$, corresponding to a molecular formula $C_{27}H_{28}O_{14}$ and -0.38 ppm of *m/z* error (Figure S19). The marker M2 (t_R = 1.21 min) was classified as the 3-hydroxy-3-methylglutaryl (HMG) moiety and *O*-glucoside with MS/MS fragmentation patterns (*m/z* 431 and 269) corresponding to the loss of the HMG (*m/z* 144) and *O*-glucoside (*m/z* 162) fragments. Thus, the M2 marker was tentatively identified as apigenin 7-*O*-[6″-*O*-(3-hydroxy-3-methylglutaryl)]β-D-glucopyranoside [57].

Multivariate analysis clearly differentiated the samples into two clusters depending on their activities: a group of 384 extracts without cytotoxicity (Group I) and a group of 52 extracts (Group II) showing less than 50% cell viability (Figure 3). These results correlated with different metabolites and active components. *C*-glycosylation or *O*-glycosylation of apigenin may affect its metabolism, and in turn, affect both its anticancer potential and biological benefits. Previous studies on flavonoids have demonstrated that the sugars of *C*- and/or *O*-glycosides have potent anticancer properties [58,59]. These results are consistent with those published elsewhere, which also found that these metabolites were the major compounds present in the gamma-irradiated mutant lines of *D. nobile* × *candidum*.

3. Materials and Methods

3.1. General

UPLC-QToF MS was performed using a Waters ACQUITY UPLC I-Class system combined with a Vion IMS QToF mass spectrometer (Waters, Milford, MA, USA), equipped with an ACQUITY UPLC BEH C18 column (2.1 mm × 100 mm i.d., 1.7 μm; Waters). All data were processed using UNIFI software (v1.9, Waters). A (^{60}Co)-irradiator (150 TBq capacity; AECL, Ottawa, ON, Canada) was used for gamma irradiation. Compounds **1-12** were isolated from Dendrobii Herba (the stems of the *Dendrobium* species), as described in our previous study [40]. The standard compounds, dendrobine (**13**), batatasin III (**16**), and lusianthridin (**17**) were purchased from Wuhan ChemFaces Biochemical Co., Ltd. (Hubei, China). Shihunine (**14**) and tristin (**15**) were obtained from Wuhan ChemNorm Biotech Co., Ltd. (Hubei, China) and Chengdu Biopurify Phytochemicals Ltd. (Chengdu, China), respectively. All other chemicals and solvents used in this study were of analytical grade (J. T. Baker, Phillipsburg, NJ, USA).

3.2. Plant Materials

Dendrobii Herba stems (CK PHARM Co., Ltd., Seoul, Korea) were purchased from the Jewondang herb shop in Jeongup-si, Jeollabuk-do, Korea. *D. nobile*, *D. candidum*, the hybrid *D. nobile × candidum*, and its gamma-irradiated mutant lines were verified as authentic and grown by Dr. Sang Hoon Kim (Korea Atomic Energy Research Institute). The hybrids (*D. nobile × candidum*) were irradiated with a single dose of γ-rays (20 Gy) emitted from a labeled cobalt (^{60}Co) source (150 TBq capacity; AECL) for 24 h at the Korea Atomic Energy Research Institute, Jeongeup-si, Korea. Approximately 1400 gamma-irradiated mutant lines derived from *D. nobile × candidum* were grown in a greenhouse under the same conditions as *D. nobile*, *D. candidum*, and *D. nobile × candidum* for two years (2017–2019). Two to three stems of these plant materials with good growth and yield (approximately 400 in total) were randomly collected in November 2019. The collected stems were dried using an air-drying method (air-conditioned at a temperature of 40 °C for 72 h). Voucher specimens of *D. nobile* (accession no. RB033), *D. candidum* (accession no. RB030), and the hybrid *D. nobile × candidum* (accession no. RB038) were deposited at the Advanced Radiation Technology Institute, Korea Atomic Energy Research Institute.

3.3. Sample Preparation

Dried stems of Dendrobii Herba, *D. nobile*, *D. candidum*, the hybrid *D. nobile × candidum*, and its gamma-irradiated mutant lines were chopped into small pieces. One gram of each stem was extracted with 30 mL of methanol using an ultrasonic bath for 60 min and was evaporated to produce the methanol extract. Each dried methanol extract (1 mg) was dissolved in 1 mL of methanol for UPLC-QTof MS analysis. Compounds **1-17** were dissolved in methanol at a concentration of approximately 0.1 mg/mL. This sample solution was filtered through a 0.20-μm polyvinylidene fluoride filter for chromatographic analysis. For cytotoxicity evaluation, each dried methanol extract (1 mg) was initially dissolved in dimethyl sulfoxide at a concentration of 100 mg/mL. Subsequently, the final concentration was diluted to 1.5625, 3.125, 6.25, 12.5, 25, 50, and 100 μg/mL using a minimum essential medium (MEM; Corning, Manassas, VA, USA) for Dendrobii Herba, *D. nobile*, *D. candidum*, and the hybrid *D. nobile × candidum*, respectively, and 50 μg/mL for the gamma-irradiated mutant lines of the hybrid.

3.4. UPLC-QTof MS Analysis

The methanol extracts of Dendrobii Herba, *D. nobile*, *D. candidum*, the hybrid, *D. nobile × candidum*, and compounds (**1-17**) were analyzed using a Waters ACQUITY UPLC I-Class system combined with a Vion IMS QToF mass spectrometer (Waters). Each sample (1000 ppm, 2 μL) was injected into an ACQUITY UPLC BEH C18 column (2.1 mm × 100 mm i.d., 1.7 μm; Waters). The temperature of the column oven was maintained at 35 °C. The flow rate was 0.4 mL/min using a mobile phase comprising 0.1% formic acid in water (*v/v*; solvent A) and 0.1% formic acid in acetonitrile (*v/v*; solvent B). Gradient elution was carried out as follows: 0–1.0 min, 23% B; 1.0–3.0 min, 23–27% B; 3.0–5.5 min, 27–37% B; 5.5–8.0 min, 37–62% B; 8.0–8.3 min, 62–100% B; 8.3–10.0 min, 100% B; 10.0–10.3 min, 100–23% B; 10.3–12.0 min, 23% B. The mass spectrometer was operated in negative or positive ion mode with the following parameters: source temperature, 110 °C; desolvation temperature, 350 °C; capillary voltage,

2300 V; cone voltage, 40 V; cone gas flow, 50 L/h; flow rate of desolvation gas (N_2), 800 L/h; mass scan range, 100–1500 Da; scan time, 0.1 s. The full scan data, MS/MS spectra, accurate mass, and elemental composition were calculated using UNIFI software (Waters).

3.5. Cytotoxicity Assay

Human pharynx squamous carcinoma FaDu cells were purchased from the Korean Cell Line Bank (Seoul, Korea). These cells were cultured in a minimum essential medium (MEM; Corning, Manassas, VA, USA) supplemented with 10% heat-inactivated FBS (Hyclone, Logan, UT, USA) in a humidified incubator with 5% CO_2 at 37 °C. To determine the viability of FaDu cells, a CCK-8 assay kit (Dojindo, Kumamoto, Japan) was used according to the manufacturer's protocol. Briefly, FaDu cells were seeded into 96-well plates at a density of 2×10^4 cells/mL and incubated at 37 °C for 24 h. After incubation, the cultured FaDu cells were treated with the indicated concentration of four *Dendrobium* extracts (1.5625−100 μg/mL) and the extracts of mutant lines (50 μg/mL) for 72 h. Thereafter, 10 μL of CCK-8 reagent was added to the cultured FaDu cells and then incubated for a further 4 h, after which the absorbance was measured at 450 nm using a SPARK® multimode microplate reader (Tecan, Männedorf, Switzerland). Afterward, the 50% inhibitory concentration (IC_{50}) values were calculated from a dose-response analysis performed using GraphPad Prism software (GraphPad Software, La Jolla, CA, USA).

3.6. Chemometric Data Analysis

The raw mass data were normalized to the total intensity (area) and analyzed using Progenesis® QI software v2.4 (Waters). The parameters included a retention time range of 1.0−8.0 min, a mass range from 100 Da to 1,500 Da, and a mass tolerance of 0.01 Da. The isotopic data were excluded, the noise elimination level was 10, and the mass and retention time windows were 0.04 min and 0.1 min, respectively. After creating a suitable processing method, the dataset was processed using the Create dataset window. The resulting two-dimensional matrix for the measured mass values and intensities for each sample was further exported to EZinfo software v3.0.3 (Waters) using both unsupervised principal component analysis and supervised OPLS-DA.

4. Conclusions

In conclusion, the chemical compositions of Dendrobii Herba, *D. nobile*, *D. chrysanthum*, and their hybrid, *D. nobile × candidum*, were analyzed via UPLC-QToF MS and were identified by comparing the ESI QToF MS data of the isolates (**1-13**) from Dendrobii Herba and the standards (**14-17**). The results showed that the chemical compositions of each sample were different, demonstrating that each sample from different species could be well-distinguished, and the quality of the different samples could be determined through this method. The distribution and content of the constituents of each sample were estimated to be related to their cytotoxic activities against FaDu cells. In addition, UPLC-QToF MS combined with multivariate analysis was used to analyze the chemical profiles of the gamma-irradiated mutant lines derived from *D. nobile × candidum*. The results showed a clear separation of different groups of the mutant lines and the distinguished marker metabolites (M1 and M2), according to the two clusters. These markers were tentatively identified as apigenin 6-C-α-L-rhamnopyranoside-8-C-[6''-O-(3-hydroxy-3-methylglutaryl)]-β-D-glucopyranoside and apigenin 7-O-[6''-O-(3-hydroxy-3-methylglutaryl)]-β-D-glucopyranoside, respectively. Based on the cytotoxicity screening results, the samples were also divided into two groups (384 inactive samples and 52 active samples), where these markers corresponded to the two groups contributing to the correlation of the inactive and active extracts, respectively. This study was an attempt to select superior and differentiated mutant lines based on biochemical analyses. Therefore, these results will serve as a reference for future investigations of the mutation mechanism of the gamma-irradiated mutant lines and their quality evaluation for improved mutant selection.

Supplementary Materials: The following are available online at https://www.mdpi.com/article/10.3390/plants10071376/s1, Figure S1: ESI QTof MS spectrum of gigantol (peak 1), Figure S2: ESI QTof MS spectrum of (1*R*,2*R*)-1,7-hydroxy-2,8-methoxy-2,3-dihydrophenanthrene-4(1*H*)-one (peak 2), Figure S3: ESI QTof MS spectrum of tristin (peak 3), Figure S4: ESI QTof MS spectrum of (−)-syringaresinol (peak 4), Figure S5: ESI QTof MS spectrum of lusianthridin (peak 5), Figure S6: ESI QTof MS spectrum of 2,7-dihydroxy-phenanthrene-1,4-dione (peak 6), Figure S7: ESI QTof MS spectrum of densiflorol B (peak 7), Figure S8: ESI QTof MS spectrum of denthyrsinin (peak 8), Figure S9: ESI QTof MS spectrum of moscatilin (peak 9), Figure S10: ESI QTof MS spectrum of lusianthridin dimer (peak 10), Figure S11: ESI QTof MS spectrum of batatasin III (peak 11), Figure S12: ESI QTof MS spectrum of ephemeranthol A (peak 12), Figure S13: ESI QTof MS spectrum of thunalbene (peak 13), Figure S14: ESI QTof MS spectrum of dehydroorchinol (peak 14), Figure S15: ESI QTof MS spectrum of dendrobine (peak 15), Figure S16: ESI QTof MS spectrum of shihunine (peak 16), Figure S17: ESI QTof MS spectrum of 1,5,7-trimethoxy-2-phenanthrenol (peak 17), Figure S18: ESI QTof MS spectrum of differential metabolite (M1), Figure S19: ESI QTof MS spectrum of differential metabolite (M2), Figure S20: Cytotoxicities of the methanol extracts of Dendrobii Herba, *D. nobile*, *D. candidum*, the hybrid, *D. nobile* × *candidum* against FaDu cells, Table S1: Cell viabilities of the methanol extracts of the stems of 436 mutant lines of *D. nobile* × *candidum* at a concentration of 50 μg/mL in FaDu cells.

Author Contributions: Conceptualization, A.-R.H. and H.-W.R.; methodology, A.-R.H., C.-H.J. (Chang-Hyun Jin), C.-H.J. (Chan-Hun Jung), S.-H.K. and H.-W.R.; software, B.N., H.-J.J., Y.-R.K. and M.-J.H.; validation, A.-R.H., C.-H.J. (Chang-Hyun Jin), K.-B.K. and H.-W.R.; formal analysis, B.N., H.-J.J., Y.-R.K. and C.-H.J. (Chang-Hyun Jin); investigation, B.N., H.-J.J., Y.-R.K., C.-H.J. (Chang-Hyun Jin), K.-B.K. and S.-H.K.; resources, C.-H.J. (Chan-Hun Jung) and S.-H.K.; data curation, B.N., H.-J.J., A.-R.H., K.-B.K. and H.-W.R.; writing—original draft preparation, A.-R.H. and H.-W.R.; writing—review and editing, A.-R.H., M.-J.H., K.-B.K. and H.-W.R.; visualization, B.N., H.-J.J., M.-J.H., S.-H.K. and H.-W.R.; supervision, A.-R.H. and H.-W.R.; project administration, A.-R.H., J.-B.K. and H.-W.R.; funding acquisition, J.-B.K. and H.-W.R. All authors have read and agreed to the published version of the manuscript.

Funding: This research was funded by the Korea Atomic Energy Research Institute (KAERI), grant number 523310-21 and the KRIBB Research Initiative Program funded by the Ministry of Science and ICT (MSIT).

Institutional Review Board Statement: Not applicable.

Informed Consent Statement: Not applicable.

Data Availability Statement: Not applicable.

Conflicts of Interest: The authors declare no conflict of interest.

References

1. Lam, Y.; Ng, T.B.; Yao, R.M.; Shi, J.; Xu, K.; Sze, S.; Zhang, K.Y. Evaluation of chemical constituents and important mechanism of pharmacological biology in *Dendrobium* plants. *Evid. Based Complement. Alternat. Med.* **2015**, *2015*, 841752. [CrossRef] [PubMed]
2. Gu, C.; Zhang, X.; Wu, L.; Jiang, X.; Huang, L. Quality evaluation of *Dendrobium* based on ultra-performance liquid chromatography (UPLC) and chemometrics. *J. Appl. Pharm. Sci.* **2017**, *7*, 17–23. [CrossRef]
3. Ng, T.B.; Liu, J.; Wong, J.H.; Ye, X.; Sze, S.C.W.; Tong, Y.; Zhang, K.Y. Review of research on *Dendrobium*, a prized folk medicine. *Appl. Microbiol. Biot.* **2012**, *93*, 1795–1803. [CrossRef] [PubMed]
4. Li, L.-S.; Lu, Y.-L.; Nie, J.; Xu, Y.-Y.; Zhang, W.; Yang, W.-J.; Gong, Q.-H.; Lu, Y.-F.; Lu, Y.; Shi, J.-S. *Dendrobium nobile* Lindl alkaloid, a novel autophagy inducer, protects against axonal degeneration induced by Aβ25-35 in hippocampus neurons in vitro. *CNS Neurosci. Ther* **2017**, *23*, 329–340. [CrossRef]
5. Wang, Q.; Gong, Q.; Wu, Q.; Shi, J. Neuroprotective effects of *Dendrobium* alkaloids on rat cortical neurons injured by oxygen-glucose deprivation and reperfusion. *Phytomedicine* **2010**, *17*, 108–115. [CrossRef] [PubMed]
6. Meng, C.-W.; He, Y.-L.; Peng, C.; Ding, X.-J.; Guo, L.; Xiong, L. Picrotoxane sesquiterpenoids from the stems of *Dendrobium nobile* and their absolute configurations and angiogenesis effect. *Fitoterapia* **2017**, *121*, 206–211. [CrossRef]
7. Morita, H.; Fujiwara, M.; Yoshida, N.; Kobayashi, J. New picrotoxinin-type and dendrobine-type sesquiterpenoids from *Dendrobium* snowflake 'Red Star'. *Tetrahedron* **2000**, *56*, 5801–5805. [CrossRef]
8. Yang, L.; Zhang, C.; Yang, H.; Zhang, M.; Wang, Z.; Xu, L. Two new alkaloids from *Dendrobium chrysanthum*. *Heterocycles-Sendai Inst. Heterocycl. Chem.* **2005**, *65*, 633–636.
9. Liu, G.-Y.; Tan, L.; Cheng, L.; Ding, L.-S.; Zhou, Y.; Deng, Y.; He, Y.-Q.; Guo, D.-L.; Xiao, S.-J. Dendrobine-type alkaloids and bibenzyl derivatives from *Dendrobium findlayanum*. *Fitoterapia* **2020**, *142*, 104497. [CrossRef]

10. Song, T.-H.; Chen, X.-X.; Lee, C.K.-F.; Sze, S.C.-W.; Feng, Y.-B.; Yang, Z.-J.; Chen, H.-Y.; Li, S.-T.; Zhang, L.-Y.; Wei, G. Dendrobine targeting JNK stress signaling to sensitize chemotoxicity of cisplatin against non-small cell lung cancer cells in vitro and in vivo. *Phytomedicine* **2019**, *53*, 18–27. [CrossRef]

11. Charoenrungruang, S.; Chanvorachote, P.; Sritularak, B.; Pongrakhananon, V. Gigantol, a bibenzyl from *Dendrobium draconis*, inhibits the migratory behavior of non-small cell lung cancer cells. *J. Nat. Prod.* **2014**, *77*, 1359–1366. [CrossRef]

12. Bhummaphan, N.; Pongrakhananon, V.; Sritularak, B.; Chanvorachote, P. Cancer stem cell–suppressing activity of chrysotoxine, a bibenzyl from *Dendrobium pulchellum*. *J. Pharmacol. Exp. Ther.* **2018**, *364*, 332–346. [CrossRef]

13. Pinkhien, T.; Petpiroon, N.; Sritularak, B.; Chanvorachote, P. Batatasin III inhibits migration of human lung cancer cells by suppressing epithelial to mesenchymal transition and FAK-AKT signals. *Anticancer Res.* **2017**, *37*, 6281–6289.

14. Chanvorachote, P.; Kowitdamrong, A.; Ruanghirun, T.; Sritularak, B.; Mungmee, C.; Likhitwitayawuid, K. Anti-metastatic activities of bibenzyls from *Dendrobium pulchellum*. *Nat. Prod. Commun.* **2013**, *8*, 1934578X1300800127. [CrossRef]

15. Pai, H.-C.; Chang, L.-H.; Peng, C.-Y.; Chang, Y.-L.; Chen, C.-C.; Shen, C.-C.; Teng, C.-M.; Pan, S.-L. Moscatilin inhibits migration and metastasis of human breast cancer MDA-MB-231 cells through inhibition of Akt and Twist signaling pathway. *J. Mol. Med.* **2013**, *91*, 347–356. [CrossRef]

16. Gong, Y.-Q.; Fan, Y.; Wu, D.-Z.; Yang, H.; Hu, Z.-B.; Wang, Z.-T. In vivo and in vitro evaluation of erianin, a novel anti-angiogenic agent. *Eur. J. Cancer.* **2004**, *40*, 1554–1565. [CrossRef]

17. Lee, E.; Han, A.-R.; Nam, B.; Kim, Y.-R.; Jin, C.H.; Kim, J.-B.; Eun, Y.-G.; Jung, C.-H. Moscatilin induces apoptosis in human head and neck squamous cell carcinoma cells via JNK signaling pathway. *Molecules.* **2020**, *25*, 901. [CrossRef]

18. Zhang, X.; Xu, J.-K.; Wang, J.; Wang, N.-L.; Kurihara, H.; Kitanaka, S.; Yao, X.-S. Bioactive bibenzyl derivatives and fluorenones from *Dendrobium nobile*. *J. Nat. Prod.* **2007**, *70*, 24–28. [CrossRef]

19. Zhou, X.-M.; Zheng, C.-J.; Gan, L.-S.; Chen, G.-Y.; Zhang, X.-P.; Song, X.-P.; Li, G.-N.; Sun, C.-G. Bioactive phenanthrene and bibenzyl derivatives from the stems of *Dendrobium nobile*. *J. Nat. Prod.* **2016**, *79*, 1791–1797. [CrossRef]

20. Yang, D.; Liu, L.-Y.; Cheng, Z.-Q.; Xu, F.-Q.; Fan, W.-W.; Zi, C.-T.; Dong, F.-W.; Zhou, J.; Ding, Z.-T.; Hu, J.-M. Five new phenolic compounds from *Dendrobium aphyllum*. *Fitoterapia* **2015**, *100*, 11–18. [CrossRef]

21. Zhang, X.; Xu, J.-K.; Wang, N.-L.; Kurihara, H.; Yao, X. Antioxidant phenanthrenes and lignans from *Dendrobium nobile*. *J. Chin. Pharm. Sci.* **2008**, *17*, 314–318.

22. Kim, J.H.; Oh, S.-Y.; Han, S.-B.; Uddin, G.M.; Kim, C.Y.; Lee, J.K. Anti-inflammatory effects of *Dendrobium nobile* derived phenanthrenes in LPS-stimulated murine macrophages. *Arch. Pharmacal Res.* **2015**, *38*, 1117–1126. [CrossRef] [PubMed]

23. Hwang, J.S.; Lee, S.A.; Hong, S.S.; Han, X.H.; Lee, C.; Kang, S.J.; Lee, D.; Kim, Y.; Hong, J.T.; Lee, M.K. Phenanthrenes from *Dendrobium nobile* and their inhibition of the LPS-induced production of nitric oxide in macrophage RAW 264.7 cells. *Bioorg. Med. Chem. Lett.* **2010**, *20*, 3785–3787. [CrossRef] [PubMed]

24. Ito, M.; Matsuzaki, K.; Wang, J.; Daikonya, A.; Wang, N.-L.; Yao, X.-S.; Kitanaka, S. New phenanthrenes and stilbenes from *Dendrobium loddigesii*. *Chem. Pharm. Bull.* **2010**, *58*, 628–633. [CrossRef]

25. Bhummaphan, N.; Petpiroon, N.; Prakhongcheep, O.; Sritularak, B.; Chanvorachote, P. Lusianthridin targeting of lung cancer stem cells via Src-STAT3 suppression. *Phytomedicine* **2019**, *62*, 152932. [CrossRef]

26. Yang, H.; Sung, S.H.; Kim, Y.C. Antifibrotic phenanthrenes of *Dendrobium nobile* stems. *J. Nat. Prod.* **2007**, *70*, 1925–1929. [CrossRef]

27. Wattanathamsan, O.; Treesuwan, S.; Sritularak, B.; Pongrakhananon, V. Cypripedin, a phenanthrenequinone from *Dendrobium densiflorum*, sensitizes non-small cell lung cancer H460 cells to cisplatin-mediated apoptosis. *J. Nat. Med.* **2018**, *72*, 503–513. [CrossRef]

28. Zhao, G.-Y.; Deng, B.-W.; Zhang, C.-Y.; Cui, Y.-D.; Bi, J.-Y.; Zhang, G.-G. New phenanthrene and 9, 10-dihydrophenanthrene derivatives from the stems of *Dendrobium officinale* with their cytotoxic activities. *J. Nat. Med.* **2018**, *72*, 246–251. [CrossRef]

29. Chen, X.-J.; Mei, W.-L.; Zuo, W.-J.; Xeng, J.-B.; Guo, Z.-K.; Song, X.-Q.; Dai, H.-F. A new antibacterial phenanthrenequinone from *Dendrobium sinense*. *J. Asian Nat. Prod. Res.* **2013**, *15*, 67–70. [CrossRef]

30. Zhang, G.-N.; Zhong, L.-Y.; Bligh, S.A.; Guo, Y.-L.; Zhang, C.-F.; Zhang, M.; Wang, Z.-T.; Xu, L.-S. Bi-bicyclic and bi-tricyclic compounds from *Dendrobium thyrsiflorum*. *Phytochemistry* **2005**, *66*, 1113–1120. [CrossRef]

31. Kyokong, N.; Muangnoi, C.; Thaweesest, W.; Kongkatitham, V.; Likhitwitayawuid, K.; Rojsitthisak, P.; Sritularak, B. A new phenanthrene dimer from *Dendrobium palpebrae*. *J. Asian Nat. Prod. Res.* **2019**, *21*, 391–397. [CrossRef]

32. Zhang, X.; Liu, H.-W.; Gao, H.; Han, H.-Y.; Wang, N.-L.; Wu, H.-M.; Yao, X.-S.; Wang, Z. Nine new sesquiterpenes from *Dendrobium nobile*. *Helv. Chim. Acta* **2007**, *90*, 2386–2394. [CrossRef]

33. Zhao, W.; Ye, Q.; Dai, J.; Martin, M.-T.; Zhu, J. Allo-aromadendrane-and picrotoxane-type sesquiterpenes from *Dendrobium moniliforme*. *Planta Med.* **2003**, *69*, 1136–1140.

34. Park, B.-Y.; Oh, S.-R.; Ahn, K.-S.; Kwon, O.-K.; Lee, H.-K. (-)-Syringaresinol inhibits proliferation of human promyelocytic HL-60 leukemia cells via G1 arrest and apoptosis. *Int. Immunopharmacol.* **2008**, *8*, 967–973. [CrossRef]

35. Zhou, C.; Xie, Z.; Lei, Z.; Huang, Y.; Wei, G. Simultaneous identification and determination of flavonoids in *Dendrobium officinale*. *Chem. Cent. J.* **2018**, *12*, 40. [CrossRef]

36. Poudel, M.R.; Chand, M.B.; Karki, N.; Pant, B. Antioxidant activity and total phenolic and flavonoid contents of *Dendrobium amoenum* Wall. ex Lindl. *Bot. Orient. J. Plant. Sci.* **2015**, *9*, 20–26. [CrossRef]

37. Zhao, Y.; Son, Y.-O.; Kim, S.-S.; Jang, Y.-S.; Lee, J.-C. Antioxidant and anti-hyperglycemic activity of polysaccharide isolated from *Dendrobium chrysotoxum* Lindl. *J. Biochem. Mol. Biol.* **2007**, *40*, 670–677. [CrossRef]

38. Liang, J.; Wu, Y.; Yuan, H.; Yang, Y.; Xiong, Q.; Liang, C.; Li, Z.; Li, C.; Zhang, G.; Lai, X. *Dendrobium officinale* polysaccharides attenuate learning and memory disabilities via anti-oxidant and anti-inflammatory actions. *Int. J. Biol. Macromol.* **2019**, *126*, 414–426. [CrossRef]

39. Xia, L.; Liu, X.; Guo, H.; Zhang, H.; Zhu, J.; Ren, F. Partial characterization and immunomodulatory activity of polysaccharides from the stem of *Dendrobium officinale* (Tiepishihu) in vitro. *J. Funct. Foods* **2012**, *4*, 294–301. [CrossRef]

40. Nam, B.; Ryu, S.M.; Lee, D.; Jung, C.-H.; Jin, C.H.; Kim, J.-B.; Lee, I.-S.; Han, A.-R. Identification of two new phenanthrenes from Dendrobii Herba and their cytotoxicity towards human hypopharynx squamous carcinoma cell (FaDu). *Molecules* **2019**, *24*, 2339. [CrossRef]

41. Ali, H.; Ghori, Z.; Sheikh, S.; Gul, A. Effects of gamma radiation on crop production. In *Crop Production and Global Environmental Issues*; Hakeem, K., Ed.; Springer: Cham, Switzerland, 2016; pp. 27–78.

42. Jankowicz-Cieslak, J.; Mba, C.; Till, B.J. Mutagenesis for crop breeding and functional genomics. In *Biotechnologies for Plant Mutation Breeding*; Jankowicz-Cieslak, J., Tai, T., Kumlehn, J., Till, B., Eds.; Springer: Cham, Switzerland, 2017; pp. 3–18.

43. Jo, Y.D.; Kim, Y.-S.; Ryu, J.; Choi, H.-I.; Kim, S.W.; Kang, H.S.; Ahn, J.-W.; Kim, J.-B.; Kang, S.-Y.; Kim, S.H. Deletion of carotenoid cleavage dioxygenase 4a (CmCCD4a) and global up-regulation of plastid protein-coding genes in a mutant chrysanthemum cultivar producing yellow petals. *Sci. Hortic.* **2016**, *212*, 49–59. [CrossRef]

44. Hong, M.J.; Kim, D.Y.; Nam, B.; Ahn, J.-W.; Kwon, S.-J.; Seo, Y.W.; Kim, J.-B. Characterization of novel mutants of hexaploid wheat (*Triticum aestivum* L.) with various depths of purple grain color and antioxidant capacity. *J. Sci. Food Agric.* **2019**, *99*, 55–63. [CrossRef] [PubMed]

45. Yang, J.; Han, X.; Wang, H.-Y.; Yang, J.; Kuang, Y.; Ji, K.-Y.; Yang, Y.; Pang, K.; Yang, S.-Z.; Qin, J.-X.; et al. Comparison of metabolomics of *Dendrobium officinale* in different habitats by UPLC-Q-TOF-MS. *Biochem. Syst. Ecol.* **2020**, *89*, 104007. [CrossRef]

46. Wang, Y.-H.; Avula, B.; Abe, N.; Wei, F.; Wang, M.; Ma, S.-C.; Ali, Z.; Elsohly, M.A.; Khan, I.A. Tandem Mass Spectrometry for Structural Identification of Sesquiterpene Alkaloids from the Stems of *Dendrobium nobile* Using LC-QToF. *Planta Med.* **2016**, *82*, 662–670. [CrossRef]

47. Tao, Y.; Cai, H.; Li, W.; Cai, B. Ultrafiltration coupled with high-performance liquid chromatography and quadrupole-time-of-flight mass spectrometry for screening lipase binders from different extracts of *Dendrobium officinale*. *Anal. Bioanal. Chem.* **2015**, *407*, 6081–6093. [CrossRef]

48. Zha, X.-Q.; Luo, J.-P.; Wei, P. Identification and classification of *Dendrobium candidum* species by fingerprint technology with capillary electrophoresis. *S. Afr. J. Bot.* **2009**, *75*, 276–282. [CrossRef]

49. Li, Y.; Wang, C.-L.; Guo, S.-X.; Yang, J.-S.; Xiao, P.-G. Two new compounds from *Dendrobium candidum*. *Chem. Pharm. Bull.* **2008**, *56*, 1477–1479. [CrossRef]

50. Xu, Y.-Y.; Wang, L.-Y.; Huang, B.; Xie, X.-L.; Wu, Q.; Shi, J.-S. Comparison of contents of polysaccharides and alkaloids in *Dendrobium* from different harvest time. *Chin. Pharm. J.* **2014**, *29*, 288–291.

51. Chen, H.; Li, X.; Xu, Y.; Lo, K.; Zheng, H.; Hu, H.; Wang, J.; Lin, Y. Study on the polar extracts of *Dendrobium nobile, D. officinale, D. loddigesii*, and *Flickingeria fimbriata*: Metabolite identification, content evaluation, and bioactivity assay. *Molecules* **2018**, *23*, 1185. [CrossRef]

52. Song, J.I.; Kang, Y.J.; Yong, H.; Kim, Y.C.; Moon, A. Denbinobin, a phenanthrene from *Dendrobium nobile*, inhibits invasion and induces apoptosis in SNU-484 human gastric cancer cells. *Oncol. Rep.* **2012**, *27*, 813–818.

53. Zhang, Y.; Zhang, Q.; Xin, W.; Liu, N.; Zhang, H. Nudol, a phenanthrene derivative from *Dendrobium nobile*, induces cell cycle arrest and apoptosis and inhibits migration in osteosarcoma cells. *Drug Des Dev. Ther.* **2019**, *13*, 2591–2601. [CrossRef]

54. Sun, J.; Guo, Y.; Fu, X.; Wang, Y.; Liu, Y.; Huo, B.; Sheng, J.; Hu, X. *Dendrobium candidum* inhibits MCF-7 cells proliferation by inducing cell cycle arrest at G2/M phase and regulating key biomarkers. *OncoTargets Ther.* **2016**, *9*, 21–30.

55. Rahman, S.; Haq, F.U.; Ali, A.; Khan, M.N.; Shah, S.M.Z.; Adhikhari, A.; El-Seedi, H.R.; Musharraf, S.G. Combining untargeted and targeted metabolomics approaches for the standardization of polyherbal formulations through UPLC–MS/MS. *Metabolomics* **2019**, *15*, 1–11. [CrossRef]

56. Liu, M.; Liu, Q.; Liu, Y.-L.; Hou, C.-Y.; Mabry, T.J. An acylated flavone C-glycoside from *Glycyrrhiza eurycarpa*. *Phytochemistry* **1994**, *36*, 1089–1090. [CrossRef]

57. König, G.M.; Wright, A.D.; Keller, W.J.; Judd, R.L.; Bates, S.; Day, C. Hypoglycaemic activity of an HMG-containing flavonoid glucoside, chamaemeloside, from *Chamaemelum nobile*. *Planta Med.* **1998**, *64*, 612–614. [CrossRef]

58. Smiljkovic, M.; Stanisavljevic, D.; Stojkovic, D.; Petrovic, I.; Vicentic, J.M.; Popovic, J.; Grdadolnik, S.G.; Markovic, D.; Sankovic-Babice, S.; Glamoclija, J.; et al. Apigenin-7-*O*-glucoside versus apigenin: Insight into the modes of anticandidal and cytotoxic actions. *EXCLI J.* **2017**, *16*, 795–807.

59. Czemplik, M.; Mierziak, J.; Szopa, J.; Kulma, A. Flavonoid C-glucosides derived from *Flax straw* extracts reduce human breast cancer cell growth in vitro and induce apoptosis. *Front. Pharmacol.* **2016**, *31*, 282. [CrossRef] [PubMed]

plants

MDPI

Communication

Phytochemical Analysis of the Fruits of Sea Buckthorn (*Hippophae rhamnoides*): Identification of Organic Acid Derivatives

Yong Hoon Lee [1,†], Hee Joo Jang [1,†], Kun Hee Park [2], Seon-Hee Kim [3], Jung Kyu Kim [4], Jin-Chul Kim [5], Tae Su Jang [6,*] and Ki Hyun Kim [1,*]

1. School of Pharmacy, Sungkyunkwan University, Suwon 16419, Korea; yhl2090@naver.com (Y.H.L.); dhn01200@naver.com (H.J.J.)
2. Department of Food Science and Biotechnology, Sungkyunkwan University, Suwon 16419, Korea; soske@skku.edu
3. Sungkyun Biotech Co., Ltd., Suwon 16419, Korea; seonhee31@gmail.com
4. School of Chemical Engineering, Sungkyunkwan University, Suwon 16419, Korea; legkim@skku.edu
5. Natural Product Informatics Research Center, KIST Gangneung Institute of Natural Products, Gangneung 25451, Korea; jckim@kist.re.kr
6. College of Medicine, Dankook University, Cheonan 31116, Korea
* Correspondence: jangts@dankook.ac.kr (T.S.J.); khkim83@skku.edu (K.H.K.); Tel.: +82-41-550-1476 (T.S.J.); +82-31-290-7700 (K.H.K.)
† Correspondence: authors contributed equally to this study.

check for updates

Citation: Lee, Y.H.; Jang, H.J.; Park, K.H.; Kim, S.-H.; Kim, J.K.; Kim, J.-C.; Jang, T.S.; Kim, K.H. Phytochemical Analysis of the Fruits of Sea Buckthorn (*Hippophae rhamnoides*): Identification of Organic Acid Derivatives. *Plants* **2021**, *10*, 860. https://doi.org/10.3390/plants10050860

Academic Editors: Narendra Singh Yadav and Jong Seong Kang

Received: 5 April 2021
Accepted: 23 April 2021
Published: 24 April 2021

Publisher's Note: MDPI stays neutral with regard to jurisdictional claims in published maps and institutional affiliations.

Abstract: *Hippophae rhamnoides* L. (Elaeagnaceae), commonly known as "Sea buckthorn" and "Vitamin tree", is a spiny deciduous shrub whose fruit is known for its nutritional composition, such as vitamin C, and is consumed as a dietary supplement worldwide. As part of our ongoing efforts to identify structurally new and bioactive constituents from natural resources, the phytochemical investigation of the extract of *H. rhamnoides* fruits led to the isolation of one malate derivative (**1**), five citrate derivatives (**2–6**), and one quinate derivative (**7**). The structures of the isolated compounds were elucidated by analysis of 1D and 2D nuclear magnetic resonance (NMR) spectroscopic data and high-resolution electrospray ionization (HR-ESI) liquid chromatography–mass spectrometry (LC/MS) data. Three of the citrate derivatives were identified as new compounds: (*S*)-1-butyl-5-methyl citrate (**3**), (*S*)-1-butyl-1′-methyl citrate (**4**), and (*S*)-1-methyl-1′-butyl citrate (**6**), which turned out to be isolation artifacts. The absolute configurations of the new compounds were established by quantum chemical electronic circular dichroism (ECD) calculation, which is an informative tool for verifying the absolute configuration of organic acid derivatives. The isolated compounds **1–7** were evaluated for their stimulatory effects on osteogenesis. Compounds **1**, **3**, **4**, **6**, and **7** stimulated osteogenic differentiation up to 1.4 fold, compared to the negative control. These findings provide experimental evidence that active compounds **1**, **3**, **4**, **6**, and **7** induce the osteogenesis of mesenchymal stem cells and activate bone formation

Keywords: *Hippophae rhamnoides*; Elaeagnaceae; citrate derivatives; nuclear magnetic resonance (NMR); electronic circular dichroism (ECD)

1. Introduction

Hippophae rhamnoides L. (Elaeagnaceae), known as "Sea buckthorn" and "Vitamin tree", is a spiny deciduous shrub. It has historically been used for medicinal purposes worldwide, especially in Central and Southeastern Asia. In China, its pharmacological use was revealed more than a thousand years ago through the records of Sibu Yidian from the Tang Dynasty and Jing Zhu Ben Cao from the Qing dynasty [1]. Sea buckthorn oil was listed in the Russian Pharmacopeia in the 1950s and used as an anti-inflammatory and wound healing agent in Russia [2]. Its bark and fruits have also been used as Tibetan folk medicine to treat pulmonary disorders, cough, fever, and tumors [3]. For the past 50 years, several medicinal preparations based on *H. rhamnoides* have been clinically used to treat radiation damage, oral inflammation, burns, and gastric ulcers, and more than 300 preparations have been reported in the literature [4].

H. rhamnoides bears yellowish or orange berries, which are commonly found as a powdered drink mix in markets as a functional food and are used to make fruit sauce and wine [5]. Vitamin C is a major nutritional component of *H. rhamnoides* fruits [6], the level of which exceeds that of lemons and oranges. Previous pharmacological studies of *H. rhamnoides* have reported that its extracts exhibited therapeutic properties, including anti-platelet effects via the inhibitory mechanism of thrombin-activated platelets and antimicrobial effects against *S. aureus* and *C. albicans* through the inhibition of adhesion and biofilm formation [7,8]. In addition, previous phytochemical investigations on *H. rhamnoides* have demonstrated the presence of diverse types of bioactive substances, including vitamins (A, E, K, riboflavin, and folic acid), carotenoids, phytosterols, minerals, organic acids, polyunsaturated fatty acids, and amino acids [9,10]. Because of the nutritional importance of *H. rhamnoides* fruits, many research groups have attempted to investigate the phytochemical constituents of these fruits, in which flavonoids, proanthocyanidins, alkaloids, and phenolic acids have been identified [11,12]. Nevertheless, previous findings suggested that only few studies have been carried out on the chemical constituents of *H. rhamnoides* fruits, despite their health benefits.

In this context, as a part of continuous research programs to discover structurally new bioactive compounds from diverse natural sources [13–19], our group has focused on the potential bioactive composition of *H. rhamnoides* fruits. In our recent study of *H. rhamnoides* fruits, we identified six chemical compounds, including one citric acid derivative, two flavonoids, one phenolic compound, and two megastigmane derivatives, which were evaluated for their anti-inflammatory effects by determining LPS-induced NO production in RAW 264.7 cells [20]. As a result, we found that a citric acid derivative, 1,5-dimethyl citrate, effectively inhibited LPS-induced NO production and simultaneously showed anti-inflammatory effects by inhibiting the expression of pro-inflammatory mediators, iNOS and COX-2, and the activity of pro-inflammatory cytokines, IL-6 and TNF-α [20]. Our recent findings provide experimental evidence that the organic acid derivative present in *H. rhamnoides* fruits could function as a promising bioactive substance. In the present study, further phytochemical investigation of the extract of *H. rhamnoides* fruits was performed to identify potential bioactive organic acid constituents based on recent preliminary findings. Phytochemical analysis of the BuOH-soluble fraction derived from the extract led to the isolation of seven organic acid derivatives: one malate derivative (**1**), five citrate derivatives (**2–6**), three of which were identified as new compounds, and one quinate derivative (**7**). The structures of the isolated organic acid derivatives (**1–7**) were determined by analysis of 1D and 2D nuclear magnetic resonance (NMR) spectroscopic data and high-resolution electrospray ionization (HR-ESI) liquid chromatography–mass spectrometry (LC/MS). In addition, the absolute configurations of the new compounds were established using quantum chemical electronic circular dichroism (ECD) calculations. Herein, we describe the isolation and structural elucidation of the organic acid derivatives **1–7** and the evaluation of their biological activities for stimulatory effects on osteogenesis.

2. Results and Discussion

2.1. Isolation of Compounds

The aqueous extract powder of *H. rhamnoides* fruits was suspended for solvent partitioning in water and then fractionated with four organic solvents, namely, hexane, dichloromethane, ethyl acetate, and *n*-butanol. LC/MS-based analysis combined with our in-house built UV library, and TLC analysis of the solvent-partitioned fractions suggested that the BuOH-soluble fraction contained the majority of the organic acid derivatives. Phytochemical analysis of the BuOH-soluble fraction was performed using repeated column chromatography with silica gel 60, RP-C$_{18}$ silica gel, Diaion HP-20, Sephadex LH-20, and high-performance liquid chromatography (HPLC). The final semi-preparative HPLC separation afforded seven organic acid derivatives (**1–7**) (Figure 1).

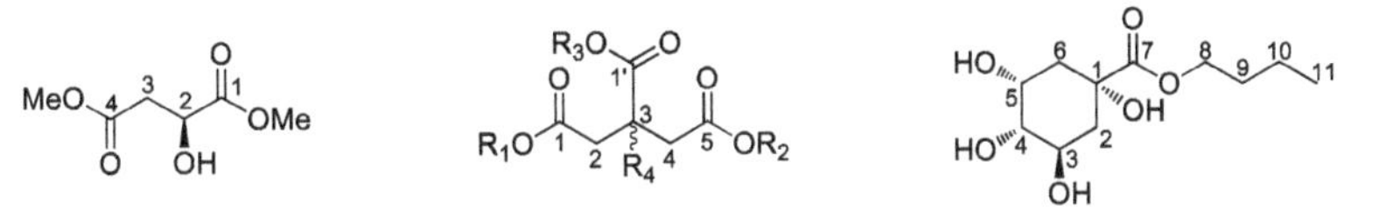

	R_1	R_2	R_3	R_4
2	Me	H	Me	α-OH
3	Butyl	Me	H	β-OH
4	Butyl	H	Me	β-OH
5	Me	Me	Butyl	OH
6	Me	H	Butyl	β-OH

Figure 1. Chemical structure of compounds **1–7**.

2.2. Elucidation of Compound Structures

Compound **3** was isolated as a yellowish gum. The molecular formula was determined to be $C_{11}H_{18}O_7$ from the protonated molecular ion peak $[M + H]^+$ at m/z 263.1122 (calculated for $C_{11}H_{19}O_7$, 263.1131) in the positive-ion mode of HRESIMS (Figure S1). The ^1H NMR spectrum (Figure S2) of **3** (Table 1) showed the presence of characteristic signals of an oxygenated methylene group at δ_H 4.08 (2H, t, J = 6.5 Hz), a methoxy group at δ_H 3.66 (3H, s), two pairs of relatively deshielded methylene groups overlapped at δ_H 2.84 and 2.92, another two pairs of methylene groups at δ_H 1.61 (2H, m) and 1.40 (2H, m), and a terminal methyl group at δ_H 0.94 (3H, t, J = 7.5 Hz). The ^{13}C NMR data of **3** (Table 1) assigned with the assistance of the HSQC experiment (Figure S4) confirmed the 11 carbon signals, which corresponded to three carboxyl carbons (δ_C 170.3, 170.5, and 175.3), two oxygenated carbons (δ_C 64.2 and 72.9), four sets of methylene groups (δ_C 18.6, 30.2, 42.7, and 42.7), one methoxy-oriented carbon (δ_C 50.7), and one methyl carbon (δ_C 12.5). Based on this evidence, it was predicted that compound **3** was a citric acid derivative, since the characteristic NMR data of **3** was similar to that of the citric acid derivatives [21,22]. The gross structure of **3** was determined by 2D NMR experiments (^1H-^1H COSY; Figure S3 and HMBC; Figure S5). The HMBC correlations of H_2-2/C-1, H_2-2/C-1', H_2-2/C-4, H_2-4/C-1', H_2-4/C-2, and H_2-4/C-5 verified the citric acid moiety. Furthermore, the HMBC correlation of 5-OCH$_3$ with C-5 confirmed that the methoxy group was attached at C-5 of the citric acid moiety, and that of H_2-1" with C-1, along with ^1H-^1H cross-peaks between H_2-1"/ H_2-2"/ H_2-3"/ H_3-4", revealed that a butyl group was attached at the other end of C-1 (Figure 2). Thus, the planar structure of **3** was established, as shown in Figure 2. Finally, the absolute configuration at C-3 was assigned using quantum chemical ECD calculations [23–25]. Two possibilities with absolute configurations of 3*R* and 3*S* were calculated. The weighted calculated ECD spectrum (green line) of **3***S* (3*S*) agreed well (although not identical) with the experimental curve of **3**, compared to that (red line) of **3***R* (3*R*) (Figure 3), thus determining the absolute configuration of C-3 as *S*. The chemical structure of **3**, including its absolute configuration, was elucidated as (*S*)-1-butyl-5-methyl citrate.

Compound **4** was obtained as a white amorphous powder. The molecular formula was determined to be the same as that of compound **3**, $C_{11}H_{18}O_7$, from HRESIMS data (Figure S6), which showed a protonated molecular ion peak $[M + H]^+$ at m/z 263.1119 (calculated for $C_{11}H_{19}O_7$, 263.1131). The ^1H (Figure S7) and ^{13}C NMR data (Figure S9) of **4** (Table 1) were also almost identical to those of compound **3**, except for apparent differences in the chemical shifts responsible for the citric acid moiety—the ^1H chemical shift of the methoxy group (δ_H 3.66 in **3**, δ_H 3.75 in **4**). Detailed analysis of the 2D NMR spectra (^1H-^1H COSY; Figure S8 and HMBC; Figure S10) assigned the gross planar structure of **4** (Figure 2), where the key HMBC correlation of 1'-OCH$_3$ with C-1' was observed, indicating that the methoxy group was attached at C-1' in compound **4** (Figure 2). Finally, to confirm the absolute configuration of C-3 in **4**, the computationally calculated ECD data of the two enantiomers, *3R*- and *3S*-isomers, were compared with the experimental ECD data of **4** (Figure 4). The ECD data of **4** showed a good fit with the calculated ECD spectrum (green line) of **4S** (*3S*) (Figure 4). This finding allowed us to assign the structure of **4** as (*S*)-1-butyl-1′-methyl citrate.

Table 1. ^{1}H (850 MHz) and ^{13}C (212.5 MHz) NMR data of compounds **3**, **4**, and **6** in CD$_3$OD [a].

Position	3		4		6	
	δ_H	δ_C	δ_H	δ_C	δ_H	δ_C
1	-	170.2	-	170.0	-	170.5
2	2.92 m [b] 2.84 m [b]	42.7 [b]	2.92 d (15.5) 2.76 d (15.5)	42.2 [b]	2.93 d (15.5) 2.79 d (15.5)	43.1
3	-	72.9	-	73.2	-	73.2
4	2.92 m [b] 2.84 m [b]	42.7 [b]	2.88 d (15.5) 2.72 d (15.5)	42.2 [b]	2.85 d (15.5) 2.70 d (15.5)	42.8
5	-	170.5	-	172.6	-	173.9 [b]
1′	-	175.3	-	174.1	-	173.9 [b]
1′-OCH$_3$	-	-	3.75 s	51.5	-	-
1-OCH$_3$	-	-	-	-	3.67 s	50.2
5-OCH$_3$	3.66 s	50.7	-	-	-	-
1″	4.08 t (6.5)	64.2	4.07 t (6.5)	64.2	4.18 t (6.5)	64.9
2″	1.61 m	30.2	1.61 m	30.2	1.67 m	30.0
3″	1.40 m	18.6	1.39 m	18.6	1.43 m	18.6
4″	0.94 t (7.5)	12.5	0.94 t (7.5)	12.5	0.97 t (7.5)	12.5

[a] Coupling constants (in Hz) are given in parentheses. Assignments based on ^{1}H-^{1}H COSY, HSQC, and HMBC.
[b] Overlapped.

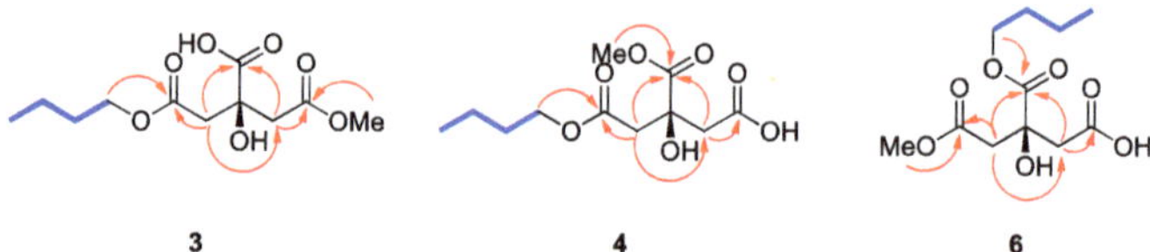

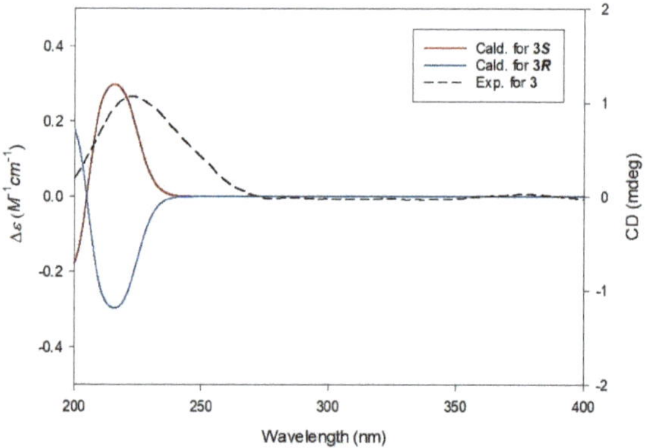

Figure 2. Key ^{1}H-^{1}H COSY (▬) and HMBC (⌒) correlations of compounds **3**, **4**, and **6**.

Figure 3. Experimental and calculated ECD spectra of **3**.

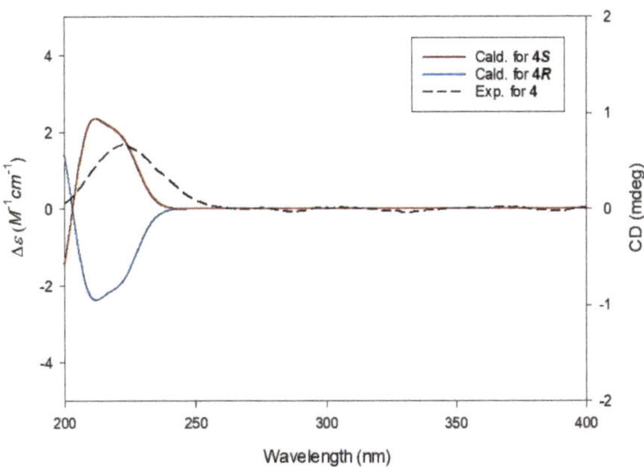

Figure 4. Experimental and calculated ECD spectra of **4**.

Compound **6**, obtained as a white amorphous powder, was determined to have the molecular formula $C_{11}H_{18}O_7$, similar to those of compounds **3** and **4**, from the protonated molecular ion peak [M + H]$^+$ at m/z 263.1145 (calculated for $C_{11}H_{19}O_7$, 263.1131) in the positive-ion mode of HRESIMS (Figure S11). The ^{1}H (Figure S12) and ^{13}C NMR data (Figure S14) (Table 1) of **6** were also quite similar to those of compounds **3** and **4**, with the only noticeable differences being that the ^{13}C chemical shift of C-1 was deshielded to δ_C 173.9, compared to these two compounds (δ_C 170.2 in **3**, δ_C 170.0 in **4**), and the ^{1}H chemical shift of H-1″ was deshielded to δ_H 4.18, compared to **3** and **4** (δ_H 4.08 in **3**, δ_H 4.07 in **4**). This finding suggested the possibility that the butyl group in **6** was attached to the different carboxyl groups of compounds **3** and **4**. The gross planar structure of **6** was verified by 2D NMR analysis (^{1}H-^{1}H COSY; Figure S13 and HMBC; Figure S15) (Figure 2). The key HMBC correlation between 1-OCH$_3$ and C-1 supported the presence of the methoxy group at C-1, and the location of the butyl group was determined to be C-1′ by the key HMBC correlation of H-1″ and C-1′ (Figure 2). Finally, the absolute configuration of **6** was assigned using quantum chemical ECD calculations. As shown in Figure 5, the experimental ECD curve of **6** is consistent with the calculated ECD spectrum (green line) of **6***S* (3*S*). Therefore, the chemical structure of **6** was characterized as (*S*)-1-methyl-1′-butyl citrate.

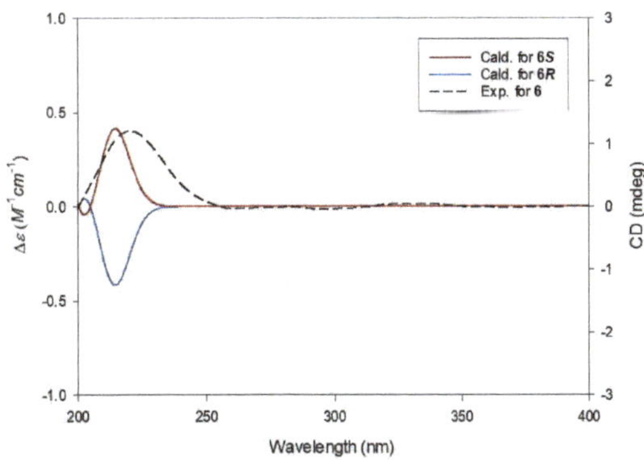

Figure 5. Experimental and calculated ECD spectra of **6**.

The structures of the known compounds (Figure 1) were determined to be (*S*)-dimethyl malate (**1**) [26], (*S*)-5,1′-dimethyl citrate (**2**) [21], 1′-butyl-1,5-dimethyl citrate (**5**) [22], and butyl quinate (**7**) [27] by spectroscopic methods, including ^{1}H and ^{13}C NMR spectra, by comparing their spectroscopic data with those previously reported in the literature and MS data obtained from LC/MS analysis. To the best of our knowledge, compounds **2**, **5**, and **7** are reported from *H. rhamnoides* for the first time in this study.

In the meantime, it appeared likely that compounds **3–7**, including the new compounds, arose from the addition of a butyl group to the corresponding carboxylic acid during the experimental procedure, because *n*-BuOH was used as the organic solvent for solvent partitioning. In order to verify whether compounds **3–7** were genuine natural compounds or artifacts, the aqueous extract of *H. rhamnoides* fruits and fractions derived from the extract were subjected to LC/MS analysis, both alone and co-injected with each compound. As a result, there was no peak with molecular ions corresponding to compounds **3–7** in the crude aqueous extract, whereas all the compounds were detected in the BuOH-soluble fraction, suggesting that the compounds with the butyl group may be artifacts produced by the addition of a butyl group during solvent partition with *n*-BuOH.

2.3. Evaluation of Biological Activities of Compounds 1–7

Mesenchymal stem cells (MSCs) in the bone marrow are pluripotent cells that are known to differentiate into osteocytes as well as adipocytes. Since microenvironmental changes cause alterations in the regulation of gene expression in MSC differentiation, where alterations in the expression of the related genes might disturb the balance between osteoprogenitor and adipocyte progenitor cells in osteoporosis patients [28–30], a therapy that is able to adjust the gene expression in MSCs would be promising in the management of postmenopausal osteoporosis. To determine the therapeutic effects of the isolated compounds **1–7** in promoting osteogenesis, all the compounds were examined for their effects on the differentiation of murine MSCs into osteoblasts. The murine mesenchymal stem cell line C3H10T1/2 was treated with 20 µM of the compounds during osteogenesis, and the differentiated cells were stained to indicate alkaline phosphatase (ALP) production, which is considered a distinctive marker of osteoblast differentiation [31]. As a result, compounds **1–7** slightly stimulated differentiation of mesenchymal stem cells into osteocytes (Figure 6). Although all the compounds did not show the superior reaction of ALP staining as in the oryzativol A-treated one, the positive control group with compounds **1–7** showed more than 1.2 times the bone differentiation-promoting effects compared to the negative control group. In particular, compounds **1**, **3**, **4**, **6**, and **7** clearly stimulated the differentiation of osteogenic premature cells up to 1.4 times compared to the negative control group (Figure 6).

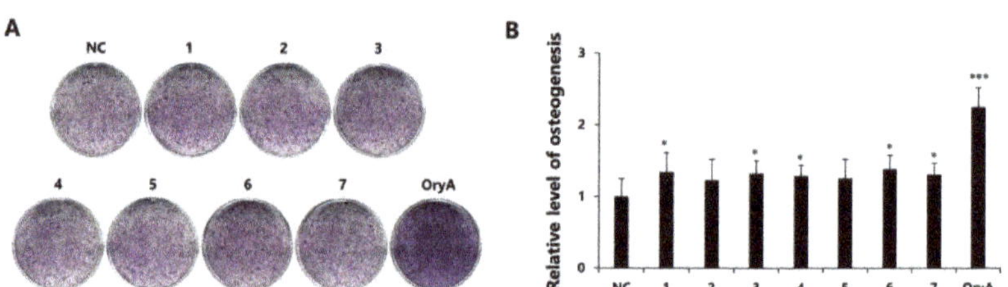

Figure 6. The effects of compounds **1–7** on the differentiation of mesenchymal stem cells (MSCs) toward osteoblasts. (**A**) Stimulatory effect of compounds **1–7** on osteogenic differentiation of MSC. Fully-differentiated C3H10T1/2 cells were stained with alkaline phosphatase (ALP) at 9 d after osteogenic differentiation with 20 µM of compounds **1–7**. (**B**) ALP enzyme activity was measured in osteogenically differentiated C3H10T1/2 cells treated with compounds **1–7**. The values were calculated relatively by setting the untreated negative control to 1. Oryzativol A (5 µM) was added to the experimental set as a positive control and marked as OryA. * denotes $p < 0.05$ and *** denotes $p < 0.001$.

3. Materials and Methods

3.1. General Experimental Procedures

Optical rotations were measured using a JASCO P-2000 polarimeter (JASCO, Easton, MD, USA). Ultraviolet (UV) spectra were acquired on an Agilent 8453 UV-visible spectrophotometer (Agilent Technologies, Santa Clara, CA, USA). Electronic circular dichroism (ECD) spectra were measured using a Jasco J-1500 spectropolarimeter (Jasco). Nuclear magnetic resonance (NMR) spectra were recorded using a Bruker AVANCE III (Bruker). HRESIMS spectra were recorded on an Agilent 1290 Infinity II series with a 6545 LC/Q-TOF mass spectrometer (Agilent Technologies). Medium-pressure liquid chromatography (MPLC) was performed on a Smart Flash AKROS (Yamazen, Osaka, Japan) using an analytical Universal ODS-SM 120 Å column (3.0 × 20.0 cm, 50 µm) (Yamazen) and Universal Premium Silica gel 60 Å column (2.3 × 12.3 cm, 30 µm) (Yamazen). Preparative and semi-preparative HPLC were performed on a Waters 1525 binary HPLC pump with a Waters 996 photodiode array detector (Waters Corporation, Milford, CT, USA) using an analytical Agilent Eclipse XDB-C18 column (250 × 21.2 mm, 7 µm) and a Phenomenex Luna Phenyl-hexyl 100 Å column (250 × 10 mm, 10 µm), respectively. LC/MS analysis was performed on an Agilent 1200 series HPLC system with a diode array detector and a 6130 Series ESI mass spectrometer using an analytical Kinetex C_{18} 100 Å column (100 mm × 2.1 mm i.d., 5 µm; flow rate: 0.3 mL/min) (Phenomenex). All HRESIMS data were obtained using an Agilent G6545B quadrupole time-of-flight (Q-TOF) mass spectrometer (Agilent Technologies) with an Agilent EclipsePlus C_{18} column (2.1 mm × 50 mm i.d., 1.8 µm; flow rate: 0.3 mL/min) maintained at 20 °C. Silica gel 60 (230–400 mesh; Merck, Darmstadt, Germany), RP-C_{18} silica gel (Merck, 230–400 mesh), silica Sep-Pak Vac 6 cc cartridges (Waters), and Diaion HP-20 (Mitsubishi Chemical Co. Ltd., Tokyo, Japan) were used for column chromatography. Sephadex LH-20 (Pharmacia, Uppsala, Sweden) was used as the packing material for molecular sieve column chromatography. Thin-layer chromatography (TLC) was performed using precoated silica gel F_{254} plates and RP-C_{18} F_{254s} plates (Merck), and spots were detected under UV light or by heating after spraying with anisaldehyde-sulfuric acid.

3.2. Plant Material

Aqueous extract powder of *H. rhamnoides* fruits was purchased from Korea Beauty and Healthcare Co., Ltd. in October 2018. The material was authenticated by one of the authors (K. H. K.). A voucher specimen of the material (VT-2018) was deposited in the herbarium of the School of Pharmacy, Sungkyunkwan University, Suwon, Korea.

3.3. Extraction and Isolation

Aqueous extract powder (270 g) of *H. rhamnoides* fruits was suspended in distilled water (700 mL) and then sequentially partitioned with hexane (HX), dichloromethane (MC), ethyl acetate (EA), and *n*-butanol (BuOH). Four layers with different polarities were obtained: HX-soluble (2.7 g), MC-soluble (3.4 g), EA-soluble (7.8 g), and BuOH-soluble (25.2 g) fractions. The BuOH-soluble fraction (25.2 g) was subjected to open-column chromatography on a Diaion HP-20 column and eluted stepwise with distilled water (3 L) and 100% methanol (MeOH). The respective fractions were evaporated to dryness. The resulting MeOH fraction (9.7 g) was chromatographed on silica gel with a gradient solvent system of MC/MeOH (10:1 → 1:1), yielding five subfractions (A–E). Subfraction B (1.2 g) was subjected to an RP-C_{18} column with a gradient solvent system of MeOH/H_2O (15:85 → 100:0), which afforded six subfractions (B1–B6). Subfraction B2 (393.3 mg) was separated by MPLC on a Yamazen Universal Premium silica gel column with MC/MeOH (99:1 → 95:5) to obtain two subfractions (B21 and B22). Subfraction B21 (219.0 mg) was purified by preparative HPLC (MeOH/H_2O, 30:70 → 80:20 → 100:0) to obtain four subfractions (B211–B214). Subfraction B212 (58.5 mg) was purified using semi-preparative HPLC (MeCN/H_2O, 5:95) to furnish compounds **1** (t_R 29.4 min, 2.2 mg) and **2** (t_R = 43.2 min, 7.4 mg). Subfraction B4 (356.1 mg) was fractionated by preparative

HPLC (MeOH/H$_2$O, 70:30 → 100:0) to yield six subfractions (B41–B46). Subfraction B45 (39.3 mg) was purified using semi-preparative HPLC (MeOH/H$_2$O, 48:52), and compound **6** (t_R = 43.1 min, 2.8 mg) was obtained. Subfraction B46 (58.0 mg) was purified using semi-preparative HPLC (MeCN/H$_2$O, 25:75), and compounds **4** (t_R = 35.8 min, 3.1 mg) and **5** (t_R 71.5 min, 1.8 mg) were obtained. Subfraction C (1.2 g) was fractionated by MPLC using an ODS-SM column with MeOH/ H$_2$O (30:70 → 100:0) to yield five subfractions (C1–C5). Subfraction C2 (338.8 mg) was chromatographed using silica Sep-Pak with an isocratic solvent system of MC/MeOH (5:1) to give four subfractions (C21–C24). Subfraction C21 (21.9 mg) was further purified using semi-preparative HPLC (MeOH/H$_2$O, 35:65) to yield compound **7** (t_R = 36.9 min, 7.5 mg). Subfraction D (1.4 g) was separated by preparative HPLC (MeOH/H$_2$O, 25:75 → 90:10 → 100:0), and two subfractions were obtained (D1 and D2). Subfraction D2 (446.2 mg) was fractionated by preparative HPLC again with a gradient solvent system of MeOH/H$_2$O (70:30 → 85:15), which afforded three subfractions (D21–D23). Subfraction D23 (140.1 mg) was purified using semi-preparative HPLC (MeOH/H$_2$O, 40:60 → 50:50) to yield compound **3** (t_R = 50.8 min, 5.7 mg).

3.3.1. (*S*)-1-Butyl-5-Methyl Citrate (**3**)

Yellowish gum; $[\alpha]_D^{25}$ +6.0 (*c* 0.50, MeOH); UV (MeOH) λ_{max} (log ε) 200 (3.9), 220 (1.5) nm; ECD (MeOH, 0.89% w/v) λ_{max} ($\Delta\varepsilon$) 217 (+1.58) nm; ^{1}H (850 MHz) and ^{13}C (212.5 MHz) NMR data, see Table 1; ESIMS (positive-ion mode) m/z 263.1 [M + H]$^+$; HRESIMS (positive-ion mode) m/z 263.1122 [M + H]$^+$ (calcd for C$_{11}$H$_{19}$O$_7$, 263.1131).

3.3.2. (*S*)-1-Butyl-1′-Methyl Citrate (**4**)

White amorphous powder; $[\alpha]_D^{25}$ −5.4 (*c* 1.65, MeOH); UV (MeOH) λ_{max} (log ε) 201 (3.8), 220 (1.4) nm; ECD (MeOH, 0.94% w/v) λ_{max} ($\Delta\varepsilon$) 222 (+0.72) nm; ^{1}H (850 MHz) and ^{13}C (212.5 MHz) NMR data, see Table 1; ESIMS (positive-ion mode) m/z 263.1 [M + H]$^+$; HRESIMS (positive-ion mode) m/z 263.1119 [M + H]$^+$ (calcd for C$_{11}$H$_{19}$O$_7$, 263.1131).

3.3.3. (*S*)-1-Methyl-1′-Butyl Citrate (**6**)

White amorphous powder; $[\alpha]_D^{25}$ −24.3 (*c* 1.41, MeOH); UV (MeOH) λ_{max} (log ε) 198 (3.9), 218 (1.5) nm; ECD (MeOH, 0.92% w/v) λ_{max} ($\Delta\varepsilon$) 217 (+1.71) nm; ^{1}H (850 MHz) and ^{13}C (212.5 MHz) NMR data, see Table 1; ESIMS (positive-ion mode) m/z 263.1 [M + H]$^+$; HRESIMS (positive-ion mode) m/z 263.1145 [M + H]$^+$ (calcd for C$_{11}$H$_{19}$O$_7$, 263.1131).

3.4. Computational Analysis

All conformers proposed in this study were acquired through the MacroModel (version 2019-3, Schrödinger LLC) module with mixed torsional/low-mode sampling, implemented with the MMFF94 force field. All searches were initially set in the gas phase, with a 10 kJ/mol energy window limit and a maximum of 10,000 steps to thoroughly explore all potential conformers. The Polak–Ribiere conjugate gradient protocol was established with 10,000 maximum iterations and a 0.001 kJ (mol Å)$^{-1}$ convergence threshold on the rms gradient to minimize conformers [23–25]. Conformers proposed in this study (within 5 kJ/mol found in the MMFF force field) were selected for geometry optimization by Tmolex 4.3.1, with the DFT settings of B3-LYP/6-31+G(d,p). ECD calculations for **3R/3S**, **4R/4S**, and **6R/6S** conformers were performed at identical theoretical levels and basis sets. The calculated ECD spectra were simulated by superimposing each transition, where σ is the width of the band at a height of 1/e. ΔE_i and R_i are the excitation energies and rotatory strengths for transition *i*, respectively. In this study, the value of σ was 0.10 eV. The excitation energies and rotational strengths for ECD spectra were calculated based on the Boltzmann populations of the conformers, and ECD visualization was carried out using SigmaPlot 14.0.

$$\Delta\epsilon(E) = \frac{1}{2.297 \times 10^{-39}} \frac{1}{\sqrt{2\pi\sigma}} \sum_A^i \Delta E_i R_i e^{[-(E-\Delta E_i)^2/(2\sigma)^2]}$$

3.5. Cell Culture and Differentiation

Mouse mesenchymal stem cell line C3H10T1/2 cells were cultured in Dulbecco's modified Eagle's medium (DMEM), 10% fetal bovine serum (FBS), and 1% penicillin-streptomycin (P/S) at 37 °C in a 5% CO_2 incubator. For osteogenic differentiation, C3H10T1/2 cells were exposed to DMEM (5% FBS, 1% P/S) media with 10 mM β-glycerophosphate and 50 μg/mL ascorbic acid for 9 d. During osteogenic differentiation, 20 μM of compounds **1–7** were added to the cells, and 5 μM of oryzativol A was used as the positive control.

To investigate the degree of differentiation into osteocytes, cells were stained for alkaline phosphatase (ALP) after 9 days of osteogenic differentiation. After fixation with 4% formaldehyde solution, cells were exposed to 0.4 mg/mL of nitroblue tetrazolium and 0.2 mg/mL of 5-bromo-4chloro-3′-indolyphosphate in alkaline phosphatase (AP) buffer (100 mM Tris-HCl, pH 9.5; 100 mM NaCl, 5 mM $MgCl_2$) in the dark for 4 h. Next, the reaction was stopped with 5 mM EDTA. To measure APL activity, cell lysates were collected from the differentiated osteogenic cells and treated with *p*-nitrophenyl phosphate (*p*-NPP) solution (Alkaline Phosphatase Assay Kit (ab83369; Abcam, Cambridge, MA, USA). Color changes of the samples were measured at 405 nm with a SpectraMax M2 Microplate Spectrophotometer.

3.6. Statistical Analysis

Each sample was tested in triplicate, and the test was repeated three times. Data are presented as the mean ± standard deviation (S.D.). Differences between the control and experimental groups were analyzed using a two-tailed unpaired Student's *t*-test; statistical significance was defined as $p < 0.05$.

4. Conclusions

In this study, phytochemical examination of the extracts of *H. rhamnoides* fruits resulted in the isolation of seven organic acid derivatives: one malate derivative (**1**), five citrate derivatives (**2–6**), and one quinate derivative (**7**). The structure of the isolates was elucidated by analysis of 1D and 2D NMR data and HR-ESIMS data as well as quantum chemical ECD calculations. Three of the citrate derivatives were identified as new compounds: (*S*)-1-butyl-5-methyl citrate (**3**), (*S*)-1-butyl-1′-methyl citrate (**4**), and (*S*)-1-methyl-1′-butyl citrate (**6**), which turned out to be isolation artifacts. To the best of our knowledge, compounds **2**, **5**, and **7** are reported for the first time from *H. rhamnoides* in this study. An osteogenesis-promoting activity bioassay in C3H10T1/2 cells demonstrated that compounds **1–7** showed little more than 1.2 times the bone differentiation promoting effect, compared to the negative control group. In particular, compounds **1**, **3**, **4**, **6**, and **7** clearly stimulated the differentiation of osteogenic premature cells up to 1.4 times, compared to the negative control group. Although the stimulatory effect of the active compounds on osteogenic differentiation was far behind that of the positive control, oryzativol A, compounds **1**, **3**, **4**, **6**, and **7** apparently helped to differentiate the mesenchymal stem cells into osteocytes.

Supplementary Materials: 1D and 2D NMR, and HR-ESIMS of compounds **3**, **4**, and **6** are available online at https://www.mdpi.com/article/10.3390/plants10050860/s1, Figure S1: HR-ESIMS data of **3**, Figure S2: ^{1}H NMR spectrum of **3** (CD_3OD, 700 MHz), Figure S3: ^{1}H-^{1}H COSY spectrum of **3** (CD_3OD), Figure S4: HSQC spectrum of **3** (CD_3OD), Figure S5: HMBC spectrum of **3** (CD_3OD), Figure S6: HR-ESIMS data of **4**, Figure S7: ^{1}H NMR spectrum of **4** (CD_3OD, 700 MHz), Figure S8: ^{1}H-^{1}H COSY spectrum of **4** (CD_3OD), Figure S9: HSQC spectrum of **4** (CD_3OD), Figure S10: HMBC spectrum of **4** (CD_3OD), Figure S11: HR-ESIMS data of **6**, Figure S12: ^{1}H NMR spectrum of **6** (CD_3OD, 700 MHz), Figure S13: ^{1}H-^{1}H COSY spectrum of **6** (CD_3OD), Figure S14: HSQC spectrum of **6** (CD_3OD), Figure S15: HMBC spectrum of **6** (CD_3OD), Table S1: Gibbs free energies and Boltzmann distribution of conformers **3S** (**3S**), Table S2: Gibbs free energies and Boltzmann distribution of conformers **4S** (**3S**), Table S3: Gibbs free energies and Boltzmann distribution of conformers **6S** (**3S**).

Author Contributions: Conceptualization, T.S.J. and K.H.K.; formal analysis, Y.H.L., H.J.J., K.H.P., S.-H.K., J.K.K., and J.-C.K.; investigation, Y.H.L., H.J.J., and K.H.P.; writing—original draft preparation, Y.H.L., H.J.J., S.-H.K., and K.H.K.; writing—review and editing, J.K.K. and K.H.K.; visualization, H.J.J. and S.-H.K.; supervision, T.S.J. and K.H.K.; project administration, T.S.J. and K.H.K.; funding acquisition, T.S.J. and K.H.K. All authors have read and agreed to the published version of the manuscript.

Funding: This work was supported by a grant from the National Research Foundation of Korea (NRF), funded by the Korean government (MSIT) (grant number: 2019R1A5A2027340) and a Korea Institute of Science and Technology intramural research grant (2E30641). This research was also supported by the Ministry of Trade, Industry and Energy (MOTIE), Korea Institute for Advancement of Technology (KIAT) through the Program for Smart Specialization Infrastructure Construction.

Conflicts of Interest: The authors declare no conflict of interest.

References

1. Mingyu, X.; Xiaoxuan, S.; Jinhua, C. The medicinal research and development of seabuckthorn. *J. Water Soil Conserv.* **1991**, 1–11.
2. Panossian, A.; Wagner, H. From traditional to evidence-based use of *Hippophae rhamnoides* L.: Chemical composition, experimental, and clinical pharmacology of sea buckthorn berries and leaves extracts. In *Evidence and Rational Based Research on Chinese Drugs*; Springer: Vienna, Austria, 2013; pp. 181–236.
3. Watanabe, T.; Rajbhanddari, K.R.; Malla, K.J.; Yahara, S. A handbook of medicinal plants of Nepal. *Banko Janakari* **2005**, *15*, 106–107.
4. Chauhan, A.S.; Negi, P.S.; Ramteke, R.S. Antioxidant and antibacterial activities of aqueous extract of Seabuckthorn (*Hippophae rhamnoides*) seeds. *Fitoterapia* **2007**, *78*, 590–592. [CrossRef]
5. Suryakumar, G.; Gupta, A. Medicinal and therapeutic potential of Sea buckthorn (*Hippophae rhamnoides* L.). *J. Ethnopharmacol.* **2011**, *138*, 268–278. [CrossRef]
6. Li, T.S.; Schroeder, W.R. Sea buckthorn (*Hippophae rhamnoides* L.): A multipurpose plant. *HortTechnology* **1996**, *6*, 370–380. [CrossRef]
7. Skalski, B.; Kontek, B.; Rolnik, A.; Olas, B.; Stochmal, A.; Żuchowski, J. Anti-platelet properties of phenolic extracts from the leaves and twigs of *Elaeagnus rhamnoides* (L.) A. Nelson. *Molecules* **2019**, *24*, 3620. [CrossRef]
8. Różalska, B.; Sadowska, B.; Żuchowski, J.; Więckowska-Szakiel, M.; Budzyńska, A.; Wójcik, U.; Stochmal, A. Phenolic and nonpolar fractions of *Elaeagnus rhamnoides* (L.) A. Nelson, extracts as virulence modulators-in vitro study on bacteria, fungi, and epithelial cells. *Molecules* **2018**, *23*, 1498. [CrossRef] [PubMed]
9. Beveridge, T.; Li, T.S.; Oomah, B.D.; Smith, A. Sea buckthorn products: Manufacture and composition. *J. Agri. Food Chem.* **1999**, *47*, 3480–3488. [CrossRef] [PubMed]
10. Yang, B.; Kallio, H.P. Fatty acid composition of lipids in sea buckthorn (*Hippophaë rhamnoides* L.) berries of different origins. *J. Agri. Food Chem.* **2001**, *49*, 1939–1947. [CrossRef] [PubMed]
11. Giuffrida, D.; Pintea, A.; Dugo, P.; Torre, G.; Pop, R.M.; Mondello, L. Determination of Carotenoids and their Esters in Fruits of Sea Buckthorn (*Hippophae rhamnoides* L.) by HPLC-DAD-APCI-MS. *Phytochem. Anal.* **2012**, *23*, 267–273. [CrossRef] [PubMed]
12. Zheng, R.X.; Xu, X.D.; Tian, Z.; Yang, J.S. Chemical constituents from the fruits of *Hippophae rhamnoides*. *Nat. Prod. Res.* **2009**, *23*, 1451–1456. [CrossRef]
13. Lee, S.; Lee, D.; Ryoo, R.; Kim, J.C.; Park, H.B.; Kang, K.S.; Kim, K.H. Calvatianone, a Sterol Possessing a 6/5/6/5-Fused Ring System with a Contracted Tetrahydrofuran B-Ring, from the Fruiting Bodies of *Calvatia nipponica*. *J. Nat. Prod.* **2020**, *83*, 2737–2742. [CrossRef]
14. Lee, S.R.; Kang, H.S.; Yoo, M.J.; Yi, S.A.; Beemelmanns, C.; Lee, J.C.; Kim, K.H. Anti-adipogenic Pregnane Steroid from a Hydractinia-associated Fungus, *Cladosporium sphaerospermum* SW67. *Nat. Prod. Sci.* **2020**, *26*, 230–235.
15. Lee, S.; Ryoo, R.; Choi, J.H.; Kim, J.H.; Kim, S.H.; Kim, K.H. Trichothecene and tremulane sesquiterpenes from a hallucinogenic mushroom *Gymnopilus junonius* and their cytotoxicity. *Arch. Pharm. Res.* **2020**, *43*, 214–223. [CrossRef] [PubMed]
16. Trinh, T.A.; Park, E.J.; Lee, D.; Song, J.H.; Lee, H.L.; Kim, K.H.; Kim, Y.; Jung, K.; Kang, K.S.; Yoo, J.E. Estrogenic activity of sanguiin H-6 through activation of estrogen receptor α coactivator-binding site. *Nat. Prod. Sci.* **2019**, *25*, 28–33. [CrossRef]
17. Ha, J.W.; Kim, J.; Kim, H.; Jang, W.; Kim, K.H. Mushrooms: An Important Source of Natural Bioactive Compounds. *Nat. Prod. Sci.* **2020**, *26*, 118–131.
18. Yu, J.S.; Li, C.; Kwon, M.; Oh, T.; Lee, T.H.; Kim, D.H.; Ahn, J.S.; Ko, S.K.; Kim, C.S.; Cao, S. Herqueilenone a, a unique rearranged benzoquinone-chromanone from the hawaiian volcanic soil-associated fungal strain *Penicillium herquei* FT729. *Bioorg. Chem.* **2020**, *105*, 104397. [CrossRef] [PubMed]
19. Yu, J.S.; Park, M.; Pang, C.; Rashan, L.; Jung, W.H.; Kim, K.H. Antifungal phenols from *Woodfordia uniflora* collected in Oman. *J. Nat. Prod.* **2020**, *83*, 2261–2268. [CrossRef]
20. Baek, S.C.; Lee, D.; Jo, M.S.; Lee, K.H.; Lee, Y.H.; Kang, K.S.; Yamabe, N.; Kim, K.H. Inhibitory effect of 1,5-dimethyl citrate from sea buckthorn (*Hippophae rhamnoides*) on lipopolysaccharide-induced inflammatory response in RAW 264.7 Mouse Macrophages. *Foods* **2020**, *9*, 269. [CrossRef]

21. Takeuchi, Y.; Nagao, Y.; Toma, K.; Yoshikawa, Y.; Akiyama, T.; Nishioka, H.; Abe, H.; Harayama, T.; Yamamoto, S. Synthesis and siderophore activity of vibrioferrin and one of its diastereomeric isomers. *Chem. Pharm. Bull.* **1999**, *47*, 1284–1287. [CrossRef]
22. Vereshchagin, A.L.; Anikina, E.V.; Syrchina, A.I.; Lapin, M.F.; Azin, L.A.; Semenov, A.A. Chemical investigation of the bitter substances of the fruit of *Lonicera caerulea*. *Chem. Nat. Compd* **1989**, *25*, 289–292. [CrossRef]
23. Lee, S.R.; Seok, S.; Ryoo, R.; Choi, S.U.; Kim, K.H. Macrocyclic trichothecene mycotoxins from a deadly poisonous mushroom, *Podostroma cornu-damae*. *J. Nat. Prod.* **2019**, *82*, 122–128. [CrossRef] [PubMed]
24. Baek, S.C.; Lee, B.S.; Yi, S.A.; Yu, J.S.; Lee, J.; Ko, Y.J.; Pang, C.; Kim, K.H. Discovery of dihydrophaseic acid glucosides from the florets of *Carthamus tinctorius*. *Plants* **2020**, *9*, 858. [CrossRef]
25. Rischer, M.; Lee, S.R.; Eom, H.J.; Park, H.B.; Vollmers, J.; Kaster, A.K.; Shin, Y.H.; Oh, D.C.; Kim, K.H.; Beemelmanns, C. Spirocyclic cladosporicin A and cladosporiumins I and J from a *Hydractinia*-associated *Cladosporium sphaerospermum* SW67. *Org. Chem. Front.* **2019**, *6*, 1084–1093. [CrossRef]
26. Jin, T.Y.; Shen, T.; Zhou, M.X.; Li, A.L.; Feng, D.; Zheng, B.; Gong, J.; Sun, J.; Li, L.; Xiang, L. Chemical constituents from *Portulaca oleracea* and their bioactivities. *J. Chin. Pharm. Sci.* **2016**, *25*, 898–905.
27. Yang, Y.B.; Li, X.; Yang, Q.; Wu, Z.J.; Sun, L.N. Study on chemical constituents of *Papaya rugosa*. *Acad. J. Second Mil. Medi. Univ.* **2009**, 1195–1198. [CrossRef]
28. Meyer, M.B.; Benkusky, N.A.; Sen, B.; Rubin, J.; Pike, J.W. Epigenetic Plasticity Drives Adipogenic and Osteogenic Differentiation of Marrow derived Mesenchymal Stem Cells. *J. Biol. Chem.* **2016**, *291*, 17829–17847. [CrossRef]
29. Ciuffreda, M.C.; Malpasso, G.; Musarò, P.; Turco, V.; Gnecchi, M. Protocols for in vitro Differentiation of Human Mesenchymal Stem Cells into Osteogenic, Chondrogenic and Adipogenic Lineages. *Methods Mol. Biol.* **2016**, *1416*, 149–158.
30. Yi, S.A.; Lee, J.; Park, S.K.; Kim, J.Y.; Park, J.W.; Lee, M.G.; Nam, K.H.; Park, J.H.; Oh, H.; Kim, S.; et al. Fermented ginseng extract, BST204, disturbs adipogenesis of mesenchymal stem cells through inhibition of S6 kinase 1 signaling. *J. Ginseng Res.* **2020**, *44*, 58–66. [CrossRef]
31. Kang, M.H.; Lee, S.J.; Lee, M.H. Bone remodeling effects of Korean Red Ginseng extracts for dental implant applications. *J. Ginseng Res.* **2020**, *44*, 823–832. [CrossRef]

Article

Differences in the Inhibitory Specificity Distinguish the Efficacy of Plant Protease Inhibitors on Mouse Fibrosarcoma

Sonia Yoo Im [1], Camila Ramalho Bonturi [1], Adriana Miti Nakahata [2], Clóvis Ryuichi Nakaie [3], Arnildo Pott [4], Vali Joana Pott [4] and Maria Luiza Vilela Oliva [1],*

[1] Departamento de Bioquímica, Universidade Federal de São Paulo (UNIFESP), São Paulo 04044-020, SP, Brazil; yoo.sonia@gmail.com (S.Y.I.); camilabntr@gmail.com (C.R.B.)

[2] Laboratório de Genômica e Biologia Molecular, Centro Internacional de Pesquisa (CIPE) do A.C.Camargo Cancer Center, São Paulo 01508-010, SP, Brazil; adriana_miti@yahoo.com.br

[3] Departamento de Biofísica, Universidade Federal de São Paulo (UNIFESP), São Paulo 04044-020, SP, Brazil; cnakaie@unifesp.br

[4] Departamento de Biologia, Universidade Federal de Mato Grosso do Sul (UFMS), Cidade Universitária, Campo Grande 79070-900, MS, Brazil; arnildo.pott@gmail.com (A.P.); vali.pott@gmail.com (V.J.P.)

* Correspondence: mlvoliva@unifesp.br; Tel.: +55-(11)-5576-4445

Abstract: Metastasis, the primary cause of death from malignant tumors, is facilitated by multiple protease-mediated processes. Thus, effort has been invested in the development of protease inhibitors to prevent metastasis. Here, we investigated the effects of protease inhibitors including the recombinant inhibitors rBbKI (serine protease inhibitor) and rBbCI (serine and cysteine inhibitor) derived from native inhibitors identified in *Bauhinia bauhinioides* seeds, and EcTI (serine and metalloprotease inhibitor) isolated from the seeds of *Enterolobium contortisiliquum* on the mouse fibrosarcoma model (lineage L929). rBbKI inhibited 80% of cell viability of L929 cells after 48 h, while EcTI showed similar efficacy after 72 h. Both inhibitors acted in a dose and time-dependent manner. Conversely, rBbCI did not significantly affect the viability of L929 cells. Confocal microscopy revealed the binding of rBbKI and EcTI to the L929 cell surface. rBbKI inhibited approximately 63% of L929 adhesion to fibronectin, in contrast with EcTI and rBbCI, which did not significantly interfere with adhesion. None of the inhibitors interfered with the L929 cell cycle phases. The synthetic peptide RPGLPVRFESPL-NH$_2$, based on the BbKI reactive site, inhibited 45% of the cellular viability of L929, becoming a promising protease inhibitor due to its ease of synthesis.

Keywords: fibrosarcoma; metastasis; natural products; plants; protease inhibitors; tumor cells

Citation: Yoo Im, S.; Ramalho Bonturi, C.; Miti Nakahata, A.; Ryuichi Nakaie, C.; Pott, A.; Pott, V.J.; Vilela Oliva, M.L. Differences in the Inhibitory Specificity Distinguish the Efficacy of Plant Protease Inhibitors on Mouse Fibrosarcoma. *Plants* **2021**, *10*, 602. https://doi.org/10.3390/plants10030602

Academic Editor: Jong Seong Kang

Received: 16 February 2021
Accepted: 17 March 2021
Published: 23 March 2021

Publisher's Note: MDPI stays neutral with regard to jurisdictional claims in published maps and institutional affiliations.

1. Introduction

Sarcomas are a heterogeneous group of malignant neoplasms that originate from mesenchymal cells and mostly affect soft tissues [1,2]. Fibrosarcoma is a rare malignant neoplasm that develops in connective tissue from fibroblasts and occurs in adults and children. The appearance of fibrosarcoma is more frequent in adults between the ages of 20 and 50, while its incidence is rare in children [3]. Fibrosarcoma is generally more aggressive in adults than in infants [1,3]. In general, fibrosarcoma presents as a palpable, painless mass (77% of cases), impairing early diagnosis. Further, fibrosarcoma is often falsely diagnosed as a traumatic hematoma or simple muscle strain due to a large number of existing histopathological subtypes [1,2]. The standard treatment for sarcomas includes surgical removal combined with radiation therapy. However, the protocol for marginal surgical resection, although of great importance, still has no medical consensus. The same is true for radiotherapy, in which a defined protocol is lacking for the duration or time of exposure, either before or after surgery [4,5]. Although sarcomas are rare tumors, the advanced stage is difficult to treat, with a survival rate below 57% and a higher incidence of recurrence [1,6]. Adjuvant therapies for surgery and radiotherapy, such as chemotherapy and target therapy, are also needed, mainly in high-grade fibrosarcoma [1]. Chemotherapeutic drugs are often combined with target therapy or immunotherapy to increase the response of unresectable or metastatic sarcomas [6,7]. However, surgery can result in limb amputation, causing physical and mental damage to patients. In addition to high recurrence rates of fibrosarcoma, available treatments have serious side effects, such as heart failure and severe or fatal hepatotoxicity.

Proteases play a fundamental role in numerous biological processes, including cell cycle progression, replication of genetic material, the immune response, cell adhesion, proliferation, migration, acting in signaling biomolecules, and apoptotic processes, all of which interfere with tumor progression. These numerous functions make them essential targets in the study of various pathophysiological processes, such as cancer [8]. In fibrosarcoma, a variety of proteases have been related to tumor progression, metastasis, and response to chemotherapy [8,9]. In this view, protease inhibitors are instruments used for the study and development of new therapeutic strategies.

Many protease inhibitors have been isolated from plants, mainly from legume seeds [10–14]. One aspect of interest is to investigate the relationship between the structure, specificity, and selectivity of different proteases and their enzymatic activity, which is an important factor in the development of drugs such as enzymes that affect the processes of cancer progression [8,9,15].

Studies with protease inhibitors have demonstrated that their cytostatic/cytotoxic activity may be the result of direct inhibition of matrix proteolysis or indirect inhibition of proteolytic cascade activation, thus preventing the spread of tumor cells [8]. Their effectiveness has been demonstrated in various studies, including in breast cancer cell lines [15] and the control of asthma in mice [16]. In particular, inhibitors derived from *Bauhinia bauhinioides*, BbKI (human plasma kallikrein inhibitor) and BbCI (cruzipain inhibitor), were investigated in models of inflammation and reperfusion [10,17], venous and arterial thrombosis of mice [13], a prostate cancer model [18], cell invasion and angiogenesis in different cell lines [12], a pulmonary emphysema model in mice [14], and as a potential insecticidal agent [19]. EcTI, a trypsin inhibitor derived from *Enterolobium contortisiliquuum*, was investigated in gastric cancer [11], triple-negative breast cancer [20], glioblastoma [21,22], the reduction of inflammation and pulmonary remodeling in induced lung inflammation [23], and in an asthma model [24]. In addition, EcTI was investigated in non-tumorigenic cells such as human mesenchymal cells [25] and showed no effects, demonstrating the potential use of these compounds as therapeutic targets. Considering that in fibrosarcoma the tumor establishment is mediated by multiple actions of proteases that cooperate with their differences in specific action [10,26], an aspect of interest in this study was to investigate the effects of plants inhibitors, with different specificities, that may help to understand the proliferative activity of fibrosarcoma. Thus, here we demonstrated the effects of EcTI and the recombinant inhibitors rBbCI and rBbKI on mouse fibrosarcoma cells.

2. Results

2.1. Inhibitor Purification and Characterization

Native and recombinant inhibitors were obtained in sufficient amounts for the experiments. Inhibitor purity was confirmed by electrophoresis and reverse-phase chromatography as published previously by Martins-Oliveira et al. [14], de Paula et al. [11], and Almeida-Reis et al. [27], and their inhibitory properties are shown in Table 1. Inhibitory activity of recombinant and native forms of the protein BbCI were maintained, as cruzain, or even improved, as human neutrophil elastase (HNE) and cruzipain. HNE activity has been upregulated in many cancers and frequently correlated to poor clinical outcomes [28]. Recombinant BbKI inhibited trypsin, human plasma kallikrein (PKa), and porcine pancreatic kallikrein, as the native protein. PKa inhibition has been related to cell growth, angiogenesis, invasion, and metastasis of a variety of cancers [29].

Table 1. Inhibitory properties of BbCI, rBbCI, BbKI, rBbKI, and EcTI (K_{iapp} values are shown in nM).

Enzyme	BbCI	rBbCI	BbKI	rBbKI	EcTI
Trypsin	φ	φ	20.00	28.00	0.88
Chymotrypsin	φ	φ	26.00	ND	1.11
Plasmin	φ	φ	330.00	ND	9.36
HNE	5.30	1.70	φ	φ	55.0
Factor Xa	φ	ND	φ	ND	φ
Thrombin	φ	ND	φ	ND	ND
PKa	φ	φ	2.40	2.00	6.15
PoPK	φ	φ	200.00	900.00	φ
Cathepsin G	160.00	ND	φ	φ	ND
Cathepsin L	0.22	ND	φ	ND	ND
Cruzain	0.30	0.30	φ	φ	ND
Cruzipain	1.30	1.20	φ	φ	ND

HNE: human neutrophil elastase; PKa: human plasma kallikrein; PoPK: porcine pancreatic kallikrein, φ: indicates a lack of measurable inhibition; ND: not determined.

The recombinant forms of BbKI and BbCI facilitated obtaining these inhibitors in high amounts since *B. bauhinioides* grows in a region of Brazil with limited access, whereas *E. contortisiliquum* is widely distributed throughout Brazil. EcTI and BbKI are both serine protease inhibitors that contain arginine at the P1 position of the reactive site [10,26]; however, specificity differences demonstrated by these proteins made them interesting candidates for cellular studies, especially those of tumor origin.

2.2. Effects of rBbCI, rBbKI, and EcTI on the L929 Fibrosarcoma Lineage

2.2.1. Effects of Inhibitors on Cell Viability

We investigated the effects of these inhibitors on fibrosarcoma cells (L929) since they have not yet been evaluated in malignant sarcoma cell lines and may be important tools in the study to understand the cellular processes involved in this type of tumor.

The effects of rBbCI, rBbKI, and EcTI (1.0–50 μM) on cell viability were evaluated at different incubation times (24, 48, and 72 h). rBbKI inhibited L929 cell viability in a dose and time-dependent manner (Figure 1). After 24 h incubation, rBbKI (50 μM) was able to reduce the viability of L929 cells by 50% (Figure 1A). Interestingly, inhibition of cell viability promoted by rBbKI was more effective after 48 h incubation, with an approximate inhibition of 80% (Figure 1B), which is quite promising. EcTI also demonstrated concentration-dependent inhibitory action, with approximately 80% inhibition occurring at a concentration of 50 μM after 72 h (Figure 1C). On the other hand, rBbCI did not significantly affect the viability of L929 cells under the analyzed conditions.

2.2.2. Effects of Inhibitors on Cell Adhesion

Cell adhesion is an essential process that can determine migration and generation of invasion process. Therefore, we analyzed the effect of the protease inhibitors on L929 cells coated with fibronectin, an important molecule that promote cell migration, invasion, and lung metastasis in soft-tissue sarcomas [30]. Two parameters were analyzed to establish ideal conditions for the adhesion assay with the fibrosarcoma cell line: the number of cells (1×10^5, 5×10^5, 1×10^6, 5×10^6, 1×10^7 cell/mL) and the incubation time (2 h, 3 h, 4 h, and 5 h). The results indicated that the optimal response for the experimental procedures was obtained using 1×10^6 cells in 2 h (Figure 2A,B).

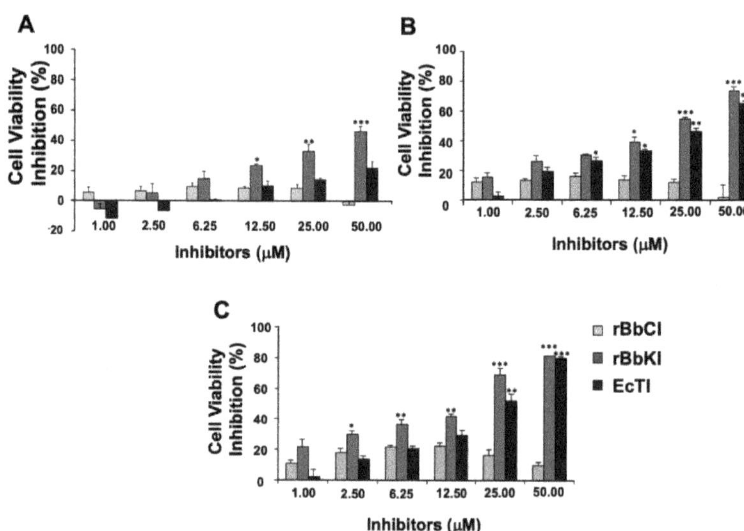

Figure 1. Action of rBbCI, rBbKI, and EcTI on the cell viability of L929 cells. L929 cells were pre-incubated for 24 h, increasing concentrations of rBbCI, rBbKI, and EcTI were added to the 96-well plates, and cells were analyzed after incubation for (**A**) 24 h, (**B**) 48 h, (**C**), or 72 h. Each bar represents the mean ± standard deviations of three repetitions. (* p <0.05, ** p <0.005, *** p <0.0001; One way-ANOVA, follow Tukey's Multiple Comparison Test).

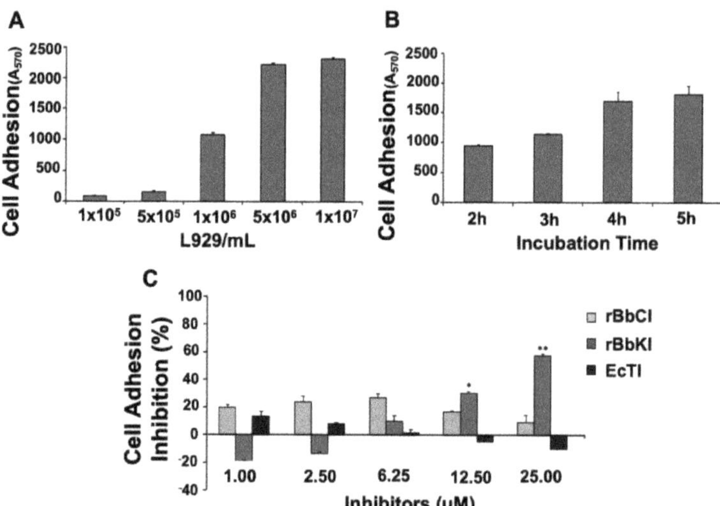

Figure 2. Action of rBbCI, rBbKI, and EcTI on cell adhesion of L929 cells. (**A**) Increasing concentrations of L929 were added to the 24-well plates previously coated with fibronectin and analyzed after 2 h of incubation. (**B**) 1×10^6 L929 cells were added to 24-well plates previously coated with fibronectin and analyzed at different incubation periods (2, 3, 4, 5 h). (**C**) L929 cells and increasing concentrations of rBbCI, rBbKI, and EcTI were pre-incubated for 15 min at room temperature and added to 24-well plates previously coated with fibronectin. Each bar represents the mean ± standard deviations of three repetitions. (* p < 0.05, ** p < 0.005; One way-ANOVA, follow Tukey's Multiple Comparison Test).

We then investigated the effect of rBbCI, rBbKI, and EcTI (1.0–25 µM) on the adhesion of 1×10^6 L929/mL to fibronectin after 2 h of incubation. The results demonstrated a dose-dependent inhibitory effect of rBbKI, which reached approximately 65% inhibition at a concentration of 25 µM, although a slight increase in cell attachment is noticed in the initial treatment (20%). On the other hand, rBbCI and EcTI did not significantly inhibit the adhesion of L929 to fibronectin under the conditions analyzed (Figure 2C).

2.2.3. Effects of Inhibitors on the Cell Cycle

Cancers have an exacerbated cell proliferation, controlled by the aberrant activity of important molecules of the cell cycle. Hence, cell cycle inhibitors are attractive for cancer therapies. For that, we investigated the effect of rBbCI, rBbKI, and EcTI (6.25 and 25 µM) on the cell cycle progression. The effects of the protease inhibitors were analyzed on the L929 cell at different incubation times (24 h—Figure 3, 48 h—Figure S1 and 72 h—Figure S2). None of the inhibitors provoked significant changes in the phases of the L929 cell cycle under the conditions analyzed.

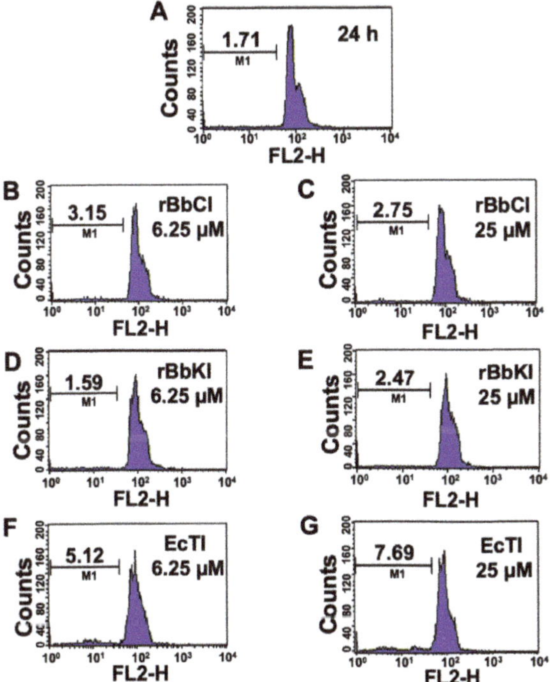

Figure 3. Effect of rBbCI, rBbKI, and EcTI on L929 cell cycle after 24 h incubation. L929 cells treated with (**A**) control, (**B**) 6.25 µM rBbCI, (**C**) 25 µM rBbCI, (**D**) 6.25 µM rBbKI, (**E**) 25 µM rBbKI, (**F**) 6.25 µM EcTI, and (**G**) 25 µM EcTI. M1 = fragmented cells.

2.2.4. Confocal Microscopy

The main difficulty of compounds with therapeutic action is access to the inside of cancer cells. The plasma membrane is the first barrier formed by the cell against the entry of chemical compounds, making it difficult for antitumor molecules to access. Therefore, the search for molecular targets that have intracellular action is important. Once the inhibitory action of rBbKI and EcTI against L929 cells was observed in cell viability and cell adhesion, we analyzed the interaction of these proteins with the cell surface. For these experiments, proteins were labeled with AlexaFluor 488, which covalently binds to primary amines present in the protein structure and emits green fluorescence when excited with an argon laser at 488 nm (blue light), which is detected by fluorescence microscopy. Thus, L929 cells were incubated with previously labeled BbKI or EcTI (Figures 4A and 5A).

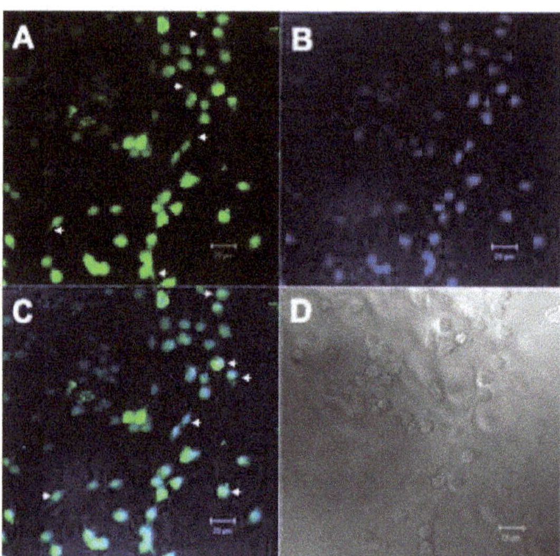

Figure 4. Confocal Microscopy Analysis of L929 with BbKI-AlexaFluor488. (**A**) L929 cells marked with BbKI conjugated with AlexaFluor 488 (green), showing cell surface localization (arrowheads), (**B**) L929 nuclei stained with 4′,6-Diamidino-2-phenylindole dihydrochloride, DAPI (blue), (**C**) overlay of images with BbKI nuclear localization (arrowheads) in L929 cells and (**D**) differential interference contrast (DIC) microscopy of L292 cells. Scale bar: 20 μm.

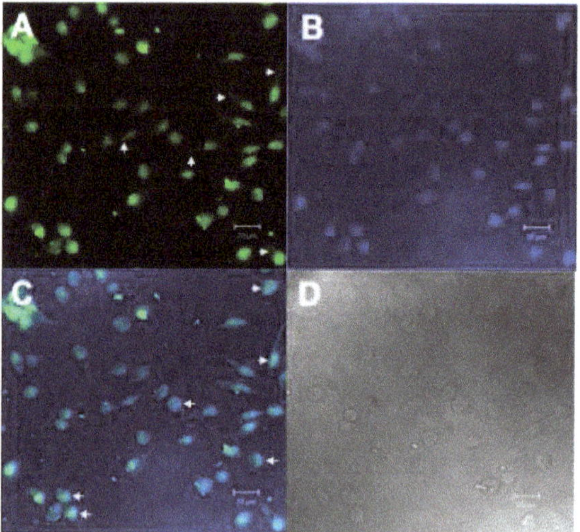

Figure 5. Confocal Microscopy Analysis of L929 with EcTI-AlexaFluor488. (**A**) L929 cells marked with EcTI conjugated with AlexaFluor 488 (green), showing cell surface localization (arrowheads), (**B**) L929 nuclei stained with 4′,6-Diamidino-2-phenylindole dihydrochloride, DAPI (blue), (**C**) overlay of images with EcTI nuclear localization (arrowheads) in L929 cells and (**D**) differential interference contrast (DIC) microscopy of L292 cells. Scale bar: 20 μm.

The fluorescent dye DAPI was used to mark the location of the cell nucleus (Figures 4B and 5B). Localization of the fluorescent-conjugated inhibitors indicated they bound to the cell surface and nucleus of L929 (Figures 4C and 5C), which is relevant.

2.2.5. Effect of the BbKI-Derived Peptide on L929 Cell Viability

Employing the synthetic peptide RPGLPVRFESPL-NH$_2$ derived from the BbKI reactive site to act on serine protease activity [31], we investigated the efficacy of BbKI peptide (0.02–1.0 mg/mL) on L929 cell viability at different incubation times (24 h—Figure 6A and 48 h—Figure 6B). The peptide inhibited cell viability in a dose-dependent manner, with maximum inhibition between 40% and 50%. These inhibitory levels did not change after 48 h, indicating the stability of the peptide.

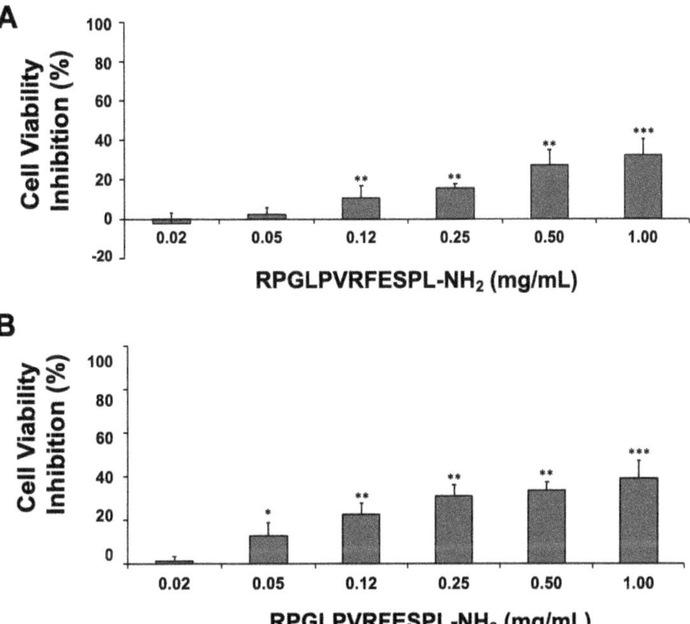

Figure 6. Action of the RPGLPVRFESPL-NH$_2$ peptide on L929 cell viability after 24 and 48 h. L929 cells were pre-incubated for 24 h, increasing concentrations of the synthetic peptide were added to the 96-well plates, and cells were analyzed after incubation for (**A**) 24 h and (**B**) 48 h. Each bar represents the mean ± standard deviations of three repetitions. (* $p < 0.05$, ** $p < 0.005$, *** $p < 0.0001$; One way-ANOVA, follow Tukey's Multiple Comparison Test).

3. Discussion

The pharmacological properties of protease inhibitors have been characterized and studied for decades in different biological models for therapeutic applications [8,9]. Their effectiveness may be due to direct inhibition of extracellular matrix (ECM) proteolysis or indirect inhibition of proteolytic cascade activation, thus preventing the spread of tumor cells, ECM degradation, and inhibition of tumor progression [8].

In the present study, we explored different inhibitory characteristics of plant protease inhibitors derivates: rBbCI, rBbKI, and EcTI by analyzing their efficacy against events elicited in the spread of tumors, such as proliferation, adhesion, and the cell cycle of mouse fibrosarcoma cells (L929). EcTI and BbKI are both serine protease inhibitors that contain arginine at the P1 position of the reactive site [10,26]; however, specificity differences demonstrated by these proteins made them interesting candidates for cellular studies, especially those of tumor origin. rBbKI and EcTI concentrations were inversely correlated with L929 cell viability, and no effect was detected with rBbCI treatment. Reduction in fibrosarcoma viability is interesting and relevant due to aggressivity and high recurrence rates of fibrosarcoma.

We determined that the action of EcTI was slower than that of rBbKI, inhibiting approximately 50% of cell viability after 24 h incubation, while EcTI reached a maximum of 20% inhibition in this period. However, after 48 h and especially 72 h, the EcTI inhibitory activity was high and similar (at 50 μM) to rBbKI. These differences probably reside in the inhibitory properties of each protease inhibitor. The action of EcTI and rBbKI must be important not only for inhibiting the activity of serine proteases, such as trypsin and chymotrypsin but also for inhibiting activation of a proteolytic cascade in which these enzymes are involved. In this case, the inhibitory activity of rBbKI against tissue kallikreins distinguishes it from EcTI. Tissue kallikreins are involved in cellular processes that lead to the activation of fundamental enzymes and receptors in the metabolism of these cells, such as protease-activated receptors [32,33]. BbKI can inhibit tissue kallikreins with K_{iapp} in the nM range, and this activity may be responsible for the faster inhibitory action of rBbKI compared with EcTI. Together with plasminogen activators, both BbKI and EcTI inhibit plasmin to form a powerful mechanism for generating proteolytic activity, which is necessary for tumor growth, metastasis, and angiogenesis. Plasminogen (the plasmin zymogen) plays an important role in metastasis as a primary tumor spreader [34,35]. Interestingly, BbCI, which is also present in the seeds of *B. bauhinioides* and exhibits 84% primary structure similarity with BbKI [36], demonstrates a distinguished inhibitory specificity by inhibiting cysteine protease cathepsin L and serine proteases elastase and cathepsin G [17]. Although increasing cathepsin L-like activity may be correlated with malignant tumors [37,38] and elastase [28], rBbKl did not affect L929 cell viability, proliferation, or cell adhesion. Thus, the inhibitory specificity of rBbKI may have impacted its toxicity towards fibrosarcoma cells.

Once rBbKI affected cell viability after 24 h and this effect was maintained after 48 h and 72 h, we investigated whether the peptide similar to the region of the reactive inhibitor site would be responsible for the effect of rBbKI on these cells. Although the greater potency of the protein compared with the peptide suggests other parts of the inhibitor structure may contribute complementary action, the difference may also be attributed to the change in specificity, as demonstrated by Cagliari et al. [31]. The peptide was not able to inhibit trypsin, and rBbKI is a potent inhibitor of this enzyme.

Degradation of extracellular matrix and adhesion is an essential process to the establishment of cancer cells. The role of fibronectin contributes to tumor malignancy, metastasis, and patients' poor prognosis [30]. The protease inhibitor rBbKI was effective in reduce L929 cell adhesion to fibronectin, demonstrating the relevance of this protein in decreased cell viability and cell adhesion.

For a compound to have potentiated therapeutic action, it is necessary to access the cell via contact with proteins, lipids, or saccharides of the plasma membrane. In this view, molecules that have intracellular action are important. The hypothesis that the plant inhibitors could penetrate the cell was already demonstrated [20]. In the L929 line, the fluorescence microscopy showed rBbKI and EcTI interaction with the cell and nucleus surface, indicated by the green stain (Alexafluor 488), distributed along L929 cell and by blue stain (DAPI), dispersed by nuclear area demonstrating that rBbKI and EcTI were internalized. Further studies need to be conducted to investigate the consequences of cell signaling besides the effects on cell viability and adhesion.

Despite progress made with the use of these inhibitors, much remains unknown. Inhibitors may indirectly or directly signal or activate different components that induce cell death, providing prospects for studies to assess the importance of proteases involved in this process. The positive results obtained with the inhibitor fragment in this study indicate that the structures of these proteins warrant further exploration, such as cell signaling and in vivo experiments involved in this type of pathology.

4. Materials and Methods

4.1. Seeds

The seeds of *Bauhinia bauhinioides* (specimen number: 4665, CGMS 28770) and *Enterolobium contortisiliquum* (Vell.) Morong (specimen number: 10254, CGMS 56403) were collected in the region of Corumbá, Brazil, and identified by Dr. Vali Joana Pott (Brazilian Agricultural Research Corporation (Embrapa), Campo Grande, Brazil).

The species *B. bauhinioides* (Mart.) Macbr. belongs to the family Fabaceae. It is popularly known as cow's hoff due to the shape of its leaves, which consist of two leaflets [35]. The species *E. contortisiliquum*, also belonging to the family Fabaceae, is a tree that grows over 20 m high, popularly known as ear pod due to its ear-shaped pods [25].

4.2. Purification of Native Inhibitors

The inhibitors from *B. bauhinioides* seeds (BbKI and BbCI) were purified according to the methodology described by Nakahata et al. [12]. The method described by de Paula et al. [11] was followed to obtain the EcTI inhibitor from *E. contortisiliquum* seeds. Some process modifications are detailed below.

4.3. Extraction of B. bauhinioides and E. contortisiliquum Inhibitors

Following the procedures shown in Figure 7A–C, *B. bauhinioides* and *E. contortisiliquum* seeds were crushed in a mill, and the resulting powder was homogenized at a 1:40 (w/v) ratio with 0.15 M NaCl in a blender. The material was centrifuged at 4000 rpm for 15 min at 4 °C and the supernatant was then heated to 56 °C for approximately 15 min before being slowly cooled in an ice bath. Proteins were precipitated with cold acetone, under slow and constant stirring at 4 °C, until reaching a final concentration of 80% (v/v). After the 30 min precipitation period, the material was centrifuged at 4000 rpm for 20 min at 4 °C. The protein precipitate was then vacuum-dried to evaporate residual acetone, solubilized in 0.15 M NaCl, followed by centrifugation under the same conditions described above.

4.4. Ion Exchange Chromatography on DEAE-Sephadex

The protein precipitate, after conductivity and pH adjustments, was applied to a DEAE-Sephadex A-50 column (3.0 × 13.0 cm, Amersham Biosciences, Piscataway, NJ, USA), previously equilibrated with 0.1 M Tris/HCl buffer (pH 8.0). The column was washed with equilibrium buffer until effluent Abs280 > 0.03. The bound protein was eluted with 0.1 M Tris/HCl buffer (pH 8.0) containing 0.15 M NaCl, followed by 0.1 M Tris/HCl buffer (pH 8.0) containing 0.3 M NaCl. Chromatography was performed under a constant flow of 30 mL/h, fractionated into 2 mL aliquots. Fractions with Abs280 > 0.05 were pooled. Pooled active fractions were dialyzed at 4 °C.

4.5. Affinity Chromatography on Trypsin-Sepharose

Dialyzed, pooled fractions containing inhibitory activity were chromatographed on a trypsin-Sepharose column (10.0 mL of resin) equilibrated in 0.1 M Tris/HCl buffer (pH 8.0). The column was washed with equilibration buffer until the effluent Abs280 > 0.03; non-bound material was collected. The column was subsequently washed with 0.1 M Tris/HCl buffer (pH 8.0) containing 0.15 M NaCl. The bound inhibitor was then eluted by acidification with 0.5 M KCl/HCl (pH 2.0) and the collected fractions (1.0 mL/min) were immediately neutralized by adding 1.0 M Tris/HCl solution (pH 9). Protein elution was followed by detecting Abs280, and inhibitory activity against trypsin (bound material) or HNE (human neutrophil elastase, non-bound material, in the case of BbCI) was followed by hydrolysis of Bz-Arg-pNan or MeO-Suc-Ala-Ala-Pro-Val-pNan substrates, as described below.

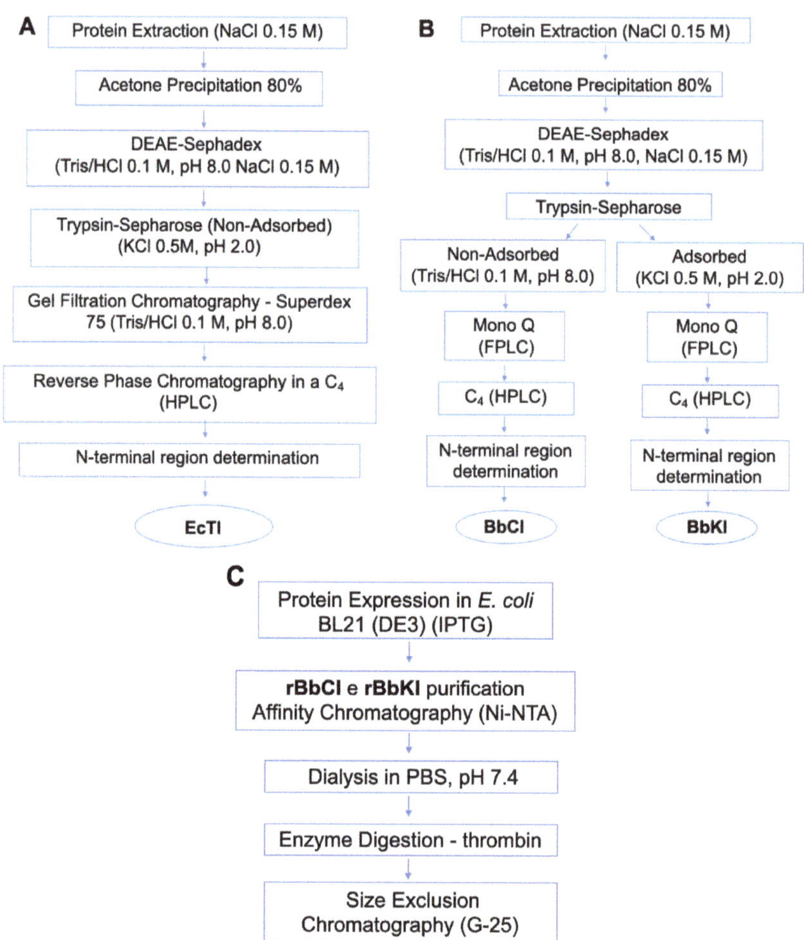

Figure 7. Purification scheme for (**A**) *Enterolobium contortisiliquum* trypsin inhibitor (EcTI), (**B**) *Bauhinia bauhinioides* cruzipain inhibitor (BbCI) and kallikrein inhibitor (BbKI), and (**C**) recombinant protein from BbCI (rBbCI) and BbKI (rBbKI).

4.6. Fast Protein Liquid Chromatography on Mono Q

B. bauhinioides materials obtained from the trypsin-Sepharose affinity column (non-bound and bound material) were purified by fast protein liquid chromatography (FPLC) to remove possible contaminants. The Mono Q column (Amersham Biosciences, GE Healthcare, Amersham, UK) was equilibrated in 0.05 M Tris/HCl buffer (pH 8.0), and elution was performed using a gradient (0–0.5 M NaCl) with 0.05 M Tris/HCl buffer (pH 8.0) containing 0.5 M NaCl, at a flow rate of 0.5 mL/min. Fractions with inhibitory activity against HNE and trypsin, respectively, were pooled, and then lyophilized for purity analysis and enzyme inhibition studies.

4.7. Recombinant rBbCI and rBbKI Inhibitors

4.7.1. Cloning

Cloning and purification of recombinant inhibitors rBbKI and rBbCI were conducted according to procedures described by Araújo et al. [39]. In summary, the genes encoding BbKI and BbCI were cloned by RT-PCR using degenerate primers, which were synthesized from protein sequences previously described by Oliva et al. [10] and de Oliveira et al. [17], respectively. The amplified product was inserted into the p-GEM T vector (Promega, Madison, WI, USA) for sequencing. Once sequences were confirmed, internal primers were synthesized to perform 5′ and 3′ RACE (Rapid Amplification of cDNA Ends). Both BbKI and BbCI were initially synthesized as pre-peptides with the following structure: 19 amino acid residues in the N-terminal region, 164 residues corresponding to the mature peptide, and 10 residues in the C-terminal region.

Sequences corresponding to the mature peptide were subcloned into the pET28a expression vector (Novagen, Madison, WI, USA). rBbKI and rBbCI inhibitors were produced by heterologous expression of a fusion protein consisting of a 6-residue histidine tail with a thrombin cleavage site between them. Transfected *E. coli* BL21 (DE3) (Merck KGaA, Darmstadt, Germany) produced inhibitors in large quantities, which were purified as described below.

4.7.2. Expression of rBbCI and rBbKI Inhibitors

E. coli BL21 (DE3) colonies harbouring rBbKI and rBbCI were inoculated in 5 mL lysogeny broth (LB) with 50 µg/mL kanamycin, with agitation at 150 rpm for 12 h at 37 °C. The inoculum was then transferred to 500 mL LB containing 50 µg/mL kanamycin, agitated at 37 °C until Abs_{600} = 0.40–0.50. The culture was then induced for 3 h to express rBbCl and rBbKl inhibitors by adding 0.2 mM isopropyl β-d-1-thiogalactopyranoside (IPTG). At the end of the induction period, the material was centrifuged at 4000 rpm for 20 min at 4 °C, resuspended in 10 mL 0.1 M Tris/HCl buffer (pH 8.0) containing 0.15 M NaCl, and stored at −70 °C until use.

4.8. Purification of rBbKI and rBbCI Inhibitors

4.8.1. Affinity Chromatography and Cleavage of the Fusion Peptide

Transfected *E. coli* were lysed by sonication (8 pulses of 30 s each) in an ice bath, followed by centrifugation at 10,000 rpm for 10 min at 4 °C. The supernatant was subjected to affinity chromatography using a Ni-NTA Superflow Column (2 mL resin; Qiagen, Hilden, Germany) previously equilibrated in 0.1 M Tris/HCl buffer (pH 8.0) containing 0.15 M NaCl. The column was washed with equilibration buffer and non-bound material was collected. Bound proteins were eluted with 0.1 M Tris/HCl buffer (pH 8.0) using an imidazole gradient (0.01 M, 0.1 M, 0.25 M, and 0.50 M). Elution of the rBbKI and rBbCI inhibitors was followed by detecting Abs_{280}. Fractions eluted with 0.1 M Tris/HCl buffer (pH 8.0) containing 0.25 M and 0.50 M imidazole were dialyzed in sodium phosphate buffer (diluted PBS solution) at pH 7.4. Subsequently, samples were incubated with thrombin (0.5 U/mg protein) for 4 h at 18 °C and immediately subjected to size-exclusion chromatography.

4.8.2. Size Exclusion Chromatography

Fractions containing native inhibitors obtained from trypsin-Sepharose chromatography, as well recombinant forms obtained by thrombin cleavage, were purified by size exclusion chromatography using the Äkta system (GE Healthcare, Chicago, IL, USA). A Superdex 75 10/300 GL column (Amersham Biosciences, GE Healthcare, Amersham, UK) was equilibrated with 0.1 M Tris/HCl buffer (pH 8.0), and previously calibrated with different molecular mass proteins (ovalbumin, 45 kDa; carbonic anhydrase, 30 kDa; soybean trypsin inhibitor, 20 kDa; cytochrome C, 12.4 kDa). Chromatography was performed under a constant flow of 0.5 mL/min, fractionated in 1 mL aliquots. Inhibitor elution was followed by detecting absorbance at 280 nm.

4.8.3. Reverse-Phase Chromatography (HPLC System)

Fractions containing inhibitory activity were concentrated by lyophilization and purified by reverse-phase chromatography (C4 column 15 × 0.5 cm; Beckman Ultrasphere, Lake Forest, CA, USA) equilibrated in 0.1% trifluoroacetic acid (TFA) in water. Elution was performed using an acetonitrile gradient (90% (*v/v*), 0–100% in 0.1% TFA in water), under a constant flow of 0.7 mL/min. Sample homogeneity was confirmed by SDS-PAGE (data not shown).

4.8.4. Enzymatic Assays

Enzymes, substrates and specific buffers were used to assess substrate hydrolysis by the proteases (Table 2). The substrates used in each experiment demonstrated optimum specificity for the enzyme tested [12]. Chromogenic substrates derived from p-nitroanilide (Calbiochem, Darmstadt, Germany) were employed for highly sensitive photometric detection of p-nitroaniline released after enzymatic hydrolysis, using a SpectraCount spectrophotometer (Hewlett-Packard, Palo Alto, CA, USA) at 405 nm. Substrates were initially diluted in dimethyl sulfoxide (DMSO) and further dilutions were performed in an appropriate buffer for each assay. Fluorogenic substrates (Calbiochem) derived from 7-amino-4-methylcoumarin (AMC) were diluted in dimethylformamide (DMF); hydrolysis was monitored at 380 nm (excitation) and 460 nm (emission) using a FluoroCount spectrofluorometer (Hewlett-Packard). Assays using substrate Abz-X-EDDnp (diluted 1:1 (v/v) in DMF and water) were performed in an F-2000 spectrofluorometer (Hitachi, Tokyo, Japan) with hydrolysis detection at 320 nm (excitation) and 420 nm (emission).

Table 2. Enzymes, substrates and buffer used to verify substrate hydrolysis by serino proteases and inhibitory activity.

Enzyme	Substrate	Buffer
trypsin (20 μL, 0.41 μM) * NPGB	Bz-Arg-pNan (25 μL, 10 mM)	Tris/HCl 0.1 M, pH 8.0, CaCl$_2$ 0.02% (v/v)
HNE (20 μL, 0.21 μM) (* α$_1$-anti-trypsin)	MeO-Suc-Ala-Ala-Pro-Val-pNan (25 μL, 1.1 mM)	Tris/HCl 0.1 M, pH 7.0, NaCl 0.5 M
PKa (20 μL, 0.84 μM) (* EcTI)	H-D-Pro-Phe-Arg-pNan (25 μL, 5 mM)	Tris/HCl 0.05 M, pH 8.0
PoPK (30 μL, 0.16 nM) (* aprotinin)	H-D-Pro-Phe-Arg-AMC (30 μL, 5 mM)	Tris/HCl 0.1 M, pH 9.0, plus albumin 0.1% (v/v)
chymotrypsin (40 μL, 0.88 μM) (* EcTI)	Suc-Phe-pNan (20 μL, 20 mM)	Tris/HCl 0.1 M, pH 8.0, CaCl$_2$ 0.02% (v/v)
plasmin (25 μL, 0.028 μM) (* BvTI)	H-D-Val-Leu-Lys-pNan (20 μL, 9 mM)	Tris/HCl 0.1 M, pH 7.4, NaCl 0.2 M
thrombin (10 μL, 0.267 μM) (* rhodinin)	H-D-Phe-L-Pip-L-Arg-pNan (20 μL, 2 mM)	Tris/HCl 0.05 M, pH 8.0
factor Xa (30 μL, 0.467 μM) (* BuXI)	Boc-Ile-Glu-Gly-Arg-AMC (60 μL, 6 mM)	Tris/HCl 0.05 M, pH 8.0
cathepsin G (30 μL, 0.25 μM) (* α$_1$-anti-trypsin)	N-Suc-Ala-Ala-Pro-Phe-pNan (25 μL, 1 mM)	Tris/HCl 0.05 M, pH 7.0, NaCl 0.5 M

* inhibitor used for enzyme titration. HNE: human neutrophil elastase; PKa: human plasma kallikrein; PoPK: porcine pancreatic kallikrein.

Apparent inhibition constants were determined by calculating the dissociation constant values of the enzyme-inhibitor complex (K_{iapp}), following the model proposed by Morrison (1989) [40]. Enzymatic kinetics were calculated using the GraFit© Version 3.0 (Erithacus Software Ltd., Horley, UK).

4.8.5. Hydrolysis of Substrates by Serine Proteases and Determination of Inhibitory Activity

Serine protease activity was assessed on specific substrates, using enzymes that were pre-incubated at 37 °C with different inhibitor concentrations in an appropriate buffer. The substrate was added (250 μL final volume) after 10 min (Table 2), incubated for 20–30 min (depending on the enzyme) at 37 °C, and the reaction was interrupted with 40 μL 40% acetic acid. Substrate hydrolysis was monitored by detecting Abs$_{405}$ from released p-nitroaniline, or fluorescence at 380 nm (excitation) and 460 nm (emission wavelengths) from AMC release. Inhibitory activity was calculated by determining the residual enzyme activity in the assays. Inhibitor concentrations were calculated assuming 1:1 reaction stoichiometry. This methodology was also used to identify inhibitory activity during the purification processes. Experiments were performed in triplicate.

4.9. Hydrolysis of Z-Phe-Arg-AMC by Cysteine Proteases and Determination of Inhibitory Activity

Cysteine protease concentrations were obtained through titrations with egg cystatin [41]. Inhibition of cathepsin L, cruzipain, or cruzain was determined through residual enzyme activity on the substrate Z-Phe-Arg-AMC. Enzymes were activated by incubation for 10 min at 37 °C in 0.1 M Na_2HPO_4 buffer (pH 6.3) containing 10 mM EDTA, 400 mM NaCl, and 2 mM DTT. In typical experiments performed with the same activation buffer, cathepsin L (18 nM), cruzipain (18 nM), or cruzain (3.2 nM) was pre-incubated with increasing concentrations of purified inhibitor for 10 min at 37 °C, and then the substrate Z-Phe-Arg-AMC (0.3 mM) was added. Substrate hydrolysis was monitored by detecting fluorescence at 380 nm (excitation) and 460 nm (emission) wavelengths, using an F-2000 spectrofluorometer (Hitachi). The increase in fluorescence was continuously recorded for 10 min. Experiments were performed in triplicate. Residual activity was determined by comparing enzymatic hydrolysis curves in the presence and absence of the inhibitor.

4.9.1. Determination of rBbKI and EcTI Concentrations

Increasing amounts of rBbKI and EcTI were pre-incubated for 10 min at 37 °C in 0.05 M Tris/HCI buffer (pH 8.0) containing 0.02% (v/v) $CaCl_2$ with trypsin previously titrated with 4-guanidinobenzoic acid 4-nitrophenylester hydrochloride (NPGB). The substrate Bz-Arg-pNan (25 uL, 10.0 mM) was then added to the pre-incubated sample (250 mL final volume). Substrate hydrolysis was monitored by detecting Abs405 of the released p-nitroaniline. Inhibitor concentrations were calculated assuming 1:1 reaction stoichiometry [10].

4.9.2. Determination of rBbCI Concentration

Increasing amounts of rBbCI were pre-incubated for 10 min at 37 °C in 0.1 M Tris/HCI buffer (pH 7.0) containing 0.5 M NaCl and 17 nM HNE. Substrate MeO-Suc-Ala-Ala-Pro-Val-pNan (25 µL, 11.0 mM) was then added to the pre-incubated sample (250 mL final volume). Substrate hydrolysis was monitored by detecting Abs_{405} of the released p-nitroaniline. Inhibitor concentrations were calculated assuming 1:1 reaction stoichiometry.

4.9.3. Synthetic Peptide RPGLPVRFESPL-NH$_2$

The peptide Arg-Pro-Gly-Leu-Pro-Val-Arg-Phe-Glu-Ser-Pro-Leu-NH$_2$ (RPGLPVRFESPL-NH$_2$), derived from the reactive site of BbKI, was previously described by Oliva et al. [10] and Cagliari et al. [31]. The peptide was synthesized at the Departamento de Biofísica, UNIFESP (São Paulo, Brazil).

4.9.4. Cultivation Conditions for the L929 Cell Line

The mouse fibrosarcoma strain L929 was maintained in RPMI medium (pH 7.4) supplemented with 10% fetal bovine serum (FBS; Invitrogen, Carlsbad, CA, USA), penicillin (100 U/mL), and streptomycin (100 µg/mL) at 37 °C under 5% CO_2 until cells reached 80–90% confluence, as described below.

For the maintenance of cell stocks, the medium was initially removed from the culture of confluent cells, which were subsequently washed once with PBS buffer (pH 7.4) containing 140 mM NaCl, 1.7 mM KH_2PO_4, and 2.7 mM KCl. Cells were detached using a 2.5% trypsin solution for approximately 1 min. Cells were then carefully resuspended several times using a serological pipette and transferred to a tube, followed by centrifugation at 2000 rpm for 3 min, at 25 °C. Finally, cells were resuspended in RPMI medium (pH 7.4) with 10% FBS, and 5×10^4 cells were transferred to a new bottle containing RPMI medium (pH 7.4) supplemented with 10% FBS. The culture medium was renewed every 2–3 days. Cells were used for experiments when they reached approximately 80–90% confluence.

4.9.5. Cell Viability

L929 cells were washed with sterile PBS buffer (pH 7.4), removed from the culture bottle with trypsin solution (2.5%, v/v), centrifuged, and suspended in RPMI (pH 7.4) with 10% FBS. Cells were then counted in a Neubauer chamber. Cells (1×10^4 cells/100 µL/well) were incubated at 37 °C under 5% CO_2 for 24 h in RPMI medium (pH 7.4) with 10% FBS to allow adhesion.

rBbCI, rBbKI, and EcTI inhibitors at concentrations ranging from 1.0–50 µM were diluted in previously filtered (0.22 µm; Millipore, Billerica, MA, USA) RPMI medium (pH 7.4) with 10% FBS. Inhibitors were added to the L929 cell-containing wells and incubated at 37 °C under 5% CO_2 for 24, 48, and 72 h. The synthetic peptide (RPGLPVRFESPL-NH$_2$) was diluted first in DMSO (final concentration 1% DMSO) and then diluted in RPMI medium (pH 7.4) with 10% FBS at concentrations ranging from 0.02–1.0 mg/mL. Synthetic peptide was added to the wells, and incubated at 37 °C under 5% CO_2 for 24 and 48 h. At the end of each incubation period, 10 µL MTT (5 mg/mL) was added to each well, and cells were incubated for 2 h at 37 °C under 5% CO_2. The culture medium was then removed, and isopropanol was added for an additional 20 min at 37 °C. Finally, Abs_{620} was measured using a SpectraCount spectrophotometer (Hewlett-Packard). Assays were performed in triplicate for each concentration of inhibitor and peptide.

4.9.6. Cell Adhesion Assay

Cell adhesion assays were performed according to the method described by de Paula et al. [11], with some modifications. Briefly, 24-well plates were pre-coated with 10 µg/250 µL/well fibronectin diluted in PBS (pH 7.4), and incubated at 37 °C under 5% CO_2 for 2 h or overnight at 4 °C. The protein-binding site not yet covered by fibronectin was blocked by incubation with 250 µL/well 1% BSA in sterile PBS (pH 7.4) for 1 h at 37 °C. After excess BSA was removed, the plates were washed 3× with PBS (pH 7.4).

L929 cells were removed from the culture bottle using trypsin solution (2.5% v/v), followed by washing with sterile PBS (pH 7.4). Cells were then centrifuged at 2000 rpm for 3 min, at 25 °C and suspended in RPMI (pH 7.4). Cells (1×10^6 cells/100 µL/well) and 1.0–25 µM inhibitor (200 µL/well) diluted in sterile RPMI (pH 7.4) were pre-incubated for 15 min at room temperature, added to the cell adhesion plates, and incubated at 37 °C under 5% CO_2 for 2 h. At the end of the incubation period, non-adherent cells were removed by washing 3× with PBS (pH 7.4). Methanol was added to the wells and incubated for 5 min at 25 °C to fix the adhered cells. Subsequently, the plates were washed 3× with PBS (pH 7.4) and stained with 1% toluidine blue solution in 1% borax for 5 min. Wells were washed 4× with PBS (pH 7.4), and pigment retained by cells was released with a 1% SDS solution. Absorbance at 570 nm (A570) was measured using a SpectraCount spectrophotometer (Hewlett-Packard). Assays were performed in triplicate for each concentration of inhibitor.

4.9.7. Cell Cycle Assay

L929 cells (1×10^4 cells/mL) were diluted in RPMI (pH 7.4) and cultured in 60×10 mm plates for 6 h at 37 °C under 5% CO_2. The medium was then removed and cells were maintained in RPMI medium (pH 7.4) supplemented with 10% FBS for 24 h. rBbCI, rBbKI, and EcTI inhibitors were added at concentrations of 6.25 and 25 µM, followed by incubation for 24, 48, and 72 h at 37 °C under 5% CO_2. After washing with PBS (pH 7.4), the medium and adhered cells were removed and pooled. Samples were centrifuged at 2000 rpm for 4 min at 25 °C, the supernatant was discarded, 5 mL PBS (pH 7.4) was added to the precipitate, and the sample was again centrifuged. Cells were then resuspended in 0.5 mL PBS (pH 7.4), and 0.5 mL absolute ethanol was added slowly under constant agitation. This material was stored at 4 ° C until use when it was again centrifuged, and the ethanol was carefully removed. The cells were then washed with 1 mL PBS (pH 7.4), and centrifuged at 2000 rpm for 4 min. The cells were resuspended in 100 µL PBS (pH 7.4) and 400 µL of propidium iodide (50 µg/mL) [25]. Cells were incubated for 30 min at 4 °C in the absence of light and immediately analyzed using a flow cytometer (FACS; Becton-Dickinson, San Diego, CA, USA).

4.10. Analysis of Cell Interaction with Inhibitors by Confocal Microscopy

4.10.1. Covalent Conjugation of Inhibitors to Fluorescent Dye

EcTI and BbKI inhibitors (1 mg/mL) were solubilized in 200 µL 0.1 M sodium bicarbonate (pH 9.0). AlexaFluor 488 (1 mg/mL) was solubilized in 100 µL DMSO. Incubation of each inhibitor with AlexaFluor 488 (1:15 protein/fluorophore molar ratio) was carried out in a nitrogen atmosphere at 25 °C in the absence of light, and under constant agitation for 3 h. After incubation, the solution containing the protein and fluorophore was subjected to size exclusion chromatography on a G-25 gel against 0.03 M HEPES buffer (pH 7.4) to eliminate excess fluorophore [11].

4.10.2. Cell Labeling for Confocal Microscopy

Glass coverslips (10 mm) in a 24-well plate were covered with 0.1 mg/mL poly-D-lysine in sterile PBS (pH 7.4), and maintained at 37 °C for 2 h. After this period, excess poly D-lysine solution was removed and the coverslips were dried at room temperature under laminar flow for 4 h [19].

Subsequently, 1 mL RPMI medium (pH 7.4) supplemented with 10% FBS containing 1×10^5 L929 cells, was placed carefully on each coverslip and cultured at 37 °C for 24 h. The cells were then washed with PBS (pH 7.4) at 4 °C and fixed with 2% paraformaldehyde (v/v) solution. After 30 min, the cells were washed with PBS (pH 7.4) containing 0.1 M glycine, and incubated for 1 h with 40 µg/mL AlexaFluor 488-conjugated inhibitors. Then, the cells were incubated for 15 min with 25 µg/mL DAPI diluted 1:400 with PBS (pH 7.4) plus 0.01% (w/v) saponin. Then, the coverslips were washed with PBS (pH 7.4) and fixed with 5 µL Fluoromont-G for analysis by confocal microscopy. The solutions were kept in an ice bath throughout the cell labeling process. Analysis of the slides was performed using a Zeiss Axiophot fluorescence microscope and a LSM 510 laser scanning confocal microscope (Zeiss, Wetzlar, Germany).

4.11. Statistical Analysis

Statistical significance was determined by One-Way ANOVA followed by Tukey's post-test to compare the means between independent groups. All experiments were performed in triplicate. Reproducible results were obtained, and representative data are shown. The values of * $p \leq 0.05$, ** $p \leq 0.005$ or *** $p \leq 0.0005$ were accepted as significant.

Supplementary Materials: The following are available online at https://www.mdpi.com/2223-7747/10/3/602/s1, Figure S1: Effect of rBbCI, rBbKI, and EcTI on L929 cell cycle after 48 h incubation, and Figure S2: Effect of rBbCI, rBbKI, and EcTI on L929 cell cycle after 72 h incubation.

Author Contributions: The manuscript was written through the contributions of all authors. Conceptualization, S.Y.I. and M.L.V.O.; methodology, S.Y.I. and A.M.N.; writing—original draft preparation, S.Y.I., C.R.B.; writing—review and editing, C.R.B., M.L.V.O., A.P. and V.J.P.; formal analysis, C.R.N. All authors have read and agreed to the published version of the manuscript.

Funding: This work was supported by Fundação de Amparo à Pesquisa do Estado de São Paulo (FAPESP) [2017/07972-9, 2017/06630-7 and 2019/22243-9]; Coordenação de Aperfeiçoamento de Pessoal de Nível Superior—Brasil (CAPES)—Finance Code 001, and Conselho Nacional de Desenvolvimento Científico e Tecnológico (CNPq) [301721/2016-5].

Institutional Review Board Statement: Not applicable.

Informed Consent Statement: Not applicable.

Data Availability Statement: There is no data.

Conflicts of Interest: The authors declare no conflict of interest.

References

1. Morrison, B.A. Soft Tissue Sarcomas of the Extremities. *Bayl. Univ. Med Cent. Proc.* **2003**, *16*, 285–290. [CrossRef] [PubMed]
2. Gore, M.R. Treatment, outcomes, and demographics in sinonasal sarcoma: A systematic review of the literature. *BMC Ear Nose Throat Disord.* **2018**, *18*, 4. [CrossRef] [PubMed]
3. Folpe, A.L. Fibrosarcoma: A review and update. *Histopathology* **2013**, *64*, 12–25. [CrossRef] [PubMed]
4. Ferguson, W.S. Advances in the adjuvant treatment of infantile fibrosarcoma. *Expert Rev. Anticancer. Ther.* **2003**, *3*, 185–191. [CrossRef]
5. Hoefkens, F.; Dehandschutter, C.; Somville, J.; Meijnders, P.; Van Gestel, D. Soft tissue sarcoma of the extremities: Pending questions on surgery and radiotherapy. *Radiat. Oncol.* **2016**, *11*, 136. [CrossRef]
6. Soofiyani, S.R.; Kazemi, T.; Lotfipour, F.; Hosseini, A.M.; Shanehbandi, D.; Hallaj-Nezhadi, S.; Baradaran, B. Gene therapy with IL-12 induced enhanced anti-tumor activity in fibrosarcoma mouse model. *Artif. Cells Nanomed. Biotechnol.* **2016**, *44*, 1988–1993. [CrossRef]
7. FDA: U.S. Food and Drug Administration. Center for Drug Evaluation and Research. Available online: https://www.cancer.gov/about-cancer/treatment/drugs/soft-tissue-sarcoma (accessed on 2 March 2021).
8. Liyanage, C.; Fernando, A.; Batra, J. Differential roles of protease isoforms in the tumor microenvironment. *Cancer Metastasis Rev.* **2019**, *38*, 389–415. [CrossRef]
9. Turk, B.; Turk, D.; Turk, V. Protease signalling: The cutting edge. *EMBO J.* **2012**, *31*, 1630–1643. [CrossRef]
10. Oliva, M.L.V.; Santomauro-Vaz, E.M.; Andrade, S.A.; Juliano, M.A.; Pott, V.J.; Sampaio, M.U.; Sampaio, C.A.M. Synthetic Peptides and Fluorogenic Substrates Related to the Reactive Site Sequence of Kunitz-Type Inhibitors Isolated from Bauhinia: Interaction with Human Plasma Kallikrein. *Biol. Chem.* **2001**, *382*, 109–113. [CrossRef]
11. de Paula, C.A.A.; Coulson-Thomas, V.J.; Ferreira, J.G.; Maza, P.K.; Suzuki, E.; Nakahata, A.M.; Nader, H.B.; Sampaio, M.U.; Oliva, M.L.V. Enterolobium contortisiliquum Trypsin Inhibitor (EcTI), a Plant Proteinase Inhibitor, Decreases in Vitro Cell Adhesion and Invasion by Inhibition of Src Protein-Focal Adhesion Kinase (FAK) Signaling Pathways. *J. Biol. Chem.* **2012**, *287*, 170–182. [CrossRef]
12. Nakahata, A.M.; Mayer, B.; Neth, P.; Hansen, D.; Sampaio, M.U.; Oliva, M.L.V. Blocking the Proliferation of Human Tumor Cell Lines by Peptidase Inhibitors from Bauhinia Seeds. *Planta Medica* **2013**, *79*, 227–235. [CrossRef] [PubMed]
13. Brito, M.V.; De Oliveira, C.; Salu, B.R.; Andrade, S.A.; Malloy, P.M.; Sato, A.C.; Vicente, C.P.; Sampaio, M.U.; Maffei, F.H.; Oliva, M.L.V. The Kallikrein Inhibitor from Bauhinia bauhinioides (BbKI) shows antithrombotic properties in venous and arterial thrombosis models. *Thromb. Res.* **2014**, *133*, 945–951. [CrossRef] [PubMed]
14. Martins-Olivera, B.T.; Almeida-Reis, R.; Theodoro-Júnior, O.A.; Oliva, L.V.; Nunes, N.N.D.S.; Olivo, C.R.; de Brito, M.V.; Prado, C.M.; Leick, E.A.; Martins, M.D.A.; et al. The Plant-DerivedBauhinia bauhinioidesKallikrein Proteinase Inhibitor (rBbKI) Attenuates Elastase-Induced Emphysema in Mice. *Mediat. Inflamm.* **2016**, *2016*, 5346574. [CrossRef]
15. Lubkowski, J.; Durbin, S.V.; Silva, M.C.C.; Farnsworth, D.; Gildersleeve, J.C.; Oliva, M.L.V.; Wlodawer, A. Structural analysis and unique molecular recognition properties of a Bauhinia forficata lectin that inhibits cancer cell growth. *FEBS J.* **2016**, *284*, 429–450. [CrossRef] [PubMed]
16. Bortolozzo, A.S.S.; Rodrigues, A.P.D.; Arantes-Costa, F.M.; Saraiva-Romanholo, B.M.; De Souza, F.C.R.; Brüggemann, T.R.; De Brito, M.V.; Ferreira, R.D.S.; Correia, M.T.D.S.; Paiva, P.M.G.; et al. The Plant Proteinase Inhibitor CrataBL Plays a Role in Controlling Asthma Response in Mice. *BioMed Res. Int.* **2018**, *2018*, 9274817. [CrossRef] [PubMed]
17. De Oliveira, C.; Santana, L.; Carmona, A.; Cezari, M.; Sampaio, M.; Sampaio, C.; Oliva, M. Structure of Cruzipain/Cruzain Inhibitors Isolated from Bauhinia bauhinioides Seeds. *Biol. Chem.* **2001**, *382*, 847–852. [CrossRef]
18. Ferreira, J.G.; Diniz, P.M.M.; de Paula, C.A.A.; Lobo, Y.A.; Paredes-Gamero, E.J.; Paschoalin, T.; Nogueira-Pedro, A.; Maza, P.K.; Toledo, M.S.; Suzuki, E.; et al. The Impaired Viability of Prostate Cancer Cell Lines by the Recombinant Plant Kallikrein Inhibitor. *J. Biol. Chem.* **2013**, *288*, 13641–13654. [CrossRef]
19. Ferreira, R.; Brito, M.; Napoleão, T.; Silva, M.; Paiva, P.; Oliva, M. Effects of two protease inhibitors from Bauhinia bauhinoides with different specificity towards gut enzymes of Nasutitermes corniger and its survival. *Chemosphere* **2019**, *222*, 364–370. [CrossRef]
20. Lobo, Y.A.; Bonazza, C.; Batista, F.P.; Castro, R.A.; Bonturi, C.R.; Salu, B.R.; Sinigaglia, R.D.C.; Toma, L.; Vicente, C.M.; Pidde, G.; et al. EcTI impairs survival and proliferation pathways in triple-negative breast cancer by modulating cell-glycosaminoglycans and inflammatory cytokines. *Cancer Lett.* **2020**, *491*, 108–120. [CrossRef]
21. Bonturi, C.R.; Motaln, H.; Silva, M.C.C.; Salu, B.R.; De Brito, M.V.; Costa, L.D.A.L.; Torquato, H.F.V.; Nunes, N.N.D.S.; Paredes-Gamero, E.J.; Turnšek, T.L.; et al. Could a plant derived protein potentiate the anticancer effects of a stem cell in brain cancer? *Oncotarget* **2018**, *9*, 21296–21312. [CrossRef]
22. Bonturi, C.R.; Silva, M.C.C.; Motaln, H.; Salu, B.R.; Ferreira, R.D.S.; Batista, F.P.; Correia, M.T.D.S.; Paiva, P.M.G.; Turnšek, T.L.; Oliva, M.L.V. A Bifunctional Molecule with Lectin and Protease Inhibitor Activities Isolated from Crataeva tapia Bark Significantly Affects Cocultures of Mesenchymal Stem Cells and Glioblastoma Cells. *Molecules* **2019**, *24*, 2109. [CrossRef] [PubMed]
23. Theodoro-Júnior, O.A.; Righetti, R.F.; Almeida-Reis, R.; Martins-Oliveira, B.T.; Oliva, L.V.; Prado, C.M.; Saraiva-Romanholo, B.M.; Leick, E.A.; Pinheiro, N.M.; Lobo, Y.A.; et al. A Plant Proteinase Inhibitor from Enterolobium contortisiliquum Attenuates Pulmonary Mechanics, Inflammation and Remodeling Induced by Elastase in Mice. *Int. J. Mol. Sci.* **2017**, *18*, 403. [CrossRef] [PubMed]

24. Rodrigues, A.P.D.; Bortolozzo, A.S.S.; Arantes-Costa, F.M.; Saraiva-Romanholo, B.M.; De Souza, F.C.R.; Brüggemann, T.R.; Santana, F.P.R.; De Brito, M.V.; Bonturi, C.R.; Nunes, N.N.D.S.; et al. A plant proteinase inhibitor from Enterolobium contortisiliquum attenuates airway hyperresponsiveness, inflammation and remodeling in a mouse model of asthma. *Histol. Histopathol.* **2018**, *34*, 537–552. [PubMed]
25. Nakahata, A.M.; Mayer, B.; Ries, C.; De Paula, C.A.A.; Karow, M.; Neth, P.; Sampaio, M.U.; Jochum, M.; Oliva, M.L.V. The effects of a plant proteinase inhibitor from Enterolobium contortisiliquum on human tumor cell lines. *Biol. Chem.* **2011**, *392*, 327–336. [CrossRef] [PubMed]
26. Batista, I.F.; Oliva, M.L.V.; Araujo, M.S.; Sampaio, M.U.; Richardson, M.; Fritz, H.; Sampaio, C.A. Primary structure of a Kunitz-type trypsin inhibitor from Enterolobium contortisiliquum seeds. *Phytochemistry* **1996**, *41*, 1017–1022. [CrossRef]
27. Almeida-Reis, R.; Theodoro-Junior, O.A.; Oliveira, B.T.M.; Oliva, L.V.; Toledo-Arruda, A.C.; Bonturi, C.R.; Brito, M.V.; Lopes, F.D.T.Q.S.; Prado, C.M.; Florencio, A.C.; et al. Plant Proteinase Inhibitor BbCI Modulates Lung Inflammatory Responses and Mechanic and Remodeling Alterations Induced by Elastase in Mice. *BioMed Res. Int.* **2017**, *2017*, 8287125. [CrossRef]
28. Lerman, I.; Hammes, S.R. Neutrophil elastase in the tumor microenvironment. *Steroids* **2018**, *133*, 96–101. [CrossRef]
29. Koumandou, V.L.; Scorilas, A. Evolution of the Plasma and Tissue Kallikreins, and Their Alternative Splicing Isoforms. *PLoS ONE* **2013**, *8*, e68074. [CrossRef]
30. Shi, K.; Wang, S.-L.; Shen, B.; Yu, F.-Q.; Weng, D.-F.; Lin, J.-H. Clinicopathological and prognostic values of fibronectin and integrin $\alpha v \beta 3$ expression in primary osteosarcoma. *World J. Surg. Oncol.* **2019**, *17*, 1–12. [CrossRef]
31. Cagliari, C.I.; De Caroli, F.P.; Nakahata, A.M.; Araújo, M.S.; Nakaie, C.R.; Sampaio, M.U.; Sampaio, C.A.M.; Oliva, M.L.V. Action of Bauhinia bauhinioides synthetic peptides on serine proteinases. *Biochem. Biophys. Res. Commun.* **2003**, *311*, 241–245. [CrossRef]
32. Figueroa, C.D.; Molina, L.; Bhoola, K.D.; Ehrenfeld, P. Overview of tissue kallikrein and kallikrein-related peptidases in breast cancer. *Biol. Chem.* **2018**, *399*, 937–957. [CrossRef] [PubMed]
33. Oikonomopoulou, K.; Hansen, K.K.; Saifeddine, M.; Vergnolle, N.; Tea, I.; Diamandis, E.P.; Hollenberg, M.D. Proteinase-mediated cell signalling: Targeting proteinase-activated receptors (PARs) by kallikreins and more. *Biol. Chem.* **2006**, *387*, 677–685. [CrossRef] [PubMed]
34. Palumbo, J.S.; Talmage, K.E.; Liu, H.; La Jeunesse, C.M.; Witte, D.P.; Degen, J.L. Plasminogen supports tumor growth through a fibrinogen-dependent mechanism linked to vascular patency. *Blood* **2003**, *102*, 2819–2827. [CrossRef] [PubMed]
35. Nyberg, P.; Ylipalosaari, M.; Sorsa, T.; Salo, T. Trypsins and their role in carcinoma growth. *Exp. Cell Res.* **2006**, *312*, 1219–1228. [CrossRef]
36. Zhou, D.; Hansen, D.; Shabalin, I.G.; Gustchina, A.; Vieira, D.F.; De Brito, M.V.; Araújo, A.P.U.; Oliva, M.L.V.; Wlodawer, A. Structure of BbKI, a disulfide-free plasma kallikrein inhibitor. *Acta Crystallogr. Sect. F Struct. Biol. Commun.* **2015**, *71*, 1055–1062. [CrossRef] [PubMed]
37. Han, M.-L.; Zhao, Y.-F.; Tan, C.-H.; Xiong, Y.-J.; Wang, W.-J.; Wu, F.; Fei, Y.; Wang, L.; Liang, Z.-Q. Cathepsin L upregulation-induced EMT phenotype is associated with the acquisition of cisplatin or paclitaxel resistance in A549 cells. *Acta Pharmacol. Sin.* **2016**, *37*, 1606–1622. [CrossRef] [PubMed]
38. Zhao, Y.; Shen, X.; Zhu, Y.; Wang, A.; Xiong, Y.; Wang, L.; Fei, Y.; Wang, Y.; Wang, W.; Lin, F.; et al. Cathepsin L-mediated resistance of paclitaxel and cisplatin is mediated by distinct regulatory mechanisms. *J. Exp. Clin. Cancer Res.* **2019**, *38*, 1–13. [CrossRef]
39. Araújo, A.P.U.; Hansen, D.; Vieira, D.F.; De Oliveira, C.; Santana, L.A.; Beltramini, L.M.; Sampaio, C.A.; Sampaio, M.U.; Oliva, M.L.V. Kunitz-type Bauhinia bauhinioides inhibitors devoid of disulfide bridges: Isolation of the cDNAs, heterologous expression and structural studies. *Biol. Chem.* **2005**, *386*, 561–568. [CrossRef]
40. Morrison, J.F. The slow-binding and slow, tight-binding inhibition of enzyme-catalysed reactions. *Trends Biochem. Sci.* **1982**, *7*, 102–105. [CrossRef]
41. Anastasi, A.; Brown, M.A.; Kembhavi, A.A.; Nicklin, M.J.H.; Sayers, C.A.; Sunter, D.C.; Barrett, A.J. Cystatin, a protein inhibitor of cysteine proteinases. Improved purification from egg white, characterization, and detection in chicken serum. *Biochem. J.* **1983**, *211*, 129–138. [CrossRef]

Article

Sustainable Processing of Floral Bio-Residues of Saffron (*Crocus sativus* L.) for Valuable Biorefinery Products

Stefania Stelluti, Matteo Caser *, Sonia Demasi and Valentina Scariot

Department of Agricultural, Forest and Food Sciences, University of Torino, Largo Paolo Braccini 2, 10095 Grugliasco, Italy; stefania.stelluti@unito.it (S.S.); sonia.demasi@unito.it (S.D.); valentina.scariot@unito.it (V.S.)
* Correspondence: matteo.caser@unito.it; Tel.: +39-011-670-8935

Abstract: Tepals constitute the most abundant bio-residues of saffron (*Crocus sativus* L.). As they are a natural source of polyphenols with antioxidant properties, they could be processed to generate valuable biorefinery products for applications in the pharmaceutical, cosmetic, and food industries, becoming a new source of income while reducing bio-waste. Proper storage of by-products is important in biorefining and dehydration is widely used in the herb sector, especially for highly perishable harvested flowers. This study aimed to deepen the phytochemical composition of dried saffron tepals and to investigate whether this was influenced by the extraction technique. In particular, the conventional maceration was compared with the Ultrasound Assisted Extraction (UAE), using different solvents (water and three methanol concentrations, i.e., 20%, 50%, and 80%). Compared to the spice, the dried saffron tepals showed a lower content of total phenolics (average value 1127.94 ± 32.34 mg GAE 100 g^{-1} DW) and anthocyanins (up to 413.30 ± 137.16 mg G3G 100 g^{-1} DW), but a higher antioxidant activity, which was measured through the FRAP, ABTS, and DPPH assays. The HPLC-DAD analysis detected some phenolic compounds (i.e., ferulic acid, isoquercitrin, and quercitrin) not previously found in fresh saffron tepals. Vitamin C, already discovered in the spice, was interestingly detected also in dried tepals. Regarding the extraction technique, in most cases, UAE with safer solvents (i.e., water or low percentage of methanol) showed results of phenolic compounds and vitamin C similar to maceration, allowing an improvement in extractions by halving the time. Thus, this study demonstrated that saffron tepals can be dried maintaining their quality and that green extractions can be adopted to obtain high yields of valuable antioxidant phytochemicals, meeting the requirement for a sustainable biorefining.

Keywords: dried tepals; total phenolic content; total anthocyanin content; antioxidant activity; vitamin C; ultrasound assisted extraction; biorefining

---

Citation: Stelluti, S.; Caser, M.; Demasi, S.; Scariot, V. Sustainable Processing of Floral Bio-Residues of Saffron (*Crocus sativus* L.) for Valuable Biorefinery Products. *Plants* **2021**, *10*, 523. https://doi.org/10.3390/plants10030523

Received: 1 February 2021
Accepted: 8 March 2021
Published: 11 March 2021

1. Introduction

Saffron (*Crocus sativus* L.), of the Iridaceae family, is a geophyte widely cultivated for its red-scarlet stigmas that, once dried, form the most expensive spice in the world ($40–50 g^{-1}, [1]). Saffron spice has nutritional and phytochemical compounds such as terpenes, phenols, and vitamin C, with antioxidant [2,3] and therapeutic [4,5] properties.

Saffron, being a triploid ($2n = 3x = 24$) sterile species, is propagated through underground clonal corms [6]. In Mediterranean climate regions flowering occurs for few weeks from early to late autumn and is mainly controlled by seasonal thermoperiodicity, soil water content, and corm size [7–9]. There are one or several flowers per saffron plant, even up to 12 [6]. Flowers have a perianth of six violet tepals, three stamens, and a style culminating in three stigmas [10]. About 110 to 300 flowers are needed to obtain 1g of spice [11,12] and tepals constitute the most abundant bio-residue (~80% of total flower mass [13]).

In Persian traditional medicine, saffron tepals are considered antispasmodic, stomachic, antitumor, antidepressant, and curative of anxiety [14]. They contain proteins, fibers, fats, ashes, minerals, polyphenols, and present high antioxidant activity [14,15]. To our knowledge, vitamin C has never been detected in saffron tepals so far.

Phenolic compounds are a class of secondary metabolites commonly found in plants in which play various biological activities, such as defense against biotic and abiotic stresses, UV filters, attraction for pollinators, fruit dispersion, plant growth, and allelopathy [16–20]. These molecules are the main source of plant antioxidants that, by quenching ROS and reactive nitrogen species (RNS) [21], control oxidative stress, which is associated with several age-related diseases such as cancers, inflammations, and neurodegenerative disorders. This class include phenolic acids and flavonoids [22].

Phenolic acids include hydroxybenzoic and hydroxycinnamic acids and are the main phenolic compounds produced by plants [17]. They can enhance the organoleptic, nutritional, and antioxidant properties of food and are ancestors for the bioactive compounds used in pharmaceutic-cosmetic and food industries [17].

Flavonoids comprise flavonols, flavan-3-ols (such as catechins), and anthocyanins, which are the water-soluble pigments responsible for the pink-orange, red, and blue color range of flowers and fruits [22,23]. Flavonoids have antioxidant, antifungal, antibacterial, and antiviral properties and are used in the agricultural, food, and pharmaceutical-cosmetic industries [23]. In particular, anthocyanins are authorized natural food colorants in Europe with the code E-163 regardless the plant source, being considered a group of harmless substances [24].

Vitamin C is ubiquitous in plants [25]. It regulates several physiological functions, such as photosynthesis, seed germination, floral inductions, and senescence. Ascorbate is the reduced and physiologically active form, while dehydro-ascorbate is the oxidized form. The regenerative nature makes ascorbate a powerful antioxidant molecule, involved in the enzymatic and non-enzymatic defense system against oxidative stress [26]. Vitamin C is an essential dietary element since it cannot be synthetized by humans and a severe lack of its intake leads to scurvy disease [25].

The selection of suitable solvents and techniques is a basic factor to get a high extraction yield of phytochemicals from plant material [27]. Water, organic solvents, and their combination are often used to extract phenolics and vitamin C [2,3,27–29]. Methanol is commonly chosen for phenolic compounds due to its similar polarity, small dimension, and low density [30]. In a previous work, Caser et al. [15] compared conventional maceration with ultrasound assisted extraction (UAE) on fresh saffron tepals using water and different concentrations of methanol (20%, 50%, and 80%) as solvents. Because harvested flowers are very perishable, dehydration is a common practice to storage and preserve their quality since it inhibits enzymatic activity and limits microbial contamination [31].

This study aimed to investigate the phytochemical composition and antioxidant activity of dried saffron tepals and select the more effective and sustainable extractions. Conventional maceration was compared to the modern UAE technique that, allowing the use of safer solvents and shorter time, can reduce the energy cost of the extraction process [31,32]. A comparison of the results with a previous work performed on fresh saffron bio-residues [15] was also possible. According to the Directive 2008/122/EC, avoid waste production and use it as a resource is encouraged in the food waste management and an eco-sustainable processing is required in biorefining. As dried saffron tepals can have food, pharmaceutical, and cosmetic applications [13], investing in this by-product may be a promising approach to minimize losses [13] and potentially increase further the economic value of this crop [32].

2. Results

2.1. Qualitative and Antioxidant Properties of Dried Tepals Extracts

Total phenolic content (TPC), total anthocyanin content (TAC), and antioxidant activity by means of three assays, i.e., FRAP, ABTS, and DPPH, were evaluated for all phytoextracts (Table 1). Statistical comparisons between solvents for both extraction techniques separately were provided in Supplementary Table S1.

Table 1. Total phenolic content (TPC), total anthocyanin content (TAC), and antioxidant activity measured with the FRAP, ABTS, and DPPH assays, in dried tepal extracts obtained through maceration (M) and Ultrasound Assisted Extraction (UAE) techniques, and the solvents water or methanol at three concentrations (20%—Met20, 50%—Met50, and 80%—Met80).

Extraction		TPC (mgGAE 100 g^{-1} DW)	TAC (mgG3G 100 g^{-1} DW)	FRAP (mmolFe^{2+} Kg^{-1} DW)	ABTS (µmolTE g^{-1} DW)	DPPH (µmolTE g^{-1} DW)
M	Water	1142.27 ± 43.52	345.04 ± 132.47 a,b	571.54 ± 3.21 a	13.82 ± 0.72	15.56 ± 2.29
M	Met20	1123.53 ± 59.86	268.13 ± 26.76 a,b	506.73 ± 13.85 b,c	14.20 ± 0.60	17.83 ± 2.46
M	Met50	1106.45 ± 9.17	300.39 ± 15.02 a,b	535.83 ± 10.30 a,b	14.62 ± 0.29	24.52 ± 2.55
M	Met80	1166.96 ± 33.15	249.13 ± 11.97 a,b	511.72 ± 22.49 a,b,c	14.29 ± 0.32	24.17 ± 1.53
UAE	Water	1150.63 ± 11.23	413.30 ± 137.16 a	556.90 ± 11.91 a,b	12.76 ± 0.81	23.55 ± 3.60
UAE	Met20	1113.27 ± 46.11	178.39 ± 34.03 b	460.05 ± 35.55 c	13.39 ± 1.46	19.35 ± 4.83
UAE	Met50	1066.89 ± 26.36	277.09 ± 49.06 a,b	506.68 ± 21.80 b,c	15.10 ± 0.38	24.58 ± 1.46
UAE	Met80	1153.49 ± 22.74	231.70 ± 30.19 a,b	513.67 ± 21.12 a,b,c	14.34 ± 0.81	21.03 ± 1.81
p		ns	0.01413 *	0.0002608 ***	ns	ns

Values of mean and standard deviation are reported for each variable. Statistical comparisons were performed using ANOVA for TPC, TAC, and FRAP or the non-parametric Kruskal–Wallis test for DPPH and ABTS ($p > 0.05$ in Levene's test). Values with the same letter are not statistically different at $p < 0.05$, according to Tukey's post-hoc test or Dunn's post-hoc test; * $p < 0.05$; *** $p < 0.001$; ns = not significant.

TAC and FRAP were significantly affected by the extractions. TPC showed an average value of 1127.94 ± 32.34 mg GAE 100 g^{-1} DW. TAC varied significantly among the different extractions, ranging from 178.39 ± 34.03 mg G3G 100 g^{-1} DW using UAE with 20% methanol to 413.30 ± 137.16 mg G3G 100 g^{-1} DW using UAE with water, with an average value of 282.90 ± 71.82 mg G3G 100 g^{-1} DW.

The assays revealed that all extracts exhibited antioxidant activity. FRAP showed significant differences ranging from 460.05 ± 35.55 mmol Fe^{2+} Kg^{-1} DW for UAE with 20% methanol to 571.54 ± 3.21 mmol Fe^{2+} Kg^{-1} DW for maceration with water. ABTS and DPPH had an average value of 14.07 ± 0.73 µmol TE g^{-1} DW and 21.32 ± 3.45 µmol TE g^{-1} DW, respectively.

2.2. Correlation Analysis

The correlation between TPC, TAC, and the antioxidant activity of the phytoextracts assessed with the FRAP, ABTS, and DPPH assays were statistically evaluated (Table 2).

Table 2. Correlation analysis between total phenolic content (TPC), total anthocyanin content (TAC), and antioxidant activity measured with the assays FRAP, ABTS, and DPPH, as regards maceration (M) and UAE. Pearson or Kendall correlation coefficients are reported for *p*-values < 0.05.

	TPC	TAC	FRAP	ABTS
M				
FRAP	ns	ns	/	
ABTS	ns	ns	ns	/
DPPH	ns	ns	ns	ns
UAE				
FRAP	ns	0.86 ***	/	ns
ABTS	−0.69 *	ns	ns	/
DPPH	−0.45 *	ns	ns	ns

p-values < 0.05 show statistically significant correlations (* $p < 0.05$; *** $p < 0.001$; ns = not significant).

Significant correlations were only found in the case of UAE, between TPC and both ABTS (−0.69) and DPPH (−0.45), and between TAC and FRAP (0.86).

2.3. Bioactive Compounds from HPLC Analysis

The bioactive compounds extracted were as follows: ferulic acid (cinnamic acid); ellagic acid (benzoic acid); hyperoside, isoquercitrin, quercitrin, and rutin (flavonols); epicatechin (catechin); and vitamin C (Table 3).

Table 3. Extraction yield (mg 100 g^{-1} DW) of the compounds obtained from dried saffron tepals expressed as mg 100 g^{-1} of dried weight (DW), using the maceration (M) and Ultrasound Assisted Extraction (UAE) techniques and the solvents water or methanol at three concentrations (20%, Met20; 50%, Met50; 80%, Met80; v:v). Quantifications were obtained through the HPLC-DAD analysis.

Extractions		Cinnamic Acids	Benzoic Acids	Flavonols				Catechins	Vitamin C
		Ferulic Acid	Ellagic Acid	Hyperoside	Isoquercitrin	Quercitrin	Rutin	Epicatechin	
M	Water	0.00 ± 0.00 b	7.67 ± 3.69 a,b,c	4.35 ± 1.04 c	0.31 ± 0.22 c,d	0.00 ± 0.00 b	8.52 ± 3.91 c	0.00 ± 0.00 b	29.61 ± 6.05 a
M	Met20	1.83 ± 0.31 a	4.43 ± 4.15 c,d	5.61 ± 0.52 a,b,c	0.22 ± 0.12 c,d	6.33 ± 5.27 a	0.32 ± 0.31 d	0.00 ± 0.00 b	33.72 ± 0.89 a
M	Met50	9.65 ± 2.62 a	0.00 ± 0.00 e	5.85 ± 4.31 b,c	4.36 ± 3.49 a,b,c	9.27 ± 3.47 a	0.00 ± 0.00 d	0.00 ± 0.00 b	0.00 ± 0.00 b
M	Met80	0.00 ± 0.00 b	1.32 ± 0.33 d,e	23.93 ± 15.51 a,b,c	7.82 ± 3.09 a	6.53 ± 0.29 a	37.61 ± 2.22 a	0.00 ± 0.00 b	0.00 ± 0.00 b
UAE	Water	0.00 ± 0.00 b	8.53 ± 8.45 b,c,d	11.58 ± 4.09 a,b,c	0.00 ± 0.00 d	0.00 ± 0.00 b	28.24 ± 4.83 a,b	0.00 ± 0.00 b	26.68 ± 4.71 a
UAE	Met20	0.00 ± 0.00 b	26.74 ± 10.80 a,b	9.68 ± 6.77 a,b,c	6.46 ± 5.03 a,b	0.00 ± 0.00 b	13.46 ± 10.25 b,c	16.62 ± 15.89 a	29.17 ± 2.31 a
UAE	Met50	0.00 ± 0.00 b	28.39 ± 4.32 a	27.26 ± 4.29 a	5.57 ± 1.90 a,b,c	7.07 ± 5.12 a	7.24 ± 1.35 c	0.00 ± 0.00 b	0.00 ± 0.00 b
UAE	Met80	0.00 ± 0.00 b	23.51 ± 5.11 a,b	24.77 ± 2.25 a,b	0.00 ± 0.00 d	0.00 ± 0.00 b	9.10 ± 2.17 c	4.22 ± 2.90 a	0.00 ± 0.00 b
p		0.001802 **	2.235×10^{-7} ***	0.004662 **	0.005466 **	0.005407 **	1.452×10^{-10} ***	0.001995 **	0.003143 **

Values of mean ± standard deviation are reported. Statistical comparisons were performed using ANOVA (for ellagic acid, hyperoside, and rutin) or Kruskal–Wallis test (for the other compounds, $p < 0.05$ in Shapiro–Wilk's test). Letters indicate statistical differences between the different extractions for each extracted compound. Values with the same letter are not statistically different at $p < 0.05$, according to Tukey's or Dunn's post-hoc test. ** $p < 0.01$; *** $p < 0.001$; ns = not significant.

Overall, the yield of phytochemicals was significantly affected by the different extractions ($p < 0.05$).

As regards ferulic acid, 9.65 ± 2.62 mg 100 g^{-1} DW was extracted through maceration with methanol at 50%, showing no significant differences to 20%. It was not obtained by the other extractions.

The yield of ellagic acid ranged from 1.32 ± 0.33 mg 100 g^{-1} DW for maceration with 80% methanol to 28.39 ± 4.32 mg 100 g^{-1} DW for UAE with 50% methanol, which was not significantly different to maceration with water and UAE with both 20% and 80% methanol. No extraction was performed by maceration with 50% methanol.

Hyperoside was obtained by all extractions. Its values varied from 4.35 ± 1.04 mg 100 g^{-1} DW regarding maceration with water to 27.26 ± 4.29 mg 100 g^{-1} DW for UAE with 50% methanol, which was not significantly different to the other methanol concentrations and maceration with both 20% and 80% methanol.

Isoquercitrin ranged from 0.22 ± 0.12 mg 100 g^{-1} DW for maceration with 20% methanol to 7.82 ± 3.09 mg 100 g^{-1} DW as regards maceration with 80% methanol, that not significantly differed to maceration with 80% methanol and UAE with both 20% and 50% methanol. It was not extracted by UAE with both water and 80% methanol.

Quercitrin was only achieved by means of maceration with all methanol levels and UAE with 50% methanol. Maceration with 50% methanol gave 9.27 ± 3.47 mg 100 g^{-1} DW, without significant differences with the other extractions.

The yield of rutin varied from 0.32 ± 0.31 mg 100 g^{-1} DW as regards maceration with 20% methanol to 37.61 ± 2.22 mg 100 g^{-1} DW for maceration with 80% methanol, which was not significantly different to UAE with water.

The value of epicatechin resulted 16.62 ± 15.89 for UAE with 20% methanol, without significant differences to UAE with 80% methanol. It was not attained by all other extractions.

Regarding vitamin C, 33.72 ± 0.89 mg 100 g^{-1} DW was gain through maceration with 20% methanol without significant differences to maceration with water and UAE with both water and 20% methanol. No extraction was achieved by all other extractions.

The phytochemical profile of the extracts depended on the extractions (Table 2). For example, all the molecules except epicatechin were obtained using maceration with 20% methanol, while UAE with 20% methanol extracted all the chemicals except ferulic acid and quercitrin. However, no specific method–solvent combination performed a significant higher yield of these compounds altogether (Figure 1).

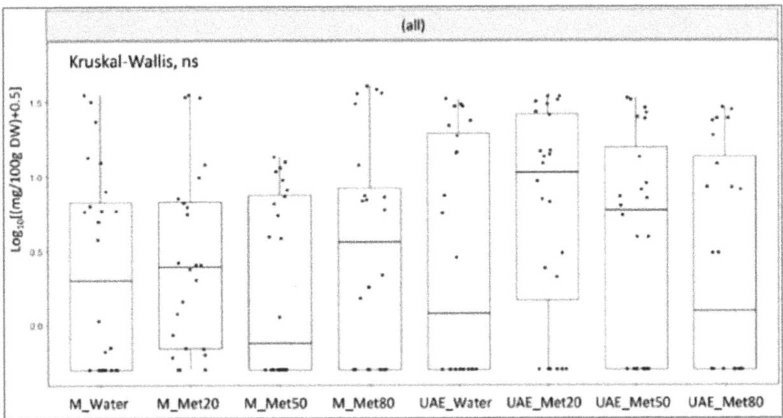

Figure 1. Effects of the extraction methods maceration (M) and Ultrasound Assisted Extraction (UAE) with the solvents water (M Water; UAE Water) and three concentrations of methanol in water (20%, Met20; 50%, Met50; 80% Met80; v:v) on the yield of all the compounds extracted and analyzed with HPLC-DAD. The extraction yield (mg 100 g^{-1} of dried weight, DW) is expressed as log transformation. Statistical comparisons were performed using the Kruskal–Wallis test; ns = not significant.

3. Discussion

Conventional maceration and UAE are the most common extraction methods employed to analyze the antioxidant molecules of saffron tepals (Table 4). Maceration requires a high consumption of organic solvents and long extraction times [33]. UAE is a modern and green method based on ultrasonic waves that, by increasing the permeability of plant cells through acoustic cavitation of the wall, improve the penetration of the solvent and the release of phytocompounds. UAE can allow the use of non-toxic solvents, to work at room temperature, shorter extraction times compared to maceration [31,34–36], which needs the addition of heat to reduce the times [32], and a lower energy cost [32].

Table 4. Some representative studies of extraction methods and solvents to extract bioactive compounds from saffron tepals in different states.

Tepals State	Extraction Methods	Solvents	Bioactive Compounds	References
Fresh	Maceration, UAE	Water and methanol at 20%, 50% and 80%	Phenolic content (ellagic acid; hyperoside; rutin; epicatechin); Anthocyanin content	[15]
Dried	Soxhlet extraction	Hexane; dichloromethane; ethanol	Phenolic and Flavonoid content	[37]
Dried	Maceration	Ethanol 25%, 50%, 75%	Anthocyanin content	[38]
Dried	Maceration	Ethyl acetate; methanol	Phenolic and Flavonoid content	[39]
Dried	Maceration	Ethanol 70%	Phenolic content (kaempferol 3-O-sophoroside-7-O-glycoside, quercetin 3,4-di-O-glycoside, kaempferol di-glycoside, kaempferol 3-O-glycoside)	[40]
Dried	Maceration	Methanol	Phenolic and Flavonoid content	[41]
Dried	Maceration	Acidified (HCl) ethanol; sulfur water solution	Phenolic and Anthocyanin content (cyanidin 3,5-diglucosides; pelargonidin 3- and 5-glucosides; delphinidin di-glucosides; pelargonidin 3-glucosides; petunidin)	[42]

Table 4. Cont.

Tepals State	Extraction Methods	Solvents	Bioactive Compounds	References
Dried	Maceration, Enzyme-assisted extraction	Acidified (HCl) ethanol; Enzymatic water (Pectinex)	Antocyanin content (cyanidin 3,5-diglycosides; pelargonidin 3- and 3,5-glycosides; delphinidin 3-glycosides; petunidin)	[43]
Freeze-dried	Maceration, UAE	Acidified (HCl) methanol; Acidified (HCl) ethanol 70%; Methanol 98% in formic acid	Phenolic, Flavonoid, Anthocyanin, and Crocetin esters content	[44]
Dried	Maceration, UAE	Acidified (HCl) deuterated methanol; Trifluoroacetic acid in acetonitrile 50%	Flavonoids (kaempferol di-hexoside; kaempferol 3-O-glucoside; kaempferol 3,4′-di-O-glucoside; kaempferol 3-O-β-sophoroside; kaempferol tri-hexoside; kaempferol di-hexosides; quercetin 3,4′-di-O-glucoside; isorhamnetin 3,4′-di-O-glucoside); Anthocyanins (delphinidin 3,5-di-O-β-glucoside; delphinidin 3-O-glucoside; petunidin 3,5-di-O-β-glucoside; petunidin 3-O-glucoside)	[45]
Dried	UAE	Ethanol 50%	Flavonoids (quercetin 3-O-sophoroside; kaempferol 3-O-sophoroside; kaempferol 3-O-glucoside)	[46]
Freeze-dried	Maceration, UAE	Methanol 50%; n-hexane; Methanol/KOH 20%	Flavonoid content (kaempferol 3-O-sophoroside-7-O-glucoside; kaempferol 3,7-O-diglucoside; quercetin 3,7-O-diglucoside; isorhamnetin 3,7-O-diglucoside; kaempferol 3-O-sophoroside; isorhamnetin 3-O-sophoroside; quercetin 3-O-glucoside; kaempferol 3-O-rutinoside; isorhamnetin 3-O-rutinoside; kaempferol 3-O-glucoside; kaempferol 3-O-(6??- acetyl-glycoside)-7-O-glycoside; isorhamnetin 3-O-glucoside; kaempferol 3-O-sophoroside-7-O-rhamnoside; kaempferol 3-O-(6??-acetyl-galactoside) or 3-O-(6??-acetyl- glucoside); kaempferol 3-O-(6??- acetyl-galactoside) or 3-O-(6??-acetyl- glucoside); quercetin 3-O-glucoside-7-O-rhamnoside; isorhamnetin 3-O-glucoside-7-O-rhamnoside; kaempferol 3-O-glucoside-7-O-rhamnoside); Anthocyanin content (delphinidin 3,7-O-diglucoside; petunidin 3,7-O-diglucoside; delphinidin 3-O-glucoside; petunidin 3-O-glucoside; malvidin O-glucoside); Lutein diesters	[32]
Dried	Microwave-assisted extraction	Acidified (HCl) ethanol	Anthocyanin content	[46]

All the extracts of dried saffron tepals presented phenolic compounds (Table 1). As showed in Table 4, other authors reported their presence in dried and fresh tepals.

The total phenolic content (TPC) was not affected by the extractions (Table 1), meaning that UAE had similar performances to maceration in half the time. This result was consistent with the absence of a better extraction to obtain all the phytochemicals analyzed with the HPLC (Figure 1). Using extraction conditions similar to this study Caser et al. [15] found that for fresh saffron tepals 80% methanol gave higher TPC. Thus, dehydrated tepals may allow the use of more ecological solvents for the extraction of phenolic compounds. UAE with water could be selected to obtain non-toxic phytoextracts to be added to foods [13].

The technique and the solvent used can influence the extraction performance, but other factors can also affect the concentration of phytocompounds for saffron, such as provenance [47], conditions of growth [2,3], and storage [48]. Compared to the average value of TPC (1127.94 ± 32.34 mg GAE 100 g^{-1} DW), Lahmass et al. [39] attained higher TPC from saffron dried tepals using maceration with both pure methanol (~9400 mg GAE 100 g^{-1} DW) and ethyl acetate (~4200 mg GAE 100 g^{-1} DW), while a lower TPC (~342 mg GAE 100 g^{-1} DW) was found by Amir et al. [49] using UAE with pure methanol. The values of TPC were lower compared to those found in saffron spice (1819.5 mg GAE 100 g^{-1} DW) by Caser et al. [2], but higher than those obtained from some traditional Chinese medicinal plants [50], such as *Lycopus lucidus* (~930 mg GAE 100 g^{-1} DW for methanolic extract of aerial parts), *Smilax glabra* (~740 mg GAE 100 g^{-1} DW for methanolic extract of root), *Plantago asiatica* (~850 mg GAE 100 g^{-1} DW for aqueous extract of seeds), *Lobelia chinensis* (~600 mg GAE 100 g^{-1} DW for aqueous extract of whole plant), *Lithospermum erythrorhizon* (~920 mg GAE 100 g^{-1} DW for root extract), *Dianthus superbus* (~690 mg GAE 100 g^{-1} DW for aqueous extract of aerial parts), *Curcuma longa* (~800 mg GAE 100 g^{-1} DW for rhizome methanolic extract), and *Zingiber officinale* (~740 mg GAE 100 g^{-1} DW for rhizome methanolic extract), as well as from some common vegetables and fruits [50], such as spinach (*Spinacia oleracea*, ~900 mg GAE 100 g^{-1} DW for methanolic extract) and lettuce (*Lactuca sativa*, ~780 mg GAE 100 g^{-1} DW for methanolic extract).

The total anthocyanins content (TAC) had an average value of 282.90 ± 71.82 mg G3G 100 g^{-1} DW (Table 1). Even in the case of anthocyanins, UAE achieved a yield similar to maceration in half the extraction time. A higher result was seen for UAE with water (413.30 ± 137.16 mg G3G 100 g^{-1} DW), showing a percentage variation from the average yield of +46%. This is in contrast with the results obtained from fresh saffron tepals [15], for which 80% methanol performed better meaning that even in the case of anthocyanins dehydration may allow the use of greener solvents. The values of TAC were lower than those obtained from saffron tepals by Lotfi et al. [43], which get 507.5 mg G3G 100 g^{-1} DW using maceration with acidified ethanol. TAC was also lower than that found in saffron spice (1867.3 mg G3G 100 g^{-1} DW) by Caser et al. [2], but greater compared to the contents of some highly pigmented vegetables [51], such as purple cauliflower (*Brassica oleracea* var. botrytis, 201 mg G3G 100g^{-1} DW) and red cabbage (*Brassica oleracea* var. capitata f. rubra, 199 mg G3G 100 g^{-1} DW), achieved through maceration with 0.1% HCl (*v/v*) in 80% methanol. Grape (*Vitis vinifera*) is one of the most common sources of these molecules [52] having a TAC ranging from 75.64 to 414.95 mg G3G 100 g^{-1} DW depending on the cultivars [53], which is comparable in its maximum values to that of our study. Since plant waste are usually used for the commercial preparation of these pigments [54], saffron tepals in both dried and fresh [15] forms could be considered a good candidate for their extraction.

All assays evaluated, namely, FRAP, ABTS, and DPPH, revealed the presence of antioxidant activity in the extracts with average values of 520.39 ± 34.48 mmol Fe^{2+} Kg^{-1} DW for FRAP, 14.07 ± 0.73 μmol TE g^{-1} DW for ABTS, and 21.32 ± 3.45 μmol TE g^{-1} DW for DPPH (Table 1). Maceration with water obtained higher antioxidant activity measured with the FRAP method, which was not significantly different to UAE with water. This was in tune with that found by Caser et al. [2] for fresh saffron tepals. Interestingly, the antioxidant activity was higher compared to that found in saffron spice by Caser et al. [2] (338.2 mmol Fe^{2+} Kg^{-1} DW for FRAP and 4.6 μmol TE g^{-1} DW for ABTS) and Caser et al. [15] (4.64 ± 0.50 μmol TE g^{-1} DW for ABTS). The ABTS values were greater than those achieved from some traditional Chinese medicinal plants [50], such as *Chrysanthemum indicum* (3.03 μmol TE g^{-1} DW for methanolic extract of inflorescence), *Artemisia annua* (6.29 μmol TE g^{-1} DW for aqueous extract of aerial parts), *Campsis radicans* (4.462 μmol TE g^{-1} DW for methanolic extract of flower), and also compared to extracts obtained from some common vegetables and fruits [50], such as wild cabbage (*Brassica oleracea*, 1.01 μmol TE g^{-1} DW) and tomato (*Solanum lycopersicum*, 1.49 μmol TE g^{-1} DW). The DPPH values were similar to those found in mengkudu (*Morinda citrifolia*, up to ~23 μmol TE g^{-1} DW) by Thoo et al. [55].

The correlations among TPC, TAC, and the antioxidant activity were assessed (Table 2). Significant correlations were only found in the case of UAE. Even though the FRAP, ABTS, and DPPH assays have been extensively applied, they do have some limitations which can cause an imprecise estimation [56]. Since the antioxidant activity of plant extracts mainly depends on the content of phenolic compounds, it was expected a strong positive correlation among TPC and the assays [50,56–58]. However, TPC was either not correlated (FRAP) or negatively correlated (−0.69 ABTS and −0.45 DPPH) with the antioxidant capacity. No significant correlations among TPC and antioxidant activity was already found in extracts [59]. The negative correlation, already reported by other authors [59,60], suggests they might have opposite behavior when subjected to different solvents. The antioxidant capacity of samples may be influenced by extraction technique, solvent, pH, and metal ions [60]. Furthermore, synergistic and antagonistic interactions among the antioxidants in the extracts may interfere with the correlation, or there can be some non-phenolic molecules which can react with the Folin–Ciocalteu reagent without being free radical scavengers [59].

TAC was positively correlated with FRAP (0.86) (Table 2), indicating that anthocyanins contributed more to antioxidant capacity than TPC.

Positive correlations among the antioxidant assays were assumed [56,58], since they are all based on the same principle of electron transfer [56]. However, other authors reported the absence of significant correlation [55,59]. This might be explained by the fact that plants extracts can contain antioxidants which may react in different ways [59].

The heterogeneity of the yields achieved for each molecule (Table 3) would explain the absence of a single better extraction for these compounds altogether (Figure 1). Overall, except for ferulic acid and quercitrin, UAE get results higher (for ellagic acid and epicatechin) or comparable (for vitamin C) to maceration using greener solvents, i.e., water (for hyperoside and rutin) or lower methanol concentrations (20% methanol for isoquercitrin) in half the extraction time (Table 3). A unique ideal extraction does not exist, but the choice should fall on that most environmentally and economically convenient.

The bioactive compounds here identified and quantified (Table 3) deepen the knowledge on the molecular composition of dried saffron tepals. The extracts showed the same molecules (ellagic acid, hyperoside, rutin, and epicatechin) as those found in fresh tepals by Caser et al. [15] using an analysis protocol similar to this study. In addition, in the dried tepals were also detected ferulic acid, isoquercitrin, quercitrin, and vitamin C. In Caser et al. [15] the absence in fresh tepals of the chemicals here obtained might be attributed to the drying procedure [61,62], which can facilitate extractions by disrupting cell walls and causing the formation of cavities and intercellular spaces [63]. Vitamin C and rutin (yield up to 37.61 ± 2.22 mg 100 g^{-1} DW) showed the highest yields among all the compounds analyzed with HPLC. Vitamin C (up to 33.72 ± 0.89 mg 100 g^{-1} DW) was lower than in the spice (73.2 mg 100 g^{-1}; Caser et al. [15]), but higher or comparable than in some freeze-dried vegetables and fruits [64], such as cabbage (*Brassica oleracea* var. capitata, up to 28.3 mg 100 g^{-1}), tomato (*Solanum lycopersicum*, up to 10.1 mg 100 g^{-1}), apple Fuji (*Malus domestica* 'Fuji', up to 3.5 mg 100 g^{-1}), mango (*Mangifera indica*, up to 18.6 mg 100 g^{-1}), pineapple (*Ananas comosus*, 5.0 mg 100 g^{-1}), and pomelo (*Citrus maxima*, up to 31.7 mg 100 g^{-1}).

4. Materials and Methods

4.1. Plant Material

Saffron tepals were provided by the company "Lo Zafferano del Monviso" located in Martiniana Po, CN (Italy—44°23′ N 7°33′ E). Saffron plants were cultivated in open field selecting corms with horizontal diameters of 2.5 to 3.5 cm, which were planted in August 2018. The flowers were harvested in October—November 2018 and immediately air-dried. In the laboratory of the Department of Agricultural, Forest, and Food Sciences (DISAFA) of the University of Turin (Italy), the dried tepals were grinded in liquid nitrogen and stored at $-80\,°C$ until use.

4.2. Extraction Methods

Two extraction procedures, i.e., maceration and UAE, with deionized water or 20%, 50%, and 80% methanol in deionized water (v:v) as solvents were applied. The extractions were conducted at room temperature (ca. 21 °C) and using a powdered sample–solvent ratio of 1:50 g ml^{-1}.

As regards maceration, the samples were soaked in each solvent type and kept into a glass tube under stirring (1000 rpm) in the dark for 30 min, whereas for UAE the sample tubes were inserted into an ultrasonic extractor (Sarl Reus, Drap, France) using a frequency of 23 kHz for 15 min. The extracted solutions were all filtered with one-layer of filter paper (Whatman No. 1, Maidstone, UK) and then with a 0.45 μm PVDF syringe filter (CPS Analitica, Milano, Italy). The extracts were then stored at $-20\,^{\circ}C$ for analyses. All the extractions were carried out in triplicate for each solvent and method used.

4.3. Spectrophotometric Analysis

All the analysis were conducted in three replicates and performed as reported by Caser et al. [15].

4.3.1. Total Phenolic Content

Total phenolic content was estimated using the Folin-Ciocalteu method. In each plastic tube 200 μL of each phytoextract were mixed with 1000 μL of diluted (1:10) Folin–Ciocalteu reagent. After 10 min of incubation at dark and room temperature, 800 μL of Na_2CO_3 7.5% (*w/v*) were added to each tube. The samples were incubated at dark and room temperature for 30 min. The absorbance at 765 nm was measured by means of a UV–Vis spectrophotometer (Cary 60 UV-Vis, Agilent Technologies, Santa Clara, CA, USA). The results were expressed as mg of gallic acid equivalents (GAE) per 100 g of dry weight (mg GAE $100\,g^{-1}$ DW).

4.3.2. Total Anthocyanin Content

Total anthocyanins content was measured using the pH-differential method. Buffer solution at pH 1 (4.026 g KCl + 12.45 mL HCl 37% in a 1 L water volume) was added to 500 μL of phytoextract reaching 5 mL in each flask. The same was made in a second flask using a buffer solution at pH 4.5 (32.82 g $C_2H_3NaO_2$ + 18 mL $C_2H_4O_2$ in a 1 L water volume). The samples were left in the dark at room temperature for 20 min. The absorbance of both flasks was read at 515 and 700 nm at a UV–Vis spectrophotometer (Cary 60 UV-Vis, Agilent Technologies, Santa Clara, CA, USA). The total anthocyanin content was calculated using the following formula:

$$[(A \times sample\ dilution\ factor \times 1000)/(molar\ absorptivity \times 1)]$$

where A is [(Absorbance 515 nm $-$ Absorbance 700 nm) at pH 1] $-$ [(Absorbance 515 nm $-$ Absorbance 700 nm) at pH 4.5]. The results were expressed as milligrams of cyanidin 3-O-glucoside (C3G) per 100 g of dry weight (mg C3G $100\,g^{-1}$ DW).

4.3.3. Antioxidant Activity

The antioxidant activity was determined using the following methods:

- Ferric ion reducing antioxidant power (FRAP) method. The FRAP solution was produced by mixing a buffer solution at pH 3.6 ($C_2H_3NaO_2$ + $C_2H_4O_2$ in water), 2,4,6-tripyridyltriazine (TPTZ, 10 mM in HCl 40 mM), and $FeCl_3 \cdot 6H_2O$ (20 mM). Then, 90 μL of deionized water and 900 μL of the FRAP reagent were added to 30 μL of phytoextract in each plastic tube. The samples were left at 37 $^{\circ}C$ for 30 min and the absorbance was read at 595 nm at a spectrophotometer (Cary 60 UV-Vis, Agilent Technologies, Santa Clara, CA, USA). Results were expressed as millimoles of ferrous iron equivalents per kilogram of dry weight (mmol Fe^{2+} Kg^{-1} DW).

- The 2,2′-azinobis (3-ethylbenzothiazoline-6-sulfonic acid) (ABTS) method. The ABTS radical cation (ABTS·) was obtained by the reaction of 7.0 mM ABTS solution with 2.45 mM $K_2S_2O_8$ solution. The solution was incubated for 12–16 h in the dark at room temperature and then diluted with distilled water up to read an absorbance of 0.70 ($\pm$0.02) at 734 nm. 500 μL of diluted ABTS· was added to 15 μL of phytoextract and, after incubation in the dark at room temperature for 10 min, the absorbance was measured at 734 nm by means of a spectrophotometer (Cary 60 UV-Vis, Agilent

Technologies, Santa Clara, CA, USA). The ABTS radical-scavenging activity was calculated as

$$[(Abs0 - Abs1/Abs0) \times 100]$$

where *Abs0* is the absorbance of the control (solution without phytoextract) and *Abs1* is the absorbance of the sample. The results were expressed as μmol of Trolox equivalents per gram of dry weight (μmol TE g^{-1} DW).

- The 2,2-diphenyl-1-picrylhydrazyl (DPPH) radical scavenging method. To obtain 100 μM of DPPH radical cation (DPPH·) 2 mg of DPPH were mixed up with 50 mL of MeOH, up to have an absorbance of 1.000 (±0.005) at 515 nm. Then, 1.5 mL of diluted DPPH· was added to 20 μL of phytoextract and the reaction was left in the dark at room temperature for 30 min. The absorbance was read at 515 nm at a spectrophotometer (Cary 60 UV-Vis, Agilent Technologies, Santa Clara, CA, USA). The DPPH radical-scavenging activity was calculated as

$$[(Abs0 - Abs1/Abs0) \times 100]$$

where *Abs0* is the absorbance of the control (solution without phytoextract) and *Abs1* is the absorbance of the sample. Results were expressed as μmol of Trolox equivalents per gram of dry weight (μmol TE g^{-1} DW).

4.4. HPLC Analysis

Qualitative and quantitative analyses of the phytoextracts were carried out using an Agilent 1200 High-Performance Liquid Chromatography coupled with an Agilent UV–Vis diode array detector (Agilent Technologies, Santa Clara, CA, USA). The chromatographic separation was made with a Kinetex C18 column (4.6 × 150 mm^2, 5 μm, Phenomenex, Torrance, CA, USA) using several mobile phases and recording UV spectra at different wavelengths [65], as described in Table 5.

Table 5. HPLC methods and conditions.

Methods	Classes of Interest	Stationary Phase	Mobile Phase	Wavelength (nm)
A	Cinnamic acids, Flavonols	KINETEX—C18 column (4.6 × 150 mm^2, 5 μm)	A: 10 mM KH_2PO_4/H_3PO_4, pH = 2.8 B: CH_3CN	330
B	Benzoic acids, catechins	KINETEX—C18 column (4.6 × 150 mm^2, 5 μm)	A: $H_2O/CH_3OH/HCOOH$ (5:95:0.1 *v/v/v*), pH = 2.5 B: $CH_3OH/HCOOH$ (100:0.1 *v/v*)	280
C	Vitamin C	KINETEX—C18 column (4.6 × 150 mm^2, 5 μm)	A: 5 mM $C_{16}H_{33}N(CH_3Br/50$ mM KH_2PO_4, pH = 2.5 B: CH_3OH	261, 348

Elution conditions. Method A, gradient analysis: 5% B to 21% B in 17 min + 21% B in 3 min (2 min conditioning time); flow: 1.5 mL min^{-1}; Method B, gradient analysis: 3% B to 85% B in 22 min + 85% B in 1 min (2 min conditioning time); flow: 0.6 mL min^{-1}; Method C, isocratic analysis: ratio of phase A and B: 95:5 in 10 min (5 min conditioning time); flow: 0.9 mL min^{-1}.

Each compound was determined comparing the retention times and UV spectra with the standards under the same chromatographic conditions as reported by Donno et al. [57]. The standards, purchased from Sigma-Aldrich (Saint Louis, MO, USA), were the following: flavonols (hyperoside, isoquercitrin, quercetin, quercitrin, and rutin), catechins (catechin and epicatechin), benzoic acids (ellagic and gallic acids), cinnamic acids (caffeic, chlorogenic, coumaric, and ferulic acids), and vitamin C (ascorbic and dehydroascorbic acids). All the analyses were performed in three replicates.

4.5. Statistical Analysis

All data were log transformed before the statistical analysis.

As regards the HPLC-DAD analysis, the limit of quantitation (LOQ) [57] was added when the treatments had data for only one or two replicates out of three, while the value 0 was added for the treatments without data.

The data were analyzed with the R-studio software to identify the statistically supported differences between the different extractions. Significant mean differences were verified with one-way ANOVA ($p < 0.05$) and Tukey's post-hoc test after checking the data for normality and homoscedasticity through Shapiro–Wilk's test ($p > 0.05$) and Levene's test ($p < 0.05$), respectively. The compounds non respecting the ANOVA assumptions were analyzed with Kruskal–Wallis non-parametric test ($p < 0.05$) and the post-hoc Dunn's comparison test.

As regards the data resulting from the spectrophotometric analysis, statistical comparisons were performed using ANOVA for TPC, TAC, and FRAP and Kruskal–Wallis test for DPPH and ABTS ($p > 0.05$ in Levene's test). Regarding the data resulting from the HPLC analysis, ANOVA was made for ellagic acid, hyperoside, and rutin, while Kruskal–Wallis test for the other compounds ($p < 0.05$ in Shapiro–Wilk's test).

A correlation analysis between the content of total phenolics (TPC) and anthocyanins (TAC), and the assays FRAP, ABTS, and DPPH used to measure the antioxidant activity of the extracts was made with the R-studio software. The Pearson correlation was adopted when the variables were normally distributed according to the Shapiro–Wilk's test ($p > 0.05$), otherwise the non-parametric Kendall correlation.

5. Conclusions

Processing food wastes and by-products to generate high-value products for industrial application has been attractive great economic and scientific interest [66,67]. The sustainable processing of biomass, including that resulting from agriculture, into valuable and safe biobased products falls within the definition of biorefining [68]. The use of renewable resources can help farmers to expand their sources of income and create new job opportunities while reducing bio-waste and the environmental footprint in a circular bioeconomy perspective [69].

The economic use of a by-product is related to its intrinsic, nutritional, or utility value. The dried saffron tepals are largely produced after stigma are separated from the flowers and are a natural source of antioxidant compounds. This study demonstrated that the phytochemical composition and the antioxidant capacity of extracts of dried saffron tepals is influenced by the extraction technique. In most cases, UAE with water or low percentage of methanol showed results of phenolic compounds and vitamin C similar to maceration, allowing an improvement in extractions by halving the time. Compared to the spice, the dried saffron tepals showed a lower content of total phenolics and anthocyanins, but a higher antioxidant activity.

To produce waste-derived products, bio-waste shall either be separated and recycled at source or collected separately before undergoing treatments, which gives storage an important role (Directive 2008/122/EC). The cost of biomass and its storage are some essential factors for the economic viability of biorefineries [67]. The drying technique has always been widely used in the herbs sector to stabilize and preserve plant material, and in particular flowers, which have high perishability due to the elevated moisture content [31]. The drying treatment can preserve tepals until their possible use in food or pharmacological-cosmetic industries, for example in the preparation of soaps and cosmetic products where they can act as antioxidants or colorants in place of synthetic excipients, meeting the demand of the cosmetic sector to use products of natural origin [70].

This study allowed to compare the results with those previously attained for fresh tepals by Caser et al. [15], which performed a similar extraction protocol. This work has highlighted that, after drying, saffron tepals retain bioactive phenolics. The dehydration procedure may improve extractions as safer solvents gave higher yield of phenolic compounds. Furthermore, the HPLC-DAD analysis detected some phenolic compounds (i.e., ferulic acid, isoquercitrin, and quercitrin) not previously found in fresh tepals. Vitamin C, already discovered in the spice, was interestingly detected also in dried tepals.

Taken together, these results assess that floral bio-residues of saffron can be sustainably processed to obtain high yields of valuable phytochemicals with potential applications in the pharmaceutical, cosmetic, and food industries, meeting the requirement for a sustainable biorefining and opening new possibilities for this by-product to become an important income source.

Supplementary Materials: The following are available online at https://www.mdpi.com/2223-7747/10/3/523/s1, Table S1: Sustainable Processing of Floral Bio-Residues of Saffron (*Crocus sativus* L.) for Valuable Biorefinery Products.

Author Contributions: Conceptualization, V.S.; methodology, M.C. and S.D.; formal analysis, M.C. and S.D.; investigation, S.S.; data curation, S.S.; writing—original draft preparation, S.S.; writing—review and editing, M.C., S.D., and V.S.; supervision, V.S.; funding acquisition, V.S. All authors have read and agreed to the published version of the manuscript.

Funding: This research was funded by the project titled "SaffronALP—Lo zafferano di montagna: tecniche sostenibili per una produzione di qualità"—Fondazione Cassa di Risparmio di Torino (RF = 2017.1966) and by the program Interreg V-A Francia Italia Alcotra "Attività innovative per lo sviluppo della filiera transfrontaliera del fiore edule–Antea" n. 1139.

Institutional Review Board Statement: Not applicable.

Informed Consent Statement: Not applicable.

Data Availability Statement: *Crocus sativus* L. tepals were kindly provided by the company "Lo Zafferano del Monviso" located in Martiniana Po, CN (Italy—44°23' N 7°33' E).

Acknowledgments: The authors acknowledge the company "Lo Zafferano del Monviso" for providing dried saffron tepals, Michele Lonati for statistical assistance, Marie-Jeanne Welter for helping in spectrophotometric analyses, and Dario Donno for chromatographic analyses.

Conflicts of Interest: The authors declare no conflict of interest. The funders had no role in the design of the study; in the collection, analyses, or interpretation of data; in the writing of the manuscript, or in the decision to publish the results.

References

1. Khan, M.; Hanif, M.A.; Ayub, M.A.; Jilani, M.I.; Chatha, S.A.S. Saffron. *Med. Plants South Asia* **2020**, 587–600. [CrossRef]
2. Caser, M.; Demasi, S.; Victorino, Í.M.M.; Donno, D.; Faccio, A.; Lumini, E.; Bianciotto, V.; Scariot, V. Arbuscular Mycorrhizal Fungi Modulate the Crop Performance and Metabolic Profile of Saffron in Soilless Cultivation. *Agronomy* **2019**, *9*, 232. [CrossRef]
3. Caser, M.; Victorino, Í.M.M.; Demasi, S.; Berruti, A.; Donno, D.; Lumini, E.; Bianciotto, V.; Scariot, V. Saffron Cultivation in Marginal Alpine Environments: How AMF Inoculation Modulates Yield and Bioactive Compounds. *Agronomy* **2018**, *9*, 12. [CrossRef]
4. Melnyk, J.P.; Wang, S.; Marcone, M.F. Chemical and biological properties of the world's most expensive spice: Saffron. *Food Res. Int.* **2010**, *43*, 1981–1989. [CrossRef]
5. Bagur, M.J.; Salinas, G.L.A.; Jiménez-Monreal, A.M.; Chaouqi, S.; Llorens, S.; Martínez-Tomé, M.; Alonso, G.L. Saffron: An Old Medicinal Plant and a Potential Novel Functional Food. *Molecules* **2017**, *23*, 30. [CrossRef] [PubMed]
6. Gresta, F.; Lombardo, G.M.; Siracusa, L.; Ruberto, G. Saffron, an alternative crop for sustainable agricultural systems. A review. *Agron. Sustain. Dev.* **2008**, *28*, 95–112. [CrossRef]
7. Halevy, A.H. recent advances in control of flowering and growth habit of geophytes. *Acta Hortic.* **1990**, 35–42. [CrossRef]
8. Kumar, R.; Singh, V.; Devi, K.; Sharma, M.; Singh, M.; Ahuja, P. State of Art of Saffron (*Crocus sativus* L.) Agronomy: A Comprehensive Review. *Food Rev. Int.* **2008**, *25*, 44–85. [CrossRef]
9. Molina, R.; Valero, M.; Navarro, Y.; Guardiola, J.; García-Luis, A. Temperature effects on flower formation in saffron (*Crocus sativus* L.). *Sci. Hortic.* **2005**, *103*, 361–379. [CrossRef]
10. Peter, K.V. *Handbook of Herbs and Spices Vol. 1: Series in Food Science*; Woodhead Publishing: Philadelphia, PA, USA, 2012; ISBN 978-0-85709-567-1.
11. Mottaghipisheh, J.; Sourestani, M.M.; Kiss, T.; Horváth, A.; Tóth, B.; Ayanmanesh, M.; Khamushi, A.; Csupor, D. Comprehensive chemotaxonomic analysis of saffron crocus tepal and stamen samples, as raw materials with potential antidepressant activity. *J. Pharm. Biomed. Anal.* **2020**, *184*, 113183. [CrossRef]

12. Ashktorab, H.; Soleimani, A.; Singh, G.; Amr, A.; Tabtabaei, S.; Latella, G.; Stein, U.; Akhondzadeh, S.; Solanki, N.; Gondré-Lewis, M.C.; et al. Saffron: The Golden Spice with Therapeutic Properties on Digestive Diseases. *Nutrients* **2019**, *11*, 943. [CrossRef]

13. Moratalla-López, N.; Bagur, M.J.; Lorenzo, C.; Salinas, M.E.M.-N.R.; Alonso, G.L. Bioactivity and Bioavailability of the Major Metabolites of Crocus sativus L. Flower. *Molecules* **2019**, *24*, 2827. [CrossRef]

14. Hosseini, A.; Razavi, B.M.; Hosseinzadeh, H. Saffron (*Crocus sativus*) petal as a new pharmacological target: A review. *Iran. J. Basic Med. Sci* **2018**, *21*, 1091–1099. [PubMed]

15. Caser, M.; Demasi, S.; Stelluti, S.; Donno, D.; Scariot, V. *Crocus sativus* L. Cultivation in Alpine Environments: Stigmas and Tepals as Source of Bioactive Compounds. *Agronomy* **2020**, *10*, 1473. [CrossRef]

16. Panche, A.N.; Diwan, A.D.; Chandra, S.R. Flavonoids: An overview. *J. Nutr. Sci.* **2016**, *5*, e47. [CrossRef] [PubMed]

17. Kumar, N.; Goel, N. Phenolic acids: Natural versatile molecules with promising therapeutic applications. *Biotechnol. Rep.* **2019**, *24*, e00370. [CrossRef] [PubMed]

18. Caser, M.; Demasi, S.; Caldera, F.; Dhakar, N.K.; Trotta, F.; Scariot, V. Activity of *Ailanthus altissima* (Mill.) Swingle Extract as a Potential Bioherbicide for Sustainable Weed Management in Horticulture. *Agronomy* **2020**, *10*, 965. [CrossRef]

19. Demasi, S.; Caser, M.; Vanara, F.; Fogliatto, S.; Vidotto, F.; Negre, M.; Trotta, F.; Scariot, V. Ailanthone from Ailanthus altissima (Mill.) Swingle as potential natural herbicide. *Sci. Hortic.* **2019**, *257*, 108702. [CrossRef]

20. Caser, M.; Chitarra, W.; D'Angiolillo, F.; Perrone, I.; Demasi, S.; Lovisolo, C.; Pistelli, L.; Pistelli, L.; Scariot, V. Drought stress adaptation modulates plant secondary metabolite production in Salvia dolomitica Codd. *Ind. Crop. Prod.* **2019**, *129*, 85–96. [CrossRef]

21. Aryee, A.N.; Agyei, D.; Akanbi, T.O. Food for Oxidative Stress Relief: Polyphenols. In *Encyclopedia of Food Chemistry*; Elsevier: Amsterdam, The Netherlands, 2018; pp. 392–398.

22. Pires, T.C.P.; Barros, L.; Santos-Buelga, C.; Ferreira, I.C. Edible flowers: Emerging components in the diet. *Trends Food Sci. Technol.* **2019**, *93*, 244–258. [CrossRef]

23. Brodowska, K.M. European Journal of Biological Research Natural flavonoids: Classification, potential role, and application of flavonoid analogues. *Eur. J. Biol. Res.* **2017**, *7*, 108–123. [CrossRef]

24. Santos-Buelga, C.; González-Paramás, A.M. Anthocyanins. In *Encyclopedia of Food Chemistry*; Elsevier: Amsterdam, The Netherlands, 2019; pp. 10–21.

25. Granger, M.; Eck, P. Dietary Vitamin C in Human Health. *Adv. Food Nutr. Res.* **2018**, *83*, 281–310. [CrossRef] [PubMed]

26. Akram, N.A.; Shafiq, F.; Ashraf, M. Ascorbic acid-a potential oxidant scavenger and its role in plant development and abiotic stress tolerance. *Front. Plant. Sci.* **2017**, *8*, 613. [CrossRef] [PubMed]

27. Garavand, F.; Rahaee, S.; Vahedikia, N.; Jafari, S.M. Different techniques for extraction and micro/nanoencapsulation of saffron bioactive ingredients. *Trends Food Sci. Technol.* **2019**, *89*, 26–44. [CrossRef]

28. Donno, D.; Mellano, M.G.; Prgomet, Ž.; Cerutti, A.K.; Beccaro, G.L. Phytochemical characterization and antioxidant activity evaluation of mediterranean medlar fruit (*Crataegus azarolus* L.): Preliminary study of underutilized genetic resources as a potential source of health-promoting compound for food supplements. *J. Food Nutr. Res.* **2017**, *56*, 18–31.

29. Khoddami, A.; Wilkes, M.A.; Roberts, T.H. Techniques for Analysis of Plant Phenolic Compounds. *Molecules* **2013**, *18*, 2328–2375. [CrossRef] [PubMed]

30. Aliaño-González, M.J.; Espada-Bellido, E.; Ferreiro-González, M.; Carrera, C.; Palma, M.; Ayuso, J.; Álvarez, J.A.; Barbero, G.F. Extraction of Anthocyanins and Total Phenolic Compounds from Açai (Euterpe oleracea Mart.) Using an Experimental Design Methodology. Part 2: Ultrasound-Assisted Extraction. *Agronomy* **2020**, *10*, 326. [CrossRef]

31. Zhao, L.; Fan, H.; Zhang, M.; Chitrakar, B.; Bhandari, B.; Wang, B. Edible flowers: Review of flower processing and extraction of bioactive compounds by novel technologies. *Food Res. Int.* **2019**, *126*, 108660. [CrossRef]

32. Medina-Torres, N.; Ayora-Talavera, T.; Espinosa-Andrews, H.; Sánchez-Contreras, A.; Pacheco, N. Ultrasound Assisted Extraction for the Recovery of Phenolic Compounds from Vegetable Sources. *Agronomy* **2017**, *7*, 47. [CrossRef]

33. Goupy, P.; Vian, M.A.; Chemat, F.; Caris-Veyrat, C. Identification and quantification of flavonols, anthocyanins and lutein diesters in tepals of Crocus sativus by ultra performance liquid chromatography coupled to diode array and ion trap mass spectrometry detections. *Ind. Crop. Prod.* **2013**, *44*, 496–510. [CrossRef]

34. A Review on the Extraction Methods Use in Medicinal Plants, Principle, Strength and Limitation. *Med. Aromat. Plants* **2015**, *4*, 196. [CrossRef]

35. Da Porto, C.; Natolino, A. Extraction kinetic modelling of total polyphenols and total anthocyanins from saffron floral bio-residues: Comparison of extraction methods. *Food Chem.* **2018**, *258*, 137–143. [CrossRef] [PubMed]

36. Ameer, K.; Shahbaz, H.M.; Kwon, J.-H. Green Extraction Methods for Polyphenols from Plant Matrices and Their Byproducts: A Review. *Compr. Rev. Food Sci. Food Saf.* **2017**, *16*, 295–315. [CrossRef]

37. Wali, A.F.; Alchamat, H.A.A.; Hariri, H.K.; Hariri, B.K.; Menezes, G.A.; Zehra, U.; Rehman, M.U.; Ahmad, P. Antioxidant, Antimicrobial, Antidiabetic and Cytotoxic Activity of Crocus sativus L. Petals. *Appl. Sci.* **2020**, *10*, 1519. [CrossRef]

38. Khazaei, K.M.; Jafari, S.M.; Ghorbani, M.; Kakhki, A.H.; Sarfarazi, M. Optimization of Anthocyanin Extraction from Saffron Petals with Response Surface Methodology. *Food Anal. Methods* **2015**, *9*, 1993–2001. [CrossRef]

39. Lahmass, I.; Lamkami, T.; Delporte, C.; Sikdar, S.; Van Antwerpen, P.; Saalaoui, E.; Megalizzi, V. The waste of saffron crop, a cheap source of bioactive compounds. *J. Funct. Foods* **2017**, *35*, 341–351. [CrossRef]

40. Sánchez-Vioque, R.; Santana-Méridas, O.; Polissiou, M.; Vioque, J.; Astraka, K.; Alaiz, M.; Herraiz-Peñalver, D.; Tarantilis, P.A.; Girón-Calle, J. Polyphenol composition and in vitro antiproliferative effect of corm, tepal and leaf from Crocus sativus L. on human colon adenocarcinoma cells (Caco-2). *J. Funct. Foods* **2016**, *24*, 18–25. [CrossRef]

41. Jadouali, S.M.; Atifi, H.; Mamouni, R.; Majourhat, K.; Bouzoubaâ, Z.; Laknifli, A.; Faouzi, A. Chemical characterization and antioxidant compounds of flower parts of Moroccan crocus sativus L. *J. Saudi Soc. Agric. Sci.* **2019**, *18*, 476–480. [CrossRef]

42. Lotfi, L.; Kalbasi-Ashtari, A.; Hamedi, M.; Ghorbani, F. Effects of sulfur water extraction on anthocyanins properties of tepals in flower of saffron (*Crocus sativus* L). *J. Food Sci. Technol.* **2013**, *52*, 813–821. [CrossRef]

43. Lotfi, L.; Kalbasi-Ashtari, A.; Hamedi, M.; Ghorbani, F. Effects of enzymatic extraction on anthocyanins yield of saffron tepals (*Crocos sativus*) along with its color properties and structural stability. *J. Food Drug Anal.* **2015**, *23*, 210–218. [CrossRef] [PubMed]

44. Serrano-Díaz, J.; Sanchez, A.M.; Maggi, L.; Martínez-Tomé, M.; García-Diz, L.; Murcia, M.A.; Alonso, G.L. Increasing the Applications ofCrocus sativusFlowers as Natural Antioxidants. *J. Food Sci.* **2012**, *77*, C1162–C1168. [CrossRef]

45. Cusano, E.; Consonni, R.; Petrakis, E.A.; Astraka, K.; Cagliani, L.R.; Polissiou, M.G. Integrated analytical methodology to investigate bioactive compounds in Crocus sativus L. flowers. *Phytochem. Anal.* **2018**, *29*, 476–486. [CrossRef]

46. Jafari, S.M.; Khazaei, K.M.; Assadpour, E. Production of a natural color through microwave-assisted extraction of saffron tepal's anthocyanins. *Food Sci. Nutr.* **2019**, *7*, 1438–1445. [CrossRef]

47. Ghanbari, J.; Khajoei-Nejad, G.; Van Ruth, S.M. Effect of saffron (Crocus sativus L.) corm provenance on its agro-morphological traits and bioactive compounds. *Sci. Hortic.* **2019**, *256*, 108605. [CrossRef]

48. Moratalla-López, N.; Lorenzo, C.; Chaouqi, S.; Sánchez, A.M.; Alonso, G.L. Kinetics of polyphenol content of dry flowers and floral bio-residues of saffron at different temperatures and relative humidity conditions. *Food Chem.* **2019**, *290*, 87–94. [CrossRef] [PubMed]

49. Goli, S.A.H.; Mokhtari, F.; Rahimmalek, M. Phenolic Compounds and Antioxidant Activity from Saffron (*Crocus sativus* L.) Petal. *J. Agric. Sci.* **2012**, *4*, 4. [CrossRef]

50. Cai, Y.; Luo, Q.; Sun, M.; Corke, H. Antioxidant activity and phenolic compounds of 112 traditional Chinese medicinal plants associated with anticancer. *Life Sci.* **2004**, *74*, 2157–2184. [CrossRef]

51. Li, H.; Deng, Z.; Zhu, H.; Hu, C.; Liu, R.; Young, J.C.; Tsao, R. Highly pigmented vegetables: Anthocyanin compositions and their role in antioxidant activities. *Food Res. Int.* **2012**, *46*, 250–259. [CrossRef]

52. Solymosi, K.; Latruffe, N.; Morantmanceau, A.; Schoefs, B. Food colour additives of natural origin. In *Colour Additives for Foods and Beverages*; Elsevier: Amsterdam, The Netherlands, 2015; pp. 3–34.

53. Ribeiro, L.; Ribani, R.; Francisco, T.; Soares, A.; Pontarolo, R.; Haminiuk, C. Profile of bioactive compounds from grape pomace (*Vitis vinifera* and *Vitis labrusca*) by spectrophotometric, chromatographic and spectral analyses. *J. Chromatogr. B* **2015**, *1007*, 72–80. [CrossRef]

54. Markakis, P. Anthocyanins as Food Additives. In *Anthocyanins as Food Colors*; Elsevier: Amsterdam, The Netherlands, 1982; pp. 245–253.

55. Thoo, Y.Y.; Ho, S.K.; Liang, J.Y.; Ho, C.W.; Tan, C.P. Effects of binary solvent extraction system, extraction time and extraction temperature on phenolic antioxidants and antioxidant capacity from mengkudu (*Morinda citrifolia*). *Food Chem.* **2010**, *120*, 290–295. [CrossRef]

56. Huang, H.; Sun, Y.; Lou, S.; Li, H.; Ye, X. In vitro digestion combined with cellular assay to determine the antioxidant activity in Chinese bayberry (*Myrica rubra* Sieb. et Zucc.) fruits: A comparison with traditional methods. *Food Chem.* **2014**, *146*, 363–370. [CrossRef] [PubMed]

57. Donno, D.; Beccaro, G.; Mellano, M.; Cerutti, A.; Bounous, G. Goji berry fruit (*Lycium* spp.): Antioxidant compound fingerprint and bioactivity evaluation. *J. Funct. Foods* **2015**, *18*, 1070–1085. [CrossRef]

58. Babbar, N.; Oberoi, H.S.; Uppal, D.S.; Patil, R.T. Total phenolic content and antioxidant capacity of extracts obtained from six important fruit residues. *Food Res. Int.* **2011**, *44*, 391–396. [CrossRef]

59. Terpinc, P.; Čeh, B.; Ulrih, N.P.; Abramovič, H. Studies of the correlation between antioxidant properties and the total phenolic content of different oil cake extracts. *Ind. Crop. Prod.* **2012**, *39*, 210–217. [CrossRef]

60. Kalinowska, M.; Gryko, K.; Wróblewska, A.M.; Jabłońska-Trypuć, A.; Karpowicz, D. Phenolic content, chemical composition and anti-/pro-oxidant activity of Gold Milenium and Papierowka apple peel extracts. *Sci. Rep.* **2020**, *10*, 1–15. [CrossRef] [PubMed]

61. Bilek, S.E. The effects of time, temperature, solvent: Solid ratio and solvent composition on extraction of total phenolic compound from dried olive (*Olea europaea* L.) leaves. *GIDA* **2010**, *35*, 411–416.

62. Safaiee, P.; Taghipour, A.; Vahdatkhoram, F.; Movagharnejad, K. Extraction of phenolic compounds from *Mentha aquatica*: The effects of sonication time, temperature and drying method. *Chem. Pap.* **2019**, *73*, 3067–3073. [CrossRef]

63. Liu, Z.; Esveld, E.; Vincken, J.-P.; Bruins, M.E. Pulsed Electric Field as an Alternative Pre-treatment for Drying to Enhance Polyphenol Extraction from Fresh Tea Leaves. *Food Bioprocess. Technol.* **2019**, *12*, 183–192. [CrossRef] [PubMed]

64. Srivastava, R.; Ahmed, H.; Dixit, R.; Dharamveer; Saraf, S. Crocus sativus L.: A comprehensive review. *Pharmacogn. Rev.* **2010**, *4*, 200–208. [CrossRef]

65. Donno, D.; Mellano, M.; Prgomet, Z.; Beccaro, G. Advances in Ribes x nidigrolaria Rud. Bauer & A. Bauer fruits as potential source of natural molecules: A preliminary study on physico-chemical traits of an underutilized berry. *Sci. Hortic.* **2018**, *237*, 20–27. [CrossRef]

66. Laufenberg, G.; Kunz, B.; Nystroem, M. Transformation of vegetable waste into value added products:(A) the upgrading concept;(B) practical implementations. *Bioresour. Technol.* **2003**, *87*, 167–198. [CrossRef]
67. Philippini, R.R.; Martiniano, S.E.; Ingle, A.P.; Marcelino, P.R.F.; Silva, G.M.; Barbosa, F.G.; Dos Santos, J.C.; Silva, S.S. Agroindustrial Byproducts for the Generation of Biobased Products: Alternatives for Sustainable Biorefineries. *Front. Energy Res.* **2020**, *8*, 1–23. [CrossRef]
68. Motola, V.; De Bari, I.; Pierro, N.; Giocoli, A. *Bioeconomy and Biorefining Strategies in the EU Member States and Beyond—Reference Year 2018*; IEA Bioenergy: Rome, Italy, 2018; ISBN 978-1-910154-60-1.
69. European Commission. *A Sustainable Bioeconomy for Europe: Strengthening the Connection between Economy, Society and the Environment*; European Commission: Brussels, Belgium, 2018; ISBN 9789279941450.
70. Bruusgaard-Mouritsen, M.A.; Johansen, J.D.; Zachariae, C.; Kirkeby, C.S.; Garvey, L.H. Natural ingredients in cosmetic products—A suggestion for a screening series for skin allergy. *Contact Dermat.* **2020**, *83*, 251–270. [CrossRef]

Article

Dibenzocyclooctadiene Lignans in Plant Parts and Fermented Beverages of *Schisandra chinensis*

Woo Sung Park [1,†], Kyung Ah Koo [1,†], Ji-Yeong Bae [2,3,†], Hye-Jin Kim [1], Dong-Min Kang [1], Ji-Min Kwon [1], Seung-Mann Paek [1], Mi Kyeong Lee [4], Chul Young Kim [5,*] and Mi-Jeong Ahn [1,*]

1   College of Pharmacy and Research Institute of Pharmaceutical Sciences, Gyeongsang National University, Jinju 52828, Korea; pws8822@gmail.com (W.S.P.); kk408842@gmail.com (K.A.K.); black200203@gmail.com (H.-J.K.); dongminkang71@gmail.com (D.-M.K.); ellusian458@gmail.com (J.-M.K.); million@gnu.ac.kr (S.-M.P.)
2   College of Pharmacy, Jeju National University, Jeju 63243, Korea; jybae@jejunu.ac.kr
3   Interdisciplinary Graduate Program in Advanced Convergence Technology & Science, Jeju National University, Jeju 63243, Korea
4   College of Pharmacy, Chungbuk National University, Cheongju 28160, Korea; mklee@chungbuk.ac.kr
5   College of Pharmacy, Institute of Pharmaceutical Science and Technology, Hanyang University, Ansan 15588, Korea
*   Correspondence: chulykim@hanyang.ac.kr (C.Y.K.); amj5812@gnu.ac.kr (M.-J.A.); Tel.: +82-31-400-5809 (C.Y.K.); +82-55-772-2425 (M.-J.A.)
†   These authors contributed equally to this work.

Citation: Park, W.S.; Koo, K.A.; Bae, J.-Y.; Kim, H.-J.; Kang, D.-M.; Kwon, J.-M.; Paek, S.-M.; Lee, M.K.; Kim, C.Y.; Ahn, M.-J. Dibenzocyclooctadiene Lignans in Plant Parts and Fermented Beverages of *Schisandra chinensis*. Plants **2021**, 10, 361. https://doi.org/10.3390/plants10020361

Academic Editor: Jong Seong Kang

Received: 21 January 2021
Accepted: 10 February 2021
Published: 13 February 2021

Abstract: The fruit of *Schisandra chinensis*, Omija, is a well-known traditional medicine used as an anti-tussive and anti-diarrhea agent, with various biological activities derived from the dibenzocyclooctadiene-type lignans. A high-pressure liquid chromatography-diode array detector (HPLC-DAD) method was used to determine seven lignans (schisandrol A and B, tigloylgomisin H, angeloylgomisin H, schisandrin A, B, and C) in the different plant parts and beverages of the fruit of *S. chinensis* grown in Korea. The contents of these lignans in the plant parts descended in the following order: seeds, flowers, leaves, pulp, and stems. The total lignan content in Omija beverages fermented with white sugar for 12 months increased by 2.6-fold. Omija was fermented for 12 months with white sugar, brown sugar, and oligosaccharide/white sugar (1:1, *w/w*). The total lignan content in Omija fermented with oligosaccharide/white sugar was approximately 1.2- and 1.7-fold higher than those fermented with white sugar and brown sugar, respectively. A drink prepared by immersion of the fruit in alcohol had a higher total lignan content than these fermented beverages. This is the first report documenting the quantitative changes in dibenzocyclooctadiene-type lignans over a fermentation period and the effects of the fermentable sugars on this eco-friendly fermentation process.

Keywords: *Schisandra chinensis*; Omija; dibenzocyclooctadiene lignans; seed; flower; fermented beverage

1. Introduction

Schisandra chinensis (Turcz.) Baill. (family Schisandraceae) is a deciduous, dioecious, woody vine plant also known as five-flavor-fruit in East Asia [1]. The fruit called Omija in Korea has been used in the treatment of diseases related to the gastrointestinal tract, respiratory failure, cardiovascular diseases, body fatigue, and weakness [2]. Diverse components of this fruit, such as lignans, terpenes, essential oils, anthocyanins, and polysaccharides have demonstrated anticancer, neuroprotective, anti-inflammatory, antioxidative, hepatoprotective, immunomodulatory, and antiviral activities [3]. Among them, dibenzocyclooctadiene-type lignans are well known to show various biological activities that could prevent cancer, hepatotoxicity, and nephrotoxicity in humans [4–6]. Schisandrin A–C, gomisin A–H, and schisandrol A and B are the major lignans that are known to exhibit the biological activities described above [6–8]. Recently, a study reported on the addition of Omija fruit to ale-type beer at different points in the brewing process in an attempt to develop an Omija fruit beverage with high antioxidant capacity [9].

Thousands of tons of schisandra berries are traded annually in Asia. They are currently commercially consumed as functional health foods in the form of beverages, jams, capsules, and powders. In general, only the fruit is beneficially consumed, and the other parts of the plant are discarded. While previous studies have focused mainly on the content of various chemical components, including lignans, in the fruit of *S. chinensis*, and in the roots or branches in some cases [10–16], there is no report on the difference in the lignan content of the plant parts, including the flower.

The fermentation of foods has a variety of benefits, including improving the flavor and long-term storage as well as increasing or producing beneficial nutrients and healthy ingredients. The raw berry of *S. chinensis* is converted primarily to a fermented beverage with sugar, as the most common form for consumption in Korea. In general, people make this beverage at home and consume the beverage after 6 or 12 months or a much longer period of fermentation. Fermentation with lactic acid bacteria or with only sugar results in changes in various biological activities, such as the anti-oxidative, anti-inflammatory, anti-hypertensive, and anti-cholesterol activities [17–21]. To the best of our knowledge, this is the first report describing the effects of the fermentation process, fermentation period, and fermenting materials on the lignan content of Omija beverages and drink.

In this study, the content of seven major dibenzocyclooctadiene-type lignans (schisandrol A (schisandrin or wuweizisu A), schisandrol B (gomisin A), tigloylgomisin H, angeloylgomisin H, schisandrin A (deoxyschisandrin), schisandrin B (gomisin N or wuweizisu B or γ-schisandrin), and schisandrin C) in the stem, leaf, pulp, seed, and flower of *S. chinensis* is determined. The content of these compounds in Omija beverages made from the whole fruit and fermented with white sugar for different fermentation times is also determined. The change in the lignan content is compared for fruit beverages fermented with three kinds of sugars (white, brown, and oligosaccharide/white) and ethanol.

2. Results and Discussion

2.1. Lignan Content in Plant Parts of S. chinensis

Calibration curves were constructed by analyzing a standard mixture containing the seven lignans (Figure 1) at various concentration levels and plotting the peak area against the concentration of each reference standard (Table 1). The curves showed good linearity, and the correlation coefficients were between 0.997 and 0.999 for all the compounds over the concentration ranges of quantification. The recovery of seven lignans was assessed by spiking samples with high and low concentrations (50 and 10 ng, respectively) of each reference compound. The average recoveries were between 84.9% and 105.7% (n = 3). The limits of detection (LOD) were determined by serial dilution based on a signal-to-noise (S/N) ratio of 3:1 (Table 1). Limits of quantification (LOQ) were 1 ng for the seven lignans. The peak purity was determined using a photodiode array detector. In addition, each peak in the absorption spectrum was compared with the characteristic peak of the corresponding standard compound. The precision was determined from the intra-day and inter-day repeatability, and the relative standard deviations were below 19%.

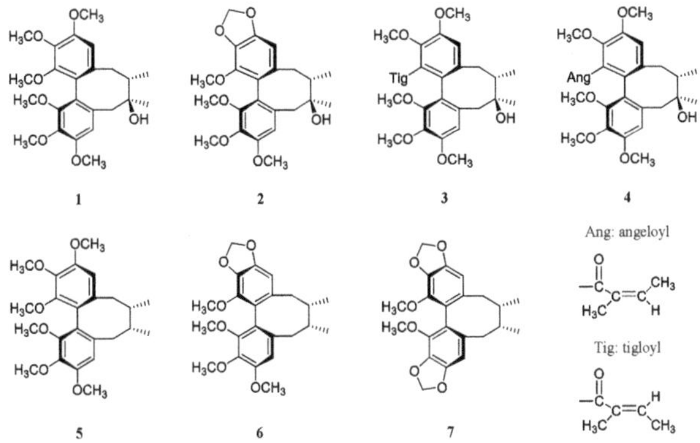

Figure 1. Chemical structures of seven dibenzocyclooctadiene-type lignans (1–7). **1**, schisandrol A; **2**, schisandrol B; **3**, tigloylgomisin H; **4**, angeloylgomisin H; **5**, schisandrin A; **6**, schisandrin B; **7**, schisandrin C.

Table 1. Linear ranges and correlation coefficients of calibration curves.

Lignans	Range (µg/mL)	Slope (a) [a]	Intercept (b) [b]	Regression (r^2)	Limits of Detection (LOD) (ng)
Schisandrol A (**1**)	0.05–50	226.12	−65.534	0.9974	~0.3
Schisandrol B (**2**)	0.05–12.5	241.27	−23.306	0.9982	~0.3
Tigloylgomisin H (**3**)	0.05–25	174.57	−22.081	0.9995	~0.3
Angeloylgomisin H (**4**)	0.05–25	179.53	−22.650	0.9993	~0.3
Schisandrin A (**5**)	0.05–50	195.70	−74.614	0.9986	~0.3
Schisandrin B (**6**)	0.05–50	251.32	−96.463	0.9985	~0.3
Schisandrin C (**7**)	0.05–12.5	520.59	39.137	0.9986	~0.3

[a,b] Slope and intercept are represented by a and b, respectively, in the Y= ax + b linear model. Y is peak area and x is concentration.

The variation in the content of seven lignans in the various parts of *S. chinensis* was evaluated by HPLC-DAD (Figure 2). The plant parts included the stem, leaf, flower, pulp, and seed. The total lignan content was highest for the seeds, with a value of 47.42 ± 2.81 mg/g·DW (Table 2). This value was more than eight times higher than the lignan contents in the other plant parts. Notably, the flower had the second highest lignan content, with a value of 5.62 ± 0.33 mg/g·DW. The most commonly used part, the pulp, had a similar lignan content (3.20 ± 0.44 mg/g·DW) to that of the leaves (3.67 ± 0.21 mg/g·DW), and a lower content than the flower. The lignan content was lowest in the stem, with a value of 2.22 ± 0.19 mg/g·DW. This result is consistent with a previous report that the leaf of *S. chinensis* may also be a good source of lignans [11].

Schisandrol A (**1**) was the most abundant of the seven lignans in the flower, pulp, and seed, while the content of schisandrol B (**2**) and schisandrin B (**6**) was highest in the stem and leaf, respectively. Schisandrol A (**1**) and schisandrin B (**6**) were found to be the major lignans present in the pulp and seed of the Omija fruit. The content of these compounds in the seed was ten times higher than that in the other plant parts (Table 2). Tigloylgomisin H (**3**) was not detected in the pulp.

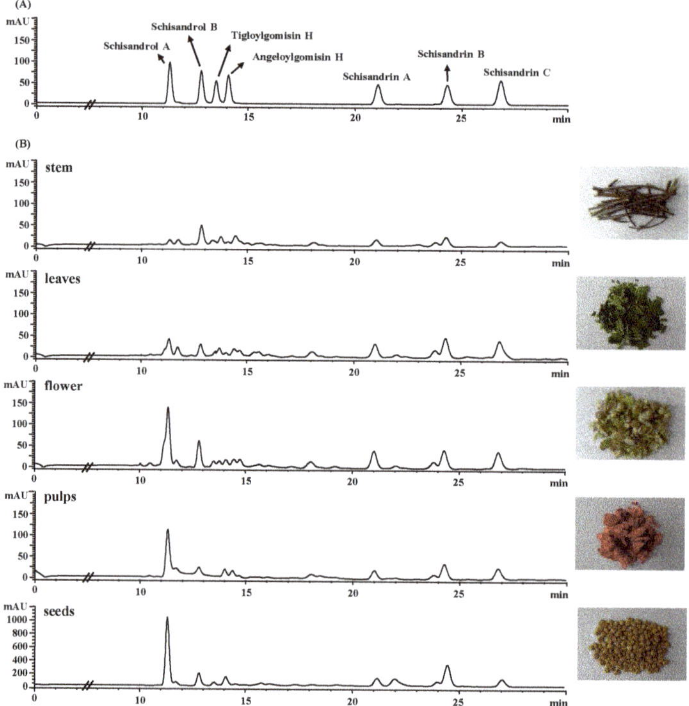

Figure 2. HPLC chromatograms of standard lignans (**A**) and sevens lignans in various parts of *S. chinensis* (**B**) measured at 254 nm.

Table 2. Contents of seven lignans in plant parts of *S. chinensis*.

Parts	1	2	3	4	5	6	7	Total
Stem	0.28 ± 0.05 [f]	0.57 ± 0.04 [c]	0.28 ± 0.02 [c]	0.15 ± 0.02 [e]	0.36 ± 0.04 [f]	0.47 ± 0.03 [f]	0.12 ± 0.00 [f]	2.22 ± 0.19 [e]
Leaf	0.73 ± 0.06 [e]	0.38 ± 0.02 [d]	0.27 ± 0.01 [c]	0.18 ± 0.01 [e]	0.78 ± 0.04 [d]	0.91 ± 0.05 [c]	0.43 ± 0.02 [c]	3.67 ± 0.21 [b]
Flower	1.85 ± 0.12 [c]	0.57 ± 0.02 [c]	0.29 ± 0.01 [c]	0.36 ± 0.02 [c]	1.08 ± 0.07 [c]	1.09 ± 0.06 [d]	0.39 ± 0.03 [d]	5.62 ± 0.33 [c]
Pulp	1.33 ± 0.24 [d]	0.26 ± 0.05 [e]	ND	0.28 ± 0.04 [d]	0.46 ± 0.03 [e]	0.66 ± 0.04 [e]	0.23 ± 0.03 [e]	3.20 ± 0.44 [d]
Seed	21.0 ± 1.00 [a]	4.02 ± 0.36 [a]	1.66 ± 0.28 [a]	3.96 ± 0.32 [a]	4.43 ± 0.24 [a]	10.6 ± 0.52 [a]	1.74 ± 0.09 [a]	47.42 ± 2.81 [a]
Fruit	11.9 ± 0.54 [b]	2.27 ± 0.23 [b]	0.89 ± 0.18 [b]	2.26 ± 0.18 [b]	2.60 ± 0.18 [b]	6.02 ± 0.36 [b]	1.04 ± 0.05 [b]	26.99 ± 1.7 [b]

Data are expressed as the mean (the average value of content for dry weight, mg/g DW) and SD of three independent experiments. Different letters in the same column indicate significant differences ($p < 0.05$) between the values. ND means not detected. **1**, schisandrol A; **2**, schisandrol B; **3**, tigloylgomisin H; **4**, angeloylgomisin H; **5**, schisandrin A; **6**, schisandrin B; **7**, schisandrin C.

The leaf displayed a higher lignan content than the stem. This result is different from a previous report indicating that the main branch or side branch had a higher lignan content than the leaf [10]. This discrepancy could be ascribed to the difference in the collection season and region, the diameter of the stem, and the habitat [6]. In fact, the stem of the side branch that is usually discarded after pruning was used in this study. It has been reported that the dried seeds have the highest content of schisandrin (schisandrol A, 9.46 mg/g·DW) and total lignan (25.97 mg/g·DW) among the skin, pulp, and seeds of the fruit [12,13]. That report is consistent with the results of this study. However, the total lignan content of the seed in our study was approximately twice as high as that found in previously reported. This discrepancy may be ascribed to the different samples and preparation methods. Soxhlet extraction at 80 °C for 3 h or supercritical CO_2 was used in the previous reports [12,13]. In the seeds collected in Europe, the content of gomisin N (**6**) was comparable to or half of that of schisandrol A (**1**) [13].

In previous reports, the lignan content of the fruit showed similar patterns in the samples collected or cultivated in Asia and Europe [7,13–15]. The content of schisandrol A (**1**) was the highest, followed by the content of gomisin N (= schisandrin B) (**6**) and schisandrol B (**2**) in order. This tendency in the lignan content of the berry was also shown in direct analysis in real time ionization source coupled with quadrupole orbitrap mass spectrometry as well as in our study [16]. The total lignan contents (%, weight/dry fruit) were in the range of 1.3 to 2.8% in the variable experimental conditions, and the value in this study was 2.7%. From these results, it can be suggested that ultrasonic extraction with methanol at room temperature is the best method for extraction of lignan compounds in the Omija berry.

2.2. Lignan Content of Omija Beverage during Fermentation Process

To investigate the changes in the content of seven lignans during the fermentation of the Omija beverage, the most popular food product of this fruit, the raw fruit with skin, pulp, and seed was fermented with white sugar for 12 months. The changes in the lignan composition were monitored at 3, 7, 10, and 12 months during the year-long fermentation process. The total content of the seven lignans after 12 months increased to a maximum value of 2.6 times compared to that after 3 months (Table 3). The lignan content increased rapidly up to 7 months and increased gradually thereafter up to 12 months. Six lignans, excluding angeloylgomisin H (**4**), increased as the fermentation period increased. Angeloylgomisin H (**4**) increased up to 7 months and decreased considerably by 12 months.

Table 3. Contents of seven lignans in an Omija beverage fermented with white sugar during different fermentation times.

Period (Month, m)	1	2	3	4	5	6	7	Total
3 m	7.01 ± 1.30 [c]	1.65 ± 0.29 [c]	1.03 ± 0.13 [c]	2.34 ± 0.44 [bc]	0.90 ± 0.16 [b]	51.71 ± 5.14 [b]	0.65 ± 0.10 [a]	65.30 ± 7.55 [b]
7 m	10.42 ± 1.10 [b]	2.39 ± 0.07 [b]	1.63 ± 0.14 [b]	4.33 ± 0.40 [a]	1.64 ± 0.11 [a]	126.43 ± 19.53 [a]	0.69 ± 0.07 [a]	147.53 ± 21.42 [a]
10 m	11.96 ± 0.65 [a]	2.79 ± 0.29 [a]	1.68 ± 0.16 [ab]	2.49 ± 0.11 [b]	1.73 ± 0.20 [a]	133.58 ± 19.30 [a]	0.75 ± 0.07 [a]	154.97 ± 20.78 [a]
12 m	13.24 ± 2.27 [a]	2.87 ± 0.52 [a]	1.95 ± 0.30 [a]	1.96 ± 0.36 [c]	2.00 ± 0.47 [a]	147.23 ± 33.09 [a]	0.89 ± 0.23 [a]	170.14 ± 37.23 [a]

Data are expressed as the mean (the average value of content for fresh weight, µg /g FW) and SD of three independent experiments. Different letters in the same column indicate significant differences ($p < 0.05$) between the values. **1**, schisandrol A; **2**, schisandrol B; **3**, tigloylgomisin H; **4**, angeloylgomisin H; **5**, schisandrin A; **6**, schisandrin B; **7**, schisandrin C.

Thus far, there is no report on the change in the content of dibenzocyclooctadiene-type lignans in Omija beverages made from the whole fruits of *S. chinensis* with white sugar according to the fermentation period. The fermentation process with sugar of 50 Brix is the most popular method used in Korea. Our study results suggest that a fermentation period of 12 months could be considered as the optimal time for highly efficient extraction or bioconversion to bioactive compounds. Among the seven lignans, schisandrin B (**6**) showed the highest content in the Omija beverage made from fresh berries with white sugar, with a content 7.4- to 12.1-fold higher than that of schisandrol A (**1**) during the fermentation period (Table 3). In the methanol extract of dried fruit of *S. chinensis*, the content of schisandrin A (**1**) was the highest and was 2.0-fold higher than that of schisandrin B (**6**) (Table 2). Schisandrin B (gomisin N) has hepatoprotective, nephroprotective, anti-cancer, anti-aging, and anti-obesity activities [3–5]. Schisandrol B (gomisin A) and schisandrin B (gomisin G) have anti-cancer and anti-aging activities [3]. Schisandrin A inhibits dengue viral replication [22]. These lignans are known to be absorbed in rat everted gut sac, human Caco-2 cell monolayer, and in vivo rat models [23]. These reports support the fact that the major lignans in Omija beverages fermented with sugar, a commonly consumed product, would be absorbed by the body and exert various biological activities.

Meanwhile, the study of natural deep eutectic solvents (NADES) has revealed that certain mixtures of natural products, such as sugars, organic acids, and amino acids, in the solid state with the proper ratio, become liquid and enhance the solubility or stability of non-water soluble bioactive compounds [24,25]. Based on this previous report, the high concentration of sugar and organic acids in the berry of *S. chinensis* in this study might play a role in promoting dissolution of the lignan compounds.

2.3. Lignan Content in Omija Beverages and Drink from S. chinensis and Sugars

Raw Omija fruit was fermented with three kinds of fermentable sugars (45-50 Brix) or immersed in alcohol (30%, v/v) for 12 months at room temperature, and the contents of seven lignans in the Omija beverages and drink were evaluated. The three kinds of fermentable sugars were white sugar, brown sugar, and oligosaccharide/white sugar (1:1). The Omija drink prepared by immersing the fruit in alcohol had a higher total lignan content (257.37 µg/g FW) than the three Omija beverages fermented with sugars (Table 4). This result is attributed to the higher solubility of lignan compounds in alcohol than in water. Among the three Omija beverages, the Omija beverage fermented with oligosaccharide/white sugar (1:1, *w/w*) had the highest total lignan content, which was 1.2- and 1.7-fold higher than those of the beverages prepared by fermentation with white sugar and blown sugar, respectively. For the Omija drink prepared by immersion in alcohol, the content of compounds 1-4 was relatively higher than that in the Omija beverages fermented with sugars (3-5-fold). For all samples, the content of compounds 5-7 did not change significantly during the total fermentation period. For the Omija beverage fermented with brown sugar, the content of schisandrin B (**6**) and schisandrin C (**7**) was relatively low compared to that in the other beverages.

Table 4. Contents of seven lignans in Omija beverages or drink fermented with different sugars.

Sugars	1	2	3	4	5	6	7	Total
White sugar	10.34 ± 1.44 [c]	2.80 ± 0.21 [b]	1.78 ± 0.35 [c]	4.69 ± 0.23 [b]	1.93 ± 0.05 [c]	151.78 ± 14.89 [b]	0.89 ± 0.03 [b]	174.50 ± 16.00 [c]
Brown sugar	6.24 ± 0.11 [d]	1.74 ± 0.20 [c]	0.94 ± 0.01 [c]	3.52 ± 0.12 [c]	2.60 ± 0.03 [b]	112.09 ± 5.15 [c]	0.63 ± 0.07 [c]	127.76 ± 3.81 [d]
Oligosaccharide/white sugar (1:1)	16.58 ± 1.07 [b]	3.38 ± 0.60 [b]	4.03 ± 0.76 [b]	3.85 ± 0.49 [c]	2.49 ± 0.42 [b]	183.01 ± 23.89 [a]	1.01 ± 0.13 [a]	214.14 ± 18.69 [b]
Alcohol	36.66 ± 1.57 [a]	6.13 ± 0.67 [a]	5.16 ± 0.90 [a]	6.69 ± 0.42 [a]	2.76 ± 0.12 [a]	199.02 ± 16.23 [a]	0.95 ± 0.02 [a]	257.37 ± 13.31 [a]

Data are expressed as the mean (the average value of content for fresh weight, µg /g FW) and SD of three independent experiments. Different letters in the same column indicate significant differences ($p < 0.05$) between the values. **1**, schisandrol A; **2**, schisandrol B; **3**, tigloylgomisin H; **4**, angeloylgomisin H; **5**, schisandrin A; **6**, schisandrin B; **7**, schisandrin C.

After the fermentation process, the sugar content was reduced to 13.5, 13.9, and 12.1 Brix for the beverages fermented with white sugar, brown sugar, and oligosaccharide/white sugar, respectively, compared to the initial concentrations. This result is consistent with the previous result in which sucrose was detected before fermentation, but was not detected in the fermented beverage [17]. Among the three kinds of fermentable sugars used in this study, oligosaccharide/white sugar might have acted as the most efficient NADES for dissolution of the lignan compounds in the Omija berry. While there is no study on the bioconversion of dibenzocyclooctadiene-type lignans, a study on the bioconversion of cereal enterolignans by in vitro fermentation for 24 h showed that the total amount of enterolignans formed was closely correlated with the presence of hexoses and pentoses [26]. Studies on the bioconversion of plant lignans to enterolignans in the human gut were reported with various plants lignans, such as secoisolariciresinol and sesaminol [27,28]. Previous studies on the use of fructo-oligosaccharides as prebiotics showed several beneficial effects, including improving mineral absorption, decreasing serum cholesterol and triglycerides, and enhancing the growth of beneficial bacteria in the colon [29]. These reports suggest that intake of Omija beverages fermented with fructo-oligosaccharide can afford further health benefits.

This fermentation process is commonly used in Korea and is not affected by specific microbial organisms, such as yeast, bacteria, or mold, due to the osmotic effect of the high sugar concentration (about 50 Brix), which inhibits growth of the microbes; the process conditions are not controlled and specialized equipment is not required [30]. Recently, the importance of a green chemistry strategy has emerged due to the environmental sustainability and the minimalization of harmful chemical exposure [31]. Although further research is needed, this fermentation method employing a high sugar concentration is recommendable in terms of nontoxicity, eco-friendliness, and ready availability of the required systems.

3. Materials and Methods

3.1. Plant Materials

The stem, leaf, and flower of *S. chinensis* were collected in May 2016, and whole fruits were collected in August 2016, from Geochang province of South Korea. Each part of the plant was washed with tap water and dried in the shade. This plant was identified by Prof. Mi-Jeong Ahn, College of Pharmacy, Gyeongsang National University, and the voucher specimens were deposited in the herbarium of the College of Pharmacy, Gyeongsang National University (PGSC420–PGSC423). The fruit pulp was peeled off from the whole fruit to prepare the seeds separately. All samples were freeze-dried at $-55\,^{\circ}$C (FDB-5503, Operon, Kimpo, Korea) and stored at $-80\,^{\circ}$C before analysis.

3.2. Chemicals and Materials

Water and methanol used in the HPLC system were HPLC grade (Fisher Scientific Korea Ltd., Korea), and the other chemicals were of extra grade (Junsei Chemical Co. Ltd., Tokyo, Japan). Seven lignans for reference, including schisandrol A, schisandrol B, tigloylgomisin H, angeloylgomisin H, schisandrin A, schisandrin B, and schisandrin C were isolated and provided by Prof. Chul Young Kim, College of Pharmacy, Hanyang University [32].

3.3. Fermentation of the Omija Fruit with Sugars and Alcohol

First, 10 kg of raw fruit was fermented with white sugar (1:1, w/w) in a porcelain container placed outdoors. An aliquot of the resulting beverage was collected after 3, 7, 10, and 12 months and filtered. Separately, 1 kg of raw fruit was fermented for 12 months in three different ways using 1 kg each of white sugar, brown sugar, and fructo-oligosacchrides/white sugar (1:1, w/w). A transparent glass jar with a lid was used as the container for fermentation. After keeping the jar at room temperature during the indicated times, an aliquot of the sample was collected and filtered with gauze. The filtrate was kept at $-20\,^{\circ}$C before the analysis. The fermentation with alcohol was started with 1 L of commercial 30% ethanol for food processing. After 12 months, an aliquot of the sample was collected and filtered with gauze. The filtrate was used for compositional analysis of the lignans. The Brix of the sample was measured three times using a Pocket Refractometer PAL-1 (Atago Co., Ltd., Tokyo, Japan) with a range of 0.0–53.0% Brix. Refractometer calibration and sample readings were performed according to the manufacturer's instructions.

3.4. Sample Preparation

To prepare the sample, 1 g of freeze-dried pulp, seed, leaf, and stem, and 200 mg of flowers were ground and extracted under sonication with 200 mL of methanol for 2 h. The extracts were centrifuged at 3000 g at $4\,^{\circ}$C for 10 min, and the supernatant was filtered with a 0.45 μm PTFE membrane filter (Whatman, New York, NY, USA) for HPLC analysis.

Subsequently, 100 μL of filtrate obtained from each Omija beverage was diluted with 200 μL of MeOH. Thereafter, 200 μL of the diluted sample was loaded on Sep-Pak Cartridges (C18, Waters Corporation, Milford, MA, USA) and then eluted with 3 mL of MeOH. The eluent was analyzed by HPLC.

3.5. Lignan Analysis

An Agilent 1100 HPLC system (Hewlett-Packard, Waldbronn, Germany) equipped with an autosampler, a column oven, and a diode array detector was used to determine the content of the lignans. Standard solution and sample extracts were injected on a Phenomenex Luna C18 column (4.5×150 mm, 5 μm, Torrance, CA, USA) with the volume of 20 μL. The flow rate was 0.8 mL/min, and the column temperature was maintained at

30 °C. The mobile phase was a mixture of water (A) and methanol (B). Solvent (A) was kept at 90% for the first 5 min, then changed from 90% to 10% for next 30 min, and finally maintained at 90% for 5 min. The eluent was detected at wavelength 254 nm. Chemstation software (Agilent Technologies, Santa Clara, CA, USA) was used to operate this HPLC system and to manipulate the data.

Peaks of seven lignans were quantified by an external standard method. Seven lignans were separately dissolved in methanol, and serial dilution was performed to obtain final concentrations of 100, 50, 25, 12.5, 6.25, 3.13, 1.56, 0.78, 0.20, and 0.05 µg/mL. At this chromatographic condition, peaks of the standard lignans were detected at the following tR (min): 11.4 for schisandrol A, 12.9 for schisandrol B, 13.5 for tigloylgomisin H, 14.1 for angeloylgomisin H, 21.1 for schisandrin A, 24.4 for schisandrin B, and 27.0 for schisandrin C (Figures 1 and 2A). The content of lignans in the fruit was calculated from the experimental data of pericarp and seed.

3.6. Statistical Analysis

All experiments were performed in triplicate, and the data were expressed as mean $\pm$ SD (standard deviation) in Excel software. The one-way ANOVA (analysis of variance) was evaluated by the SPSS statistics 21.0 (IBM Corp, Armonk, NY, USA) using Dunnett's test. The statistically significance was considered at the value of $p < 0.05$.

4. Conclusions

The content of seven lignans in various plant parts of *S. chinensis*, including the stem, flower, leaves, seeds, and pulp, was determined using an HPLC-DAD method. The total lignan content followed the descending order of seed, flower, leaves, pulp, and stem. Analysis of seven major lignans during the fermentation of the Omija fruit with white sugar revealed that the lignan content increased by ~2.4-fold after seven months' fermentation compared to that at three months' fermentation, indicating that the lignan content changed in proportion to the fermentation period. The content of total lignans in the Omija beverage fermented with oligosaccharide/white sugar was approximately 1.2- and 1.7-fold higher than that of the Omija beverages fermented with white sugar and brown sugar, respectively. These results provide scientific background illuminating the eco-friendly food processing method using a high sugar concentration that promotes dissolution of the naturally occurring substances as well as the usage of other plant parts of *S. chinensis* as functional materials.

Author Contributions: Conceptualization, W.S.P., K.A.K. and J.-Y.B.; methodology, W.S.P., H.-J.K. and C.Y.K.; software, W.S.P. and D.-M.K.; validation, M.K.L. and H.-J.K.; formal analysis, D.-M.K., S.-M.P. and K.A.K.; resources, W.S.P., M.K.L. and C.Y.K.; writing—original draft preparation, W.S.P., J.-Y.B. and M.-J.A.; writing—review and editing, K.A.K., S.-M.P., J.-Y.B. and M.-J.A.; visualization, H.-J.K., J.-M.K. and M.K.L.; supervision, C.Y.K. and M.-J.A. All authors read and agreed to the published version of manuscript.

Funding: This research was funded by a grant from the National Research Foundation of Korea (NRF-2017R1A2B4008859) and the Basic Science Research Program of the Research Institute for Basic Sciences (RIBS) of Jeju National University through the National Research Foundation of Korea (NRF) funded by the Ministry of Education (2019R1A6A1A10072987).

Institutional Review Board Statement: Not applicable.

Informed Consent Statement: Not applicable.

Data Availability Statement: Not applicable.

Acknowledgments: In this section, you can acknowledge any support given which is not covered by the author contribution or funding sections. This may include administrative and technical support, or donations in kind (e.g., materials used for experiments).

Conflicts of Interest: The authors declare no conflict of interest.

References

1. Kim, S.-H.; Joo, M.H.; Yoo, S.H. Structural identification and antioxidant properties of major anthocyanin extracted from Omija (*Schizandra chinensis*) fruit. *J. Food Sci.* **2009**, *74*, 134–140. [CrossRef]
2. Panossian, A.; Wikman, G. Pharmacology of *Schisandra chinensis* Bail.: An overview of Russian research and uses in medicine. *J. Ethnopharmacol.* **2008**, *118*, 183–212. [CrossRef]
3. Nowak, A.; Zakłos-Szyda, M.; Błasiak, J.; Nowak, A.; Zhang, Z.; Zhang, B. Potential of *Schisandra chinensis* (Turcz.) Baill. in human health and nutrition: A review of current knowledge and therapeutic perspectives. *Nutrients* **2019**, *11*, 333. [CrossRef]
4. Nagappan, A.; Jung, D.Y.; Kim, J.-H.; Lee, H.; Jung, M.H. Gomisin N alleviates ethanol-induced liver injury through ameliorating lipid metabolism and oxidative stress. *Int. J. Mol. Sci.* **2018**, *19*, 2601. [CrossRef]
5. Park, S.Y.; Choung, S.Y. Inhibitory effect of schizandrin on nephrotoxicity of cisplatin. *Korean J. Environ. Toxicol.* **1998**, *13*, 125–131.
6. Szopa, A.; Ekiert, R.; Ekiert, H. Current knowledge of *Schisandra chinensis* (Turcz.) Baill. (Chinese magnolia vine) as a medicinal plant species: A review on bioactive components, pharmacological properties, analytical and biotechnological studies. *Phytochem. Rev.* **2017**, *16*, 195–218. [CrossRef]
7. Liu, H.; Lai, H.; Jia, X.; Zhang, Z.; Qi, Y.; Zhang, J.; Song, J.; Wu, C.; Zhang, B.; Xiao, P. Comprehensive chemical analysis of *Schisandra chinensis* by HPLC-DAD-MS combined with chemometrics. *Phytomedicine* **2013**, *20*, 1135–1143. [CrossRef]
8. Choi, Y.W.; Takamatsu, S.; Khan, S.I.; Srinivas, P.V.; Ferreira, D.; Zhao, J.; Khan, I.A. Schisandrene, a dibenzocyclooctadiene lignan from *Schisandra chinensis*: Structure-antioxidant activity relationships of dibenzocyclooctadiene lignans. *J. Nat. Prod.* **2006**, *69*, 356–359. [CrossRef]
9. Deng, Y.; Lim, J.; Nguyen, T.T.H.; Mok, I.K.; Piao, M.; Kim, D. Composition and biochemical properties of ale beer enriched with lignans from *Schisandra chinensis* Baillon (omija) fruits. *Food Sci. Biotechnol.* **2019**, *29*, 609–617. [CrossRef]
10. Ding, P.; Wang, B.; Song, X.; Li, X.-K.; Chen, T.; Liu, C. HPLC determination of six lignans in different parts of *Schisandra chinensis*. *China J. Chin. Mater. Med.* **2013**, *38*, 2078–2081.
11. Ekiert, R.J.; Szopa, A.; Ekiert, H.; Krzek, J.; Dzik, E. Analysis of lignans in *Schisandra chinensis* fruits, leaves, biomasses from in vitro cultures and food supplements. *J. Funct. Foods* **2013**, *5*, 1576–1581. [CrossRef]
12. Lee, K.S.; Lee, B.H.; Seong, B.J.; Kim, S.I.; Han, S.H.; Kim, G.H.; Park, S.B.; Kim, H.H.; Choi, T.Y. Chemical components composition on different parts of fruit in *Schisandra chinensis* Baillon. *J. Korean Soc. Food Sci. Nutr.* **2016**, *45*, 851–858. [CrossRef]
13. Slanina, J.; Táborská, E.; Lojkoá, E. Lignans in the seeds and fruits of *Schisandra chinensis* cultured in Europe. *Planta Med.* **1997**, *63*, 277–280. [CrossRef] [PubMed]
14. Nakajima, K.; Taguchi, H.; Ikeya, Y.; Endo, T.; Yosioka, I. Constituents of *Schizandra chinensis* Baill. XIII. Quantitative analysis of lignans in the fruits of *Schizandra chinensis* Baill. by high performance liquid chromatography. *Yakugaku Zasshi* **1983**, *103*, 743–749. [CrossRef]
15. Tvrdá, E.; Michalko, J.; Árvay, J.; Vukovic, N.L.; Ivanišová, E.; Ďuračka, M.; Matušíková, I.; Kačániová, M. Characterization of the Omija (*Schisandra chinensis*) extract and its effects on the bovine sperm vitality and oxidative profile during *in vitro* storage. *Evid. Based. Complement. Alternat. Med.* **2020**, 7123780. [CrossRef]
16. Guo, Y.; Chi, H.; Liu, X.; Sun, X.; Wang, Y.; Liu, S. Rapid characterization of Schisandra species by using direct analysis in real time mass spectrometry. *J. Pharm. Biomed. Anal.* **2021**, *192*, 113648. [CrossRef]
17. Cho, E.K.; Cho, H.E.; Choi, Y.J. Antioxidant and antibacterial activities, and tyrosinase and elastase inhibitory effect of fermented omija (*Schizandra chinensis* Baillon.) beverage. *J. Appl. Biol. Chem.* **2010**, *53*, 212–218. [CrossRef]
18. Hwang, K.-A.; Hwang, Y.-J.; Hwang, H.-J.; Hwang, I.-G.; Kim, Y.J. Evaluation of biological activities of fermented *Schisandra chinensis* extracts by Streptococcus thermophiles. *J. Korean Soc. Food Sci. Nutr.* **2018**, *47*, 1338–1343. [CrossRef]
19. Park, S.-C. Characteristics of fermented omija (*Schizandra chinensis* Baillon) sugar treatment extracts by *Lactobacillus* sp. *Korean J. Microbiol.* **2014**, *50*, 60–66. [CrossRef]
20. Jeong, J.H.; Jung, H.; Lee, S.R.; Lee, H.J.; Hwang, K.T.; Kim, T.Y. Anti-oxidant, antiproliferative and anti-inflammatory activities of the extracts from black raspberry fruits and wine. *Food Chem.* **2010**, *123*, 338–344. [CrossRef]
21. Hwa, S.-H.; Yeon, S.-J.; Hong, G.-E.; Cho, W.-Y.; Lee, H.-J.; Kim, J.-H.; Lee, C.-H. Evaluation of lignan compound content and bioactivity of raw omija and sugared omija in serum of Sprague Dawley rat. *Foods* **2019**, *8*, 373. [CrossRef]
22. Yu, J.S.; Wu, Y.H.; Tseng, C.K.; Lin, C.K.; Hsu, Y.C.; Chen, Y.H.; Lee, J.C. Schisandrin A inhibits dengue viral replication via upregulating antiviral interferon responses through STAT signaling pathway. *Sci. Rep.* **2017**, *7*, 45171. [CrossRef] [PubMed]
23. Yang, J.; IP, S.P.; Yeung, H.J.; Che, C. HPLC-MS analysis of *Schisandra* lignans and their metabolites in Caco-2 cell monolayer and rat everted gut sac models and in rat plasma. *Acta Pharm. Sin. B* **2011**, *1*, 46–55. [CrossRef]
24. Dai, Y.; van Spronsen, J.; Witkamp, G.J.; Verpoorte, R.; Choi, Y.H. Natural deep eutectic solvents as new potential media for green technology. *Anal. Chim. Acta* **2013**, *766*, 61–68. [CrossRef]
25. Dai, Y.; Verpoorte, R.; Choi, Y.H. Natural deep eutectic solvents providing enhanced stability of natural colorants from safflower (*Carthamus tinctorius*). *Food Chem.* **2014**, *159*, 116–121. [CrossRef]
26. Bartkiene, E.; Juodeikiene, G.; Basinskiene, L. *In vitro* fermentative production of plant lignans from cereal products in relationship with constituents of non-starch polysaccharides. *Food Technol. Biotechnol.* **2012**, *50*, 237–245.
27. Jan, K.C.; Hwang, L.S.; Ho, C.T. Biotransformation of sesaminol triglucoside to mammalian lignans by intestinal microbiota. *J. Agric. Food Chem.* **2009**, *57*, 6101–6106. [CrossRef]

28. Liu, Z.; Saarinen, N.M.; Thompson, L.U. Sesamin is one of the major precursors of mammalian lignans in sesame seed (*Sesamum indicum*) as observed *in vitro* and in rats. *Nutrition* **2006**, *136*, 906–912. [CrossRef]
29. Osama, O.I. Functional oligosaccharides: Chemicals structure, manufacturing, health benefits, applications and regulations. *J. Food Chem. Nanotechnol.* **2018**, *4*, 65–76. [CrossRef]
30. Jackson, R.S. Stuck and sluggish fermentation in Wine science principles and applications, a volume in Food Science and Technology. Chapter 7: Fermentation. In *Wine Science*, 3rd ed.; Academic Press/Elsevier Inc.: Cambridge, MA, USA, 2008; pp. 332–417.
31. Brahmachari, G. Design of organic transformations at ambient conditions: Our sincere efforts to the cause of green chemistry practice. *Chem. Rec.* **2016**, *16*, 98–123. [CrossRef]
32. Lee, H.J.; Kim, C.Y. Simultaneous determination of nine lignans using pressurized liquid extraction and HPLC-DAD in the fruits of *Schisandra chinensis*. *Food Chem.* **2010**, *120*, 1224–1228. [CrossRef]

plants

Article

Fatty Acid Derivatives Isolated from the Oil of *Persea americana* (Avocado) Protects against Neomycin-Induced Hair Cell Damage

SeonJu Park [1,†], Seo Yule Jeong [2,†], Youn Hee Nam [2], Jun Hyung Park [3], Isabel Rodriguez [2], Ji Heon Shim [2], Tamanna Yasmin [2], Hee Jae Kwak [3], Youngse Oh [3], Mira Oh [3], Kye Wan Lee [4], Jung Suk Lee [4], Do Hoon Kim [4], Yu Hwa Park [4], In Seok Moon [5], Se-Young Choung [6], Kwang Won Jeong [7], Bin Na Hong [2], Seung Hyun Kim [3,*] and Tong Ho Kang [2,*]

1 Chuncheon Center, Korea Basic Science Institute (KBSI), Chuncheon 24341, Korea; sjp19@kbsi.re.kr
2 Department of Oriental Medicine Biotechnology, College of Life Sciences and Graduate School of Biotechnology, Kyung Hee University, Gyeonggi 17104, Korea; tjdbf26@gmail.com (S.Y.J.); 01030084217@hanmail.net (Y.H.N.); isabelula3r@gmail.com (I.R.); jee1015235@gmail.com (J.H.S.); pm.tamanna.yasmin@gmail.com (T.Y.); habina22@hanmail.net (B.N.H.)
3 Yonsei Institute of Pharmaceutical Sciences, College of Pharmacy, Yonsei University, Incheon 21983, Korea; kayskaron@naver.com (J.H.P.); moon3685@naver.com (H.J.K.); oys9300@naver.com (Y.O.); purunmei2002@naver.com (M.O.)
4 R&D Center, Dongkook Pharm. Co. Ltd., Gyeonggi 16229, Korea; lkw1@dkpharm.co.kr (K.W.L.); ljs@dkpharm.co.kr (J.S.L.); kdh2@dkpharm.co.kr (D.H.K.); pyh@dkpharm.co.kr (Y.H.P.)
5 Department of Otorhinolaryngology, Yonsei University College of Medicine, Seoul 03722, Korea; ISMOONMD@yuhs.ac
6 Department of Preventive Pharmacy and Toxicology, College of Pharmacy, Kyung Hee University, Seoul 02453, Korea; sychoung@khu.ac.kr
7 Gachon Institute of Pharmaceutical Sciences, College of Pharmacy, Gachon University, Incheon 21936, Korea; kwjeong@gachon.ac.kr
* Correspondence: kimsh11@yonsei.ac.kr (S.H.K.); panjae@khu.ac.kr (T.H.K.)
† These two authors contributed equally to the work.

Citation: Park, S.; Jeong, S.Y.; Nam, Y.H.; Park, J.H.; Rodriguez, I.; Shim, J.H.; Yasmin, T.; Kwak, H.J.; Oh, Y.; Oh, M.; et al. Fatty Acid Derivatives Isolated from the Oil of *Persea americana* (Avocado) Protects against Neomycin-Induced Hair Cell Damage. *Plants* **2021**, *10*, 171. https://doi.org/10.3390/plants10010171

Received: 8 December 2020
Accepted: 10 January 2021
Published: 18 January 2021

Publisher's Note: MDPI stays neutral with regard to jurisdictional claims in published maps and institutional affiliations.

Abstract: Avocado oil is beneficial to human health and has been reported to have beneficial effects on sensorineural hearing loss (SNHL). However, the compounds in avocado oil that affect SNHL have not been identified. In this study, we identified 20 compounds from avocado oil, including two new and 18 known fatty acid derivatives, using extensive spectroscopic analysis. The efficacy of the isolated compounds for improving SNHL was investigated in an ototoxic zebrafish model. The two new compounds, namely (2*R*,4*R*,6*Z*)-1,2,4-trihydroxynonadec-6-ene and (2*R*,4*R*)-1,2,4-trihydroxyheptadecadi-14,16-ene (compounds 1 and 2), as well as compounds 7, 9, 14, 17 and 19 showed significant improvement in damaged hair cells in toxic zebrafish. These results led to the conclusion that compounds from avocado oil as well as oil itself have a regenerative effect on damaged otic hair cells in ototoxic zebrafish.

Keywords: avocado oil; fatty acids; hearing loss; zebrafish; hair cell

1. Introduction

Sensorineural hearing loss (SNHL) is a major disease that may be genetic or acquired as a consequence of disease, ototoxic drugs or chemicals, noise, or trauma, among other causes [1–3]. Around 466 million people in world have disabling hearing loss, and this number will be estimated to be over 900 million by 2050 [4]. Therefore, it is important to focus on preventing hearing loss. In a continuing study to identify phytochemicals that protect against auditory hair cell damage, we investigated the protective effects of compounds from avocado leaves on SNHL in vitro [5].

Avocado (*Persea americana* Mill.) is a fruit tree that is indigenous to tropical and subtropical regions [6]. Due to its functional properties and nutritional value, interest in the study of various parts of avocado tree such as its fruit, leaves and oil has increased. Avocado contains approximately 60% oil [7]. Avocado oil is renowned for its healing and renewing effects and reduced inflammation during the wound healing process [8]. In addition, it is a functional food that composed of a high content of unsaturated fatty acids especially oleic acid and linoleic acid (about 50% and 10% in total fatty acid), which are beneficial to human health. Among various unsaturated fatty acids, omega-3 fatty acids have previously demonstrated efficacy in preventing age-related hearing loss [9,10]. We previously reported the efficacy of avocado oil on sensorineural hearing loss in vitro and in vivo [2]. However, the compounds in avocado oil that improve SNHL have not been identified. Therefore, we aimed to isolate compounds from avocado oil and confirm their efficacy for improving hearing loss in an ototoxic zebrafish model. Thus, in a continuing project to identify phytochemicals that improve auditory hair cell function, phytochemical investigation and the biological evaluation of avocado oil led to isolation of two new compounds and 18 known fatty acid derivatives. The protecting effects of the isolated compounds against neomycin-induced hearing loss was investigated in a zebrafish model.

2. Results and Discussion

2.1. Structure Elucidation

Compound **1** was isolated as a colorless oil. The molecular formula was determined to be $C_{19}H_{38}O_3$ on the basis of HR–ESI–MS pseudo-ion at m/z 297.2775 [M + H−H$_2$O]$^+$ (calcd for $C_{19}H_{37}O_2$, 297.2788). The ^{1}H-NMR spectrum of **1** was similar to that of (2*R*,4*R*,6*E*)-1,2,4-trihydroxynonadec-6-ene [11]. There were three hydroxy proton signals at δ_H 3.52 (2H, dd, J = 5.3, 14.3 Hz), 3.81 (1H, dt, J = 4.1, 8.2 Hz) and 3.85 (1H, ddt, J = 2.4, 4.4, 9.2 Hz) (Table 1). In **1**, two *cis* olefinic protons at δ_H 5.38 and 5.39 (each 1H, overlapped) replaced the two *trans* olefinic protons at 5.45 and 5.50 (each 1H, dt, J = 6, 15 Hz). A double bond that is located at C-6 and C-7 was verified from the COSY (Figure 1), which exhibited correlations between H-5 (δ_H 2.82) and H-4 (δ_H 3.81), H-6 (δ_H 5.39); H-7 (δ_H 5.38) and H-6 (δ_H 5.39), H-8 (δ_H 2.11). The ^{13}C and HSQC spectra of **1** exhibited nineteen carbon signals, including fourteen methylenes (one oxygenated), four methines (two olefinics) and one methyl signal. Unlike the two *trans* olefinic carbons of (2*R*,4*R*,6*E*)-1,2,4-trihydroxynonadec-6-ene at δ_C 127.2 and 134.4, the signal difference between the two *cis* olefinic carbons of **1** at δ_C 129.1 and 130.9 is smaller. Thus, this allows the planar structure of **1** to be (6*Z*)-1,2,4-trihydroxynonadec-6-ene. Based on the NOE correlation between H-2 (δ_H 3.85) and H-4 (δ_H 3.81) and comparing the optical rotation of **1** with two possible synthesized stereoisomers of 1,2,4-trihydroxyheptadec-16-ene, the absolute configuration of **1** is 2*R*, 4*R*. The synthetic (2*R*,4*R*)-1,2,4-trihydroxyheptadec-16-ene has given negative optical rotation ($[\alpha]_D^{20}$: −6.4 (c 1.1, CHCl$_3$)) as compound **1**, whereas (2*S*,4*S*)-1,2,4-trihydroxyheptadec-16-ene has positive value, $[\alpha]_D^{20}$: +6.0 (c 1.0, CHCl$_3$) [12]. Thus, the structure of **1** is (2*R*,4*R*,6*Z*)-1,2,4-trihydroxynonadec-6-ene.

Compound **2** was isolated as a colorless oil and its molecular formula was confirmed to be $C_{17}H_{32}O_3$ by the HR–ESI–MS ion at m/z 267.2314 [M + H − H$_2$O]$^+$ (calcd for $C_{17}H_{31}O_2$, 267.2324). The ^{1}H NMR spectrum showed three major regions: a large number of methylenes from δ_H 1.20 to 1.60, several oxygenated protons from δ_H 3.50 to 3.80, and two double bonds from δ_H 4.90 to 6.30. The ^{13}C and HSQC spectra of **1** exhibited seventeen carbon signals, including twelve methylenes (one olefinic, one oxygenated) and five methines (three olefinic) signals. The ^{1}H-NMR spectrum of **2** exhibited similar ^{1}H signals to that of compound **1** except for the additional double bond. The most downfield ^{1}H signal (δ_H 6.32, H-16) displayed a typical exomethylene double bond. J values of 17.0 and 10.3 Hz were observed for *trans* and *cis* protons (H-17$_{trans}$ and H-17$_{cis}$), respectively. H-17$_{trans}$ (δ_H 5.07) exhibited a dd pattern with J values of 1.8 and 17.0 Hz due to H-16 and H-17$_{cis}$, while H-17$_{cis}$ (δ_H 4.93) with a ddt pattern displayed J values of 1.8 and 10.3 Hz for H-16 and H-17$_{trans}$. Additionally, another double bond was located next to the terminal double bond, which was determined by correlations between H-16 (δ_H 6.32)/H-15 (δ_H 6.03) and C-13 (δ_C 33.6)/C-14 (δ_C 136.2)/C-15 (δ_C 132.4)/C-16 (δ_C 138.7)/C-17 (δ_C 114.9) (Figure 1).

A previous study related to relative configuration was conducted by synthesizing all four stereoisomers of 1,2,4-trihydroxyheptadec-16-ene. Both compounds **2** and (2*R*,4*R*)-1,2,4-trihydroxyheptadec-16-ene exhibited negative optical rotation and thus the configurations of C-2 and C-4 were determined to be *R* [12]. Based on the evidence above, compound **2** is (2*R*,4*R*)-1,2,4-trihydroxyheptadecadi-14,16-ene.

Table 1. NMR spectroscopic data for compounds **1** and **2**.

Pos.	**1** δ_C [a, b]	**1** δ_H [a, c] (*J* in Hz)	**2** δ_C [a, b]	**2** δ_H [a, c] (*J* in Hz)
1	67.2	3.52 (dd, 5.3, 14.3)	67.2	3.50 (dd, 5.3, 14.3)
2	72.1	3.85 (ddt, 2.4, 4.4, 9.2)	72.1	3.82 (m)
3	41.2	1.55 (m) 1.69 (m)	41.2	1.54* 1.68 (dt, 4.5, 14.2)
4	71.2	3.81 (dt, 4.1, 8.2)	71.2	3.78 (m)
5	26.5	5.39 (m)	38.5	1.47 *
6	129.1	5.38 (m)	(C-6–C-12) 30.3, 30.4, 30.6, 30.7, 30.8, 30.8, 30.9	(H-6–H-12) 1.31 *
7	130.9	2.11 (m)		
8	28.2	1.34 *		
9	32.6	1.37*		
10				
11				
12	(C-10–C-16) 30.2, 30.4, 30.5, 30.7, 30.8, 30.8, 30.9,	(H-10–H-16) 1.36 *		
13			33.6	2.1 (qd, 1.6, 7.1)
14			136.2	5.71 (dt, 7.1, 15.2)
15			132.4	6.06 (dd, 10.3, 15.2)
16			138.7	6.32 (dt, 10.3, 17.0)
17	32.7	1.34 *	114.9	4.93 (dd, 1.8, 10.3) 5.07 (dd, 1.8, 17.0)
18	23.6	1.40 *		
19	14.5	0.95 (t)		

[a] Measured in methanol-d_4, [b] 100 MHz, [c] 600 MHz, * overlapped signal, assignments were done by HSQC, HMBC, COSY and NOESY.

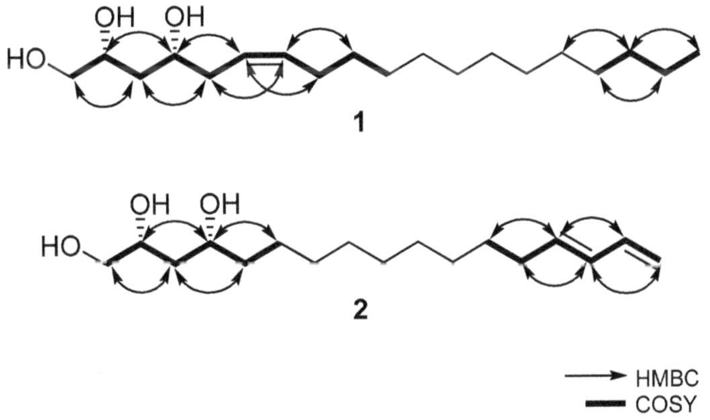

Figure 1. The key HMBC and COSY correlations of **1** and **2**.

By comparing the NMR and MS data with previously reported literature, known compounds were identified as (2*R*,4*R*)-1,2,4-trihydroxynonadecane, (2*R*,4*R*)-1,2,4-trihydroxyheptadec-16-yne, avocadene (**3–4**, **6**) [13], 4-acetoxy-1,2-dihydroxyheptadec-16-yne (**5**) [11], avocadenol A (**7**) [14], 1-palmitoleylglycerol (**8**) [15], palmitoleic acid and oleic acid (**9–10**) [16], linoleic acid (**11**) [17], *α*-linolenic acid (**12**) [18], avocadene acetate, 1-(acetyloxy)-2-hydroxy-4-heptadecanone, avocadenone acetate (**13**, **15–16**) [19], 1,4*R*-diacetoxy-2*R*-hydroxyheptadeca-16-ene (**14**) [20], persenone B (**17**) [21], avocadyne acetate (**18**) [22],

(5*E*,12*Z*)-2-hydroxy-4-oxoheneicosa-5,12-dien-1-yl acetate and (5*E*,12*Z*,15*Z*)-2-hydroxy-4-oxoheneicosa-5,12,15-trien-1-yl acetate (**19–20**) [23] (Figure 2).

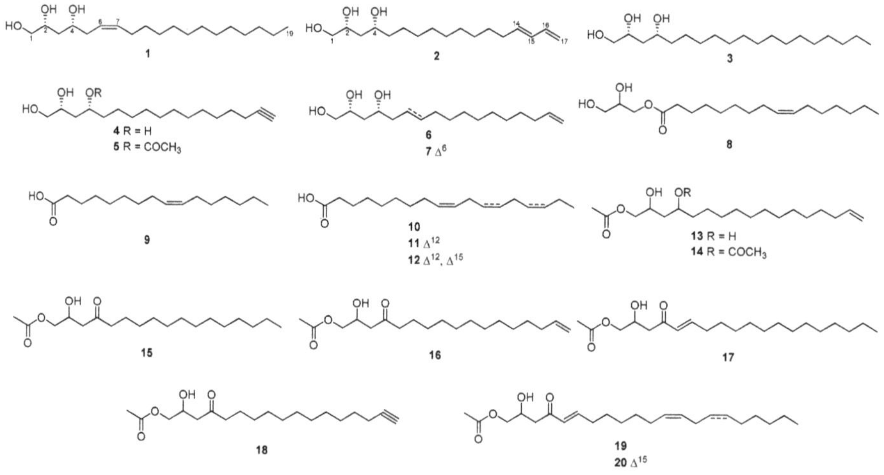

Figure 2. Chemical structures of compounds **1–20**.

2.2. Otic Hair Cell Recovery

The effects of compounds **1–20** on otic hair cell recovery were evaluated after neomycin exposure using a zebrafish model. Hair cells in zebrafish neuromast have reported apoptosis during normal turnover and after damage by aminoglycosides, similarly to the hair cells in an organ of Corti in mammals [24–30]. In addition, genes and pathways involved in apoptosis, neuronal development, and oxidation are highly conserved in zebrafish [31,32], which allows to study the effects of potential therapeutics, such as neurotrophic factors, antioxidants and antiapoptotic agents on hair cell loss. The hair cells within the otic (O1) neuromast were severely damaged by neomycin exposure (Figure 3). Neomycin significantly decreased the number of otic hair cells ($p < 0.001$). However, 1 µM of compound **1** ($p < 0.001$), **2** ($p < 0.001$), **7** ($p < 0.001$), and **9** ($p = 0.04$) exhibited significant hair cell recovery compared to the control. Furthermore, compounds **14** ($p = 0.01$), **17** ($p = 0.006$), and **19** ($p = 0.01$) led to significant recovery in hair cells compared to controls. In our previous study, avocado oil had a protective effect against neomycin-induced ototoxicity in sensory hair cells. These beneficial effects may be related to avocado oil compounds **1, 2, 7, 9, 14, 17**, and **19**, which significantly increased otic hair cell recovery after neomycin-induced ototoxicity in a zebrafish model. All those compounds except **9** are the avocado-derived fatty acids. Compound **9**, palmitoleic acid, is a monounsaturated fatty acid that can be found in a variety of animal fats, vegetable oils, and marine oils. Future studies should focus on these compounds and their mechanism of action in SNHL.

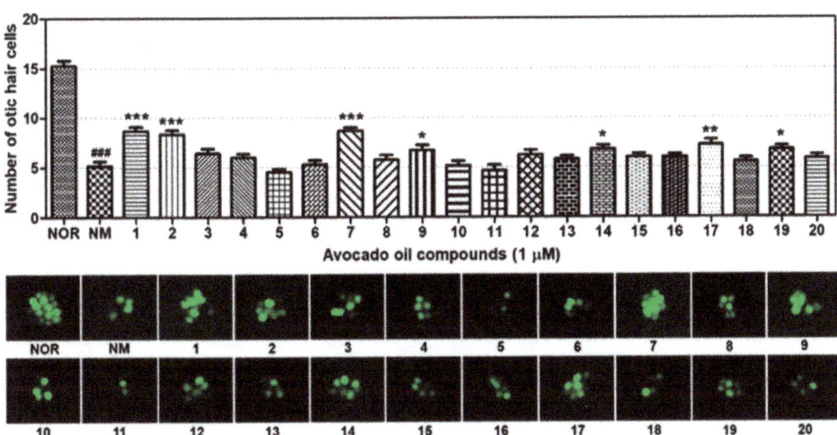

Figure 3. Otic hair cell recovery of compounds **1–20** after neomycin-induced hair cell damage. Data are presented as means ± SEM. ### $p < 0.001$ (NOR group versus NM group). * $p < 0.05$, ** $p < 0.01$ and *** $p < 0.001$ (NM versus avocado oil compounds treated groups).

3. Materials and Methods

3.1. General Experimental Procedures

The resonances for ^{1}H and ^{13}C were observed at 400 and 100 MHz, respectively, in an Agilent 400-MR-NMR spectrometer. Chemical shifts are expressed in parts per million in the scale relative to tetramethylsilane. Data processing was performed by the MestReNova ver. 6.0.2 program. HR–ESI–MS analyses and prep-HPLC were carried out by an AGILENT 6550 iFunnel Q-TOF LC/MS system and by an AGILENT 1200 HPLC system, respectively. Column chromatography was done on silica–gel (Kieselgel 60, 70–230 mesh and 230–400 mesh, Merck, Kenilworth, NJ, USA) or YMC RP-18 resins (30–50 μm, Fujisilisa Chemical Ltd., Kasugai Aichi, Japan). A pre-coated silica–gel 60 F254 (0.25 mm, Merck) and RP-18 F254S plates (0.25 mm, Merck) were used for thin layer chromatography (TLC).

3.2. Plant Material

Avocado oil that was made from whole parts of the fruit (peel, pulp and seed) was purchased from Avoplus in Morelia, Mexico, in 2018. It was manufactured by a general manufacturing process for avocado oil: grinding fully ripe fruits and undergoing centrifugal extraction. Avocado fruits were supplied by certified producers and Avoplus, the oil manufacturer, has its certification.

3.3. Extraction and Isolation

The oil of *P. americana* (60 g) was subjected to column chromatography (CC) separation over silica gel and eluted with a gradient of CHCl$_3$: MeOH (17:1, v/v) gave four smaller fractions, AVO1 (0.4 g), AVO2 (0.3 g), AVO3 (0.8 g) and AVO4 (0.8 g) and cleansed with MeOH (AVO-M). The AVO1 fraction was performed prep-HPLC with J'sphere ODS H-80 column (250 mm × 20 mm), eluting with 78% MeCN and a flow rate of 3 mL/min to yield 15 (12.7 mg), 16 (14.1 mg), 17 (18.6 mg), 18 (1.9 mg), 19 (4.2 mg) and 20 (43.3 mg). The AVO2 fraction was also applied to prep-HPLC as conditions above with 75% MeCN to yield 11 (117.4 mg), 12 (13.2 mg), and 13 (53.2 mg), whereas the AVO4 fraction was applied to silica gel CC, eluting with CHCl$_3$: MeOH (20:1, v/v) to obtain 10 (994.0 mg). The AVO-M fraction was applied to a silica gel CC with CHCl$_3$: MeOH (15: 1, v/v) and three sub-fractions, AVO-M1 (1.0 g), AVO-M2 (1.3 g) and AVO-M3 (1.8 g), were obtained. The AVO-M1 fraction was purified further using a RP-18 CC, which eluted with acetone: H$_2$O (4:1, v/v) to yield 8 (13.4 mg), 9 (23.2 mg), and 14 (44.1 mg). The AVO-M2 fraction was subjected to prep-HPLC as per the above conditions with 65% MeCN to yield 1 (2.4 mg), 2 (4.0 mg), and 4 (13.7 mg). Lastly, the AVO-M3 fraction was performed and RP-18 CC eluted with acetone: H_2O (4:1, *v/v*) to yield 3 (23.4 mg), 5 (32.7 mg), 6 (2.3 mg), and 7 (3.2 mg).

3.3.1. (2R,4R,6Z)-1,2,4-Trihydroxynonadec-6-ene (1)

Colorless oil; $[\alpha]_D^{20}$: −33.4 (c 0.02, $CHCl_3$); $C_{19}H_{38}O_3$, HR-ESI-MS *m/z*: 297.2775 $[M + H - H_2O]^+$ (calcd for $C_{19}H_{37}O_2$, 297.2788); ^1H (CD_3OD, 600 MHz) and ^{13}C NMR (CD_3OD, 100 MHz) data can be seen in Table 1.

3.3.2. (2R,4R)-1,2,4-Trihydroxyheptadecadi-14,16-ene (2)

Colorless oil; $[\alpha]_D^{20}$: −38.0 (c 0.02, $CHCl_3$); $C_{17}H_{32}O_3$, HR-ESI-MS *m/z*: 267.2314 $[M + H - H_2O]^+$ (calcd for $C_{17}H_{31}O_2$, 267.2324); ^1H (CD_3OD, 600 MHz) and ^{13}C NMR (CD_3OD, 100 MHz) data can be seen in Table 1.

3.4. Animals

Wild-type adult zebrafish (*Danio rerio*) were cultured in a zebtec stand-alone system (1500(W) × 400(D) × 2050(H) mm, WoojungBio, Inc., Suwon, Korea). Three pairs of zebrafish were set-up overnight in a spawning box and eggs were gathered 3 h after fertilization. Embryos were cultured in 0.03% of sea salt solution (Sigma-Aldrich Co., St. Louis, MO, USA) in a petri dish under a 14 h light/10 h dark cycle in an incubator at 26.5–28.5 °C until 6 days post-fertilization (dpf) when the experiments were performed as previously described [2].

3.5. Ethical Statement

All zebrafish experimental procedures were performed in accordance with standard zebrafish protocols and the Animal Care and Use Committee of Kyung Hee University [KHUASP(SE)-15-10] approved the experiments.

3.6. Neomycin-Induced Ototoxicity in a Zebrafish Model

Wild-type zebrafish larvae were sited into a 96-well plate and treated with 100 μL of 2 μM neomycin sulfate (MB Cell Co., Seoul, Korea) for an hour to induce ototoxicity as previously described [2].

To evaluate the protective effects of compounds 1–20 on otic hair cell damage, the neomycin solution was removed and zebrafish were rinsed with 0.03% sea salt solution. The zebrafish were then treated with 100 μL of 1 μM compounds 1–20. After 8 h treatment, the zebrafish were rinsed again with 0.03% sea salt solution and stained for 30 min with 0.1% YO-PRO-1 (Fisher Scientific Inc., Hampton, MA, USA), then anesthetized with 0.04% tricaine. The otic hair cells were counted under a fluorescence microscope (Olympus 1 × 70; Olympus Co., Tokyo, Japan) after visualization. Image analysis was performed by Focus Lite software (Focus Co., Daejeon, Korea).

3.7. Statistical Analysis

Data were analyzed by GraphPad Prism (version 5) statistical software package (GraphPad, San Diego, CA, USA). All data are expressed as the means ± standard error of the mean. The statistical significance of the differences between groups was determined by a paired *t*-test. *p*-value of <0.05 (*), <0.01 (**) and <0.001 (***) are considered statistically significant.

4. Conclusions

A detailed phytochemical study of avocado oil identified two previously undescribed fatty acid derivatives, (2R,4R,6Z)-1,2,4-trihydroxynonadec-6-ene and (2R,4R)-1,2,4-trihydroxyheptadecadi-14,16-ene, as well as 18 known compounds. We evaluated the protective effect of these 20 compounds against neomycin-induced ototoxicity in a zebrafish model. Our results demonstrated that new compounds 1 and 2, as well as compounds 7, 9, 14, 17, and 19 protect otic hair cells from ototoxicity.

Author Contributions: Conceptualization I.S.M., S.-Y.C., K.W.J., B.N.H., S.H.K. and T.H.K.; methodology, Y.H.N., I.R., S.Y.J., J.H.S., T.Y., K.W.L., J.S.L., D.H.K. and Y.H.P.; formal analysis, S.P., Y.H.N., I.R. and S.Y.J.; investigation, S.P., S.Y.J., J.H.P., H.J.K., Y.O., M.O. and J.H.S.; resources, K.W.L., J.S.L., D.H.K. and Y.H.P.; writing—original draft preparation, S.P., S.Y.J., Y.H.N. and I.R.; writing—review and editing, S.P., S.Y.J., Y.H.N. and I.R.; supervision, S.H.K. and T.H.K. All authors have read and agreed to the published version of the manuscript.

Funding: This work was supported by the Technology Innovation Program (10080691, Development of an individual approval health functional food and global product; protection of hearing loss and improvement of auditory function for the people with light to moderate hearing loss (26–55 dB of pure tone audiometry threshold) using avocado fraction) funded By the Ministry of Trade, Industry & Energy (MOTIE, Korea).

Institutional Review Board Statement: All zebrafish experimental procedures were carried out in accordance with standard zebrafish protocols and were approved by the Animal Care and Use Committee of Kyung Hee University (Protocol No. KHUASP(SE)-15-10).

Informed Consent Statement: Not applicable.

Data Availability Statement: The data presented in this study are available on request from the corresponding author.

Conflicts of Interest: The authors declare no conflict of interest.

References

1. Basner, M.; Babisch, W.; Davis, A.; Brink, M.; Clark, C.; Janssen, S.; Stansfeld, S. Auditory and non-auditory effects of noise on health. *Lancet* **2014**, *383*, 1325–1332. [CrossRef]
2. Nam, Y.H.; Rodriguez, I.; Jeong, S.Y.; Pham, T.N.M.; Nuankaew, W.; Kim, Y.H.; Castañeda, R.; Jeong, S.Y.; Park, M.S.; Lee, K.W. Avocado oil extract modulates auditory hair cell function through the regulation of amino acid biosynthesis genes. *Nutrients* **2019**, *11*, 113. [CrossRef] [PubMed]
3. Shargorodsky, J.; Curhan, S.G.; Curhan, G.C.; Eavey, R. Change in prevalence of hearing loss in US adolescents. *JAMA* **2010**, *304*, 772–778. [CrossRef]
4. D'Haese, P.S.; Van Rompaey, V.; De Bodt, M.; Van de Heyning, P. Severe hearing loss in the ageing population poses a global public health challenge. How can we better realise the Benefits of cochlear implantation to mitigate this crisis? *Front. Public Health* **2019**, *7*, 227. [CrossRef] [PubMed]
5. Park, S.; Nam, Y.H.; Rodriguez, I.; Park, J.H.; Kwak, H.J.; Oh, Y.; Oh, M.; Park, M.S.; Lee, K.W.; Lee, J.S. Chemical constituents of leaves of *Persea americana* (avocado) and their protective effects against neomycin-induced hair cell damage. *Rev. Bras. Farmacogn.* **2019**, *29*, 739–743. [CrossRef]
6. Di Stefano, V.; Avellone, G.; Bongiorno, D.; Indelicato, S.; Massenti, R.; Lo Bianco, R. Quantitative evaluation of the phenolic profile in fruits of six avocado (*Persea americana*) cultivars by ultra-high-performance liquid chromatography-heated electrospray-mass spectrometry. *Int. J. Food Prop.* **2017**, *20*, 1302–1312. [CrossRef]
7. Tan, C.; Tan, S.; Tan, S. Influence of geographical origins on the physicochemical properties of Hass avocado oil. *J. Am. Oil Chem. Soc.* **2017**, *94*, 1431–1437. [CrossRef]
8. de Oliveira, A.P.; Franco, E.D.; Rodrigues Barreto, R.; Cordeiro, D.P.; de Melo, R.G.; de Aquino, C.M.F.; de Medeiros, P.L.; da Silva, T.G.; Góes, A.J.; Maia, M.B. Effect of semisolid formulation of *Persea americana* Mill (avocado) oil on wound healing in rats. *Evid.-Based Complement. Altern. Med.* **2013**, *2013*, 472382. [CrossRef]
9. Gopinath, B.; Flood, V.M.; Rochtchina, E.; McMahon, C.M.; Mitchell, P. Consumption of omega-3 fatty acids and fish and risk of age-related hearing loss. *Am. J. Clin. Nutr.* **2010**, *92*, 416–421. [CrossRef]
10. Martínez-Vega, R.; Partearroyo, T.; Vallecillo, N.; Varela-Moreiras, G.; Pajares, M.A.; Varela-Nieto, I. Long-term omega-3 fatty acid supplementation prevents expression changes in cochlear homocysteine metabolism and ameliorates progressive hearing loss in C57BL/6J mice. *J. Nutr. Biochem.* **2015**, *26*, 1424–1433. [CrossRef]
11. Abe, F.; Nagafuji, S.; Okawa, M.; Kinjo, J.; Akahane, H.; Ogura, T.; Martinez-Alfaro, M.A.; Reyes-Chilpa, R. Trypanocidal constituents in plants 5. Evaluation of some Mexican plants for their trypanocidal activity and active constituents in the seeds of *Persea americana*. *Biol. Pharm. Bull.* **2005**, *28*, 1314–1317. [CrossRef] [PubMed]
12. Sugiyama, T.; Sato, A.; Yamashita, K. Synthesis of all four stereoisomers of antibacterial component of avocado. *Agric. Biol. Chem.* **1982**, *46*, 481–485. [CrossRef]
13. Oberlies, N.H.; Rogers, L.L.; Martin, J.M.; McLaughlin, J.L. Cytotoxic and insecticidal constituents of the unripe fruit of *Persea americana*. *J. Nat. Prod.* **1998**, *61*, 781–785. [CrossRef] [PubMed]

14. Lu, Y.-C.; Chang, H.-S.; Peng, C.-F.; Lin, C.-H.; Chen, I.-S. Secondary metabolites from the unripe pulp of *Persea americana* and their antimycobacterial activities. *Food Chem.* **2012**, *135*, 2904–2909. [CrossRef]

15. Coleman, B.E.; Cwynar, V.; Hart, D.J.; Havas, F.; Mohan, J.M.; Patterson, S.; Ridenour, S.; Schmidt, M.; Smith, E.; Wells, A.J Modular approach to the synthesis of unsaturated 1-monoacyl glycerols. *Synlett* **2004**, *2004*, 1339–1342. [CrossRef]

16. Batchelor, J.G.; Cushley, R.J.; Prestegard, J.H. Carbon-13 Fourier transform nuclear magnetic resonance. VIII. Role of steric and electric field effects in fatty acid spectra. *J. Org. Chem.* **1974**, *39*, 1698–1705. [CrossRef]

17. Marwah, R.G.; Fatope, M.O.; Deadman, M.L.; Al-Maqbali, Y.M.; Husband, J. Musanahol: A new aureonitol-related metabolite from a *Chaetomium* sp. *Tetrahedron* **2007**, *63*, 8174–8180. [CrossRef]

18. Lee, S.O.; Choi, S.Z.; Choi, S.U.; Ryu, S.Y.; Lee, K.R. Phytochemical constituents of the aerial parts from *Aster hispidus*. *Nat. Prod. Sci.* **2004**, *10*, 335–340.

19. Degenhardt, A.G.; Hofmann, T. Bitter-tasting and kokumi-enhancing molecules in thermally processed avocado (*Persea americana* Mill.). *J. Agric. Food Chem.* **2010**, *58*, 12906–12915. [CrossRef]

20. Lee, T.H.; Tsai, Y.F.; Huang, T.T.; Chen, P.Y.; Liang, W.L.; Lee, C.K. Heptadecanols from the leaves of *Persea americana* var. *americana*. *Food Chem.* **2012**, *132*, 921–924. [CrossRef]

21. Kim, O.K.; Murakami, A.; Nakamura, Y.; Takeda, N.; Yoshizumi, H.; Ohigashi, H. Novel nitric oxide and superoxide generation inhibitors, persenone A and B, from avocado fruit. *J. Agric. Food Chem.* **2000**, *48*, 1557–1563. [CrossRef] [PubMed]

22. Kashman, Y.; Neeman, I.; Lifshitz, A. New compounds from avocado pear. *Tetrahedron* **1969**, *25*, 4617–4631. [CrossRef]

23. Kawagishi, H.; Fukumoto, Y.; Hatakeyama, M.; He, P.; Arimoto, H.; Matsuzawa, T.; Arimoto, Y.; Suganuma, H.; Inakuma, T.; Sugiyama, K. Liver injury suppressing compounds from avocado (*Persea americana*). *J. Agric. Food Chem.* **2001**, *49*, 2215–2221. [CrossRef] [PubMed]

24. Fritzsch, B. The amphibian octavo-lateralis system and its regressive and progressive evolution. *Acta Biol. Hung.* **1988**, *39*, 305–322.

25. Kornblum, H.I.; Corwin, J.T.; Trevarrow, B. Selective labeling of sensory hair cells and neurons in auditory, vestibular, and lateral line systems by a monoclonal antibody. *J. Comp. Neurol.* **1990**, *301*, 162–170. [CrossRef]

26. Higgs, D.M.; Souza, M.J.; Wilkins, H.R.; Presson, J.C.; Popper, A.N. Age-and size-related changes in the inner ear and hearing ability of the adult zebrafish (*Danio rerio*). *J. Assoc. Res. Otolaryngol.* **2002**, *3*, 174–184. [CrossRef]

27. Higgs, D.M.; Rollo, A.K.; Souza, M.J.; Popper, A.N. Development of form and function in peripheral auditory structures of the zebrafish (*Danio rerio*). *J. Acoust. Soc. Am.* **2003**, *113*, 1145–1154. [CrossRef]

28. Williams, J.A.; Holder, N. Cell turnover in neuromasts of zebrafish larvae. *Hear. Res.* **2000**, *143*, 171–181. [CrossRef]

29. Murakami, S.L.; Cunningham, L.L.; Werner, L.A.; Bauer, E.; Pujol, R.; Raible, D.W.; Rubel, E.W. Developmental differences in susceptibility to neomycin-induced hair cell death in the lateral line neuromasts of zebrafish (*Danio rerio*). *Hear. Res.* **2003**, *186*, 47–56. [CrossRef]

30. Matsui, J.I.; Cotanche, D.A. Sensory hair cell death and regeneration: Two halves of the same equation. *Curr. Opin. Otolaryngol. Head Neck Surg.* **2004**, *12*, 418–425. [CrossRef]

31. Inohara, N.; Nunez, G. Genes with homology to mammalian apoptosis regulators identified in zebrafish. *Cell Death Differ.* **2000**, *7*, 509–510. [CrossRef] [PubMed]

32. Ton, C.; Parng, C. The use of zebrafish for assessing ototoxic and otoprotective agents. *Hear. Res.* **2005**, *208*, 79–88. [CrossRef] [PubMed]

Article

Phytochemical Analysis and Evaluation of Antioxidant and Biological Activities of Extracts from Three Clauseneae Plants in Northern Thailand

Keerati Tanruean [1], Pisit Poolprasert [1], Nakarin Suwannarach [2,3], Jaturong Kumla [2,3] and Saisamorn Lumyong [2,3,4,*]

1   Biology Program, Faculty of Science and Technology, Pibulsongkram Rajabhat University, Phitsanulok 65000, Thailand; keerati.t@psru.ac.th (K.T.); poolprasert_p@psru.ac.th (P.P.)
2   Research Center of Microbial Diversity and Sustainable Utilization, Faculty of Science, Chiang Mai University, Chiang Mai 50200, Thailand; suwan.462@gmail.com (N.S.); jaturong_yai@hotmail.com (J.K.)
3   Department of Biology, Faculty of Science, Chiang Mai University, Chiang Mai 50200, Thailand
4   Academy of Science, The Royal Society of Thailand, Bangkok 10200, Thailand
*   Correspondence: scboi009@gmail.com or saisamorn.l@cmu.ac.th; Tel.: +668-1881-3658

Abstract: This study established the DNA barcoding sequences (*mat*K and *rbc*L) of three plant species identified in the tribe Clauseneae, namely *Clausena excavata*, *C. harmandiana* and *Murraya koenigii*. The total phenolic and total flavonoid contents, together with the biological activities of the derived essential oils and methanol extracts, were also investigated. Herein, the success of obtaining sequences of these plant using two different barcode genes *mat*K and *rbc*L were 62.5% and 100%, respectively. Both regions were discriminated by around 700 base pairs and these had resemblance with those of the Clausenae materials earlier deposited in Genbank at a 99–100% degree of identity. Additionally, the use of *mat*K DNA sequences could positively confirm the identity as monophyletic. The highest total phenolic and total flavonoid content values ($p < 0.05$) were observed in the methanol extract of *M. koenigii* at 43.50 mg GAE/g extract and 66.13 mg QE/g extract, respectively. Furthermore, anethole was detected as the dominant compound in *C. excavata* (86.72%) and *C. harmandiana* (46.09%). Moreover, anethole (26.02%) and caryophyllene (21.15%) were identified as the major phytochemical compounds of *M. koenigii*. In terms of the biological properties, the *M. koenigii* methanol extract was found to display the greatest amount of antioxidant activity (DPPH; IC_{50} 95.54 µg/mL, ABTS value 118.12 mg GAE/g extract, FRAP value 48.15 mg GAE/g extract), and also revealed the highest α-glucosidase and antihypertensive inhibitory activities with percent inhibition values of 84.55 and 84.95. Notably, no adverse effects on human peripheral blood mononuclear cells were observed with regard to all of the plant extracts. Furthermore, *M. koenigii* methanol extract exhibited promise against human lung cancer cells almost at 80% after 24 h and 90% over 48 h.

Keywords: tribe Clauseneae; DNA barcode; volatile compounds; antioxidant activity; ACE inhibitory activity; anticancer activity; α-glucosidase inhibitory activity

Citation: Tanruean, K.; Poolprasert, P.; Suwannarach, N.; Kumla, J.; Lumyong, S. Phytochemical Analysis and Evaluation of Antioxidant and Biological Activities of Extracts from Three Clauseneae Plants in Northern Thailand. *Plants* **2021**, *10*, 117. https://doi.org/10.3390/plants10010117

Received: 12 December 2020
Accepted: 5 January 2021
Published: 8 January 2021

Publisher's Note: MDPI stays neutral with regard to jurisdictional claims in published maps and institutional affiliations.

1. Introduction

Oxidative stress and autoxidation are known to be harmful to the human body. The presence of these conditions can result from an imbalance between free radicals and antioxidants present in the body, which can ultimately be damaging to the human body's cells, proteins and DNA. This set of circumstances can lead to a range of health problems such as aging, diabetes and cancer, as well as certain cardiovascular and neurogenerative diseases [1–3]. To treat these disease, synthetic compounds can be used to help prevent any effects resulting from the specific mechanisms that may occur. In spite of their advantages, it has been recently reported that an overdose of chemical compounds might lead to a range of detrimental side effects and health issues including the formation of carcinogenic tissues or tumors [4]. Consequently, natural products have been recognized for their potential in the development of medicinal treatment to address many of these illnesses. In this way, the active molecules derived from plants and other natural resources have been isolated and enhanced with the intention of reducing the oxidation processes that occur in these plant substances and to development alternative and beneficial forms of treatment. The Clauseneae tribe belongs to the Rutaceae family, which is generally recognized as a citrus family of flowering plants (Angiosperms). The tribe Clauseneae is comprised of five genera, namely *Micromelum*, *Marillia*, *Glycosmis*, *Murraya* and *Clausena* [5]. The latter genus has been identified as a potent natural resource. It has been used in a variety of applications and is considered a traditional food source. It is now being considered for use in the field of pharmacology. Furthermore, there is a great need to discover and develop novel bioactive compounds that can be utilized for a range of medicinal purposes including for anticancer, antibacterial, antifungal and anti-malarial treatments [6–8]. *Clausena* is a relatively small genus of strongly scented evergreen trees with odd-pinnate leaves. It is comprised of around 30 species that are distributed throughout tropical Asia, South Asia and South East Asia [6,9] and are primarily found in India. In Thailand, the genus *Clausena* is considered an economically important group of plants with a range of potential health benefits. Specifically, *C. excavata* and *C. harmandiana* have been identified as plants within this genus that show potential in the development of natural health treatments [10,11]. Concerning the tribe Clauseneae, several species in this tribe have been acknowledged as sources of edible fruits, essential oils, herbal medicines, spices (*Murraya exotica* and *M. kwangsiensis*) and various horticultural items (*Clausena anisum-olens*, *C. lansium* and *Glycosmis parviflora*). Several species have also been recognized as potential sources of lumber and pharmaceutical compounds [12]. Additionally, *Murraya koenigii*, a popular and commonly known species, is scattered throughout different regions of the globe [13]. Recently, Balakrishnan et al. [14] speculated that *M. koenigii* was a beneficial source of various bioactive compounds such as alkaloids, polyphenols, terpenoids and flavonoids. Additionally, *M. koenigii* was also found to be capable of expressing a range of anticarcinogenic, proapoptotic, antiangiogenic, antimetastatic, immunomodulatory and antioxidant properties. These properties can be further developed as antiemetic and antidiarrheal agents. They could then potentially be used to treat diseases like dysentery as a febrifuge, blood purifier, tonic and stomachic. Furthermore, *M. koenigii* is already being used as a flavoring agent in curries and chutneys. In some cultures, *M. koenigii* is being utilized in traditional systems of medicine [15].

Over the last decades, molecular technique based on DNA sequence can be applied to the systematic identification of known and unknown plants and animals. DNA barcoding, a modern method for the rapid identification of any species, is considered as a molecular and bioinformatics tool for species differentiation, identification and discovery of new species at molecular systematic level [16]. In terms of barcoding of vascular plants, it was mostly focused on markers of chloroplast genes, several markers were assessed and with time most commonly applied combinations are *rbc*L, *mat*K, *trn*H-*psb*A, with a nuclear internal transcribed spacer 2 (ITS2) established [17–20].

As mentioned above, bioactivities, namely anticarcinogenic and antioxidant have extensively addressed in the Clauseneae. Nonetheless, many aspects among these plants such as chemical profiles, some biological properties and DNA evident are still scarce and they are needed to be explored.

Nonetheless, due to a set of resembling systematics, this tribe has been labeled with a diverse range of local names at the local level, which can mislead people and result in incidences of misidentification. Furthermore, similar plant species have been utilized for numerous alternative purposes as ingredients in food or in pharmaceuticals. This present study, therefore, has aimed to confirm the obscure major taxa of Clauseneae i.e., *Clausena excavata*, *C. harmandiana* and *Murraya koenigii* gathered from Chiang Mai, Thailand using the DNA barcoding sequences obtained from two candidates of barcoding loci (*mat*K and *rbc*L). In addition, phytochemical screening and an assessment of the biological properties of these Clauseneae plants were also carried out.

2. Results and Discussion

2.1. Genetic Variation Analysis

Botanically, three plant species in the Clauseneae tribe have been recognized as a type of strongly fragrant evergreen trees with a tall slender shape of approximately 10 m tall, and odd-pinnate narrow leaves positioned in around 10–15 pairs separated to slanting leaf-lets that are 3.5 to 7 cm long. All materials were morphologically confirmed by an expert botanist by using the Forest Trees of Southern Thailand [21]. It was determined that the specific collected materials labeled with voucher numbers PSRU-RUT001-PSRU-RUT002 could not be identified by species. However, voucher numbers PSRU-RUT003-PSRU-RUT004 and PSRU-RUT005-PSRU-RUT006 could be identified as *Clausena excavata* and *C. harmandiana*, respectively. The details of the morphological features of each representative Claseneae tree are presented in Figure 1.

Figure 1. General morphology of plants. (**A**) PSRU-RUT001 (foliage of unknown plant species), (**B**), *Clausena excavata* plant (PSRU-RUT003), (**C**) Fruits of *C. excavata*, (**D**) flowers of *C. excavata*, (**E**) *Clausena harmandiana* plant (PSRU-RUT005), (**F**) Fruits of *C. harmandiana* and (**G**) flowers of *C. harmandiana*.

Furthermore, their relevant molecular traits were assessed to identify all plant specimens. Herein, we have established the DNA barcoding sequences (*mat*K and *rbc*L) of these three species of plants to be in the tribe Clauseneae. All obtained sequences were verified by their accession numbers (*mat*K; MH187239- MH187243 and *rbc*L; MH187222-MH187229) and compared to the sequences previously deposited in Genbank using BLAST. Regarding molecular identification, genomic DNA was successfully extracted from every specimen. Two main chloroplast *mat*K and *rbc*L genes were employed for the purposes of DNA barcoding. Upon PCR amplification, fragments encompassing both *mat*K and *rbc*L were distinguished by approximately 700 base pairs. The success rates of PCR amplification of *mat*K and *rbc*L were 62.5% and 100%, respectively. Regions of similarity between the sequences were observed via the BLAST program. It was determined that these were similar to those of the *Clausena*e specimens previously deposited in Genbank at a 99–100% degree of accuracy. Based on *mat*K DNA sequences, five samples of Clauseneae (PSRU_RUT001-PSRU_RUT002, PSRU_RUT003-PSRU_RUT004 and PSRU_RUT005) were found to be closely related to *Murraya koenigii* (L.) Spreng, *Clausena excavata* Burm.f. and *Clausena harmandiana* (Pierre) Pierre ex Guillaumin, respectively. Furthermore, eight *rbc*L sequences revealed a close resemblance to *M. koenigii* (L.) Spreng (PSRU_RUT001-PSRU_RUT002), *C. excavata* Burm.f. (PSRU_RUT003-PSRU_RUT004) and *C. harmandiana* (Pierre) Pierre ex Guillaumin (PSRU_RUT005-PSRU_RUT008). Nevertheless, some *rbc*L sequences were observed to be similar to the sequences previously deposited in Genbank, namely *M. koenigii* (99.58%, complete genome), *M. paniculate* (99.58%, complete genome) and *Micromelum minutum* (99.44%, complete genome) etc. This set of circumstances resulted in a number of difficulties with regard to establishing the correct species name. Therefore, representative specimens obtained from the *mat*K DNA sequences [PSRU-RUT001 (MH187239), PSRU-RUT003 (MH187241) and PSRU-RUT005 (MH187243)] and a relevant outgroup were further assessed by phylogenic analysis using the maximum likelihood (ML) method with 1000 bootstrap replicates. The phylogenic construction revealed that each representative plant species was reciprocally monophyletic as depicted in Figure 2.

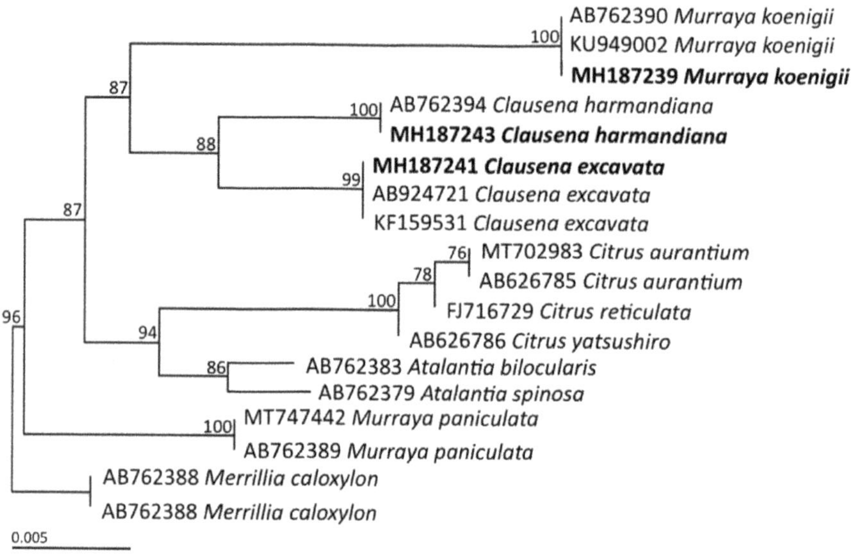

Figure 2. Phylogenetic tree derived from maximum likelihood analysis of the *mat*K gene of 18 sequences. Bar represents 0.005 substitutions per nucleotide position. The plants from this study are presented in bold.

In the past decade, DNA barcoding inferred from nuclear ribosomal internal transcribed spacer (ITS) sequence was employed for the discrimination of *M. koenigii*. It was suggested that the ITS sequence was an appropriate molecular marker for isolation of *M. koenigii* from selected plants [5]. At this time, the *mat*K gene was determined to be more reliable when applied as a DNA barcode for the identification of the designated three plant species in the tribe Clauseneae. It was noted that the *rbc*L sequences were inconsistent with the basic principles of DNA barcoding because of certain drawbacks, for instance low rates of variation and amplification, poor universality of the primer, gene deletion and a potential number of other issues as has been postulated by Shivakumar et al. [17]. Although *mat*K had a higher discriminatory power than *rbc*L, it was more difficult to amplify it across distantly related species [18]. In the same manner, this result conformed to those of previous studies conducted by Penjor et al. [19] who established the phylogenetic relationships of citrus plants and their relatives inferred from *mat*K gene sequences. It was found that use of *mat*K as a marker alone could establish the curry tree as *M. koenigii*. A combination of both markers, *mat*K + *rbc*L, could help to discriminate between a maximum number of species. However, a number of researchers have used the integration of the DNA barcode to classify and identify species, while several different combinations of DNA barcodes have been effectively put forward for different plants [20]. To attain a maximum discrimination rate among closely related species, a combination of Internal Transcribed Spacers (ITS + *mat*K + *rbc*L) has been suggested by Li et al. [22].

2.2. Phytochemical Analysis of Essential Oils

The phytochemical composition of the leaves used to obtain the essential oil of *C. excavata*, *C. harmandiana* and *M. koenigii* are shown in Table 1. A GC-MS chromatogram is presented in Figure 3. Various compounds were detected by gas chromatography. Notably, *C. harmandiana* was acknowledged as a high-performance specimen containing compounds amounting to 93.60% followed by *C. excavata* 89.86% and *M. koenigii* 88.87%. Furthermore, anethole was the most dominant compound identified in essential oil, while *C. excavata* was found to contain the highest yield at 86.72% followed by *C. harmandiana* and *M. koenigii* at 46.09% and 26.02%, respectively.

Table 1. Chemical composition of *Clausena excavata*, *Clausena harmandiana* and *Murraya koenigii* leaves essential oil.

Rt (min) [a]	Compound [b]	% Composition		
		C. excavata	*C. harmandiana*	*M. koenigii*
5.617	α-Pinene	-	1.05	12.23
5.943	Camphene	-	9.61	0.27
5.998	Sabinene	-	-	0.05
6.685	β-Terpinene	-	7.87	-
6.703	β-Pinene	-	0.88	0.46
6.952	6-Methyl- 5-hepten-2-one	-	-	0.28
7.068	β-Mycene	-	3.69	0.65
7.513	α-Phellandrene	-	-	0.63
7.693	3-Carene	0.76	2.41	0.33
8.172	*p*-Cymene	-	1.27	0.53
8.406	D-Limonene	-	7.07	-
8.342	β-Phellandrene	-	-	3.67
8.456	Eucalyptol	-	0.5	-
8.616	(E)-Ocimene	-	0.9	-
8.947	α-Ocimene	-	0.57	0.78
9.403	γ-Terpinene	-	3.57	-
9.705	(Z)-β-Terpineol	-	0.4	-
10.419	Terpinolene	0.25	1.38	0.85
10.895	3,7-dimethyl-1,6-Octadien-3-ol	-	-	0.57
11.761	*trans*-1-Methyl-4-(1-methylethyl)-2-cyclohexen-1-ol	-	0.29	-
13.986	4-Terpinenol	-	3.34	0.09
14.849	Estragole	1.23	1.68	0.38
18.805	Anethole	86.72	46.09	26.02
21.137	α-Cubebene	-	-	0.15
22.030	Ylangene	-	-	0.16
22.220	Copaene	-	-	0.31
22.400	*cis*-β-Guaiene	-	-	0.16
22.925	β-Elemene	-	-	0.41
23.067	6,10,11,11-Tetramethyl-tricyclo[6.3.0.1(2,3)]undec-7-ene	-	-	0.23
23.598	α-Gurjunene	-	-	0.70
24.202	Caryophyllene	-	0.67	21.15
24.695	Zingiberene	-	-	1.29
24.817	Spathulenol	-	-	1.25
24.978	Eudesma-4(14),11-diene	-	-	0.55
25.434	α-Caryophyllene	-	0.17	3.92
26.739	β-helmiscapene	-	-	3.81
27.105	γ-Elemene	-	0.19	-
27.126	α-Selinene	0.34	-	6.10
30.285	Spathulenol	0.56	-	-
30.445	Caryophyllene oxide	-	-	0.85
Total		**89.86**	**93.60**	**88.87**

[a] Retention time (in minutes). [b] Compounds listed in order of elution from a HP-5 MS column. "-"= not detected.

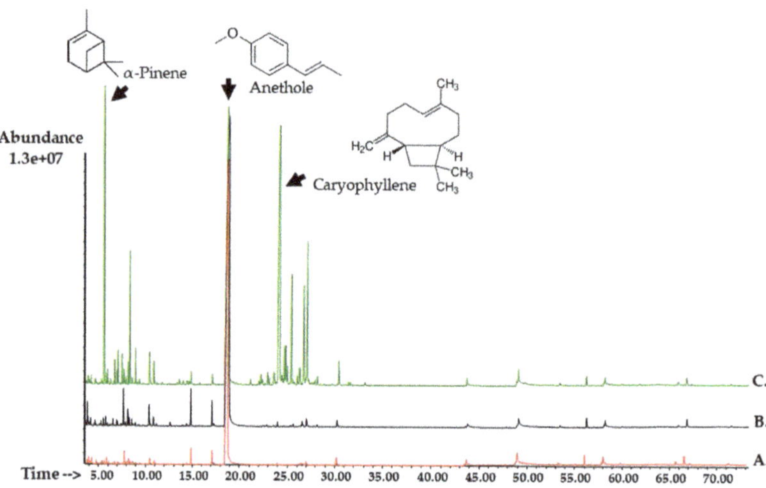

Figure 3. GC-MS Chromatogram of (**A**) *Clausena excavata* leaves in essential oil (**B**) *Clausena harmandiana* leaves in essential oil and (**C**) *Murraya koenigii* leaves in essential oil.

As determined by the results, anethole was identified as a major phytochemical compound of *C. excavata* and *C. harmandiana*. Additionally, minor compounds were observed to be present, namely Camphene (9.61%), β-Terpinene (7.87%), D-Limonene (7.07%), β-Mycene (3.69%), γ-Terpinene (3.57%), 4-Terpinenol (3.34%) and others, of which less than 3% were found in the essential oils. Cheng et al. [23] reported that the essential oil obtained from the fresh leaves of *C. excavata* contained Safrole (75.85%) and Terpinolene (17.86%) as the primary components, while Trung et al. [24] discovered that β-Caryophyllene (16.7%), Spathulenol (11.9%) and Bicyclogermacrene (7.5%) were the major constituents of the *C. excavata* plant collected in Vietnam. On the other hand, Arbab et al. [6] revealed that α-Pinene (12.23%) and Copaene (12.40%) were the major constituents present in the essential oil of *C. harmandiana*. Moreover, bioactive Carbazoles and Coumarins were observed as the major constituents of *C. excavata* and *C. harmandiana*, of which these two compounds were identified as bioactive compounds [11,25,26]. According to the results of the assessment of the essential oil of *M. koenigii*, it was found to possess Anethole (26.02%), Caryophyllene (21.15%) and α-Pinene (12.23%) as the major components. These results are in accordance with those presented in the published report of Sukkaew et al. [27], wherein the essential oil obtained from the fresh leaves of *M. koenigii* collected from Surat Thani Province, Thailand, was found to contain β-Caryophyllene (21.4%), α-Selinene (10.2%) and α-Humulene (7.1%) and α-Pinene (4.4%) as its major components. However, in this study, the chemical composition of the essential oil of *M. koenigii* was distinctly different from that of previous studies. The composition of the essential oils of the fresh leaves of *M. koenigii* that was cultivated at six locations in Peninsula Malaysia and Borneo revealed that the two major volatile metabolites were identified as β-Caryophyllene (16.6–26.6%) and α-Humulene (15.2–26.7%). The volatile substances could be categorized as sesquiterpene hydrocarbons and oxygenated sesquiterpenes as the major groups along with oxygenated monoterpenes and oxygenated diterpenes [28], Caryophyllene (9.5%), β-Myrcene (3.2%), α-Caryophyllene (2.8%), 4-Terpineol (2.8%), g-Terpinene (2.7%) and Allyl (methoxy) dimethylsilane (2.6%) [29]. Rao et al. [13] stated that the essential oil obtained from wild and cultivated *M. koenigii* leaves collected from ten Indian locations revealed different chemical compositions. For the essential oil of the wild plants, α-Pinene (55.7%) and β-Pinene (10.6%) were detected as the dominant constituents.

Furthermore, α-Pinene (13.5–35.7%) and/or β-Phellandrene (14.7–50.2%) were identified as the major components of the specimen collected from seven other locations.

Additionally, (E)-Caryophyllene (26.5%–31.5%) and α-Selinene (9.5%–10.4%) were considered the principal components of the essential oils derived from the specimens collected from two locations. The chemical compositions of the essential oil of two Chemotypes of *M. koenigii* collected from northern India were assessed in different seasons. Chemotype A contained the major compounds of α-Pinene (34.6–41.9%), Sab-inene (26.1–36.1%), (E)-Caryophyllene (2.4–5.4%) and Terpinen-4-ol (1.5–5.3%), while chemotype B contained α-Pinene (52.7–65.3%), β-Pinene (10.7–12.9%), (E)-Caryophyllene (3.1–10.3%) and Limonene (5.1–5.7%) [30]. On the other hand, the major compounds detected in the essential oil of *M. koenigii* leaves collected from southern India, namely Linalool (32.83%), Elemol (7.44%), Geranyl acetate (6.18%), Myrcene (6.12%), Allo-Ocimene (5.02%), α-Terpinene (4.9%), (E)-β-Ocimene (3.68%) and Neryl acetate (3.45%), were also identified [31]. Notably, Tripathi et al. [32] established that 3-Carene (18.52%), β-Pinene (13.57%), α-Pinene (9.38%), Linalool (5.42%), α-Eudesmol (4.55%), ρ-Cymene (3.61%), γ-Terpinene (3.48%), α-Amorphene (3.38%), Allo-Ocimene (2.75%), Sabinene (2.55%), γ-Terpinene (2.48%), Linalyl acetate (2.46%), Myrcene (2.43%) and β-Eudesmol (2.16%) were the crucial constituents of the essential oil of *M. koenigii* collected from Uttarakhand, India. Hence, variations in the compositions of the essential oils of curry leaves were found to be dependent upon season and the specific location of collection.

2.3. Total Phenolic and Total Flavonoid Contents

The total phenolic (TPC) and total flavonoid contents (TFC) in the extracts of the leaves of *C. excavata*, *C. harmandiana* and *M. koenigii* were expressed as mg gallic acid equivalent (GAE) and mg quercetin equivalent (QE) per gram extract. As is shown in Table 2, TPC ranged from 7.07 to 43.50 mg GAE/g extract, of which *M. koenigii* in the methanol extract exhibited the highest yields amounting to 43.50 mg GAE/g extract ($p < 0.05$) followed by *M. koenigii* in essential oil, *C. excavata* in methanol extract, *C. har-mandiana* in methanol extract, *C. excavata* in essential oil and *C. harmandiana* in essen-tial oil (32.28, 22.89, 19.71, 9.70 and 7.07 mg GAE/g extract, respectively). Meanwhile, TFC ranged from 16.82 to 66.13 mg QE/g extract *M. koenigii* in the methanol extract and presented the highest value at 66.13 mg QE/g extract ($p < 0.05$) followed by *M. koenigii* in essential oil, *C. harmandiana* in methanol extract, *C. excavata* in methanol extract, *C. excavata* in essential oil, and *C. harmandiana* in essential oil (50.57, 39.95, 30.89, 23.91 and 16.82 mg QE/g extract, respectively). Additionally, the phenolic compounds and flavonoids that were found in the three Clauseneae extracts, along with certain other phytoconstituents, such as phenols, steroids, saponins, quinones, alkaloids, flavonoids, tannins, carbohydrates, proteins and volatile oils, were found to be present in the Clauseneae plants [33].

Table 2. Total phenolic and total flavonoid contents of methanol extract and essential oil of *Clausena excavata*, *Clausena harmandiana* and *Murraya koenigii*.

Plant Extracts	Percent Yield (%Yield)	Total Phenolic Content (mg GAE/g Extract)	Total Flavonoid Content (mg QE/g Extract)
C. excavata methanol extract	5.52 ± 0.32a	22.89 ± 0.93c	30.89 ± 2.15d
C. excavata essential oil	0.83 ± 0.06d	9.70 ± 0.72d	23.91 ± 1.98e
C. harmandiana methanol extract	4.12 ± 0.19c	19.71 ± 0.83c	39.95 ± 0.63c
C. harmandiana essential oil	0.64 ± 0.08d	7.07 ± 0.73d	16.82 ± 1.25f
M. koenigii methanol extract	4.92 ± 0.39b	43.50 ± 4.30a	66.13 ± 1.69a
M. koenigii essential oil	0.86 ± 0.09d	36.28 ± 1.65b	50.57 ± 1.11b

Average ± standard deviations from three replicates. Different letters in the same column are considered significantly different according to Duncan's multiple comparison test ($p < 0.05$).

Antioxidant activity of the essential oils and methanol extracts of three Clauseneae plants were measured by DPPH, ABTS and FRAP assays. In terms of establishing potential antioxidant activities, the DPPH free radical scavenging ability is usually employed as the preferred method. In this situation, the antioxidant compounds in the plant extracts donate an electron or hydrogen radical to an unstable DPPH free radical and then become a stable diamagnetic molecule [34]. In this study, the DPPH free radical scavenging activity of the three Clauseneae plant extracts were expressed as IC_{50} (μg/mL). The DPPH radical scavenging activities of *C. excavata*, *C. harmandiana* and *M. koenigii* in the methanol extracts and essential oils were observed and are presented in Table 3. Specifically, the IC_{50} values ranged from 95.54 μg/mL to 2865.26 μg/mL. Furthermore, it was revealed that *M. koenigii* in the methanol extract displayed the highest radical scavenging activity at 95.54 μg/mL followed by *M. koenigii* in essential oil (167.74 μg/mL), *C. excavata* in methanol extract (904.53 μg/mL), *C. harmandiana* in methanol extract (2037.66 μg/mL), *C. excavata* in essential oil (2059.29 μg/mL) and *C. harmandiana* in essential oil (2865.26 μg/mL). There was a strong correlation between the DPPH free radical scavenging activity in terms of the total phenolic (r = 0.810, $p < 0.05$) and total flavonoid (r = 0.835, $p < 0.05$) contents of all plant extracts. Furthermore, the antioxidant capacities of the essential oil and methanol extracts of the three Clauseneae plants were investigated through the scavenging activity of the ABTS cation free radicals. With regard to the mechanism of this assay, the antioxidant compounds donated an electron and hydrogen atom to an unstable $ABTS^+$ cation radical, which could then convert to a stable ABTS radical form and then be used to evaluate the degree of antioxidant activity by reduction of the bluegreen $ABTS^+$ radical. The ABTS free radical scavenging ability of various extracts of the Clauseneae plant extracts was expressed as mg GAE/g extract. The ABTS cation scavenging activity of the methanol extract of *M. koenigii* was found to be significantly higher than the other extracts ($p < 0.05$) at a value of 118.12 mg GAE/g extract, followed by the methanol extracts of *C. excavata* (88.65 mg GAE/g extract) and *C. harmandiana* (80.11 mg GAE/g extract).

Table 3. Antioxidant activity of methanol extract and essential oil of *Clausena excavata*, *Clausena harmandiana* and *Murraya koenigii*.

Plant Extracts	DPPH Free Radical Scavenging (IC$_{50}$, ug/mL)	Antioxidant Activity ABTS Cation Free Radical Scavenging (mg GAE/g extract)	Ferric Reducing Antioxidant Power (mg GAE/g extract)
C. excavata methanol extract	904.53 ± 3.23c	88.65 ± 0.71b	16.48 ± 0.72d
C. excavata essential oil	2059.29 ± 83.13d	12.27 ± 0.1.75d	5.07 ± 0.17e
C. harmandiana methanol extract	2037.66 ± 39.23d	80.11 ± 1.01c	19.07 ± 0.55c
C. harmandiana essential oil	2865.26 ± 8.22e	12.83 ± 1.04d	5.37 ± 0.32e
M. koenigii methanol extract	95.54 ± 3.46a	118.12 ± 1.01a	48.15 ± 1.21a
M. koenigii essential oil	167.74 ± 6.97b	79.52 ± 1.01c	28.22 ± 0.94b

Average ± standard deviations from three replicates. Different letters in the same column are considered significantly different according to Duncan's multiple comparison test ($p < 0.05$).

Furthermore, the essential oil of all plants displayed less scavenging activity of the ABTS cation radical than the methanol extracts, with the exception of the essential oil extract of *M. koenigii* which possessed a degree of high ABTS cation radical that was close to that of the methanol extract of *C. harmandiana* ($p > 0.05$). Importantly, a high correlation between ABTS cation free radical scavenging activity with total phenolic (r = 0.881, $p < 0.05$) and total flavonoid (r = 0.880, $p < 0.05$) contents of all plant extracts was observed. Additionally, the antioxidant activity of Clauseneae plant extracts was investigated by ferric reducing antioxidant power (FRAP) assay based on the measurement of the ability of antioxidants to reduce ferric iron (Fe^{3+}) to its ferrous (Fe^{2+}) form. The potential antioxidant activity was then measured through its reducing ca-pacity. The FRAP values of the extracts were expressed as mg GAE/g extract. The highest FRAP value was found in the methanol extract of *M. koenigii* (48.15 mg GAE/g extract) ($p < 0.05$), while the essential oil of *C. excavata* displayed a minimum degree of antioxidant activity at 5.07 mg GAE/g extract. Additionally, a high correlation was observed between their FRAP values and their total phenolic (r = 0.823, $p < 0.05$) and total flavonoid (r = 0.875, $p < 0.05$) contents. It was further observed that the methanol extract of *M. koenigii* showed strong DPPH free radical scavenging activity, the highest degree of ABTS free radical scavenging ability, and also exhibited high reducing power. All of which indicated that the methanol extract of

M. koenigii has the potential to scavenge both neutral and cation free radicals. Moreover, a significant correlation between chemical constituents and antioxidant activity suggests that the phenolic and flavonoid contents of Clauseneae plants extracts are associated with their strong antioxidant activities. In addition, the antioxidant activities of the extracts may be a result of the level of terpene contained in the extract. A previous study conducted by Ma et al. [35] revealed that the alkenes isolated from *M. koenigii* exhibited significant antioxidant activities using DPPH free radical scavenging assay. The study conducted by Truong et al. [36] reported that the methanol extract of *Severinia buxifolia*, belonging to the Rutaceae family, showed the highest extraction yield along with the highest content of phenolics, flavonoids, alkaloids and terpenoids. Moreover, the methanol extract also possessed the highest degree of antioxidant activity indexed by the DPPH free radical scavenging assay and the highest in vitro anti-inflammatory activity. Similarly, Albaayit et al. [37] speculated that the extract of the methanol leaves of *C. excavata* displayed the highest phenolic content and antioxidant activity based on FRAP and DPPH radical scavenging activity assays. These finding support the contention that methanol is the optimal solvent for the extraction of Rutaceae plants.

The α-glucosidase inhibitory activities of *C. excavata*, *C. harmandiana* and *M. koenigii* in the methanol extract and essential oil are shown in Table 4. The results reveal that the percentage of inhibition of α-glucosidase inhibitory activities ranged from 24.33% to 84.55% and the methanol extract of *M. koenigii* displayed positive inhibitory activity at 84.55% inhibition. The antidiabetic activity of Claseneae plants has been reported, while the root ethanolic extract of *C. excavata* that contained dentatin and heptaphyllin exhibited high activity in term of α-glucosidase inhibitory [38]. Specifically, the extract of *M. koenigii* exhibited high potential in preventing diabetes [39,40].

Table 4. *Alpha*-glucosidase and antihypertensive inhibitory activity of methanol extract and essential oil of *Clausena excavata*, *Clausena harmandiana* and *Murraya koenigii*.

Plant Extracts	α-Glucosidase Inhibitory Activity (% Inhibition)	Antihypertensive Inhibitory Activity (% Inhibition)
C. excavata methanol extract	49.30 ± 1.10c	47.63 ± 1.11c
C. excavata essential oil	36.20 ± 1.49e	38.90 ± 1.05d
C. harmandiana methanol extract	41.42 ± 0.65d	58.10 ± 1.75b
C. harmandiana essential oil	24.33 ± 0.91f	31.07 ± 1.16e
M. koenigii methanol extract	84.55 ± 0.49a	84.95 ± 1.24a
M. koenigii essential oil	52.39 ± 1.16b	39.33 ± 1.13d

Average ± standard deviations from three replicates. Different letters in the same column are considered significantly different according to Duncan's multiple comparison test ($p < 0.05$).

Furthermore, the inhibition of α-glucosidase, an important enzyme that hydrolyses oligosaccharides, can reduce the risk of diabetes and obesity. Based on our results, the high potential of α-glucosidase inhibitory activity observed in the methanol extract of *M. koenigii* could be related to its high total phenolic and total flavonoid contents. This would be indicative of strong relative correlations of 0.792 and 0.804 ($p < 0.05$). These results are supported by the findings of Nagarani et al. [41], which suggested that the α-glucosidase inhibitory activity of natural substances was strongly correlated with the contents of the flavonoid and phenolic compounds. Moreover, antihypertensive inhibitory activity was investigated and the percentage ranged from 31.07% to 84.95% (Table 4). The methanol extract of *M. koenigii* recorded the highest value at 84.95%. These results revealed a strong correlation r = 0.704 ($p < 0.05$) and r = 0.727 ($p < 0.05$) between the antihypertensive inhibitory activity of all plant extracts and their total phenolic and total flavonoid contents, respectively. Moreover, the high potential of the antihypertensive (>80%) properties of plants has been reported in various plant species such as *Quercus infectoria* (Fagaceae) (93.9%), *Berberis integerrima* (Berberidaceae) (88.2%), *Crataegus microphylla* (Rosaceae) (80.9%) and *Onopordon acanthium* (Asteraceae) (80.2%) [42].

2.4. Determination of Methanol Extract and Essential Oil of Clausena excavata, Clausena harmandiana and Murraya koenigii against Human Normal Cells and Human Lung Cancer Cells

Over the past decade, the number of cancer deaths worldwide has been increasing. It is a significant cause of many of the health problems experienced by humans around the world. Cancer is followed by a range of cardiovascular diseases in terms of mortality rates. Notably, the leading cause of deaths attributed to cancer is lung cancer. Therefore, the antitumor activity of the three Clauseneae plant extracts in this study were investigated. The effect of the essential oil and methanol extracts of *C. excavata*, *C. harmandiana* and *M. koenigii* by varied concentrations (0–1 μg/mL) were observed for 24 and 48 h against human normal cells using human peripheral blood mononuclear cell (PBMC) and human lung cancer cell (A549).

The results revealed that the essential oil and methanol extracts of *C. excavata*, *C. harmandiana* and *M. koenigii* did not have any toxic effect on PBMC at a concentration of 1 μg/mL relating to the increasing number of cell viability after being tested over 24 h and 48 h (data not shown). For A549 cells assessment, the essential oil and methanol extracts of *C. excavata*, *C. harmandiana* and *M. koenigii* could effectively decrease cell viability when concentrations were increased to 1 μg/mL (Figure 4). This was especially true for *M. koenigii* in the methanol extract, which displayed high efficacy activity against A549 cells at almost 80% after 24 h and 90% over 48 h. The study conducted by Huang et al. [43] determined that carbazole alkaloids and coumarins obtained from *Clausena* plants exhibited anticancer activity. Based on the finding of our study, all the Clauseneae plant extracts exhibited high activity against human lung cancer cells. This was especially true of the methanol extract of *M. koenigii* that displayed high potential activity against A549 cells when compared with the methanol leaves and stem extracts of Marsdenia glabra Cost. In turn, this extract displayed low inhibitory activity within a range of about 15–25% at 1000 μg/mL [44]. Moreover, the study conducted by Muthumani et al. [45] revealed that the extract of *M. koenigii* displayed a high degree of potential anticancer activity in experimental animal trials.

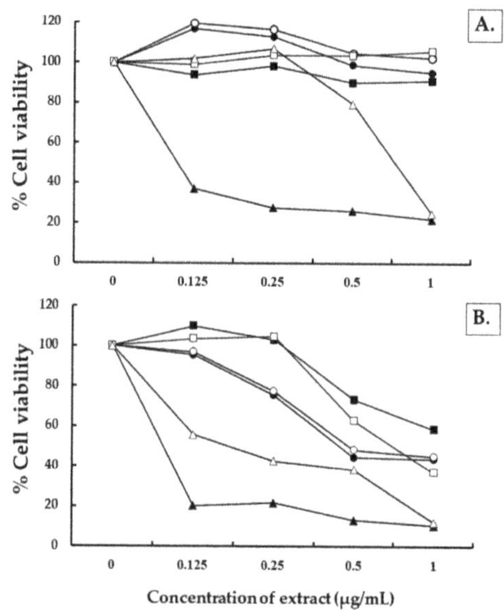

Figure 4. Effects of the extracts of *Clausena excavata* (essential oil, -○-; methanol extract, -●-), *Clausena harmandiana* (essential oil, -□-; methanol extract, -■-) and *Murraya koenigii* (essential oil, -△-; methanol extract, -▲-) against human lung cancer cells (A549) at 24 h (**A**) and 48 h (**B**).

3. Materials and Methods

3.1. Plant Materials and Chemicals

All plant samples were tentatively identified to species level by comparison with the reference materials at the herbarium of Queen Sirikit Botanic Garden, Chiang Mai, Thailand, and deposited as voucher specimens (PSRU-RUT001-008) in the collection of Faculty of Science and Technology, Pibulsongkram Rajabhat University (PSRU), Phitsanulok Province, Thailand.

Gallic acid was purchased from Merck (Germany) and folin-ciocalteu reagent was bought from BDH Chemicals Ltd. (Poole, England). The 2,2′-Diphenyl-1-picrylhydrazyl radical (DPPH) and 2,4,6-tris(2-pyridyl)-s-triazine (TPTZ) were purchased from Fluka (Steinheim, Germany). 2, 2′-Azino-bis (3-ethylbenzothiazoline-6-sulfonic acid) (ABTS) was derived from SIGMA (Oakville, ON, Canada). Intestinal acetone powder and 3-(4,5-dimethylthiazol-2-yl)-2,5-diphenyltetrazolium bromide were purchased from Sigma-Aldrich Chemical Co. (St. Louis, MO, USA). Dulbecco's modified Eagle medium and fetal bovine serum were purchased from Invitrogen Corp. (Grand Island, NY, USA). All the solvents and other chemicals were an analytical grade (A.R.).

3.2. Collection of Plant Material for Genetic Variation Analysis

3.2.1. Plant Sample Collection

The fresh leaves were individually collected from local area in Chiang Mai province, upper northern Thailand, labeled and transported in sterile plastic bags and frozen until processed for DNA extraction. The morphological details of each representative Clauseneae leaf are demonstrated as Figure 5.

Figure 5. Representative leaf images of three plant species in Claseneae; (**A**) PSRU_RUT005-006, (**B**) PSRU_RUT003-004 and (**C**) PSRU_RUT001-002.

3.2.2. DNA Extraction

Genomic DNA was isolated using the DNeasy Plant Mini Kit (QIAGEN, Germany) following the manufacturers protocol. Purity and quantity of the extracted genomic DNA was performed with a NanoDrop spectrophotometer. DNA was checked using 1.0% an agarose gel stained with stained with Redsafe™.

3.2.3. PCR Amplication

We amplified and sequenced the chloroplast DNA regions (*mat*K and *rbc*L) based on primers 1) *mat*K (*mat*K-F; 5′-TAA TTT ACG ATC AAT TCA TTC-3′ and *mat*K-R; 5′-TCT GGA GTC TTT CTT GAG CG-3′) and 2) *rbc*L (*rbc*L-F; 5′-TCA CCA CAA ACA GAA ACT AAA GC-3′ and *rbc*L-R; 5′-GGC ACA AAA TAA GAA ACG ATC TC-3′). PCR was conducted in a final reaction volume of 20 μL made up of 5× PCR Enhancer, 2 μL of 10× HF Reaction Buffer, 0.4 μL of 10 mM dNTP Mix, 0.3 μL each of the forward and reverse primers

(10 μmol/L), 0.3 μL of Long and High-Fidelity DNA Polymerase (0.75 U) (biotechrabbit, Germany), 10.7 μL of nuclear free water and at least 20 ng of the DNA template. PCR amplification was performed using a T100™ Thermal Cycler (BioRad, USA). The thermal profile used was: 94 °C for 3 min, 30 cycles of 94 °C for 1 min, 48 °C for 1 min, 72 °C for 1 min, and 72 °C for 10 min, and a final extension at 72 °C for 5 min. PCR products were verified on 1% agarose gels stained with Redsafe™ under UV light and were then purified using GenUP PCR/Gel Cleanup Kit (Biotechrabbit, Germany). Afterwards, the purified product was direct sequenced with forward and reverse primers by Macrogen, Inc (http://www.macrogen.com).

3.2.4. Alignment of Sequences and Phylogenetic Analysis

A similarity search for each sequence was verified using BLAST (https://www.ncbi.nlm.nih.gov/). The maximum likelihood (ML) methods from the MEGA (version 5.0) program [46] were used to create phylogenetic trees. The reliability of each branch was tested by bootstrap analysis with 1000 replications. Clade with bootstrap values of 70% was considered significantly supported [47]. The sequences of *Merrillia caloxyon* were used as an outgroup. All DNA sequences were finally trimmed to 718 (*mat*K) and 713 (*rbc*L) base pairs. The sequences obtained after removing the primers used for PCR amplification were deposited to NCBI-Genbank BankIt (https://www.ncbi.nlm.nih.gov/BankIt/) under accession numbers (*mat*K; MH187239-MH187243 and *rbc*L; MH187222-MH187229).

3.3. Preparation of Plant Extracts

The leaves of three plants of Clauseneae were dried at 50 °C for 72 h, grounded into small pieces and stored at room temperature. Plant materials (100 g) were individually extracted with 1000 mL of methanol (1:10) and left for 12 h at ambient temperature. The extracts were sonicated using an ultrasonicator (Crest, USA) for 30 min, filtered through Whatman no. 1 filter paper and evaporated under a vacuum at 40 °C using a rotary evaporator until dry. Dry extracts were kept at 4 °C in the dark for further study.

The essential oil was extracted from the dried leaves and stems by stream distillation with a ratio of 1:20 and partitioned with dichloromethane for 3 h. The essential oil was obtained after evaporation of the dichloromethane. After that, the essential oil was dehydrated with anhydrous sodium sulphate and stored at 4 °C for further analysis.

The methanol extract and essential oil of each plant species were performed in triplicate.

3.4. Phytochemical Analysis of Essential Oils

Phytochemical analysis of the essential oils was determined according to the previous research of Tanruean et al. [48]. The extracts were analyzed for their phytochemicals using a gas chromatography (GC) 6890 Agilent Technologies/MSD 5973 Hewlett Packard, equipped with a MS detector and an HP-5MS capillary column (bonded and cross-linked 5% phenyl-methylpolysiloxane 30 × 0.25 μm, film thickness 0.25 μm). The injector and detector temperatures were set at 270 and 280 °C, respectively. The oven temperature was set at 80 °C and held for 2 min, and then increased at a rate 10 °C/min to 120 °C and held for 4 min. The oven temperature was then in-creased at a rate 10 °C/min to 155 °C and held for 4 min, and then increased at a rate 5 °C/min to 280 °C and held for 12.50 min. The total running time was 55 min. Helium was used as a carrier gas at a flow rate of 1 mL/min. The sample (1 μL) was injected in the splitless mode. GC-MS detection of an electron ionization system with an ionization energy measurement of 70 eV was used. Injector and MS transfer line temperatures were set at 270 and 290 °C, respectively. The components were identified based on a comparison of their relative retention times and the mass spectra with W8N08 and Wiley7n libraries data of the GC-MS system.

3.5. Determination of Total Flavonoid Contents

The total flavonoid contents were determined by the method of Kaewnarin et al. [49] with a few slight modifications. The extract (0.5 mL) was mixed with 2 mL of methanol, followed by the addition 0.15 mL of 50 g/L $NaNO_2$. After 5 min, 0.15 mL of 100 g/L $AlCl_3$ was added. The reaction was mixed and incubated at ambient temperature for 15 min, and the absorbance was measured at 415 nm. Quercetin solution was used as a standard for the determination [50–52] and the results were expressed as mg QE/g extract. The data were presented as the average of the triplicate analyses.

3.6. Determination of Total Phenolic Contents

Total phenolic contents were estimated using the protocol of Thitilertdecha et al. [53] with slight modifications. The procedure involved of combining 0.25 mL of sample (1 mg/mL) with 2.5 mL of deionized water and 0.5 mL of folin-ciocalteu reagent. After 5 min, 0.5 mL of 20% (*w/v*) Na_2CO_3 was added, and the solution was incubated for 1 h at ambient temperature. Absorbance was then measured at 760 nm. Gallic acid solution was used as a standard for the determination and the results were expressed as mg GAE/g extract. The data were presented as the average of the triplicate analyses.

3.7. Determination of Antioxidant Activities

3.7.1. DPPH Free Radical Scavenging Assay

The free radical scavenging ability was determined according to the method of Gülçin et al. [54] with slight modifications. The 2,2′-diphenyl-1-picrylhydrazyl radical (DPPH•) solution in ethanol (0.1 mM, 1.5 mL) was mixed with 0.5 mL of different concentrations of each extract, and methanol was used as the control. The mixtures were well shaken and kept at ambient temperature for 30 min in the dark. The absorbance was measured at 517 nm and gallic acid was used as the comparative standard. The percent of DPPH• discoloration of the samples was calculated according to the formula:

Percent inhibition = $(A_o - A_s/A_o) \times 100$

Where A_o is the absorbance of the control (containing all reagents except the test compound), and A_s is the absorbance of the mixture containing the test compound. The test sample concentrations providing 50% inhibition (IC_{50}) were calculated from the plot of inhibition percentage against extract concentration values. The radical scavenging ability was presented IC_{50} values. The data were presented as the average of the triplicate analyses.

3.7.2. ABTS Radical Scavenging Activity

The $ABTS^{•+}$ scavenging activity was measured according to the method of Re et al. [55] with some modification. The stock solution of ABTS cation chromophore was prepared by a reaction between a 7 mM ABTS solution (100 mL) and 2.45 mM potassium persulphate (final concentration) (100 mL) and was kept in a dark place at an ambient temperature for 16 h. The $ABTS^{•+}$ solution was diluted with phosphate buffer (50 mM, pH 7.4) to an absorbance of 0.70 ± 002 at 734 nm. An aliquot of each extract (100 μL) was added to 3 mL of an $ABTS^{•+}$ solution, and the resulting mixture was then incubated for 30 min at an ambient temperature prior to measuring the absorbance at 734 nm. Gallic acid was used as a reference compound, and the $ABTS^{•+}$ scavenging activity was expressed as mg GAE/g extract. The data were presented as the average of the triplicate analyses.

3.7.3. Ferric Reducing Antioxidant Power (FRAP) Assay

The FRAP assay was determined according to the protocol of Li et al. [56] with some modifications. The FRAP reagent containing 10 mM of 2,4,6-tris(2-pyridyl)-s-triazine (TPTZ) solution in 40 mM hydrochloric acid (20 mL), 20 mM ferric (III) chloride (20 mL) and acetate buffer (5 mL, 300 mM, pH 3.6) was pre-pared freshly prior to being used. Different concentrations of each extract (0.1 mL) was mixed with the FRAP reagent (1.5 mL) and 1.4 mL of acetate buffer (300 mM, pH 3.6) and were then incubated at an ambient temperature for 30 min. The absorbance was measured at 593 nm. Gallic acid was used as a standard and FRAP value was calculated as mg GAE/g extract. The data were presented as the average of the triplicate analyses.

3.8. Determination of α-Glucosidase Inhibitory Activity

α-Glucosidase (AGH) solution was prepared from rat intestinal acetone powder by partial modification of the procedure reported by Oki et al. [57]. 100 mg of intestinal acetone powder (Sigma-Aldrich Chemical Co.; St. Louis, MO, USA) was added to 3 mL of 0.9% NaCl solution, homogenized with the sonication and kept in an ice bath. After centrifugation at 6000 rpm for 30 min at 4 °C, the resulting supernatant was kept cold and directly subjected to inhibitory assay. The method of Adisakwattana et al. [58] was used to determine AGH inhibitory assay. The assay was defined as the percent inhibition under the assay conditions, which was calculated according to the formula:

Percent inhibition = $(A_o - A_s/A_o) \times 100$

Where A_o is the absorbance of the control, and A_s is the absorbance of the mixture containing the test compound. The data were presented as the average of the triplicate analyses.

3.9. Determination of Antihypertensive Activity

The angiotensin-I-converting enzyme (ACE) inhibitory activity was evaluated by the modified method of Park et al. [59]. The sample (50 μL) was mixed with 50 μL of 25 mU/mL ACE (Sigma-Aldrich Chemical Co.; St. Louis, MO, USA) and pre-incubated at 37 °C for 10 min. Then, 6 mM hippuryl-histidyl-leucine (HHL) in 50 mM Tris with 300 mM NaCl 100 μL was added and further incubated for 30 min. The reaction was stopped by adding 200 μL of 1.0 M HCl. Hippuric acid was extracted by ethyl acetate (600 μL), followed by centrifugation at 4880 rpm for 15 min. The supernatant (200 μL) was transferred to a test tube and evaporated at 95 °C to remove the ethyl acetate. Hot distilled water (1.0 mL) was added to dissolve the hippuric acid and the absorbance was determined at 228 nm. The ACE inhibition was calculated from this equation:

Percent inhibition = $[1 - (A_s/A_o)] \times 100$

Where A_o is the absorbance of the control (containing all reagents except the test compound), and A_s is the absorbance of the mixture containing the test compound. The results of all experiments were expressed as mean ± standard deviation.

3.10. Antitumor Activity and Cell Toxicity Assay

Antitumor activity and cell toxicity assay of the extracts were determined according to the protocol of Wang et al. [60] with some modifications. Tumor cells, A549, were cultured in Dulbecco's modified Eagle medium (DMEM) until reaching 80% con-fluence. Trypan blue exclusion method was applied to determine the cell viability. In this experiment, optimum cell viability was above 98% and concentration level was adjusted for further experimentation. Human lymphocyte cells, peripheral blood mononuclear cells (PBMCs), were obtained from healthy volunteers by venipuncture and heparin was used as an anticoagulant. The blood solution was diluted with one-fold sterile phosphate buffer saline (PBS) and was centrifuged with Ficoll-Hypaque gradient centrifugation to separate PBMCs from the other specimens. Briefly, the diluted blood solution was overlaid in Ficoll-Hypaque solution and centrifuged at 1300 rpm, 25 °C for 30 min. The PBMC layer was collected, washed two times with sterile PBS and the redissolved PBMC pellets were treated with RPMI-1640 cell medium with 10% fetal bovine serum (FBS). Cell viability was determined. and the concentration level was adjusted for further experiment.

The 3-(4,5-dimethylthiazol-2-yl)-2,5-diphenyltetrazolium bromide (MTT) assay was used to investigate the cytotoxicity of the three Clauseneae plants extracts on A549 and human lymphocyte cells. Tumor cells (A549) and human lymphocyte cells were cultured in a 96-well tissue culture plate, which contained 5000 and 10,000 cells, in each well, respectively. Different concentrations of the sample solution were added to each well and they were then incubated at 37 °C in a 5% CO_2 incubator for 24 h. After that, 20 μL of MTT solution (5 mg/mL of MTT in PBS, pH 7.4) was added and the specimens were further incubated at 37 °C for another 4 h. Cell medium was drained out and the formazan dye sediment was dissolved with 100 μL of dimethyl sulfoxide (DMSO). Absorbance was measured at 540 nm and the cell viability ratio was calculated by comparing the absorbance of the wells that did not contain any sample solution.

3.11. Ethical Considerations

Concerning the cultured human cells used in this current research, approval to conduct this study was gained from the Ethics Review Committee for Research In-volving Human Research Subjects, by the human ethics committees of the Faculty of Associated Medical Sciences, Chiang Mai University, Thailand (Study Code: AM-SEC-63EM-032).

3.12. Statistical Analysis

The results of all experiments were expressed as mean $\pm$ standard deviation. Analysis of variance was performed by ANOVA procedure and Duncan's multiple comparison test was used to determine any significant differences ($p < 0.05$) identified between treatments. The correlation (r) between the two variants was analyzed using the Pearson test. The statistical analyses were performed using SPSS software (SPSS v.25 for windows; IBM Crop., Armonk, NY, USA).

4. Conclusions

It can be concluded that matK possesses a good DNA barcode for Clauseneae, especially with regard to the discrimination of *Murraya koenigii* from *Clausena excavata* and *C. harmandiana*. This is due to its ease of amplification and sequence variation. However, a combination with other DNA fragments would increase its degree of identification efficiency. Notably, the extracts obtained from *Clausena excavata*, *C. harmandiana* and *Murraya koenigii* were found to be effective as natural antioxidant, antidiabetic and anti-hypertension agents and were also found to inhibit human lung cancer cells without any side effects on human normal cells. Moreover, the methanol extract of *M. koenigii* exhibited the strongest degree of biological activities because it possessed the highest amount of containable phenolic compounds and flavonoids. Suggesting that the extracts of the three Clauseneae plants could be promising candidates as natural bioactive agents in various relevant fields, which the biological activities in animal models could be initially studied.

Author Contributions: Conceptualization, K.T., N.S. and S.L.; investigation, K.T., J.K., N.S. and P.P.; software, J.K., N.S. and P.P.; writing—original draft, K.T.; writing—review and editing, K.T., J.K., N.S., P.P. and S.L.; supervision, K.T. and S.L. All authors have read and agreed to the published version of the manuscript.

Funding: We are grateful to Chiang Mai University and Biology Program, Faculty of Science and Technology, Pibulsongkram Rajabhat University.

Institutional Review Board Statement: The study was conducted according to the guidelines of the Declaration of Helsinki, and approved by the Institutional Review Board of Faculty of Associated Medical Sciences, Chiang Mai University, Thailand (Study Code: AM-SEC-63EM-032, Date of Approval: December 30, 2020).

Informed Consent Statement: These cultured human cells are considered by IRB as exemptions.

Data Availability Statement: Data sharing not applicable.

Acknowledgments: We are grateful to Russell K. Hollis for the English proofreading of this manuscript. Special appreciation is also extended to Khanittha Punturee, Department of Medical Technology, Faculty of Associated Medical Sciences, Chiang Mai University, for the technical support in the clinical laboratory.

Conflicts of Interest: The authors declare no conflict of interest.

References

1. Pizzino, G.; Irrera, N.; Cucinotta, M.; Pallio, G.; Mannino, F.; Arcoraci, V.; Squadrito, F.; Altavilla, D.; Bitto, A. Oxidative stress: Harms and benefits for human health. *Oxid. Med. Cell Longev.* **2017**, *2017*, 8416763. [CrossRef] [PubMed]
2. Nguyen, H.C.; Lin, K.H.; Huang, M.Y.; Yang, C.M.; Shih, T.H.; Hsiung, T.C.; Lin, Y.C.; Tsao, F.C. Antioxidant activities of the methanol extracts of various parts of *Phalaenopsis* orchids with white, yellow, and purple flowers. *Not. Bot. Horti Agrobo. Cluj Napoca.* **2018**, *46*, 457–465. [CrossRef]
3. Yen, G.C.; Chen, C.S.; Chang, W.T.; Wu, M.F.; Cheng, F.T.; Shiau, D.K.; Hsu, C.L. Antioxidant activity and anticancer effect of ethanolic and aqueous extracts of the roots of *Ficus beecheyana* and their phenolic components. *J. Food Drug Anal.* **2018**, *26*, 182–192. [CrossRef] [PubMed]
4. Biparva, P.; Ehsani, M.; Hadjmohammadi, M.R. Dispersive liquid–liquid microextraction using extraction solvents lighter than water combined with high performance liquid chromatography for determination of synthetic antioxidants in fruit juice sample. *J. Food Compost. Anal.* **2012**, *27*, 87–94. [CrossRef]
5. Sitthitharworn, W.; Wirasathien, L.; Kampiranont, L.; Khomdej, P.; Waiyasilp, W. Identification of *Murraya koenigii* (L.) Spreng using DNA barcoding technique based on the ITS sequence. *Thai J. Pharm. Sci.* **2010**, *5*, 202–205.
6. Arbab, I.A.; Abdul, A.B.; Aspollah, M.; Abdullh, R.; Abdelwahab, S.I.; Ibrahim, M.Y.; Ali, L.Z. A review of traditional uses, phytochemical and pharmacological aspects of selected members of *Clausena* genus (Rutaceae). *J. Med. Plant Res.* **2012**, *6*, 5107–5118.
7. Liu, J.; Li, C.J.; Du, Y.Q.; Li, L.; Sun, H.; Chen, N.H.; Zhang, D.M. Bioactive compounds from the stems of *Clausena lansium*. *Molecules* **2017**, *22*, 2226. [CrossRef]
8. Palombo, E.A. Traditional medicinal plant extracts and natural products with activity against oral bacteria: Potential application in the prevention and treatment of oral diseases. *Evid Based Complement Alternat Med.* **2011**, *2011*, 680354. [CrossRef]
9. Peng, W.; Fu, X.; Li, Y.; Xiong, Z.; Shi, Z.; Shi, X.; Zhang, F.; Huo, G.; Li, B. Phytochemical study of stem and leaf of *Clausena lansium*. *Molecules* **2019**, *24*, 3124. [CrossRef]
10. Yenjai, C.; Sripontan, S.; Sriprajun, P.; Kittakoop, P.; Jintasirikul, A.; Tanticharoen, M.; Thebtaranonth, Y. Coumarins and carbazoles with antiplasmodial activity from *Clausena harmandiana*. *Planta Med.* **2000**, *66*, 277–279. [CrossRef]
11. Wongthet, N.; Sanevas, N.; Schinnerl, J.; Valant-Vetschera, K.; Bacher, M.; Vajrodaya, S. Chemodiversity of *Clausena excavata* (Rutaceae) and related species: Coumarins and carbazoles. *Biochem. Syst.* **2018**, *80*, 84–90. [CrossRef]
12. Mou, F.J.; Tu, T.Y.; Chen, Y.Z.; Zhang, D.X. Phylogenetic relationship of *Clauseneae* (Rutaceae) inferred from plastid and nuclear DNA data and taxonomic implication for some major taxa. *Nord. J. Bot.* **2018**, *36*, e01552. [CrossRef]
13. Rao, B.R.; Rajput, D.K.; Mallavarapu, G.R. Chemical diversity in curry leaf (*Murraya koenigii*) essential oils. *Food Chem.* **2011**, *126*, 989–994.
14. Balakrishnan, R.; Vijayraja, D.; Jo, S.H.; Ganesan, P.; Su-Kim, I.; Choi, D.K. Medicinal profile, phytochemistry, and pharmacological activities of *Murraya koenigii* and its primary bioactive compounds. *Antioxidants* **2020**, *24*, 101. [CrossRef] [PubMed]
15. Ajay, S.; Rahul, S.; Sumit, G.; Paras, M.; Mishra, A.; Gaurav, A. Comprehensive review: *Murraya koenigii* Linn. *Asian J. Pharm. Life Sci.* **2011**, *1*, 417–425.
16. Hebert, P.D.N.; Cywinska, A.; Ball, S.L.; de Waard, J.R. Biological Identifications through DNA Barcodes. *Proc. Royal Soc. B Biol. Sci.* **2003**, *270*, 313–321. [CrossRef]
17. Shivakumar, V.S.; Appelhans, M.S.; Johnson, G.; Carlsen, M.; Zimmer, E.A. Analysis of whole chloroplast genomes from the genera of the Clauseneae, the curry tribe (Rutaceae, *Citrus* family). *Mol. Phylogenet. Evol.* **2017**, *117*, 135–140. [CrossRef]
18. Hollingsworth, P.M.; Graham, S.W.; Little, D.P. Choosing and using a plant DNA barcode. *PLoS ONE* **2011**, *6*, e19254. [CrossRef]
19. Penjor, T.; Yamamoto, M.; Uehara, M.; Ide, M.; Matsumoto, N.; Matsumoto, R.; Nagano, Y. Phylogenetic relationships of *Citrus* and its relatives based on *matK* gene sequences. *PLoS ONE* **2013**, *8*, e62574. [CrossRef]
20. Duan, H.; Wang, W.; Zeng, Y.; Guo, M.; Zhou, Y. The screening and identification of DNA barcode sequences for *Rehmannia*. *Sci. Rep.* **2019**, *9*, 17295. [CrossRef]
21. Gardner, S.; Sidisunthorn, P.; Chayamarit, K. Rutaceae. In *Forest Trees of Southern Thailand*; Amarin Printing and Publishing Plc.: Bangkok, Thailand, 2018; Volume 3 (Mo-Z), pp. 1968–1969.
22. Li, D.Z.; Gao, L.M.; Li, H.T.; Wang, H.; Ge, X.J.; Liu, J.Q.; Chen, Z.D.; Zhou, S.L.; Chen, S.L.; Yang, J.B.; et al. Comparative analysis of a large dataset indicates that internal transcribed spacer (ITS) should be incorporated into the core barcode for seed plants. *Proc. Nat. Acad. Sci. USA* **2011**, *108*, 19641–19646. [PubMed]
23. Cheng, S.S.; Chang, H.T.; Lin, C.Y.; Chen, P.S.; Huang, C.G.; Chen, W.J.; Chang, S.T. Insecticidal activities of leaf and twig essential oils from *Clausena excavata* against *Aedes aegypti* and *Aedes albopictus* larvae. *Pest Manag. Sci.* **2009**, *65*, 339–343. [CrossRef] [PubMed]
24. Trung, H.D.; Thang, T.D.; Ban, P.H.; Hoi, T.M.; Dai, D.N.; Ogunwande, I.A. Terpene constituents of the leaves of five Vietnamese species of *Clausena* (Rutaceae). *Nat. Prod. Res.* **2014**, *28*, 622–630. [CrossRef] [PubMed]
25. Songsiang, U.; Thongthoom, T.; Zeekpudsa, P.; Kukongviriyapan, V.; Boonyarat, C.; Wangboonskul, J.; Yenjai, C. Antioxidant activity and cytotoxicity against cholangiocarcinoma of carbazoles and coumarins from *Clausena harmandiana*. *Sci. Asi.* **2012**, *38*, 75–81. [CrossRef]
26. Greger, H. Phytocarbazoles: Alkaloids with great structural diversity and pro-nounced biological activities. *Phytochem. Rev.* **2017**, *16*, 1095–1153. [CrossRef]

27. Sukkaew, S.; Pripdeevech, P.; Thongpoon, C.; Machan, T.; Wongchuphan, R. Volatile constituents of *Murraya koenigii* fresh leaves using headspace solid phase microextraction—Gas chromatography—Mass spectrometry. *Nat. Prod. Commun.* **2014**, *9*, 1783–1786. [CrossRef]
28. Nagappana, T.; Ramasamyb, P.; Vairappana, C.S. Chemotaxonomical markers in essential oil of *Murraya koenigi*. *Nat. Prod. Commun.* **2012**, *7*, 1375–1378.
29. Chowdhury, J.U.; Bhuiyan, M.N.I.; Yusuf, M. Chemical composition of the leaf essential oils of *Murraya koenigii* (L.) Spreng and *Murraya paniculata* (L.) Jack. *Bangladesh J. Pharmacol.* **2008**, *3*, 59–63. [CrossRef]
30. Verma, R.S.; Padalia, R.C.; Arya, V.; Chauhan, A. Aroma profiles of the curry leaf, *Murraya koenigii* (L.) Spreng. Chemotypes: Variability in north India during the year. *Ind. Crops Prod.* **2012**, *36*, 343–348. [CrossRef]
31. Rajendran, M.P.; Pallaiyan, B.B.; Selvaraj, N. Chemical composition, antibacterial and antioxidant profile of essential oil from *Murraya koenigii* (L.) leaves. *Avicenna J. Phytomed.* **2014**, *4*, 200–214.
32. Tripathi, Y.C.; Anjum, N.; Rana, A. Chemical composition and in vitro antifungal and antioxidant activities of essential oil from *Murraya koenigii* (L.) Spreng. Leaves. *Asian J. Biomed. Pharm.* **2018**, *8*, 6–13. [CrossRef]
33. Gupta, S.; Paarakh, P.M.; Gavani, U. Isolation of phytoconstituents from the leaves of *Murraya koenigii* Linn. *J. Pharm Res.* **2009**, *2*, 1313–1314.
34. Soares, J.R.; Dins, T.C.P.; Cunha, A.P.; Almeida, L.M. Antioxidant activity of some extracts of *Thymus zygis*. *Free Radic. Res.* **1997**, *26*, 469–478. [CrossRef] [PubMed]
35. Ma, Q.G.; Xu, K.; Sang, Z.P.; Wei, R.R.; Liu, W.M.; Su, Y.L.; Yang, J.B.; Wang, A.G.; Ji, T.F.; Li, L.J. Alkenes with antioxidative activities from *Murraya koenigii* (L.) Spreng. *Bioorganic Med. Chem. Lett.* **2016**, *26*, 799–803. [CrossRef]
36. Truong, D.H.; Nguyen, D.H.; Ta, N.T.A.; Bui, A.V.; Do, T.H.; Nguyen, H.C. Evaluation of the use of different solvents for phytochemical constituents, antioxidants, and in vitro anti-inflammatory activities of *Severinia buxifolia*. *J. Food Qual.* **2019**, *2019*, 8178294. [CrossRef]
37. Albaayit, S.F.A.; Abba, Y.; Abdullah, R.; Abdullah, N. Evaluation of antioxidant activity and acute toxicity of *Clausena excavata* leaves extract. *Evid. Based Complementary Altern. Med.* **2014**, *2014*, 975450. [CrossRef]
38. Thant, T.M.; Aminah, N.S.; Kristanti, A.N.; Ramadhan, R.; Aung, H.T.; Takaya, Y. Antidiabetes and antioxidant agents from *Clausena excavata* root as medicinal plant of Myanmar. *Open Chem.* **2019**, *17*, 1339–1344. [CrossRef]
39. Arulselvan, P.; Senthilkumar, G.P.; Sathish Kumar, D.; Subramanian, S. Antidiabetic effect of *Murraya koenigii* leaves on streptozotocin induced diabetic rats. *Pharmazie* **2006**, *61*, 874–877.
40. Yankuzo, H.; Ahmed, Q.U.; Santosa, R.I.; Akter, S.F.U.; Talib, N.A. Beneficial effect of the leaves of *Murraya koenigii* L. Spreng (Rutaceae) on diabetes-induced renal damage in vivo. *J. Ethnopharmacol.* **2011**, *135*, 88–94. [CrossRef]
41. Nagarani, G.; Abirami, A.; Siddhuraju, P. A comparative study on antioxidant potentials, inhibitory activities against key enzymes related to metabolic syndrome, and anti-inflammatory activity of leaf extract from different *Momordica* species. *Food Sci. Hum. Well* **2014**, *3*, 36–46. [CrossRef]
42. Sharifi, N.; Souri, E.; Ziai, S.A.; Amin, G.; Amanlou, M. Discovery of new angiotensin converting enzyme (ACE) inhibitors from medicinal plants to treat hypertension using an in vitro assay. *DARU J. Pharm. Sci.* **2013**, *21*, 74. [CrossRef] [PubMed]
43. Huang, L.; Feng, Z.L.; Wang, Y.T.; Lin, L.G. Anticancer carbazole alkaloids and coumarins from *Clausena* plants. A review. *Chin. J. Nat. Med.* **2017**, *15*, 881–888. [CrossRef]
44. Tanruean, K.; Suwannarach, N.; Choonpicharn, S.; Lumyong, S. Evaluation of phytochemical constituents and biological activities of leaves and stems of *Marsdenia glabra* Cost. *Int. Food Res. J.* **2017**, *24*, 2572–2579.
45. Muthumani, P.; Venkatraman, S.; Ramseshu, K.; Meera, R.; Devi, P.; Kameswari, B.; Eswarapriya, B. Pharmacological studies of anticancer, anti-inflammatory activities of *Murraya koenigii* L. Spreng in experimental animals. *J. Pharm. Sci. Res.* **2009**, *1*, 131–141.
46. Tamura, K.; Peterson, D.; Peterson, N.; Stecher, G.; Nei, M.; Kumar, S. MEGA5: Molecular evolutionary genetics analysis using maximum likelihood, evolutionary distance, and maximum parsimony methods. *Mol. Biol. Evol.* **2011**, *28*, 2731–2739. [CrossRef]
47. Hillis, D.M.; Bull, J.J. An empirical test of bootstrapping as a method for assessing confidence in phylogenetic analysis. *Syst. Biol.* **1993**, *42*, 182–192. [CrossRef]
48. Tanruean, K.; Kaewnarin, K.; Suwannarach, N.; Lumyong, S. Comparative evaluation of phytochemicals, and antidiabetic and antioxidant activities of *Cuscuta reflexa* grown on different hosts in northern Thailand. *Nat. Prod. Commun.* **2017**, *12*, 51–54. [CrossRef]
49. Kaewnarin, K.; Niamsup, H.; Shank, L.; Rakariyatham, N. Antioxidant and antidiabetic activities of some edible and medicinal plants. *Chiang Mai J. Sci.* **2014**, *41*, 105–116.
50. Zhishen, J.; Mengcheng, T.; Jianming, W. The determination of flavonoid contents in mulberry and their scavenging effects on superoxide radicals. *Food Chem.* **1999**, *64*, 555–559. [CrossRef]
51. Kamarul Zaman, M.A.; Azzeme, A.M.; Ramle, I.K.; Normanshah, N.; Ramli, S.N.; Shaharuddin, N.A.; Ahmad, S.; Abdullah, S.N.A. Induction, multiplication, and evaluation of antioxidant activity of *Polyalthia bullata* callus, a woody medicinal plant. *Plants* **2020**, *9*, 1772. [CrossRef] [PubMed]
52. Radulescu, C.; Buruleanu, L.C.; Nicolescu, C.M.; Olteanu, R.L.; Bumbac, M.; Holban, G.C.; Simal-Gandara, J. Phytochemical profiles, antioxidant and antibacterial activities of grape (*Vitis vinifera* L.) seeds and skin from organic and conventional vineyards. *Plants* **2020**, *9*, 1470. [CrossRef] [PubMed]

53. Thitilertdecha, N.; Teerawutgulrag, A.; Rakariyatham, N. Antioxidant and antimicrobial activities of *Nephelium lappacium* L. extracts. *LWT Food Sci. Technol.* **2008**, *41*, 2029–2035. [CrossRef]

54. Gülçın, İ.; Oktay, M.; Kıreçcı, E.; Küfrevıoğlu, Ö.İ. Screening of antioxidant and antimicrobial activities of anise (*Pimpinella anisum* L.) seed extracts. *Food Chem.* **2003**, *83*, 371–382. [CrossRef]

55. Re, R.; Pellegrini, N.; Proteggente, A.; Pannala, A.; Yang, M.; Rice- Evans, C. Antioxidant activity applying an improved ABTS radical cation decolorization assay. *Radic. Biol. Med.* **1999**, *26*, 1231–1237. [CrossRef]

56. Li, Y.; Guo, C.; Yang, J.; Wei, J.; Xu, J.; Cheng, S. Evaluation of antioxidant properties of pomegranate peel extract in comparison with pomegranate pulp extract. *Food Chem.* **2006**, *96*, 254–260. [CrossRef]

57. Oki, T.; Matsui, T.; Osajima, Y. Inhibitory effect of α-glucosidase inhibitors varies according to its origin. *J. Agric. Food Chem.* **1999**, *47*, 550–553. [CrossRef] [PubMed]

58. Adisakwattana, S.; Chantarasinlapin, P.; Thammarat, H.; Yibchok-Anun, S. A series of cinnamic acid derivatives and their inhibitory activity on intestinal α-glucosidase. *J. Enzyme Inhib. Med. Chem.* **2009**, *24*, 1194–1200. [CrossRef]

59. Park, P.J.; Je, J.Y.; Kim, S.K. Angiotensin I converting enzyme (ACE) inhibitory activity of hetero-chitooligosaccharides prepared from partially different deacetylated chitosans. *J. Agric. Food Chem.* **2003**, *51*, 4930–4934. [CrossRef]

60. Wang, P.; Henning, S.M.; Heber, D. Limitations of MTT and MTS-based assays for measurement of antiproliferative activity of green tea polyphenols. *PLoS ONE* **2010**, *5*, e10202. [CrossRef]

 *plants*

Article

Vasorelaxant Effect of *Boesenbergia rotunda* and Its Active Ingredients on an Isolated Coronary Artery

Deepak Adhikari [1,†], Dal-Seong Gong [1,†], Se Hee Oh [1], Eun Hee Sung [1], Seung On Lee [2], Dong-Wook Kim [2], Min-Ho Oak [1,*]  and Hyun Jung Kim [1,*]

[1] College of Pharmacy and Natural Medicine Research Institute, Mokpo National University, Muan-gun 58554, Korea; dpak7adh@gmail.com (D.A.); nh4011@naver.com (D.-S.G.); yuhui16@naver.com (S.H.O.); hannn0828@hanmail.net (E.H.S.)

[2] Department of Oriental Medicine Resources, Mokpo National University, Muan-gun 58554, Korea; lso6918@naver.com (S.O.L.); dbkim@mokpo.ac.kr (D.-W.K.)

* Correspondence: mhoak@mokpo.ac.kr (M.-H.O.); hyunkim@mokpo.ac.kr (H.J.K.); Tel.: +82-61-450-2681 (M.-H.O.); +82-61-450-2686 (H.J.K.)

† The authors contributed to this work equally.

Received: 4 November 2020; Accepted: 27 November 2020; Published: 1 December 2020

Abstract: Cardiovascular diseases are a major cause of death in developed countries. The regulation of vascular tone is a major approach to prevent and ameliorate vascular diseases. As part of our ongoing screening for cardioprotective natural compounds, we investigated the vasorelaxant effect of rhizomes from *Boesenbergia rotunda* (L.) Mansf. [*Boesenbergia pandurata* (Roxb.) Schltr.] used as a spice and herbal medicine in Asian countries. The methanol extract of *B. rotunda* rhizomes (BRE) exhibited significant vasorelaxation effects ex vivo at EC_{50} values of 13.4 ± 6.1 μg/mL and 40.9 ± 7.9 μg/mL, respectively, with and without endothelium in the porcine coronary artery ring. The intrinsic mechanism was evaluated by treating with specific inhibitors or activators that typically affect vascular reactivity. The results suggested that BRE induced relaxation in the coronary artery rings via an endothelium-dependent pathway involving NO-cGMP, and also via an endothelium-independent pathway involving the blockade of Ca^{2+} channels. Vasorelaxant principles in BRE were identified by subsequent chromatographic methods, which revealed that flavonoids regulate vasorelaxant activity in BRE. One of the flavonoids was a Diels-Alder type adduct, 4-hydroxypanduratin A, which showed the most potent vasorelaxant effect on porcine coronary artery with an EC_{50} of 17.8 ± 2.5 μM. Our results suggest that rhizomes of *B. rotunda* might be of interest as herbal medicine against cardiovascular diseases.

Keywords: *Boesenbergia rotunda*; Zingiberaceae; flavonoid; 4-hydroxypanduratin; vasorelaxation

1. Introduction

Cardiovascular diseases are a major cause of death in developed countries. The vascular tone plays an important role in the regulation of blood pressure. High blood pressure is a major risk factor for cardiovascular diseases such as hyperpiesia, arrhythmia, heart failure, and atherosclerosis. Therefore, the regulation of vascular tone is one of the major approaches designed to prevent and ameliorate vascular diseases [1]. Several epidemiological studies have suggested that diets rich in polyphenols including fruits and vegetables and beverages protect against cardiovascular risks [2–5]. As part of our ongoing screening for cardioprotective phytochemicals, the extract from *Boesenbergia rotunda* rhizomes was found to exhibit significant vasorelaxation effects in the coronary artery.

Boesenbergia rotunda (L.) Mansf. [*Boesenbergia pandurata* (Roxb.) Schltr.; *Kaempferia pandurata* Roxb.], commonly known as fingerroot or Chinese keys, is a perennial culinary herb used as a flavoring food ingredient belonging to the family Zingiberaceae. It is cultivated and produced in India and southeastern Asian countries such as Indonesia, Thailand, Malaysia and Southwest China [6,7]. The underground parts of this plant have been widely used as traditional medicine to treat cough, muscular pain, rheumatism, gout, infections, inflammatory lesions, and gastrointestinal disorders in Southeast Asia [7]. Intensive studies analyzing the phytochemical composition and biological evaluations of *B. rotunda* rhizomes have been investigated based on these applications [7,8]. Phytochemicals including flavanones, flavones, chalcones and prenylated flavonoids were identified as the main secondary metabolites contributing to the potential biological and medicinal effects [7,8]. The extracts and flavonoids derived from *B. rotunda* rhizomes revealed antioxidant [9], anti-inflammatory [10,11], antimicrobial [12–16], antitumor [17–19], anti-aging [20,21], skin-whitening [22], antiulcer [23] and anti-obesity [24,25] activities. However, the vasodilatory effect of *B. rotunda* rhizomes has yet to be reported. In the present study, we investigated the vasorelaxant effect of *B. rotunda* rhizome extracts and the underlying mechanism, followed by isolation and structural elucidation of the key vasodilators in the active extract.

2. Results and Discussion

2.1. Vascular Relaxantion Induced by B. rotunda Rhizomes on Coronary Artery

Dried rhizomes of *B. rotunda* were milled and extracted with MeOH at room temperature, and the filtered liquid extract was evaporated in vacuo to yield a crude extract. The effect of *B. rotuda* rhizome methanol extract (BRE) on the vascular tone was assessed using porcine coronary artery rings contracted submaximal with the thromboxane A_2 analogue, U46619. BRE induced concentration-dependent relaxation in the endothelial rings, with an EC_{50} value of 13.4 ± 6.1 µg/mL (Figure 1A,B). The endothelium-dependent relaxation was initiated at concentrations greater than 3 µg/mL and reached a near maximal value at 100 µg/mL (E_{max} = 101.8 ± 4.2%, Figure 1A,B). Next, the possibility that BRE, besides inducing vasorelaxations, also affects the contractile responses was assessed. Both 30 and 100 µg/mL of BRE also significantly inhibited the concentration-dependent contractile response to U46619 in a nonparallel fashion and suppressed their maximal responses to 103.9 ± 3.2% and 40.7 ± 10.3% respectively (versus control group 130.3 ± 5.6% compared with 80 mM KCl) in endothelium-intact artery rings (Figure 1C).

To elucidate the role of endothelium in BRE-mediated relaxation in the porcine coronary artery, the concentration responses to BRE were evaluated in endothelium-denuded rings pre-contracted by U46619. BRE caused endothelium-independent relaxation of coronary artery, with an EC_{50} of 40.9 ± 7.9 µg/mL and E_{max} of 97.8 ± 5.3% at 100 µg/mL (Figure 1A,B). The EC_{50} value of vasorelaxant BRE in endothelium-intact coronary artery pre-contracted using U46619 was significantly lower compared with that of endothelium-denuded coronary artery, suggesting that vasorelaxant effects of BRE were mediated in both an endothelium-dependent and -independent manner.

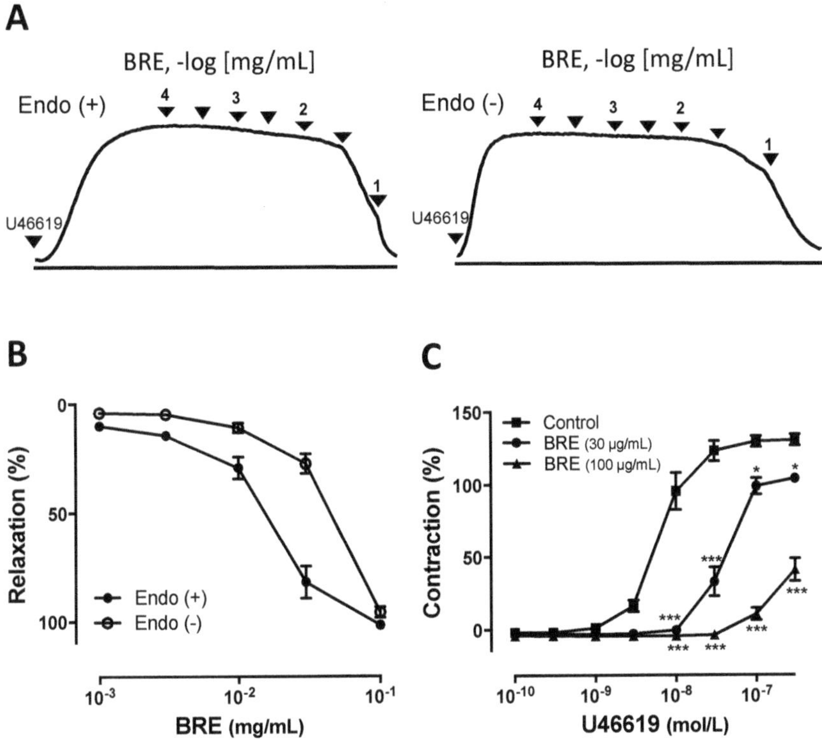

Figure 1. *Boesenbergia rotunda* rhizome methanol extract (BRE) induced concentration-dependent relaxation and prevented contraction in porcine coronary artery rings. Intact and endothelium-denuded rings were contracted with U46619 before the construction of BRE concentration-relaxation. (**A**) representative tracing; (**B**) accumulated figure; (**C**) coronary artery rings were treated with BRE at the indicated concentration before the construction of a concentration–contraction curve in response to U46619. Results are shown as mean ± SEM of 5–10 experiments. $* p < 0.05$, $*** p < 0.001$ versus control.

2.2. Mechansim of BRE-Induced Vasorelaxation

Vascular endothelium is a barrier separating circulating blood from vascular smooth muscle, and is considered an important dynamic organ regulating vascular tone by releasing various vasoactive factors such as vasodilatory factors [nitric oxide (NO), prostacyclin (PGI₂) and endothelium derived hyperpolarizing factors (EDHF) etc.] and vasoconstricting factors [thromboxane (TBXs) and endothelin etc.] [26]. Pre-incubation of endothelial rings with L-nitroarginine (L-NA, 100 µM), an inhibitor of nitric oxide synthase (NOS), significantly reduced the BRE-induced relaxation, with an EC_{50} value of 52.2 ± 12.2 µg/mL (versus 13.4 ± 6.1 µg/mL in the absence of L-NA, Figures 1B and 2A). However, indomethacin, a cyclooxygenase inhibitor did not significantly affect the BRE-induced relaxation in endothelium-intact rings (Figure 2A). It was previously reported that the release of NO-induced vasodilation via activation of soluble guanylate cyclase [27] and stimulation of Na⁺/K⁺-ATPase in aortic rings [28]. Therefore, we tested whether or not the vasorelaxant effect of BRE was affected by 1*H*-[1,2,4]oxadiazole[4,3-alpha]quinoxalin-1-one (ODQ, 10 µM), an inhibitor of guanylate cyclase, and ouabain (1 µM), an inhibitor of Na⁺/K⁺-ATPase. ODQ and ouabain significantly reduced BRE-induced relaxation, with EC_{50} values of 64.3 ± 10.1 µg/mL and 80.5 ± 13.4 µg/mL, respectively (versus 13.4 ± 6.1 µg/mL without inhibitors, Figure 2B). These results suggest that the activation of eNOS may play a major role, at least in part, in BRE-induced vasorelaxation.

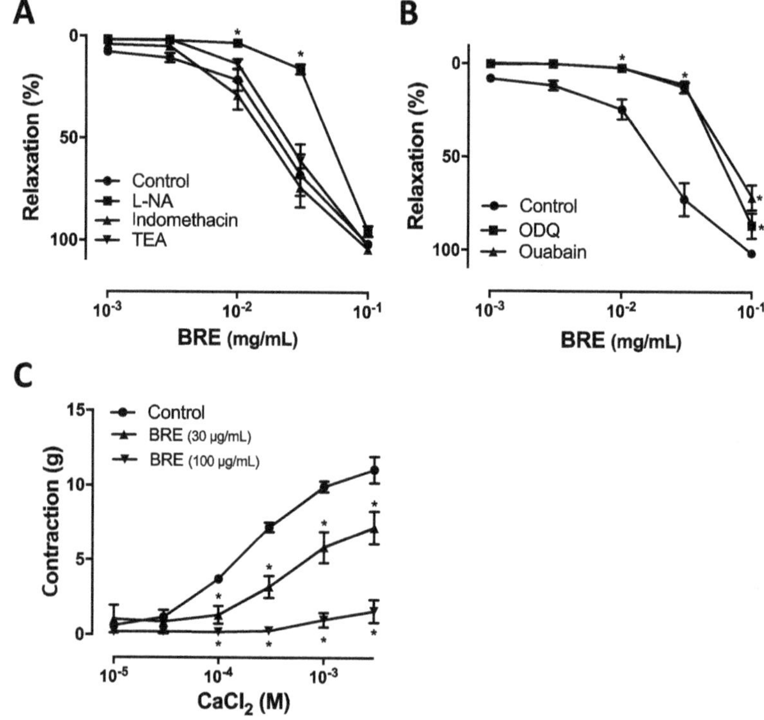

Figure 2. Cumulative concentration–response curves of BRE in coronary artery rings precontracted with U46619 in the presence of (**A**) L-NA (nitric oxide synthase inhibitor, 100 µM), indomethacin (cyclooxygenase inhibitor, 10 µM), TEA (non-selective potassium channel blocker) and (**B**) ODQ (guanylyl cyclase inhibitor, 10 µM), and ouabain (Na^+/K^+-ATPase inhibitor, 1 µM). (**C**) BRE inhibits vasoconstriction induced by calcium. Coronary artery rings were treated with BRE at the indicated concentration before drawing a concentration–contraction curve in response to $CaCl_2$. Results are shown as mean ± SEM of 5–10 experiments. * $p < 0.05$ versus control.

However, the observation that the vasorelaxant effect of BRE persisted in endothelium-denuded coronary artery ring or in those treated with L-NA suggests that BRE has a direct effect on vascular smooth muscle cells. The opening of K^+ channels in the vascular smooth muscle cells induces membrane potential hyperpolarization, which is an important mechanism of arterial dilation [29]. However, tetraethylammonium (TEA, 1 mM) which is a K^+ channel blocker, did not significantly affect the BRE-induced relaxation in the artery ring, suggesting that the vasodilating effect of BRE was not associated with the K^+ channel (Figure 2A). Another important vascular smooth muscle contraction is mediated by Ca^{2+} influx. BRE significantly and dose-dependently stimulated $CaCl_2$-induced contraction in endothelium-denuded coronary artery rings in Ca^{2+}-free medium containing 60 mM KCl. The maximal contraction induced by $CaCl_2$ (3 mM) was 11.1 ± 0.7 g, 7.2 ± 1.8 g, and 1.6 ± 0.5 g in the absence and presence of BRE (30 µg/mL and 100 µg/mL), respectively (Figure 2C). Taken together, our results suggest that BRE induces relaxation in coronary artery rings via an endothelium-dependent pathway, involving NO-cGMP, and also via an endothelium-independent pathway, involving blockade of Ca^{2+} channels.

2.3. Isolation and Structural Elucidaton of Flavonoids in the Rhizomes of B. rotunda

The BRE inducing vasorelaxation was fractionated via liquid–liquid partition based on solvent polarity to afford ethyl acetate (EtOAc), *n*-butyl alcohol (BuOH) and aqueous residue. The EC_{50} values were 15.5 ± 5.2 µg/mL and more than 100 µg/mL for EtOAc and aqueous fractions, respectively. Vascular reactivity showed that the EtOAc fraction exhibited similar vasorelaxant activity compared with that of the crude BRE (13.4 ± 6.1 µg/mL). The BuOH fraction also showed vasorelaxation effects (EC_{50} value, 35.3 ± 12.1 µg/mL), which were weaker than that of the EtOAc fraction, and residual aqueous layer was not active at the highest concentration tested (100 µg/mL). Interestingly, HPLC profiles of active fractions prepared with EtOAc and BuOH revealed the presence of flavonoids in UV spectral data of PDA, while no peaks corresponding to flavonoids were detected in the chromatogram of aqueous residue obtained under the same LC condition (Figure S1).

The EtOAc fraction was chromatographed by MPLC, and further purified using preparative HPLC to afford eight flavonoids (**1–8**) (Figure 3). Their structures were identified by spectroscopic analysis including NMR, UV, and MS, and compared with previously reported spectral data [30–35]. These flavonoids were also detected as major peaks in the HPLC chromatogram of BRE (Figure 4).

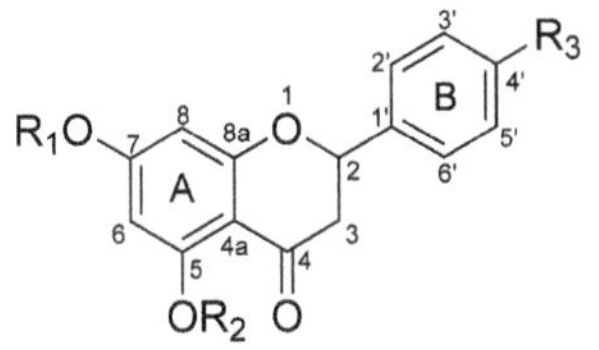

naringenin 5-methyl ether (**1**) R_1= H, R_2= CH_3, R_3= OH
alpinetin (**2**) R_1= H, R_2= CH_3, R_3= H
pinocembrin (**3**) R_1= H, R_2= H, R_3= H
pinostrobin (**6**) R_1= CH_3, R_2= H, R_3= H

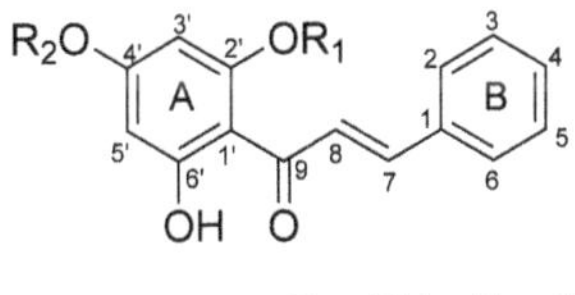

cardamonin (**4**) R_1= CH_3, R_2= H
pinostrobin chalcone (**5**) R_1= H, R_2= CH_3

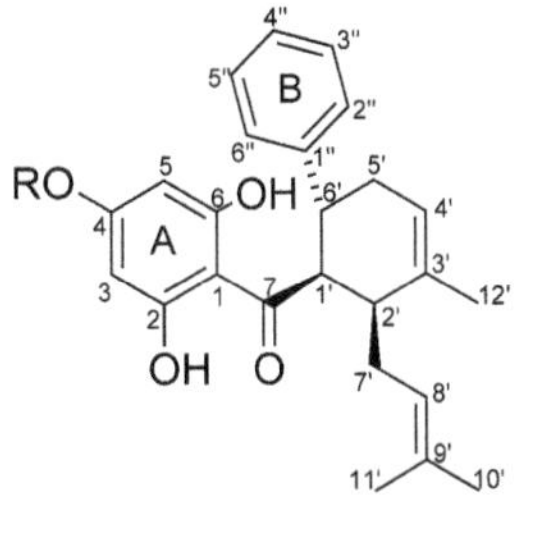

4-hydroxypanduratin A (**7**) R= H
panduratin A (**8**) R= CH_3

Figure 3. Structures of compounds isolated from *B. rotunda* rhizomes.

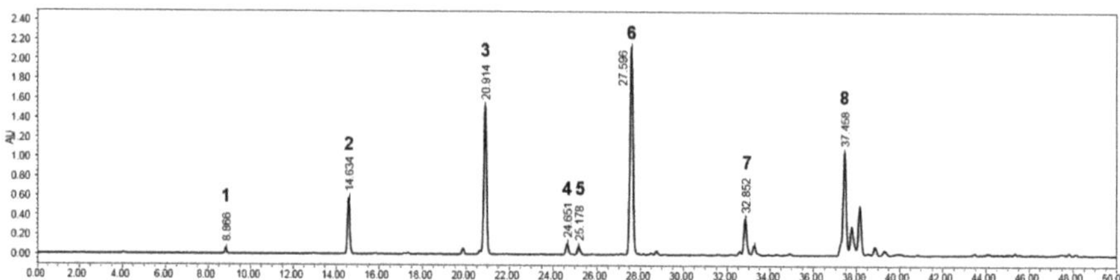

Figure 4. HPLC profile of the BRE derived from *B. rotunda* rhizomes. The chromatogram was based on the following conditions: a linear gradient solvent system of acetonitrile and 0.1% HCOOH-containing water, ranging from 20% to 90% acetonitrile, for 40 min, followed by isocratic elution with 100% acetonitrile for 10 min, on a SunFire C18 (5 μm, 4.6 × 150 mm) waters column, at a flow rate of 1.0 mL/min, and detection under 300 nm; naringenin 5-methyl ether (**1**), alpinetin (**2**), pinocembrin (**3**), cardamonin (**4**), pinostrobin chalcone (**5**), pinostrobin (**6**), 4-hydroxypanduratin A (**7**) and panduratin A (**8**).

Naringenin 5-methyl ether (**1**), alpinetin (**2**), pinocembrin (**3**), and pinostrobin (**6**) exhibited similar UV spectral patterns with maximal absorption between 284 and 288 nm in HPLC-PDA analysis, which suggested that their structures carried a flavonoid subgroup, flavanone. The ^{1}H NMR data of all these compounds revealed the presence of a 2-phenyl-4-chromanone skeleton including two aromatic ring protons (δ 5.94–6.15, d, H-6 and H-8) with typical meta-coupling constants (1.2–2.4 Hz), an oxymethine proton signal (δ 5.32–5.63, dd, H-2) with *J* values (12–13 and 3.0 Hz), and two methylene proton signals (δ 2.51–2.82 and δ 2.98–3.31, dd, H-3) with coupling constants of 16–17, 12–13 Hz and 16–17, 3.0 Hz, respectively. Their ^{13}C NMR data also displayed a 4-chromanone moiety including one carbonyl carbon (C-4), two alicyclic carbons (C-2 and C-3), and six aromatic carbons (C-5~C-8, C-4a, C-8a) as well as a phenyl moiety with six aromatic carbons (C-1′–C-6′). The aromatic protons of the ring B in **2**, **3** and **6** included five protons at δ 7.36–7.57 (H-2′–H-6′) indicating an unsubstituted B-ring, while phenyl protons in naringenin 5-methyl ether (**1**) showed a set of AA′BB′ protons as two doublets (each 2H) at δ 6.77 (C-3′,5′) and δ 7.28 (C-2′,6′) indicating para-substituted ring B. The position of each methoxy group in **1**, **2**, and **6** was identified based on long range couplings in HMBC spectra, respectively.

Cardamonin (**4**) and pinostrobin chalcone (**5**) displayed UV spectra of chalcones with essential and maximal absorption band at nearby 340 nm via HPLC-PDA analysis. NMR data of these compounds revealed that the presence of an α,β-unsaturated ketone with two aromatic rings. The NMR data of both compounds showed two trans-olefin protons (δ 8.02/8.30 and δ 7.75/7.79, H-7 and H-8, d) with a large coupling constant (15.6 Hz) correlating with one carbonyl carbon (δ 193.2 and δ 194.2, C-9). Additionally, their ^{1}H NMR spectra showed an unsubstituted phenyl ring possessing five aromatic protons at δ 7.43–7.73 (H-2–H-6), and a 1,2,4,6-tetrasubstituted benzene ring including two aromatic meta-protons at around δ 6.00 (H-3′ and H-5′). The positions of methoxy functions were respectively identified at C-2′ and C-4′ in **4** and **5** by HSQC and HMBC spectra.

4-Hydroxypanduratin A (**7**) and panduratin A (**8**) showed closely similar UV spectra with the maximal absorptions at around UV 290 nm suggesting a conjugated ketone moiety and revealed the difference in 14 mass units between two compounds. The ^{1}H NMR data of both shared five aromatic protons of a phenyl group at δ 7.05–7.22 (H-2″–C-6″) and two protons in a 2,4,6-trihydroxyphenyl group near δ 5.90 (2H, H-3,5). They also revealed the presence of a dimethylallyl group consisting of two proton signals in a methylene bridge (δ 2.08–2.29, H-7′), an olefinic proton signal (δ 4.94/4.88, respectively, H-8′) and singlets of dimethyl protons (δ 1.51–1.52, 6H, H-10′ and 11′), as well as a 3-methyl-cyclohex-3-ene moiety with trans–cis configuration ($J_{1′,6′}$ = 10.0–11.7 Hz, $J_{1′,2′}$ = 4.5–5.5 Hz) corresponding to five aliphatic proton signals at δ 1.99–4.80 (each proton of H-1′, 2′, 6′ and two proton signals of H-5′), an olefinic signal at δ 5.41/5.42 (H-4′) and an allylic methyl at δ1.77 (3H, H-12′). The ^{1}H NMR spectrum of panduratin A (**8**) displayed an additional proton signal at δ 3.74 (3H, s). These observations indicated that both compounds were closely related Diels-Alder adducts with a distinct methyl group with long range correlation to the oxygenated aromatic carbon C-4. The presence of all the isolated compounds were previously reported in the rhizomes of *B. rotunda* [8].

2.4. Vasorelaxant Activity of Flavonoids from B. rotunda

The vasorelaxant effect of the isolated compounds derived from BRE indicated that naringenin 5-methyl ether (**1**), alpinetin (**2**), pinocembrin (**3**), pinostrobin (**6**) and 4-hydroxypanduratin A (**7**) elicited significant vasodilation of coronary artery (Table 1 and Figure 5). Vascular reactivity of flavonoids (**1–8**) from BRE described herein suggests a structure-activity relationship underlying for the role and potential of flavonoid scaffold in part. Chalcones including cardamonin (**4**) and pinostrobin chalcone (**5**) showed significantly less potent vasorelaxant effects than those of flavanones including naringenin 5-methyl ether (**1**), alpinetin (**2**), pinocembrin (**3**) and pinostrobin (**6**). Additionally, with functional moieties of the flavanones (**1–3**, and **6**), the comparison of EC_{50} values indicated that the vascular reactivity is affected by not only the hydroxyl groups but also the position of a methoxyl in the A and B rings. Furthermore, the presence of hydroxylation of C-4 in ring A demonstrates the pivotal role of vasorelaxant effects induced by the Diels-Alder type adducts such as panduratin A derivatives. Panduratin A (**8**) with a methoxy function at C-4 did not alter the vascular tone, whereas 4-hydroxypanduratin A (**7**) showed the most potent vascular reactivity among the major flavonoids present in the BRE.

Table 1. Values of half maximal effective dose (EC_{50}) and the maximum effect (E_{max}) of isolated compounds from BRE.

Compound	EC_{50} (µM)	E_{max} (%)
Naringenin 5-methyl ether (**1**)	52.3 ± 10.4	100.4 ± 1.9
Alpinetin (**2**)	85.6 ± 23.9	104.6 ± 8.9
Pinocembrin (**3**)	35.1 ± 15.3	107.2 ± 3.5
Cardamonin (**4**)	166.7 ± 35.5	65.4 ± 7.4
Pinostrobin chalcone (**5**)	92.1 ± 22.7	103.7 ± 3.7
Pinostrobin (**6**)	52.0 ± 14.3	89.9 ± 2.1
4-Hydroxypanduratin A (**7**)	17.8 ± 2.5	92.7 ± 6.8
Panduratin A (**8**)	$EC_{50} \geq 1000$	23.1 ± 4.5

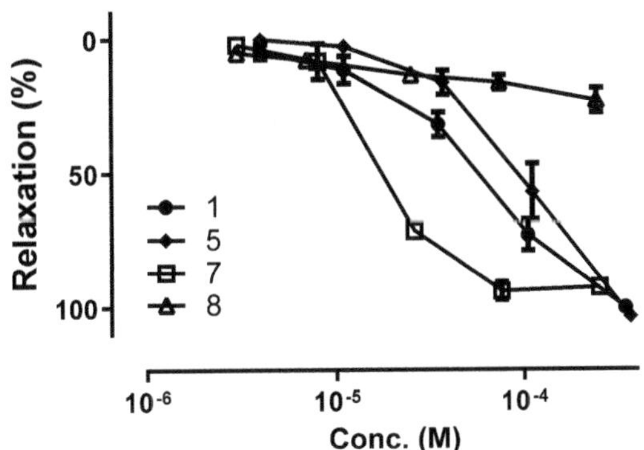

Figure 5. Concentration-response curves illustrating the vasorelaxant effect of naringenin 5-methyl ether (**1**), pinostrobin chalcone (**5**), 4-hydroxypanduratin A (**7**) and panduratin A (**8**). Results are shown as mean ± SEM of 5–10 experiments.

The study of the relationship between dietary factors and cardiovascular mortality suggested that the diet-related cardiometabolic deaths are predominantly observed in the population associated with a low intake of vegetables and fruits [36]. Polyphenols naturally exist in plants and plant products, including fruits and vegetables. Currently, more than 8000 phenolic structures are available, including more than 4000 belonging to the flavonoid class, and several hundred flavonoids are present in edible vegetables [37]. The polyphenol structure is characterized by at least a simple phenol core bearing at least one hydroxyl group and classified according to the arrangement of the carbon atoms and their substituents into two main classes. It is largely admitted that flavonoids derived from vegetables and medicinal plants exhibit cardiovascular protective effects [38]. Among eight flavonoids isolated from BRE in the present study, the vasorelaxant effect of alpinetin (**2**) isolated from *Alpinia henryi* K. Schum. [39], pinocembrin (**3**) from propolis [40,41], cardamonin (**4**) from *A. henryi* K. Schum. [39], and pinostrobin (**6**) from *Teloxys graveolens* [42] was reported previously. Consistent with our results, the vasorelaxant effect of alpinetin (**2**), pinocembrin (**3**) and cardamonin (**4**) was shown both in endothelium-dependent, which was associated with nitric oxide, and endothelium-independent, which was associated with the blockade of Ca^{2+} channels [39–42]. To the best of our knowledge, the present investigation is the first report suggesting that naringenin 5-methyl ether (**1**) and 4-hydroxypanduratin A (**7**) isolated from rhizomes of *B. rotunda* induce vasorelaxation; however, the molecular mechanism of action remains to be clarified, and cardiovascular effect and toxicity of the isolated vasoactive flavonoids need to be further investigated via in vitro and in vivo studies. Taken together, the BRE induced relaxation in coronary artery via endothelium-dependent and independent mechanisms, whose effects are, at least in part, due to constituents including that naringenin 5-methyl ether (**1**), alpinetin (**2**), pinocembrin (**3**), pinostrobin (**6**) and 4-hydroxypanduratin A (**7**). They further suggest that BRE might be of interest as a source of herbal medicine or functional food to prevent the development of cardiovascular diseases.

3. Materials and Methods

3.1. Plant Material

The dried rhizomes of *B. rotunda* originating in Indonesia were obtained from an herbal company in Suwon, Korea in 2016, and were identified by Prof. Hyun Jung Kim at the Laboratory of Pharmacognosy, College of Pharmacy, Mokpo National University. A voucher specimen was deposited at the herbarium placed in the same institute (No. P2016BRR001).

3.2. Reagents and Chemicals

9,11-Dideoxy-9α,11α-methanoepoxy prostaglandin $F_{2\alpha}$ (U46619) was supplied by Cayman Chemical (Ann Arbor, MI, USA), and indomethacin, N-ω-nitro-L-arginine (L-NA), bradykinin, ODQ, 9-(tetrahydro-2-furanyl)-9*H*-purin-6-amine (SQ22563), ouabain by Sigma-Aldrich (St. Louis, MO, USA). All chemicals were of analytical grade. HPLC grade solvents (Fisher Scientific, Hampton, NH, USA) were used for all the HPLC experiments, and extra pure grade solvents were used for extraction.

3.3. Extraction and Isolation of the Rhizomes of B. rotunda

Dried rhizomes of *B. rotunda* (600 g) were milled and extracted with MeOH (1 L × 4) at room temperature, and the filtered liquid extract was evaporated in vacuo to afford a crude extract (91.9 g). An aliquot of this extract (48.7 g) was suspended in 1 L of H_2O, subsequently partitioned with EtOAc (1 L × 4) and *n*-BuOH (1 L × 3). The EtOAc fraction (25 g) was subjected to Si gel MPLC (Biotage Isolera One, Biotage, Uppsala, Sweden) with Biotage SNAP cartridge KP-SIL 340 g using *n*-hexane and $CHCl_3$/acetone (1:1) mixtures at 60 mL/min for 60 min to obtain 20 fractions (F01-F20). The combined F03 through F05 (4.88 g) was purified by preparative HPLC (Waters 600 system, Waters, Milford, MA) with a SunFire^{TM} Prep OBD^{TM} column (5 μm, 19 × 150 mm, Waters) using acetonitrile and 0.1% HCOOH containing water (30:70 to 100% acetonitrile, 30 min) at the flow rate of 10.0 mL/min to give pinostrobin (**6**, 944.0 mg). The fractions F06 and F07 were combined, and an aliquot of a fraction (7 g) was chromatographed with Biotage SNAP Cartridge (KP-C18-HS 120 g) using a gradient solvent system of acetonitrile and water (30:70 to 100% acetonitrile) at 30 mL/min for 60 min to yield eight subfractions (SF01-SF08). Alpinetin (**2**, 94.0 mg) from SF01, pinocembrin (**3**, 69.8 mg) and cardamonin (**4**, 49.4 mg) from SF02, pinostrobin chalcone (**5**, 12.7 mg) from SF03, panduratin A (**8**, 116. 4 mg) from SF05, and 4-hydroxypanduratin A (**7**, 32.9 mg) from SF06, were obtained respectively via preparative HPLC purification with a SunFire^{TM} Prep OBD^{TM} column (5 μm, 19 × 150 mm, Waters) using acetonitrile and 0.1% HCOOH containing water (30:70 to 100% acetonitrile, 30 min) at 10.0 mL/min. F07 was further separated using the same preparative HPLC condition using acetonitrile and 0.1% HCOOH containing water (25:75 to 70:30, 15 min) at the flow rate of 10.0 mL/min to obtain naringenin 5-methyl ether (**1**, 84.0 mg). The structures of all the isolated compounds (**1–8**) were elucidated by spectroscopic analysis and compared with previously reported spectral data (Figures S2–S9) [30–35].

Naringenin 5-methyl ether (**1**) m/z 287.1 [M + H]$^+$; UV λ_{max} 284 nm; ^{1}H NMR (600 MHz, DMSO-d_6): 7.28 (2H, d, J = 8.4 Hz, H-2′,6′), 6.77 (2H, d, J = 8.4 Hz, H-3′,5′), 6.04 (1H, d, J = 1.2 Hz, H-6), 5.94 (1H, d, J = 1.2 Hz, H-8), 5.32 (1H, dd, J = 12.6, 3.0 Hz, H-2), 3.72 (3H, s, 5-OCH$_3$), 2.98 (1H, dd, J = 16.2, 12.6 Hz, H-3), 2.51 (1H, dd, J = 16.2, 3.0 Hz, H-3); ^{13}C NMR (150 MHz, DMSO-d_6): 187.7 (C-4), 164.8 (C-8a), 164.2 (C-7), 162.2 (C-5), 157.6 (C-4′), 129.4 (C-1′), 128.1 (C-2′,6′), 115.1 (C-3′,5′), 104.2 (C-4a), 95.7 (C-8), 93. 4 (C-6), 78.0 (C-2), 55.6 (5-OCH$_3$), 44.8 (C-3).

Alpinetin (**2**) m/z 271.1 [M + H]$^+$; UV λ_{max} 286 nm; ^{1}H NMR (600 MHz, DMSO-d_6): 7.49 (2H, brd, J = 7.3 Hz, H-2′,6′), 7.41 (2H, brt, J = 7.3 Hz, H-3′,5′), 7.36 (1H, brt, J = 7.3 Hz, H-4′), 6.07 (1H, d, J = 1.8 Hz, H-6), 6.00 (1H, d, J = 1.8 Hz, H-8), 5.48 (1H, dd, J = 12.3, 3.0 Hz, H-2), 3.74 (3H, s, 5-OCH$_3$), 2.98 (1H, dd, J = 16.2, 12.3 Hz, H-3), 2.62 (1H, dd, J = 16.2, 3.0 Hz, H-3); ^{13}C NMR (150 MHz, DMSO-d_6): 187.3 (C-4), 164.6 (C-7), 164.1 (C-8a), 162.2 (C-5), 139.2 (C-1′), 128.5 (C-3′,5′), 128.3 (C-4′), 126.4 (C-2′,6′), 104.4 (C-4a), 95.7 (C-8), 93. 4 (C-6), 78.0 (C-2), 55.6 (5-OCH$_3$), 44.9 (C-3).

Pinocembrin (**3**) m/z 257.1 [M + H]$^+$; UV λ_{max} 288 nm; ^{1}H NMR (600 MHz, acetone-d_6): 12.16 (1H, s, 5-O<u>H</u>), 7.57 (2H, brt, J = 7.2 Hz, H-2′,6′), 7.45 (2H, td, J = 7.2, 1.8 Hz, H-3′,5′), 7.40 (1H, tt, J = 7.2, 1.8 Hz, H-4′), 6.00 (1H, d, J = 2.4 Hz, H-6), 5.96 (1H, d, J = 2.4 Hz, H-8), 5.58 (1H, dd, J = 12.9, 3.0 Hz, H-2), 3.17 (1H, dd, J = 17.4, 12.9 Hz, H-3), 2.81 (1H, dd, J = 17.4, 3.0 Hz, H-3); ^{13}C NMR (150 MHz, acetone-d_6): 196.9 (C-4), 167.7 (C-7), 165.4 (C-5), 164.2 (C-8a), 140.1 (C-1′), 129.6 (C-3′,5′), 129.5 (C-4′), 127.4 (C-2′,6′), 103.3 (C-4a), 97.1 (C-8), 96. 0 (C-6), 80.0 (C-2), 43.7 (C-3).

Cardamonin (**4**) m/z 271.1 [M + H]$^+$; UV λ_{max} 344 nm; ^{1}H NMR (600 MHz, acetone-d_6): 8.02 (1H, d, J = 15.6 Hz, H-7), 7.75 (1H, d, J = 15.6 Hz, H-8), 7.73 (2H, brd, J = 7.2 Hz, H-2,6), 7.44 (3H, m, H-3,4,5), 6.09 (1H, d, J = 2.2 Hz, H-3′), 6.01 (1H, d, J = 2.2 Hz, H-5′), 3.98 (3H, s, 2′-OCH$_3$); ^{13}C NMR (150 MHz, acetone-d_6): 193.2 (C-9), 169.1 (C-6′), 166.3 (C-4′), 164.4 (C-2′), 142.7 (C-8), 136.5 (C-1), 131.0 (C-4), 129.9 (C-3,5), 129.3 (C-2,6), 128.6 (C-7), 106.4 (C-1′), 97.1 (C-5′), 92.4 (C-3′), 56.5 (2′-OCH$_3$).

Pinostrobin chalcone (**5**) m/z 271.1 [M + H]$^+$; UV λ_{max} 340 nm; ^{1}H NMR (600 MHz, acetone-d_6): 8.30 (1H, d, J = 15.6 Hz, H-7), 7.79 (1H, d, J = 15.6 Hz, H-8), 7.70 (2H, dd, J = 7.8, 1.8 Hz, H-2,6), 7.44 (2H, m, H-3,5), 7.43 (1H, m, H-4), 6.04 (2H, s, H-3′,5′), 3.81 (3H, s, 4′-OCH$_3$); ^{13}C NMR (150 MHz, acetone-d_6): 194.2 (C-9), 168.0 (C-4′), 166.5 (C-2′,6′), 143.5 (C-8), 137.2 (C-1), 131.6 (C-4), 130.5 (C-3,5), 129.9 (C-2,6), 129.2 (C-7), 107.0 (C-1′), 95.2 (C-3′,5′), 56.5 (4′-OCH$_3$).

Pinostrobin (**6**) m/z 271.1 [M + H]$^+$; UV λ_{max} 288 nm; ^{1}H NMR (600 MHz, DMSO-d_6). 12.11 (1H, s, 5-O<u>H</u>), 7.53 (2H, brd, J = 7.2 Hz, H-2′,6′), 7.44 (2H, brt, J = 7.2 Hz, H-3′,5′), 7.39 (1H, tt, J = 7.2, 1.2 Hz, H-4′), 6.15 (1H, d, J = 1.8 Hz, H-8), 6.10 (1H, d, J = 1.8 Hz, H-6), 5.63 (1H, dd, J = 13.2, 3.0 Hz, H-2), 3.79 (3H, s, 7-OCH$_3$), 3.31 (1H, dd, J = 17.2, 13.2 Hz, H-3), 2.82 (1H, dd, J = 17.2, 3.0 Hz, H-3); ^{13}C NMR (150 MHz, DMSO-d_6): 196.5 (C-4), 167.5 (C-7), 163.2 (C-5), 162.7 (C-8a), 138.5 (C-1′), 128.6 (C-4′), 128.5 (C-3′,5′), 126.7 (C-2′,6′), 102.7 (C-4a), 94.8 (C-6), 93. 9 (C-8), 78.6 (C-2), 55.9 (5-OCH$_3$), 42.2 (C-3).

4-Hydroxypanduratin A (**7**) m/z 393.2 [M + H]$^+$; UV λ_{max} 291 nm; ^{1}H NMR (600 MHz, acetone-d_6): 7.22 (2H, brd, J = 7.8 Hz, H-2″,6″), 7.18 (1H, brt, J = 7.8 Hz, H-3″,5″), 7.05 (1H, tt, J = 7.6, 1.2 Hz, H-4″), 5.89 (2H, s, H-3, 5), 5.41 (1H, s, H-4′), 4.94 (1H, tq, J = 7.2, 1.2 Hz, H-8′), 4.80 (1H, dd, J = 11.7, 4.5 Hz, H-1′), 3.43 (1H, td, J = 10.8, 6.6 Hz, H-6′), 2.69 (1H, q, J = 5.4 Hz, H-2′), 2.36 (1H, m, H-5′), 2.29 (1H, dt, J = 15.6, 7.2 Hz, H-7′), 2.08 (1H, m, H-7′), 1.99 (1H, ddq, J = 18.0, 4.8, 1.8 Hz, H-5′), 1.77 (3H, d, J = 1.2 Hz, H-12′), 1.52 (3H, s, H-10′), 1.51 (3H, d, J = 1.2 Hz, H-11′); ^{13}C NMR (150 MHz, acetone-d_6): 207.1 (C-7), 165.0 (C-2,4,6), 148.4 (C-1″), 138.0 (C-3′), 131.8 (C-9′), 129.0 (C-3″,5″), 128.1 (C-2″,6″),

126.3 (C-4″), 125.5 (C-8′), 121.8 (C-4′), 106.3 (C-1), 96.0 (C-3,5), 54.5 (C-1′), 43.4 (C-2′), 37.9 (C-6′), 36.9 (C-5′), 29.6 (C-7′), 26.0 (C-11′), 23.1 (C-12′), 18.1 (C-10′).

Panduratin A (**8**) m/z 407.2 [M + H]$^+$; UV λ_{max} 290 nm; ^{1}H NMR (500 MHz, CDCl$_3$): 7.22 (4H, m, H-2″,3″,5″,6″), 7.09 (1H, m, H-4″), 5.91 (2H, s, H-3, 5), 5.42 (1H, brs, H-4′), 4.88 (1H, brt, J = 6.0 Hz, H-8′), 4.72 (1H, dd, J = 11.5, 4.5 Hz, H-1′), 3.74 (4-OCH$_3$), 3.44 (1H, tdd, J = 10.0, 6.0, 1.5 Hz, H-6′), 2.64 (1H, q, J = 5.5 Hz, H-2′), 2.39 (1H, m, H-5′), 2.27 (1H, dt, J = 15.5, 7.5 Hz, H-7′), 2.08 (1H, m, H-7′), 2.02 (1H, m, H-5′), 1.77 (3H, s, H-12′), 1.52 (6H, s, H-10′,11′); ^{13}C NMR (125 MHz, CDCl$_3$): 206.6 (C-7), 165.1 (C-2,4,6), 147.1 (C-1″), 137.2 (C-3′), 131.8 (C-9′), 128.3 (C-3″,5″), 127.1 (C-2″,6″), 125.5 (C-4″), 124.3 (C-8′), 121.0 (C-4′), 105.9 (C-1), 94.1 (C-3,5), 55.3 (4-OCH$_3$), 53.7 (C-1′), 42.5 (C-2′), 37.1 (C-6′), 35.9 (C-5′), 28.9 (C-7′), 25.7 (C-11′), 22.8 (C-12′), 17.9 (C-10′).

3.4. HPLC-PDA and LC-MS Analysis

HPLC analysis of extracts was carried out in a Waters HPLC system (Waters Corporation, Milford, MA) composed of a 1525 binary pump with a column oven, a 2707 autosampler, and a 2998 photodiode array detector (210–400 nm) using a Waters SunFire C18 column (5 μm, 4.6 × 150 mm). Plant extract, fractions, and single compounds were eluted by a linear gradient system using acetonitrile and water (0.1% HCOOH), ranging from 20% A to 90% acetonitrile (for 40 min) followed by an isocratic solvent 100% acetonitrile (10 min) at the flow rate of 1.0 mL/min. UV absorption was monitored under 300 nm. LC-ESI-MS experiments were performed on Agilent 6120 single quadruple MS hyphenated to Agilent 1260 Infinity quaternary LC (Agilent Technologies, Santa Clara, CA, USA) using a Thermo Acclaim Polar Advantage II (2.2 μm, 2.1 × 100 mm) in the positive mode, and developed by a gradient solvent mixture with acetonitrile and water (0.1% HCOOH), 20% to 100% acetonitrile for 25 min at 0.3 mL/min. The MS detection was carried out using electrospray ionization (ESI) with API source and MS spectra were obtained between m/z 100–1000 in a positive mode.

3.5. Vascular Reactivity

Pig hearts were collected from the local slaughterhouse (Mokpo, Korea) and the vascular reactivity was assessed using coronary artery as indicated previously [43]. Briefly, the left anterior descending coronary arteries of porcine heart were dissected, cleaned of connective tissue, and cut into rings (4–5 mm in length) carefully. Then, porcine coronary artery rings were incubated in organ baths containing oxygenated (95% O$_2$ and 5% CO$_2$) Krebs bicarbonate solution (mmol/L; NaCl, 119; KCl, 4.7; KH$_2$PO$_4$, 1.18; MgSO$_4$, 1.18; CaCl$_2$, 1.25; NaHCO$_3$, 25; and D-glucose, 11; pH 7.4, 37 °C) to determine the changes in isometric tension. Following equilibration for 90 min under a resting tension of 5 g, the rings were contracted twice with KCl (80 mmol/L). Subsequently, the rings were pre-contracted with the thromboxane mimetic U46619 (1–60 nmol/L) to about 80% of the maximal contraction and the integrity of the endothelium was checked with bradykinin (0.3 μmol/L). After washout and a 30 min equilibration period, the rings were again contracted with U46619 before construction of a concentration–relaxation curve involving extracts and isolated compounds. In some experiments analyzing the role of endothelium-derived vasoactive factors, the rings were exposed to various inhibitors for 30 min before the addition of U46619. The inhibition of contractile response was assessed by exposing the rings to extracts for 30 min before construction of concentration-contraction curve either for U46619 or CaCl$_2$ in the presence of 40 mM KCl. The contractile response to CaCl$_2$ or U46619 was expressed as the percentage of the maximal contraction induced by KCl (80 mmol/L) in a standard Krebs solution.

3.6. Statistical Analysis

All data are expressed as mean ± SEM. Statistical analysis of the data was performed using the Student's t-test or multiway ANOVA followed by Fisher's protected least significant difference test where appropriate. A value of p < 0.05 was considered statistically significant.

4. Conclusions

This study demonstrates that *B. rotunda* (L.) Mansf. [*B. pandurata* (Roxb.) Schltr.] extracts induce relaxation in coronary artery rings via, at least in part, an endothelium-dependent pathway, involving NO-cGMP, and also via an endothelium-independent pathway, involving blockade of Ca^{2+} channels. Vasoactive flavonoids were identified from the active fraction, including naringenin 5-methyl ether (**1**), alpinetin (**2**), pinocembrin (**3**), pinostrobin (**6**) and 4-hydroxypanduratin A (**7**), which induced significant vasodilation of coronary artery. Although the exact underlying mechanism of each compounds still remains to be elucidated, our findings suggest that rhizomes of *B. rotunda* might be of interest as an herbal medicine and functional food to prevent the development of cardiovascular diseases.

Supplementary Materials: The following are available online at http://www.mdpi.com/2223-7747/9/12/1688/s1, Figure S1. Comparison of HPLC profiles for three fractions derived from the vasoactive MeOH extract of *Boesenbergia rotunda* rhizomes (BRE), Figure S2. ^{1}H and ^{13}C NMR data of naringenin 5-methyl ether (**1**), Figure S3. ^{1}H and ^{13}C NMR data of alpinetin (**2**), Figure S4. ^{1}H and ^{13}C NMR data of pinocembrin (**3**), Figure S5. ^{1}H and ^{13}C NMR data of cardamonin (**4**), Figure S6. ^{1}H and ^{13}C NMR data of pinostrobin chalcone (**5**), Figure S7. ^{1}H and ^{13}C NMR data of pinostrobin (**6**), Figure S8. ^{1}H and ^{13}C NMR data of 4-hydroxypanduratin A (**7**), Figure S9. ^{1}H and ^{13}C NMR data of panduratin A (**8**).

Author Contributions: Conceptualization, M.-H.O. and H.J.K.; methodology, D.-W.K., M.-H.O. and H.J.K.; analysis, D.A., D.-S.G., S.H.O., E.H.S. and S.O.L.; investigation, D.A. and D.-S.G.; data curation, M.-H.O. and H.J.K.; writing—original draft preparation, D.A. and D.-S.G.; writing—review and editing, D.-W.K., M.-H.O. and H.J.K.; supervision, M.-H.O. and H.J.K.; project administration, M.-H.O. and H.J.K.; funding acquisition, M.-H.O. and H.J.K. All authors have read and agreed to the published version of the manuscript.

Funding: This research was studied partly by funds of MNU Innovative Programs for National University in 2019 and supported in part by the Basic Research Program through the National Research Foundation of Korea (NRF) funded by the Ministry of Education (2018R1D1A1B07050107 and 2020R1F1A106784911).

Acknowledgments: We thank the Gwangju Center of Korea Basic Science Institute (KBSI) for running the NMR experiments. We especially appreciate Bo Young Lee, a graduate of the College of Pharmacy in Mokpo National University, for her interest in fingerroot (*B. rotunda* rhizomes), which prompted this study.

Conflicts of Interest: The authors declare no conflict of interest.

Abbreviations

BRE	*Boesenbergia rotunda* rhizomes methanol extract
HPLC-PDA	High performance liquid chromatography-photodiode array detector
LC-MS	Liquid chromatography-mass spectrometry
eNOS	Endothelial nitric oxide synthase
ODQ	1*H*- [1,2,4]oxadiazole [4,3-alpha]quinoxalin-1-one
L-NA	L-Nitroarginine
TEA	Tetraethylammonium
cGMP	Cyclic guanosine monophosphate

References

1. Li, C.; Li, J.; Weng, X.; Lan, X.; Chi, X. Farnesoid X receptor agonist CDCA reduces blood pressure and regulates vascular tone in spontaneously hypertensive rats. *J. Am. Soc. Hypertens.* **2015**, *9*, 507–516.e7. [CrossRef]
2. Renaud, S.; de Lorgeril, M. Wine, alcohol, platelets, and the French paradox for coronary heart disease. *Lancet* **1992**, *339*, 1523–1526. [CrossRef]
3. Sofi, F.; Macchi, C.; Abbate, R.; Gensini, G.F.; Casini, A. Mediterranean diet and health status: An updated meta-analysis and a proposal for a literature-based adherence score. *Public Health Nutr.* **2013**, *17*, 2769–2782. [CrossRef] [PubMed]
4. Oude Griep, L.M.; Verschuren, W.M.M.; Kromhout, D.; Ocké, M.C.; Geleijnse, J.M. Variety in fruit and vegetable consumption and 10-year incidence of CHD and stroke. *Public Health Nutr.* **2012**, *15*, 2280–2286. [CrossRef]
5. Rienks, J.; Barbaresko, J.; Nothlings, U. Association of polyphenol biomarkers with cardiovascular disease and mortality risk: A systematic review and meta-analysis of observational studies. *Nutrients* **2017**, *9*, 415. [CrossRef] [PubMed]

6.   The Plant List. Available online: http://www.theplantlist.org/tpl1.1/record/kew-221874 (accessed on 30 November 2019).
7.   Eng-Chong, T.; Yean-Kee, L.; Chin-Fei, C.; Choon-Han, H.; Sher-Ming, W.; Li-Ping, C.T.; Gen-Teck, F.; Khalid, N.; Abd Rahman, N.; Karsani, S.A.; et al. *Boesenbergia rotunda*: From ethnomedicine to drug discovery. *Evid. Based Complement. Alternat. Med.* **2012**, *2012*, 473637. [CrossRef] [PubMed]
8.   Chahyadi, A.; Hartati, R.; Wirasutisna, K.R.; Elfahmi. *Boesenbergia pandurata* Roxb., an indonesian medicinal plant: Phytochemistry, biological activity, plant biotechnology. *Procedia Chem.* **2014**, *13*, 13–37.
9.   Shindo, K.; Kato, M.; Kinoshita, A.; Kobayashi, A.; Koike, Y. Analysis of antioxidant activities contained in the *Boesenbergia pandurata* Schult. rhizome. *Biosci. Biotechnol. Biochem.* **2006**, *70*, 2281–2284. [CrossRef]
10. Tuchinda, P.; Reutrakul, V.; Claeson, P.; Pongprayoon, U.; Sematong, T.; Santisuk, T.; Taylor, W.C. Anti-inflammatory cyclohexenyl chalcone derivatives in *Boesenbergia pandurata*. *Phytochemistry* **2002**, *59*, 169–173. [CrossRef]
11. Yun, J.M.; Kwon, H.; Hwang, J.K. In vitro anti-inflammatory activity of panduratin A isolated from *Kaempferia pandurata* in raw264.7 cells. *Planta Med.* **2003**, *69*, 1102–1108.
12. Hwang, J.K.; Chung, J.Y.; Baek, N.I.; Park, J.H. Isopanduratin A from *Kaempferia pandurata* as an active antibacterial agent against cariogenic *Streptococcus mutans*. *Int. J. Antimicrob. Agents* **2004**, *23*, 377–381. [CrossRef] [PubMed]
13. Rukayadi, Y.; Lee, K.; Han, S.; Yong, D.; Hwang, J.K. In vitro activities of panduratin A against clinical *Staphylococcus* strains. *Antimicrob. Agents Chemother.* **2009**, *53*, 4529–4532. [CrossRef] [PubMed]
14. Rukayadi, Y.; Han, S.; Yong, D.; Hwang, J.K. In vitro antibacterial activity of panduratin A against enterococci clinical isolates. *Biol. Pharm. Bull.* **2010**, *33*, 1489–1493. [CrossRef] [PubMed]
15. Cheenpracha, S.; Karalai, C.; Ponglimanont, C.; Subhadhirasakul, S.; Tewtrakul, S. Anti-HIV-1 protease activity of compounds from *Boesenbergia pandurata*. *Bioorg. Med. Chem.* **2006**, *14*, 1710–1714. [CrossRef] [PubMed]
16. Kiat, T.S.; Pippen, R.; Yusof, R.; Ibrahim, H.; Khalid, N.; Rahman, N.A. Inhibitory activity of cyclohexenyl chalcone derivatives and flavonoids of fingerroot, *Boesenbergia rotunda* (L.), towards dengue-2 virus NS3 protease. *Bioorg. Med. Chem. Lett.* **2006**, *16*, 3337–3340. [CrossRef]
17. Kirana, C.; Jones, G.P.; Record, I.R.; McIntosh, G.H. Anticancer properties of panduratin A isolated from *Boesenbergia pandurata* (Zingiberaceae). *J. Nat. Med.* **2007**, *61*, 131–137. [CrossRef]
18. Yun, J.M.; Kweon, M.H.; Kwon, H.; Hwang, J.K.; Mukhtar, H. Induction of apoptosis and cell cycle arrest by a chalcone panduratin A isolated from *Kaempferia pandurata* in androgen-independent human prostate cancer cells PC3 and DU145. *Carcinogenesis* **2006**, *27*, 1454–1464. [CrossRef]
19. Isa, N.M.; Abdul, A.B.; Abdelwahab, S.I.; Abdullah, R.; Sukari, M.A.; Kamalidehghan, B.; Hadi, A.H.A.; Mohan, S. Boesenbergin A, a chalcone from *Boesenbergia rotunda* induces apoptosis via mitochondrial dysregulation and cytochrome c release in A549 cells in vitro: Involvement of HSP70 and Bcl2/Bax signalling pathways. *J. Funct. Foods* **2013**, *5*, 87–97. [CrossRef]
20. Shim, J.S.; Kwon, Y.Y.; Han, Y.S.; Hwang, J.K. Inhibitory effect of panduratin A on UV-induced activation of mitogen-activated protein kinases (MAPKs) in dermal fibroblast cells. *Planta Med.* **2008**, *74*, 1446–1450. [CrossRef]
21. Shim, J.S.; Han, Y.S.; Hwang, J.K. The effect of 4-hydroxypanduratin A on the mitogen-activated protein kinase-dependent activation of matrix metalloproteinase-1 expression in human skin fibroblasts. *J. Dermatol. Sci.* **2009**, *53*, 129–134. [CrossRef]
22. Lee, C.W.; Kim, H.S.; Kim, H.K.; Kim, J.W.; Yoon, J.H.; Cho, Y.; Hwang, J.K. Inhibitory effect of panduratin A isolated from *Kaempferia panduarata* Roxb. on melanin biosynthesis. *Phytother. Res.* **2010**, *24*, 1600–1604. [CrossRef]
23. Abdelwahab, S.I.; Mohan, S.; Abdulla, M.A.; Sukari, M.A.; Abdul, A.B.; Taha, M.M.; Syam, S.; Ahmad, S.; Lee, K.H. The methanolic extract of *Boesenbergia rotunda* (L.) Mansf. and its major compound pinostrobin induces anti-ulcerogenic property in vivo: Possible involvement of indirect antioxidant action. *J. Ethnopharmacol.* **2011**, *137*, 963–970. [CrossRef] [PubMed]
24. Kim, D.; Lee, M.S.; Jo, K.; Lee, K.E.; Hwang, J.K. Therapeutic potential of panduratin A, LKB1-dependent AMP-activated protein kinase stimulator, with activation of PPARα/δ for the treatment of obesity. *Diabetes Obes. Metab.* **2011**, *13*, 584–593. [CrossRef] [PubMed]
25. Kim, D.Y.; Kim, M.S.; Sa, B.K.; Kim, M.B.; Hwang, J.K. *Boesenbergia pandurata* attenuates diet-induced obesity by activating AMP-activated protein kinase and regulating lipid metabolism. *Int. J. Mol. Sci.* **2012**, *13*, 994–1005. [CrossRef] [PubMed]

26. Feletou, M.; Vanhoutte, P.M. Endothelium-derived hyperpolarizing factor: Where are we now? *Arterioscler. Thromb. Vasc. Biol.* **2006**, *26*, 1215–1225. [CrossRef]

27. Zeller, A.; Wenzl, M.V.; Beretta, M.; Stessel, H.; Russwurm, M.; Koesling, D.; Schmidt, K.; Mayer, B. Mechanisms underlying activation of soluble guanylate cyclase by the nitroxyl donor Angeli's salt. *Mol. Pharmacol.* **2009**, *76*, 1115–1122. [CrossRef]

28. Gupta, S.; McArthur, C.; Grady, C.; Ruderman, N.B. Role of endothelium-derived nitric oxide in stimulation of Na(+)-K(+)-ATPase activity by endothelin in rabbit aorta. *Am. J. Physiol.* **1994**, *266*, H577–H582. [CrossRef]

29. Nelson, M.T.; Quayle, J.M. Physiological roles and properties of potassium channels in arterial smooth muscle. *Am. J. Physiol.* **1995**, *268*, C799–C822. [CrossRef]

30. Jaipetch, T.; Kanghae, S.; Pancharoen, O.; Patrick, V.; Reutrakul, V.; Tuntiwachwuttikul, P.; White, A. Constituents of *Boesenbergia pandurata* (syn. *Kaempferia pandurata*): Isolation, crystal structure and synthesis of (±)-boesenbergin A. *Aust. J. Chem.* **1982**, *35*, 351–361.

31. Trakoontivakorn, G.; Nakahara, K.; Shinmoto, H.; Takenaka, M.; Onishi-Kameyama, M.; Ono, H.; Yoshida, M.; Nagata, T.; Tsushida, T. Structural analysis of a novel antimutagenic compound, 4-hydroxypanduratin A, and the antimutagenic activity of flavonoids in a Thai spice, fingerroot (*Boesenbergia pandurata* Schult.) against mutagenic heterocyclic amines. *J. Agric. Food Chem.* **2001**, *49*, 3046–3050. [CrossRef]

32. Win, N.N.; Awale, S.; Esumi, H.; Tezuka, Y.; Kadota, S. Bioactive secondary metabolites from *Boesenbergia pandurata* of Myanmar and their preferential cytotoxicity against human pancreatic cancer PANC-1 cell line in nutrient-deprived medium. *J. Nat. Prod.* **2007**, *70*, 1582–1587. [CrossRef]

33. Pandji, C.; Grimm, C.; Wray, V.; Witte, L.; Proksch, P. Insecticidal constituents from four species of the Zingiberaceae. *Phytochemistry* **1993**, *34*, 415–419. [CrossRef]

34. Tuntiwachwuttikul, P.; Pancharoen, O.; Reutrakul, V.; Byrne, L. (1'RS,2'SR,6'RS)-(2,6-dihydroxy-4-methoxyphenyl)-[3'-methyl-2'-(3''-methylbut-2''-enyl)-6'-phenyl-cyclohex-3'-enyl]methanone (panduratin A)-a constituent of the red rhizomers of a variety of *Boesenbergia pandurata*. *Aust. J. Chem.* **1984**, *37*, 449–453. [CrossRef]

35. Agrawal, P.K. *Carbon-13 NMR of Flavonoids*, 1st ed.; Elsevier Science: Amsterdam, The Netherlands, 1989.

36. Micha, R.; Penalvo, J.L.; Cudhea, F.; Imamura, F.; Rehm, C.D.; Mozaffarian, D. Association between dietary factors and mortality from heart disease, stroke, and type 2 diabetes in the United States. *JAMA* **2017**, *317*, 912–924. [CrossRef]

37. Panche, A.N.; Diwan, A.D.; Chandra, S.R. Flavonoids: An overview. *J. Nutr. Sci.* **2016**, *5*, e47. [CrossRef] [PubMed]

38. Oak, M.H.; Auger, C.; Belcastro, E.; Park, S.H.; Lee, H.H.; Schini-Kerth, V.B. Potential mechanisms underlying cardiovascular protection by polyphenols: Role of the endothelium. *Free Radic. Biol. Med.* **2018**, *122*, 161–170. [CrossRef] [PubMed]

39. Wang, Z.T.; Lau, C.W.; Chan, F.L.; Yao, X.; Chen, Z.Y.; He, Z.D.; Huang, Y. Vasorelaxant effects of cardamonin and alpinetin from *Alpinia henryi* K. Schum. *J. Cardiovasc. Pharmacol.* **2001**, *37*, 596–606. [CrossRef]

40. Zhu, X.-M.; Fang, L.-H.; Li, Y.-J.; Du, G.-H. Endothelium-dependent and -independent relaxation induced by pinocembrin in rat aortic rings. *Vascul. Pharmacol.* **2007**, *46*, 160–165. [CrossRef]

41. Li, L.; Yang, H.-G.; Yuan, T.-Y.; Zhao, Y.; Du, G.-H. Rho kinase inhibition activity of pinocembrin in rat aortic rings contracted by angiotensin II. *Chin. J. Nat. Med.* **2013**, *11*, 258–263. [CrossRef]

42. Luna-Vazquez, F.J.; Ibarra-Alvarado, C.; Camacho-Corona, M.D.R.; Rojas-Molina, A.; Rojas-Molina, J.I.; Garcia, A.; Bah, M. Vasodilator activity of compounds isolated from plants used in Mexican traditional medicine. *Molecules* **2018**, *23*, 1474. [CrossRef]

43. Sharma, K.; Lee, H.H.; Gong, D.S.; Park, S.H.; Yi, E.; Schini-Kerth, V.; Oak, M.H. Fine air pollution particles induce endothelial senescence via redox-sensitive activation of local angiotensin system. *Environ. Pollut.* **2019**, *252*, 317–329. [CrossRef] [PubMed]

Publisher's Note: MDPI stays neutral with regard to jurisdictional claims in published maps and institutional affiliations.

plants

Article

Evaluation of the Antiwrinkle Activity of Enriched Isatidis Folium Extract and an HPLC–UV Method for the Quality Control of Its Cream Products

Dan Gao [1], Chong Woon Cho [1], Cheong Taek Kim [2], Won Seok Jeong [2] and Jong Seong Kang [1,*]

1   College of Pharmacy, Chungnam National University, Daejeon 34134, Korea; gaodan521361@hotmail.com (D.G.); chongw113@naver.com (C.W.C.)
2   RNS Inc., Daejeon 34014, Korea; happilion@biorns.com (C.T.K.); zmal1329@biorns.com (W.S.J.)
*   Correspondence: kangjss@cnu.ac.kr; Tel.: +82-42-821-5928

Received: 15 October 2020; Accepted: 13 November 2020; Published: 16 November 2020

Abstract: Currently, many extracts from natural sources are added to cosmetic products for reducing facial aging and wrinkles. This study investigated the antiwrinkle activity of enriched extract of Isatidis Folium used for a novel antiwrinkle cream product. The result demonstrated that this enriched extract has excellent antiwrinkle activity by significantly inhibiting mRNA expression of matrix metalloproteinase-1, matrix metalloproteinase-3, and pro-inflammatory cytokines IL-1β and upregulating the mRNA expression of IL-4 and procollagen. Additionally, to implement effective quality control of the entire manufacturing process of antiwrinkle cream products based on the enriched extract of Isatidis Folium, the main chemical constituents of the enriched extract of Isatidis Folium was evaluated by high–performance liquid chromatography-photodiode array-tandem mass spectrometry (HPLC-PDA-ESI-MS/MS), five constituents were undisputedly confirmed. An HPLC-UV method in 15-min analysis time for quality assessment of the entire manufacturing process of antiwrinkle cream products was proposed and validated. The optimal conditions for extracting TMCA (3,4,5-trimethoxycinnamic acid) from the developed antiwrinkle cream products were determined using response surface methodology based on central composite design. The established HPLC method and optimal extract condition are suitable for routinely analyzing this novel antiwrinkle cream product.

Keywords: isatidis folium; HPLC-PDA-ESI-MS/MS; antiwrinkle activity; quality control; response surface methodology

1. Introduction

The wrinkle formation, which is the most obvious skin-aging sign, is affected by UV irradiation, exterior environment, and age [1,2]. Notably, deficiency in collagen synthesis and matrix metalloproteinase (MMP) overexpression are the main causes of wrinkles [3,4]. Collagen comprises approximately 75% of the dry weight of skin and is critically involved in the strength and structural elasticity of skin tissue [5]. Procollagen is the main raw material for collagen synthesis [6]. MMPs are enzymes that directly degrade a large number of extracellular matrix components, including collagen, elastin, and gelatin [7]. UV radiation contributes to MMP overexpression, such as MMP-1 (interstitial collagenase), which inaugurates the degradation of type I and III collagens, and MMP-3 (stromelysin 1), which necessitates the degradation of type IV and activation of pro-MMP-1 [8,9]. However, UV exposure has been suggested to induce inflammatory responses and photoaging in the skin, which stimulates the transcription of proinflammatory cytokine genes, such as IL-1β, IL-6, and IL-8 [10–12]. Therefore, evaluating specific materials for promoting procollagen synthesis and inhibitory of MMP expression could be used to screen potential compounds or extracts used in producing the antiwrinkle products.

Isatidis Folium is the dried leaves of *Isatia indigotica* Fort. Belonging to the Cruciferae family and is a well-known herb with various biological activities and is deemed a promising agent for treating skin diseases because of its rich anti-inflammation and antioxidant components [13–15]. Isatidis Folium demonstrated the best antiwrinkle effect among the prescreened plants. Consequently, using Isatidis Folium extracts as a raw material in developing effective antiwrinkle cosmetics is a promising approach in the cosmetic industry innovation.

Some previously developed analytical methods to control the Isatidis Folium quality are thin-layer chromatography, high-performance liquid chromatography-ultraviolet (HPLC-UV), HPLC-atmospheric pressure chemical ionization (APCI)-mass spectrometry (MS), HPLC-electrospray ionization (ESI)-MS, nuclear magnetic resonance, and Fourier Transform (FT)-Raman spectroscopy [16–19]. However, none of these established methods can be used to identify and quantify the bioactive compounds in our enriched Isatidis Folium extract because of many components with a high amount in the crude extract (such as alkaloids and glycosides) were removed during the purification process. Unfortunately, it is necessary to profile the chemical constituents of enriched Isatidis Folium extract and find characteristic compounds to ensure the quality and safety of the novel antiwrinkle cream products.

Therefore, this study investigated the antiwrinkle activities of Isatidis Folium by evaluating its potency in regulating MMP-1, MMP-3, procollagen, and proinflammatory cytokine gene expression using the technique of real-time reverse-transcription polymerase chain reaction (real-time RT-PCR). A new antiwrinkle cream product was developed using enriched extracts of Isatidis Folium that exhibited excellent antiwrinkle activity. We investigated the major chemical constituents of enriched Isatidis Folium extracts using HPLC-photodiode array (PDA)-tandem mass spectrometry (HPLC-PDA-MS/MS) and established an effective and rapid analytical method that allowed us to determine characteristic compound TMCA (3,4,5-trimethoxycinnamic acid) in all steps of this product production process.

2. Results and Discussion

2.1. Antiwrinkle Activity

The mRNA expression of MMP-1, MMP-3, procollagen, IL-1β, and IL-4 involved in wrinkle formation was evaluated using RT-RCR in CCD-986sk cells treated with enriched Isatidis Folium extract at 5-μg/mL concentration. Measurements proving the enriched Isatidis Folium extract effects on mRNA expression of procollagen and IL-4 in vitro showed an increase in mRNA expression of IL-4 and procollagen 2.5 and 2.4 times, respectively, compared with the vehicle group (Figure 1). Meanwhile, Figure 1 illustrates that the enriched Isatidis Folium extract can downregulate the mRNA expression of MMP-1, MMP-3, and pro-inflammatory cytokines IL-1β. These results revealed that the enriched Isatidis Folium extract has an excellent antiwrinkle efficacy, suggesting that it could be used to develop antiwrinkle cream products.

2.2. Structural Characterization of the Main Compounds from Enriched Isatidis Folium Extract

HPLC-PDA-ESI-MS/MS was adopted to tentatively identify major compounds in enriched Isatidis Folium extract. While the PDA data were effective for identifying the class of the flavonoid and phenolic acid compounds, the MS data supplied further structural characterization. The produced fragmentation of the predominant negative ions in the MS/MS mode were used to produce fragmentation ions to obtain more detailed information about the molecular masses of various compounds [20,21]. Compound identification was confirmed by comparing their UV spectra, MS spectra (Figures S1–S5, Table 1), and elution order with those in the previously reported literature of components found in Isatidis

Folium. Figure 2 shows the HPLC-PDA chromatogram monitored at 360 nm profiles of enriched Isatidis Folium extract. First, peak 5 was clearly confirmed to be TMCA by comparing their maximum UV wavelength, MS spectra and retention time with the reference compound (Figure 3). TMCA was deemed as an active ingredient in Isatidis Folium, which was suggested to have anti-inflammation, antitumor, and anticonvulsant activities [22–24]. Therefore, as the most representative component in the enriched extract, TMCA can be selected as a marker compound to evaluate this enriched extract quality. Moreover, all UV spectra with their λmax at approximately 270 and 330 nm of peaks 1–4 agree with flavonoid characterization [25]. Based on the ESI-MS/MS and literature data, peaks 1–4 were assigned as isovitexin, isovitexin-3″-O-glucopyranoside, isoscoparin, and isoscoparin-3″-O-glucopyranoside (Figure 3) [15,16].

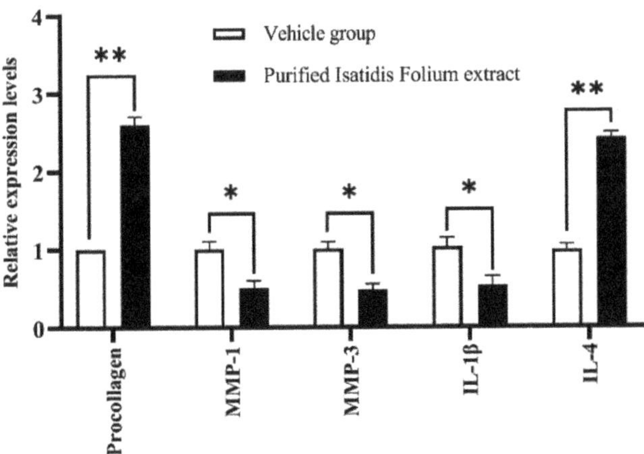

Figure 1. The effect of enriched Isatidis Folium on mRNA expression of procollagen, MMP-1, MMP-3, IL-1β, and IL-4. (* $p < 0.05$), (** $p < 0.01$) versus vehicle-treated group.

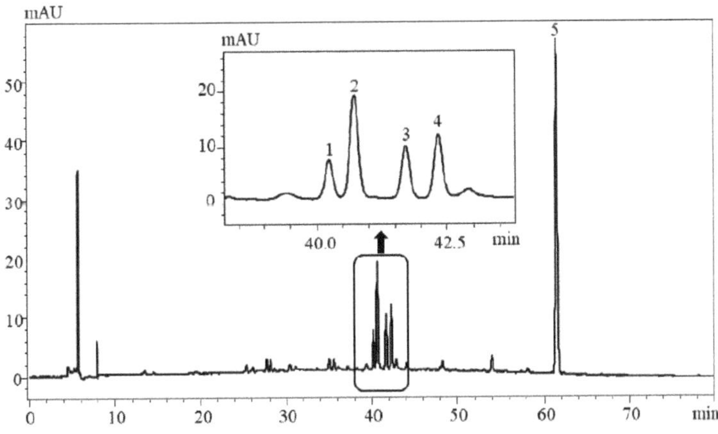

Figure 2. High–performance liquid chromatography-photodiode array (HPLC-PDA) chromatogram of detected compounds in enriched Isatidis Folium extract. HPLC conditions: column; Optimapark C_{18} (4.6 mm × 250 mm, 5 μm), elution; 0.1% formic acid aqueous solution (A) and acetonitrile-formic acid (99.9:0.1, B) at a linear-gradient elution program from 5% to 70% for 80 min, column temp; 30 °C, detection wavelength; 360 nm.

Table 1. Identification of major compounds in enriched Isatidis Folium extracts by HPLC-PDA-ESI-MS/MS.

No.	Rt [a] (min)	UV λ (nm)	Precursorion (m/z)	Production (m/z)	Molecular Formula	Identification
1	40.19	213, 269, 336	431.05 (M-H)$^-$	311.12 (M-H-C$_4$H$_9$O$_4$)$^-$	C$_{21}$H$_{20}$O$_{10}$	isovitexin
2	40.82	213, 269, 338	592.65 (M-H)$^-$	311.12 (M-Glc^b-C$_4$H$_9$O$_4$-H)$^-$ 431.00 (M-Glc-H)$^-$	C$_{27}$H$_{30}$O$_{15}$	isovitexin-3″-O-glucopyranoside
3	41.78	209, 270, 346	461.20 (M-H)$^-$	341.28 (M-H-C$_4$H$_9$O$_4$)$^-$	C$_{22}$H$_{22}$O$_{11}$	isoscoparin
4	42.35	208, 270, 348	624.03 (M-H)$^-$	341.28 (M-Glc-C$_4$H$_9$O$_4$-H)$^-$ 461.12 (M-Glc–H)$^-$	C$_{28}$H$_{32}$O$_{16}$	isoscoparin-3″-O-glucopyranoside
5	62.43	229,302	239.00 (M-H)$^+$	239.00 (M + H)$^+$	C$_{12}$H$_{14}$O$_5$	TMCA^c

[a] Rt: retention time. [b] Glc: beta-D-glucopyranosyl. [c] TMCA: 3,4,5-trimethoxycinnamic acid.

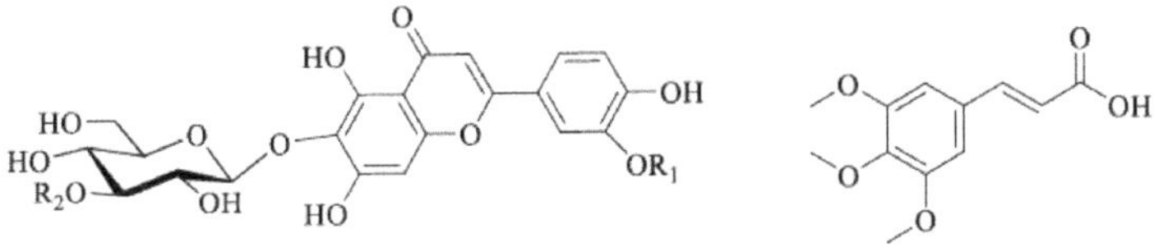

TMCA (5)

Isovitexin (1): R_1=H, R_2=H
Isovitexin 3''-O-glucopyranoside (2): R_1=H, R_2=glucosyl
Isoscoparin (3): R_1=OCH_3, R_2=H
Isoscoparin 3''-O-glucopyranoside (4): R_1=OCH_3, R_2=glucosyl

Figure 3. Chemical structures of identified compounds from enriched Isatidis Folium extract.

2.3. *Optimizing the Analytical Method*

To obtain rapid and simple chromatographic conditions, various HPLC parameters were compared and explored, including the mobile phase compositions (acetonitrile-water and methanol-water containing different buffers, such as formic acid, acetic acid, and trifluoroacetic acid), the temperature of column (25, 30, 35, and 40 °C), and the flow rate of the mobile phase (0.7, 0.9, 1.0, and 1.2 mL/min). Using the PDA for obtaining a satisfying resolution and separation, the UV spectra of TMCA were characterized. The optimal wavelength was set up at 302 nm. Certainly, a good peak shape, separation, and resolution of TMCA were obtained under the optimized conditions. HPLC chromatograms of (a) the standard solution of TMCA, (b) enriched powder of Isatidis Folium extract, (c) cream matrix, and (d) developed antiwrinkle cream products are shown in Figure 4. The retention time of TMCA was 8.94 min and no interference was observed in the chromatograms of the samples, which showed that this proposed HPLC analytical conditions would be useful and powerful for routinely determining TMCA in the production process of antiwrinkle cream products.

2.4. *Method Validation*

The proposed method was validated in terms of linearity, limit of quantitation (LOQ), limit of detection (LOD), accuracy, precision, repeatability, and recovery test. The results showed that this developed method was reliable, stable, and conducive for determining TMCA in our developed antiwrinkle cream products (Table S1).

2.4.1. Linearity, LOD, and LOQ

The linearity of the peak area versus concentration of TMCA was shown by building up a calibration curve with five concentrations under optimized analytical conditions. The linearity range of TMCA was 1.0–10.0 μg/mL. The regression equation between the peak area (Y) and the concentration (X) of TMCA was Y = 360X − 60. The correlation coefficient was 0.9996 for TMCA. The LOD and LOQ were calculated by setting up the detector response signal to noise ratio of 3:1 and 10:1, respectively. The values of LOD and LOQ were 6.08 ng/mL and 18.44 ng/mL, respectively, which reflected the high sensitivity of this novel HPLC method.

2.4.2. Accuracy and Precision

Accuracy and precision were obtained by analyzing three concentrations (low, middle, and high) of the TMCA standard five times, which were expressed as the relative standard deviation (RSD). The precision of intra-day and inter-day for the TMCA was 0.2–0.6% and 1.9–3.2%, respectively, and the accuracy of this analytical method ranged from 97.3–107.0% (Table S1).

2.4.3. Repeatability

Repeatability was used to verify the stability of the HPLC instrument after consecutive injection. In this study, repeatability was obtained by calculating the retention times and contents of TMCA in cream samples. The RSD of retention time and concentration of TMCA were 0.23% and 0.18%, respectively. The analytical method was shown to be effective and accurate.

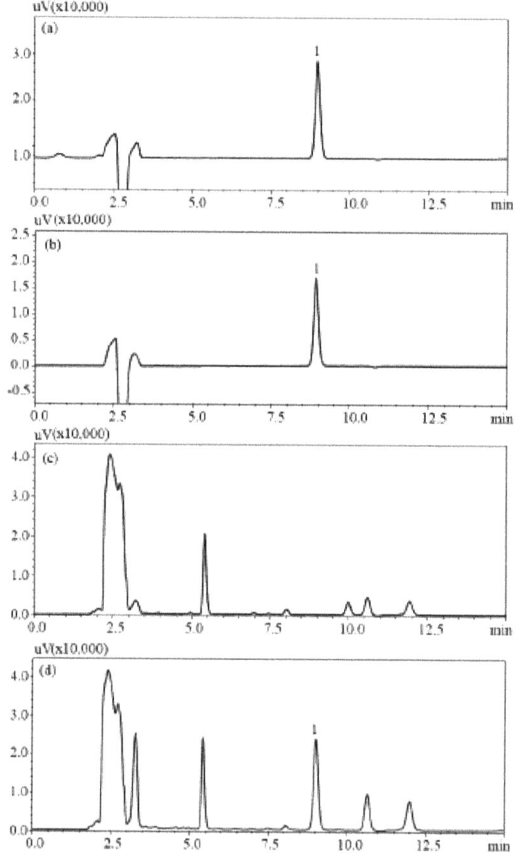

Figure 4. High–performance liquid chromatography-photodiode array (HPLC-PDA) chromatograms of (a) the standard solution of TMCA, (b) the powder of enriched Isatidis Folium extract, (c) cream matrix, and (d) developed anti-wrinkle cream products. HPLC conditions: column; Optimapark C_{18} (4.6 mm × 250 mm, 5 µm), mobile phase; 0.1% formic acid aqueous solution/acetonitrile (65:35), column temp; 30 °C, detection wavelength; 302 nm Peak 1. 3,4,5-trimethoxycinnamic acid, Retention time: 8.94 min.

2.4.4. Recoveries

The recoveries were obtained by calculating the ratio percentage as 100× (found concentration–original concentration)/spiked concentration. A certain concentration of TMCA in three concentrations (5, 6, and 7 µg/mL) was added to the cream solutions, extracted, and determined using the developed method. Acceptable recoveries of this analytical method are showing in Table S1.

2.5. Optimization of Sample Preparation for Extracting TMCA in Developed Antiwrinkle Cream Products

From the results of single-factor experiments, the extraction temperature had no significant effect on TMCA extraction ($p > 0.05$), and the two main empirical parameters of sonication time (A) and methanol-to-material ratio (B) were further optimized using response surface methodology (RSM) by central composite design (CCD). RSM is an all-powerful approach to optimize different extract conditions for obtaining the best response values [26,27]. Thirteen designed experiments were accomplished and the response values (the content of TMCA) are listed in Table 2. A quadratic model was suggested to predict the response values. The response to the content of TMCA (Y) was given by the following Equation (1):

$$Y = 29.63 - 1.24\,A - 0.917\,B + 1.10\,AB - 4.23A^2 - 3.09\,B^2, \tag{1}$$

Table 2. Central composite design matrix of two independent variables in coded (in brackets) and the content of 3,4,5-trimethoxycinnamic acid in the developed cream products.

Run	A: Sonication Time (min)	B: Methanol-to-Material Ratio (mL/g)	Content of TMCA Y (µg/g)
1	40 (0)	7 (0)	30.2
2	40 (0)	4 (−1)	27.3
3	40 (0)	7 (0)	29.7
4	40 (0)	7 (0)	29.6
5	40 (0)	10 (1)	25.2
6	50 (+1)	10 (+1)	21.3
7	40 (0)	7 (0)	29.4
8	50 (+1)	4 (−1)	20.8
9	40 (0)	7 (0)	29.7
10	30 (−1)	10 (+1)	21.9
11	50 (1)	7(0)	24.2
12	30 (−1)	4 (−1)	25.8
13	30 (−1)	7 (0)	26.0

To evaluate the statistically significant difference in this model, Fisher's test and p-value were conducted. The model F-value of 175.99 indicated that this mode is significant [28]. The p-value < 0.05 indicated that the model terms were significant. Analysis of variance (ANOVA) was performed to analyze the significance of the two independent variables and interaction terms, and the results implied that the model terms of A, B, AB, A^2, and B^2 were significant (Table S2). Conclusively, the TMCA extraction efficiency was greatly affected by sonication time and methanol-to-material ratio. Furthermore, a lack-of-fit p-value of 0.1297 indicated that the lack-of-fit was not significantly related to pure error [29].

The coefficient of determination (R^2 and adjusted-R^2) was used to evaluate the model fitting quality. The predicted R^2 of 0.9495 properly agreed with the adjusted R^2 of 0.9865, and the difference was < 0.2. Adeq Precision of 32.296 indicated an adequate signal. The coefficient of variation (CV) was selected to assess the predicted model reproducibility, which was obtained by calculating the ratio of the standard error to the mean value [30]. The CV of 1.52% suggested that the predicted model has great reproducibility. In this way, the model could be used to navigate the design space. From Figure S6a, the predicted and actual data plots were distributed near the line, which revealed that the proposed model was excellently fitted. Figure S6b shows that the residual plots were randomly scattered, which demonstrated that the variance of the experimental results was consistent for all response values [26,31].

A three-dimensional (3D) response surface was used to show the impact of each variable on the response value. Figure 5 illustrates the interactive effects of sonication time and methanol-to-material ratio on the yield of TMCA. The optimal extract conditions predicted by this excellent model are shown as follows: sonication time of 38.31 min and the methanol-to-material ratio of 6.47. To confirm the adequacy and validity of the established regression models, three parallel validation experiments were conducted under predicted conditions and the results are listed in Table S3. The matching degree between the predicted and real experimental results was 97.85%, indicating that the RSM was a reliable approach to obtain the optimum extract conditions of the TMCA from the novel antiwrinkle cream products.

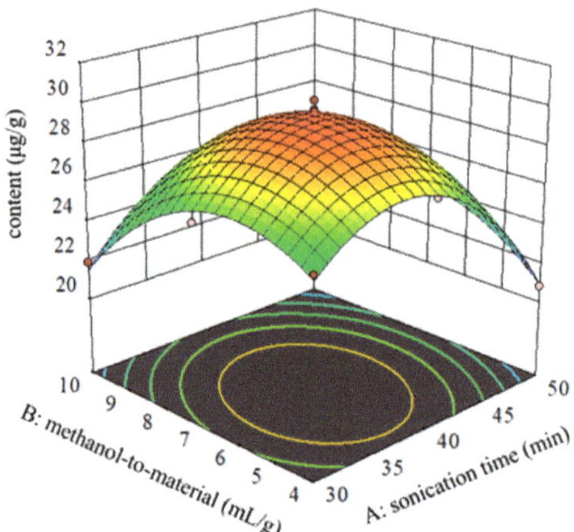

Figure 5. Response surface (3-D) plot showing the effect of sonication time (min) and methanol-to-material (mL/g) on the extraction yields of TMCA (3,4,5-trimethoxycinnamic acid). (**A**) sonication time (min); (**B**) methanol-to-material (mL/g).

2.6. Method Application to Raw Materials and Products

To evaluate the reliability of the novel HPLC method, different batches of enriched Isatidis Folium extract powder and developed antiwrinkle cream products were analyzed with the proposed method. The analyte contents were calculated using the standard calibration curves and the results are summarized in Table 3. No significant differences in TMCA content among different batches of samples ($p > 0.05$) were observed, indicating that the quality of the enriched Isatidis Folium extract and developed antiwrinkle cream products were even and stable, and the proposed method is practical for the whole industrial process for producing the antiwrinkle cream products containing the enriched Isatidis Folium extract.

Table 3. TMCA concentrations were found in different batches of the samples.

Sample	Batch Number	Concentration
The powder of enriched Isatidis Folium extract	B-1	126.32 ± 0.32 mg/g
	B-2	127.45 ± 0.21 mg/g
	B-3	126.61 ± 0.01 mg/g
Anti-wrinkle cream products	C-1	29.2 ± 0.01 µg/g
	C-2	28.9 ± 0.02 µg/g
	C-3	29.6 ± 0.01 µg/g

3. Materials and Methods

3.1. Plant Material and Reagents

The authenticated raw material samples of Isatidis Folium were purchased from the company of Hanyajaesijang (Seoul, Korea), which was produced in July 2019 from China. The reference compound of TMCA (99%) was obtained from Alfa Aesar (Alfa Aesar, Ward Hill, MA, USA). Cream matrix and novel antiwrinkle cream products were provided by RNS Inc. (Daejeon, Korea).

HPLC-grade acetonitrile and methanol were obtained from Burdick & Jackson (Muskegon, MI, USA); a buffer containing formic acid, acetic acid, and trifluoroacetic acid (MS grade) was purchased from Sigma-Aldrich (St. Louis, MO, USA). Distilled water was enriched using a Milli-Q system (Sinhan, Seoul, Korea). The primers used in this study were purchased from Bioneer Inc. (Daejeon, Korea).

3.2. Preparing Enriched Isatidis Folium Extract and Cosmetic Creams

Isatidis Folium powder (100 g) was mixed with cyclohexane (500 mL) for removing the fat-soluble compounds and soaked at room temperature for 72 h. Then, the cyclohexane extract solution was filtered with a filter paper, and residues were dried and re-extracted with methanol (500 mL) by soaking for 72 h at room temperature. Subsequently, to obtain the powder, the methanol extract was further filtered and concentrated. The extract powder was mixed 5 times with ethyl acetate solution and stirred for 24 h. The ethyl acetate extract solution was then filtered and the residue was dissolved 5 times in methanol, after which the powder of the enriched Isatidis Folium extract was obtained by vacuum concentration. Finally, the enriched Isatidis Folium extract (1 g) was added to 1,3-butylene glycol (99 mL) for making the solution that would be used in antiwrinkle cream products. The obtained 1,3-butylene glycol solution (2.5 g) was added to the cream matrix (97.5 g) for manufacturing novel antiwrinkle cream products.

3.3. Preparation of Sample and Standard Solution

The enriched Isatidis Folium extract powder (0.1 g) was dissolved in 10-mL methanol and then diluted 200 times for HPLC analysis. The developed antiwrinkle cream (2 g) was dissolved in 4-mL methanol and sonicated for 40 min at 50 °C. Furthermore, a standard stock solution (10 mg/mL) of TMCA was prepared by dissolving it in methanol.

All stock and working solutions were stored in a refrigerator at 4 °C until use. All samples and standard solutions were filtered using a 0.22-μm membrane filter before HPLC analysis.

3.4. Optimizing Experimental Parameters

The maximum yield of target compounds is crucial for pharmacological testing and production. In this study, the "one-factor-at-a-time" methodology was used to identify an acceptable level for evaluated factors. First, different extraction times (0, 10, 30, 40, 50, and 60 min) were investigated using methanol as the solvent and a solvent-to-material ratio of 4:1. Then, extraction temperatures (30 °C, 45 °C, 60 °C, 75 °C, and 80 °C) were examined using methanol as the solvent and a solvent-to-material ratio of 4:1 for 40 min. Finally, the methanol-to-material (2, 4, 7, 10, 16, and 22 mL/g) for extracting TMCA was determined using 40-min extraction time and 30 °C extraction temperature. The TMCA in novel antiwrinkle cream products (μg/g) were used as response values to establish a proper range among the studied factors. The results of these independent factors were optimized by RSM using Design-Expert software (version 12, Sta-Ease Inc., Minneapolis, MN, USA).

3.5. HPLC-PDA-ESI-MS/MS Analysis

HPLC-PDA-ESI-MS/MS analysis was performed using a ProminenceTM HPLC system (Shimadzu, Kyoto, Japan) coupled to an LCMS-8040 system equipped with an electrospray ion source (ESI) (Shimadzu, Kyoto, Japan). A linear-gradient elution program from 5% to 70% for 80 min with a flow rate of 0.5 mL/min was run for separation. The wavelength for PDA detection was set up between 200 and 400 nm. The mass spectrometer was performed in negative and positive modes, and the interface voltage was set up at −3.5 kV and 3.5 kV, respectively. Other analytical parameters were displayed as follows: a nebulizing gas flow rate = 3 L/min, a desolvation line temperature = 250 °C, a drying gas flow = 15 L/min, and a heat block temperature = 400 °C. The MS^2 data were obtained using a production survey scan in positive and negative modes.

3.6. Quality Control of Enriched Isatidis Folium Extract and Developed Antiwrinkle Cream Products

The HPLC analysis was performed on a Shimadzu HPLC system (Kyoto, Japan) equipped with a PDA detector. All samples were eluted through an Optimapark C_{18} column (4.6 mm × 250 mm, 5 μm) at a column temperature of 30 °C and a flow rate of 1 mL/min. The mobile phase comprising 0.1% formic acid in aqueous solution (solvent A) and acetonitrile (solvent B) at a ratio of 65:35

was used in the isocratic mode. PDA detection was used to conduct spectral scans over the range 200–400 nm, and detection wavelengths were monitored at 302 nm for determining and quantifying TMCA. Additionally, different validation parameters of this developed HPLC analytical method, such as precision, accuracy, LOQ, LOD, repeatability and others were assessed according to the International Council for Harmonization Guidance for Industry and Method Validation Guidelines of the Korean Ministry of Food and Drug Safety [32].

3.7. Cell Culture

Human fibroblast CCD-986sk cells were purchased from the Korean Cell Line Bank (KCLB, Seoul, Korea). CCD-986sk cells were cultured and maintained in Iscove's Modified Dulbecco's Medium (Hyclone Co., Logan, UT, USA) complemented with 10% FBS (Hyclone Co., Logan, UT, USA) and 1% antibiotic antimycotic (100 U/mL penicillin and 100 µg/mL streptomycin) at 37 °C, 5% CO_2 incubator.

3.8. Real-Time Reverse-Transcription Polymerase Chain Reaction (Real-Time RT-PCR)

CCD-986sk cells were added to a 6-well plate at a concentration of 5×10^4/well and then cultured in a 5% CO_2 incubator at 37 °C. After that, the cells were treated with the enriched Isatidis Folium extract (5 µg/mL) for 24 h. According to the manufacturer's instructions, total RNA was isolated from CCD-986sk cells with TRIzol Reagent (Cat. No: 15596018, Life Technologies, New York, NY, USA) and stored at −70 °C until use in the experiment. To analyze the mRNA template concentration, PCR was performed on a 7500 Real-time PCR system (Applied Biosystems, Foster City, CA, USA) using SYBR® Premix Ex Taq™ (Takara Bio Inc., Shiga, Japan) based on the manufacturer's instructions and valuable pieces of literature. The data were expressed as the ratio of target mRNA expression to 36B4 (vehicle) mRNA expression. The primers used in this study are summarized in Table S4. These experiments were performed in triplicate and independently duplicated at least three times.

3.9. Statistical Analysis

All statistical analysis was performed using GraphPad Prism (version 8.02, GraphPad Software Inc., La Jolla, CA, USA). * p-value < 0.05, and ** p-value < 0.01 were considered statistically significant difference. RSM was performed to predict the best extract conditions, which was performed using Design-Expert software (version 12, Sta-Ease Inc., Minneapolis, MN, USA).

4. Conclusions

In this study, the antiwrinkle activity of enriched Isatidis Folium extract was investigated, the result demonstrated that the enriched Isatidis Folium extract significantly increased the mRNA expression of procollagen and IL-4, and also decreased the mRNA expression of MMP-1, MMP-3, and pro-inflammatory cytokines IL-1β. Furthermore, the major chemical constituents of the enriched Isatidis Folium extract were analyzed using HPLC-PDA-ESI-MS/MS, and five components were conditionally identified. The representative bioactive component TMCA was then selected as a marker compound to control the quality of different samples from the enriched Isatidis Folium extract powder to the finished antiwrinkle cosmetic products. This reliable HPLC-UV method was shown to have good linearity, accuracy, precision, repeatability, and recovery. RSM was conducted to optimize the sample preparation process for TMCA extraction from novel antiwrinkle cream products. The optimal parameters were obtained as follows: sonication time of 38.31 min and the methanol-to-material ratio of 6.47.

Supplementary Materials: The following are available online at http://www.mdpi.com/2223-7747/9/11/1586/s1, Figures S1–S5: MS and MS/MS spectrum of identified components from enriched Isatidis Folium extract, Figure S6. The plot of actual values versus predicted values for extraction TMCA (**a**) and residual plot (**b**), Table S1. Validation data of TMCA, Table S2. Evaluation of CCD predicted model by ANOVA, Table S3: The response value of prediction and experiment acquired by the optimal conditions, Table S4: Primers sequences using for RT-PCR.

Author Contributions: HPLC analysis and method optimization, D.G. and C.W.C.; activity evaluation, C.T.K. and W.S.J.; research idea, J.S.K. writing—review and editing, D.G. and J.S.K. All authors have read and agreed to the published version of the manuscript.

Funding: This research received no external funding.

Conflicts of Interest: The authors declare no conflict of interest.

References

1. Mukherjee, P.K.; Maity, N.; Nema, N.K.; Sarkar, B.K. Bioactive compounds from natural resources against skin aging. *Phytomedicine* **2011**, *19*, 64–73. [CrossRef] [PubMed]

2. Trojahn, C.; Dobos, G.; Lichterfeld, A.; Blume-Peytavi, U.; Kottner, J. Characterizing facial skin ageing in humans: Disentangling extrinsic from intrinsic biological phenomena. *BioMed Res. Int.* **2015**, *2015*. [CrossRef] [PubMed]

3. Kim, M.; Park, Y.G.; Lee, H.-J.; Lim, S.J.; Nho, C.W. Youngiasides A and C isolated from *Youngia denticulatum* inhibit UVB-Induced MMP expression and promote type I procollagen production via repression of MAPK/AP-1/NF-κB and activation of AMPK/Nrf2 in HaCaT cells and human dermal fibroblasts. *J. Agric. Food Chem.* **2015**, *63*, 5428–5438. [CrossRef] [PubMed]

4. Hwang, E.; Ngo, H.T.; Seo, S.A.; Park, B.; Zhang, M.; Yi, T.-H. Protective effect of dietary *Alchemilla mollis* on UVB-irradiated premature skin aging through regulation of transcription factor NFATc1 and Nrf2/ARE pathways. *Phytomedicine* **2018**, *39*, 125–136. [CrossRef] [PubMed]

5. Peng, Y.; Wang, G.; Cao, F.; Fu, F.-F. Collection and evaluation of thirty-seven pomegranate germplasm resources. *Appl. Biol. Chem.* **2020**, *63*, 1–15. [CrossRef]

6. Liu, S.; You, L.; Zhao, Y.; Chang, X. Hawthorn polyphenol extract inhibits UVB-induced skin photoaging by regulating MMP expression and type I procollagen production in mice. *J. Agric. Food Chem.* **2018**, *66*, 8537–8546. [CrossRef]

7. Raffetto, J.D.; Khalil, R.A. Matrix metalloproteinases and their inhibitors in vascular remodeling and vascular disease. *Biochem. Pharmacol.* **2008**, *75*, 346–359. [CrossRef]

8. Pourrajab, F.; Forouzannia, S.K.; Tabatabaee, S.A. Molecular characteristics of bone marrow mesenchymal stem cells, source of regenerative medicine. *Int. J. Cardiol.* **2013**, *163*, 125–131. [CrossRef]

9. Hornebeck, W.; Maquart, F.X. Proteolyzed matrix as a template for the regulation of tumor progression. *Biomed. Pharmacother* **2003**, *57*, 223–230. [CrossRef]

10. Katiyar, S.K.; Matsui, M.S.; Elmets, C.A.; Mukhtar, H. Polyphenolic antioxidant (-)-epigallocatechin-3-gallate from green tea reduces UVB-induced inflammatory responses and infiltration of leukocytes in human skin. *Photochem. Photobiol.* **1999**, *69*, 148–153. [CrossRef]

11. Gong, M.; Zhai, X.; Yu, L.; Li, C.; Ma, X.; Shen, Q.; Han, Y.; Yang, D. ADSCs inhibit photoaging-and photocarcinogenesis-related inflammatory responses and extracellular matrix degradation. *J. Cell. Biochem.* **2020**, *121*, 1205–1215. [CrossRef] [PubMed]

12. Shin, M.H.; Lee, S.-R.; Kim, M.-K.; Shin, C.-Y.; Lee, D.H.; Chung, J.H. Activation of peroxisome proliferator-activated receptor alpha Improves aged and UV-irradiated skin by catalase induction. *PLoS ONE* **2016**, *11*, e0162628. [CrossRef] [PubMed]

13. Yu, Y.; Gong, D.; Zhu, Y.; Wei, W.; Sun, G. Quality consistency evaluation of Isatidis Folium combined with equal weight quantified ratio fingerprint method and determination of antioxidant activity. *J. Chromatogr. B* **2018**, *1095*, 149–156. [CrossRef] [PubMed]

14. Lü, H.T.; Liu, J.; Deng, R.; Song, J.Y. Preparative isolation and purification of indigo and indirubin from Folium isatidis by high-speed counter-current chromatography. *Phytochem. Anal.* **2012**, *23*, 637–641. [CrossRef]

15. Zhang, Q.; Hong, B.; Zheng, L.; Wang, X.; Cai, D. Matrix solid-phase dispersion extraction followed by HPLC-diode array detection method for the determination of major constituents in a traditional Chinese medicine *Folium isatidis* (Da-qing-ye). *J. Sep. Sci.* **2012**, *35*, 2453–2459. [CrossRef]

16. Liang, Z.; Li, B.; Liang, Y.; Su, Y.; Ito, Y. Separation and purification of two minor compounds from *Radix isatidis* by integrative MPLC and HSCCC with preparative HPLC. *J. Liq. Chromatogr. Relat. Technol.* **2015**, *38*, 647–653. [CrossRef]

17. Chen, L.; Song, F.; Liu, Z.; Zheng, Z.; Xing, J.; Liu, S. Study of the ESI and APCI interfaces for the UPLC–MS/MS analysis of pesticides in traditional Chinese herbal medicine. *Anal. Bioanal. Chem.* **2014**, *406*, 1481–1491. [CrossRef]

18. Jiang, L.; Lu, Y.; Jin, J.; Dong, L.; Xu, F.; Chen, S.; Wang, Z.; Liang, G.; Shan, X. n-Butanol extract from Folium isatidis inhibits lipopolysaccharide-induced inflammatory cytokine production in macrophages and protects mice against lipopolysaccharide-induced endotoxic shock. *Drug Des. Dev. Ther.* **2015**, *9*, 5601. [CrossRef]

19. Wu, B.; Wang, L.; Jiang, L.; Dong, L.; Xu, F.; Lu, Y.; Jin, J.; Wang, Z.; Liang, G.; Shan, X. n-butanol extract from Folium isatidis inhibits the lipopolysaccharide-induced downregulation of CXCR1 and CXCR2 on human neutrophils. *Mol. Med. Rep.* **2018**, *17*, 179–185. [CrossRef]

20. Saleem, H.; Htar, T.T.; Naidu, R.; Anwar, S.; Zengin, G.; Locatelli, M.; Ahemad, N. HPLC–PDA polyphenolic quantification, UHPLC–MS secondary metabolite composition, and in vitro enzyme inhibition potential of *Bougainvillea glabra*. *Plants* **2020**, *9*, 388. [CrossRef]

21. Torres-Vega, J.; Gómez-Alonso, S.; Pérez-Navarro, J.; Pastene-Navarrete, E. Green extraction of alkaloids and polyphenols from *Peumus boldus* leaves with natural deep eutectic solvents and profiling by HPLC-PDA-IT-MS/MS and HPLC-QTOF-MS/MS. *Plants* **2020**, *9*, 242. [CrossRef] [PubMed]

22. Zhao, Z.; Song, H.; Xie, J.; Liu, T.; Zhao, X.; Chen, X.; He, X.; Wu, S.; Zhang, Y.; Zheng, X. Research progress in the biological activities of 3, 4, 5-trimethoxycinnamic acid (TMCA) derivatives. *Eur. J. Med. Chem.* **2019**, *173*, 213–227. [CrossRef] [PubMed]

23. Rouau, X.; Cheynier, V.; Surget, A.; Gloux, D.; Barron, C.; Meudec, E.; Louis-Montero, J.; Criton, M. A dehydrotrimer of ferulic acid from maize bran. *Phytochemistry* **2003**, *63*, 899–903. [CrossRef]

24. Wang, Q.; Xiao, B.-X.; Pan, R.-L.; Liu, X.-M.; Liao, Y.-H.; Feng, L.; Cao, F.-R.; Chang, Q. An LC-MS/MS method for simultaneous determination of three Polygala saponin hydrolysates in rat plasma and its application to a pharmacokinetic study. *J. Ethnopharmacol.* **2015**, *169*, 401–406. [CrossRef]

25. Mariani, C.; Braca, A.; Vitalini, S.; De Tommasi, N.; Visioli, F.; Fico, G. Flavonoid characterization and in vitro antioxidant activity of *Aconitum anthora* L.(Ranunculaceae). *Phytochemistry* **2008**, *69*, 1220–1226. [CrossRef]

26. Gao, D.; Le Ba, V.; Rustam, R.; Cho, C.W.; Yang, S.Y.; Su, X.D.; Kim, Y.H.; Kang, J.S. Isolation of bioactive components with soluble epoxide hydrolase inhibitory activity from *Stachys sieboldii* MiQ. by ultrasonic-assisted extraction optimized using response surface methodology. *Prep. Biochem. Biotechnol.* **2020**, 1–10. [CrossRef]

27. Balli, D.; Bellumori, M.; Orlandini, S.; Cecchi, L.; Mani, E.; Pieraccini, G.; Mulinacci, N.; Innocenti, M. Optimized hydrolytic methods by response surface methodology to accurately estimate the phenols in cereal by HPLC-DAD: The case of millet. *Food Chem.* **2020**, *303*, 125393. [CrossRef]

28. Cheng, X.; Cheng, Y.; Zhang, N.; Zhao, S.; Cui, H.; Zhou, H. Purification of flavonoids from *Carex meyeriana* Kunth based on AHP and RSM: Composition analysis, antioxidant, and antimicrobial activity. *Ind. Crops Prod.* **2020**, *157*, 112900. [CrossRef]

29. Mei, Z.; Zhang, R.; Zhao, Z.; Zheng, G.; Xu, X.; Yang, D. Extraction process and method validation for bioactive compounds from *Citrus reticulata* cv. Chachiensis: Application of response surface methodology and HPLC–DAD. *Acta Chromatogr.* **2020**. [CrossRef]

30. Zhang, Y.; Yu, L.; Jin, W.; Li, C.; Wang, Y.; Wan, H.; Yang, J. Simultaneous optimization of the ultrasonic extraction method and determination of the antioxidant activities of hydroxysafflor yellow A and anhydrosafflor yellow B from safflower using a response surface methodology. *Molecules* **2020**, *25*, 1226. [CrossRef]

31. Pandey, A.; Belwal, T.; Sekar, K.C.; Bhatt, I.D.; Rawal, R.S. Optimization of ultrasonic-assisted extraction (UAE) of phenolics and antioxidant compounds from rhizomes of *Rheum moorcroftianum* using response surface methodology (RSM). *Ind. Crops Prod.* **2018**, *119*, 218–225. [CrossRef]

32. Lim, H.-S.; Choi, J.-C.; Song, S.-B.; Kim, M. Quantitative determination of carmine in foods by high-performance liquid chromatography. *Food Chem.* **2014**, *158*, 521–526. [CrossRef] [PubMed]

Publisher's Note: MDPI stays neutral with regard to jurisdictional claims in published maps and institutional affiliations.

 *plants*

Article

Bioactive Compounds from *Polygala tenuifolia* and Their Inhibitory Effects on Lipopolysaccharide-Stimulated Pro-inflammatory Cytokine Production in Bone Marrow-Derived Dendritic Cells

Le Ba Vinh [1,2,†], Myungsook Heo [1,†], Nguyen Viet Phong [2], Irshad Ali [3], Young Sang Koh [3], Young Ho Kim [1,*] and Seo Young Yang [1,*]

1   College of Pharmacy, Chungnam National University, Daejeon 34134, Korea; vinhrooney@gmail.com (L.B.V.); inyl1110@naver.com (M.H.)
2   Institute of Marine Biochemistry (IMBC), Vietnam Academy of Science and Technology (VAST), Hanoi 100000, Vietnam; ngvietphong@gmail.com
3   School of Medicine and Jeju Research Center for Natural Medicine, Jeju National University, Jeju 63243, Korea; irshad.qau200@gmail.com (I.A.); yskoh7@jejunu.ac.kr (Y.S.K.)
*   Correspondence: yhk@cnu.ac.kr (Y.H.K.); syyang@cnu.ac.kr (S.Y.Y.); Tel.: +82-42-821-5933 (Y.H.K.); +82-42-821-7321 (S.Y.Y.)
†   These authors contributed equally to this work.

Received: 4 September 2020; Accepted: 19 September 2020; Published: 20 September 2020

Abstract: The roots of *Polygala tenuifolia* Wild (Polygalaceae), which is among the most important components of traditional Chinese herbal medicine, have been widely used for over 1000 years to treat a variety of diseases. In the current investigation of secondary metabolites with anti-inflammatory properties from Korean medicinal plants, a phytochemical constituent study led to the isolation of 15 compounds (**1–15**) from the roots of *P. tenuifolia* via a combination of chromatographic methods. Their structures were determined by means of spectroscopic data such as nuclear magnetic resonance (NMR), 1D- and 2D-NMR, and liquid chromatography-mass spectrometry (LC-MS). As the obtained results, the isolated compounds were divided into two groups—phenolic glycosides (**1–9**) and triterpenoid saponins (**10–15**). The anti-inflammatory effects of crude extracts, fractions, and isolated compounds were investigated on the production of the pro-inflammatory cytokines interleukin (IL)-12 p40, IL-6, and tumour necrosis factor-α in lipopolysaccharide-stimulated bone marrow-derived dendritic cells. The IC_{50} values, ranging from 0.08 ± 0.01 to 21.05 ± 0.40 μM, indicated potent inhibitory effects of the isolated compounds on the production of all three pro-inflammatory cytokines. In particular, compounds **3–12**, **14**, and **15** showed promising anti-inflammatory activity. These results suggest that phenolic and triterpenoid saponins from *P. tenuifolia* may be excellent anti-inflammatory agents.

Keywords: *Polygala tenuifolia*; phenolic glycosides; saponins; anti-inflammatory effect; bioactive compound

1. Introduction

Chronic inflammation, which is associated with complications such as osteoarthritis and cancer, is currently the most challenging public health issue. Inflammation is at the root of several non-communicable diseases, which kill approximately 40 million people worldwide each year and account for 70% of all deaths, according to the World Health Organization. Inflammation is a complex biological response to tissue injury that can be caused by mechanical stimulation, microbial invasion, and irritants [1,2]. The inflammatory process often involves the release of biochemical mediators and can lead to the development of diseases such as cancer, rheumatoid arthritis, osteoarthritis, and diabetes mellitus [3].

Inflammatory cells produce cytokines, including interleukin (IL)-12 p40, IL-6, and tumor necrosis factor (TNF)-α. IL-12 p40, IL-6, and TNF-α play important roles in the inflammatory response, as well as cell apoptosis and transformation [1]. Many studies have focused on the inhibition of pro-inflammatory cytokines in an effort to decrease inflammation-related diseases [4–6].

Polygala tenuifolia is a herb in the Polygalaceae family, which is widely distributed in Asia. In traditional folk medicine, *P. tenuifolia* has been used for thousands of years as an expectorant and stimulant to treat bronchial asthma, chronic bronchitis, and whooping cough [7]. The major components of *P. tenuifolia* include triterpenoid saponins [8], xanthone glycosides [9], phenolic glycosides [10], and oligosaccharide ester derivatives [11]. Isolated components from *P. tenuifolia* roots exhibit diverse pharmacological effects, including anti-inflammatory [7] and anti-diabetic properties. In particular, saponin compounds (onjisaponins A, B, E, F, and G) from the roots of *P. tenuifolia* have been used to treat psychosis, whereas xanthone constituents were found to inhibit neuraminidases from influenza A viruses [12].

Natural bioactive compounds, such as phenolics, triterpene saponins, and carbohydrates, are responsible for the pharmaceutical activities of medicinal herbs [13,14]. In an effort to discover new anti-inflammatory agents in medicinal plants, we isolated 15 compounds from a methanol (MeOH) extract of *P. tenuifolia* roots using bioassay-guided and chromatographic separation methods. In addition, the anti-inflammatory activities of crude MeOH extract, dichloromethane (DCM), ethyl acetate (EtOAc), and aqueous fractions, and of isolated compounds, were examined by measuring their inhibitory effects on lipopolysaccharide (LPS)-induced expression of the pro-inflammatory cytokines IL-12 p40, IL-6, and TNF-α in bone marrow-derived dendritic cells (BMDCs). These results suggest that *P. tenuifolia* extracts and isolated compounds **3–12**, **14**, and **15** have significant anti-inflammatory effects and may be candidates for the treatment of inflammation and other related diseases.

2. Results and Discussion

Bioassay-guided fractionation of a MeOH extract of dried *P. tenuifolia* roots yielded 15 distinct compounds (**1–15**). These compounds, shown in Figures 1 and 2, are as follows: glomeratose A (**1**), 3′-*O*-(*O*-methylferuloyl)sucrose (**2**), sibiricose A5 (**3**), 4-*O*-benzoyl-3′-*O*-(*O*-methylsinapoyl)sucrose (**4**), tenuifoliside A (**5**), 6-*O*-(*O*-methyl-*p*-benzoyl)-3′-*O*-(*O*-methylsinapoyl)sucrose (**6**), arillanin A (**7**), 6,3′-di-*O*-sinapoylsucrose (**8**), tenuifoliside C (**9**), polygalasaponin XXXII (**10**), desacylsenegasaponin B (**11**), onjisaponin B (**12**), micranthoside A (**13**), polygalasaponin XXVIII (**14**), and platycodin D (**15**). Their structures were characterized by nuclear magnetic resonance (NMR) spectroscopy (^{1}H-NMR, ^{13}C-NMR, COSY, HMQC, and HMBC spectra) and liquid chromatography-mass spectrometry (LC-MS) (see Supplementary Materials), and compared with structures reported in the literature. The detailed spectroscopic data of the isolated compounds were also reported. As the obtained results, the purified compounds were divided into two groups—phenolic glycosides (**1-9**) and triterpenoid saponins (**10-15**). Many of these phenolic glycosides and triterpenoid saponins are commonly found in *Polygala* species, such as *P. arillata* [15], *P. sibirica* [16], and *P. aureocauda* [17]. However, this is the first report of compound **15** being isolated from the genus *Polygala*.

By studying how anti-inflammatory compounds inhibit the expression of inflammatory mediators such as IL-12 p40, IL-6, and TNF-α, therapeutic targets for anti-inflammatory agents can be identified for health promotion and disease prevention [6,18]. In an effort to discover new anti-inflammatory agents from medicinal plants, we extracted the roots of *P. tenuifolia* three times with MeOH. The crude extract was used to treat LPS-stimulated BMDCs and evaluate their production of cytokines. The MeOH extract of *P. tenuifolia* inhibited the production of IL-12 p40, IL-6, and TNF-α (IC_{50} = 3.38, 1.65, and 3.09 µg/mL, respectively) (Figure 3). Given the high anti-inflammatory activity of the MeOH extract, its components were further separated into DCM, EtOAc, and water fractions. As shown in Table 1, the water fraction strongly inhibited IL-12 p40, IL-6, and TNF-α production (IC_{50} = 0.94, 0.24, and 2.43 µg/mL, respectively). Therefore, the aqueous fraction was used to isolate individual active compounds. Using chromatographic techniques (silica gel column chromatography (CC); RP-C18 CC and Sephadex LH-20 columns), 15 compounds (**1–15**) were isolated from the water fraction of *P. tenuifolia* root extracts.

The isolated compounds were assessed in terms of their effects on the production of IL-12 p40. Compounds **3–10** and **12–15** greatly inhibited IL-12 p40 production, with IC_{50} values ranging from 0.08 ± 0.01 to 14.34 ± 0.03 µM, whereas compound **11** showed a moderate inhibitory effect on IL-12 p40 production (IC_{50} value of 21.05 ± 0.40) (Figure 4). Compounds **3–15** greatly inhibited IL-6 and TNF-α production, with IC_{50} values ranging from 0.24 ± 0.06 to 9.04 ± 0.05 and from 1.04 ± 0.12 to 6.34 ± 0.12 µM, respectively (Figure 5). The anti-inflammatory activities of the isolated compounds were comparable to those of the positive control, SB203580 (IC_{50} values of 5.00 ± 0.01, 3.50 ± 0.02, and 7.20 ± 0.02 µM, respectively) (Table 2). The extracts and isolated compounds were further assessed in terms of their effects on the viability of BMDCs using a colorimetric MTT (3-(4,5-dimethylthiazol-2-yl)-2,5-diphenyltetrazolium bromide) assay. Compounds **1**, **2**, and **13** showed strong cytotoxicity toward the BMDCs, whereas the other compounds displayed no notable cytotoxicity.

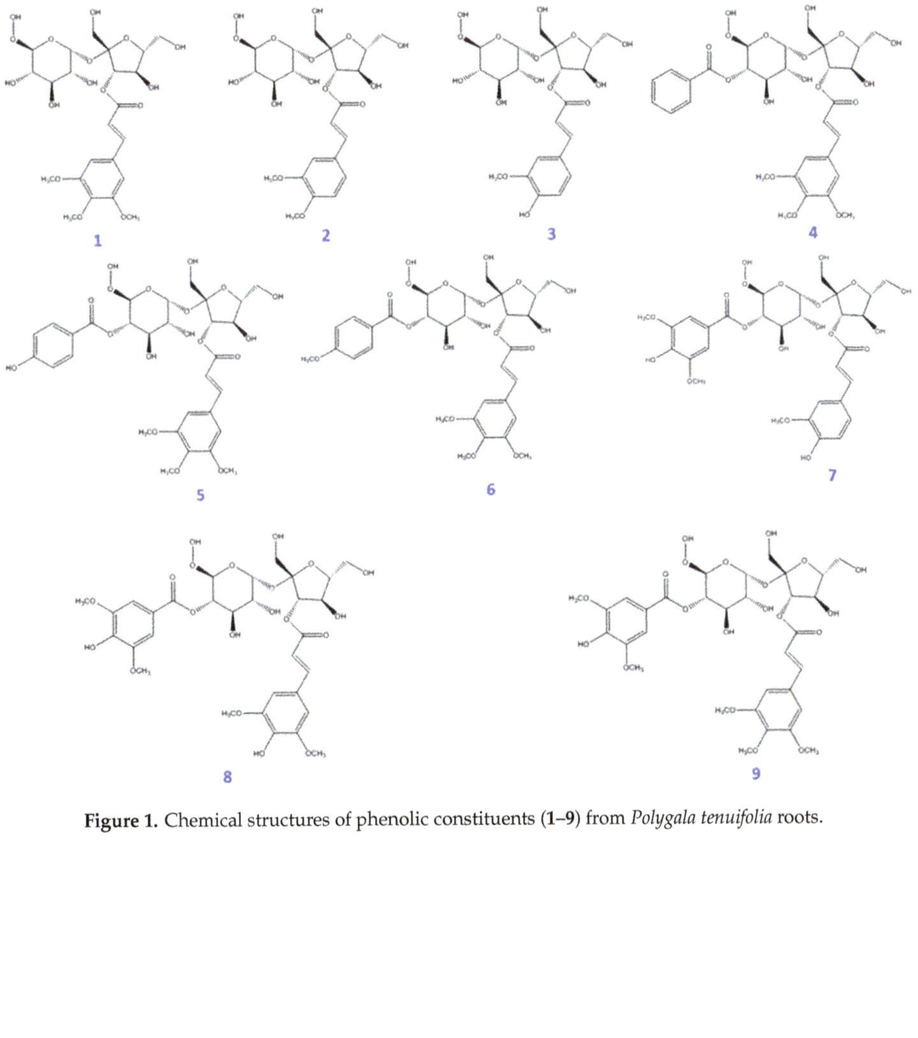

Figure 1. Chemical structures of phenolic constituents (**1–9**) from *Polygala tenuifolia* roots.

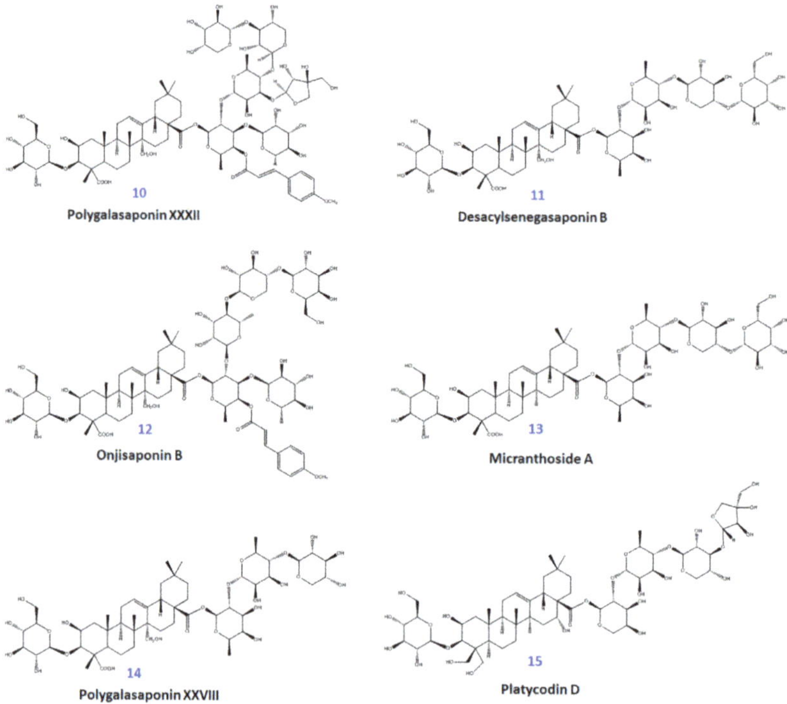

Figure 2. Chemical structures of triterpene saponin constituents (**10–15**) from *P. tenuifolia* roots.

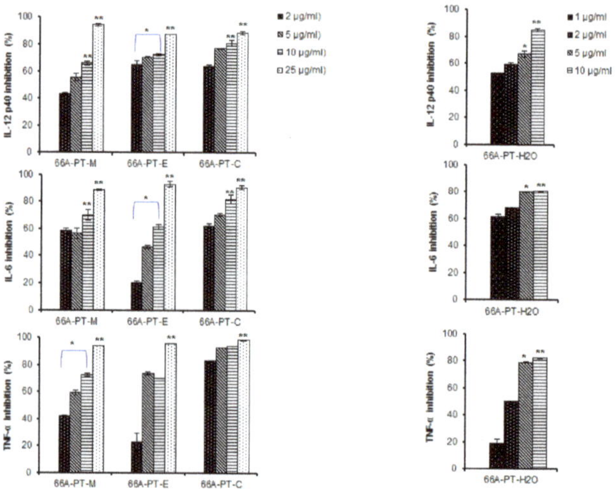

Figure 3. Effect of crude extracts (1, 2, 5, 10, and 15 µg/mL, respectively) on IL-12 p40, IL-6, and TNF-α production by LPS-stimulated bone marrow-derived dendritic cells (BMDCs). (* $P < 0.05$), (** $P < 0.01$) versus compound untreated BMDCs in the presence of LPS. BMDCs were pre-treated with respective extracts at the indicated concentrations for 1 h and then stimulated with LPS (10 ng/mL) for 18 h. Inhibitory concentrations of IL-12 p40, IL-6, and TNF-α in the culture supernatant were determined by ELISA assay.

Table 1. Anti-inflammatory effects of extracts of *P. tenuifolia* roots on LPS-stimulated bone marrow-derived dendritic cells.

	Extracts	$IC_{50} \pm SD$ (µg/mL) [a]		
		IL-12 p40	IL-6	TNF-α
1	**MeOH extract**	3.38 ± 0.02	1.65 ± 0.16	3.09 ± 0.14
2	**Water-layer**	0.94 ± 0.04	0.24 ± 0.08	2.43 ± 0.22
3	**EtOAc fraction**	0.05 ± 0.01	5.83 ± 0.26	3.92 ± 0.13
4	**CH₂Cl₂ fraction**	0.37 ± 0.10	0.89 ± 0.12	0.005 ± 0.001
	SB203580 [b]	5.00 ± 0.08	3.50 ± 0.08	7.20 ± 0.06

[a] IC_{50} values < 50 µg/mL are considered to be active. [b] Positive control

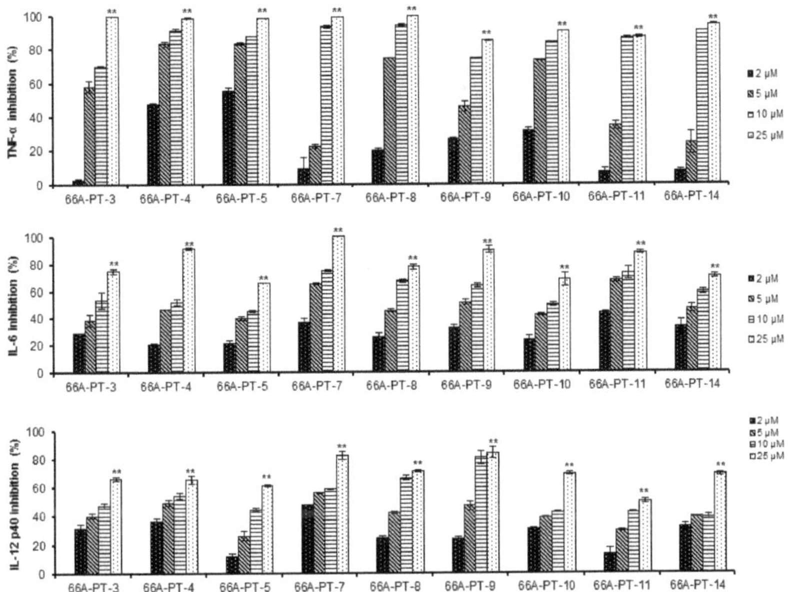

Figure 4. Effect of isolated compounds (**3–5**, **7–11**, and **14**) (2, 5, 10, and 15 µg/mL, respectively) on IL-12 p40, IL-6, and TNF-α production by LPS-stimulated BMDCs. BMDCs (bone marrow-derived dendritic cells) were pre-treated with respective compounds at the indicated concentrations for 1 h and then stimulated with LPS (10 ng/mL) for 18 h. Inhibitory concentrations of IL-12 p40, IL-6, and TNF-α in the culture supernatant were determined by ELISA assay. (* $P < 0.05$), (** $P < 0.01$) versus compound untreated BMDCs in the presence of LPS.

The structure–activity relationship (SAR) in phenolic glycosides and triterpenoid saponins may be deduced from anti-inflammatory effects. Consideration of the SAR of these triterpenoid saponins (**10-15**) suggests that the presence of the sugar units at C-3 and/or C-28 of the aglycon might play an important role in the anti-inflammatory inhibitory activity of these active compounds. In addition, the presence of substitute groups in molecules of phenolic glycosides (**1–9**), specifically –OH, –OCH₃, and –COOH also clearly affect the IC_{50} values. Hence, further study is warranted to understand the SAR between the sugar moieties in aglycon and their pharmacological properties, particularly in vivo.

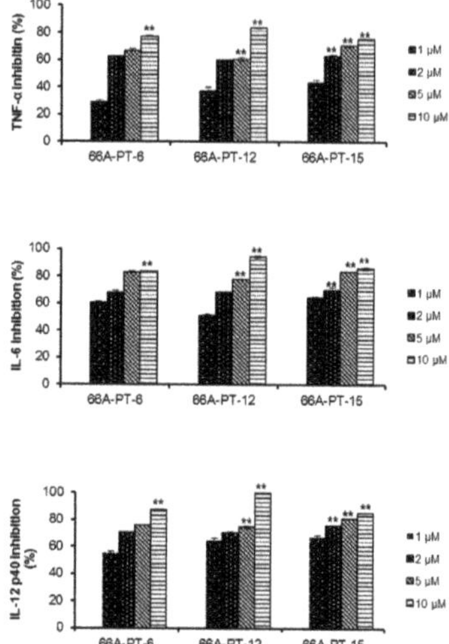

Figure 5. Effect of isolated compounds (**6**, **12**, and **15**) (1, 2, 5, and 10 µg/mL, respectively) on IL-12 p40, IL-6, and TNF-α production by LPS-stimulated BMDCs. BMDCs (bone marrow-derived dendritic cells) were pre-treated with respective compounds at the indicated concentrations for 1 h and then stimulated with LPS (10 ng/mL) for 18 h. Inhibitory concentrations of IL-12 p40, IL-6, and TNF-α in the culture supernatant were determined by ELISA assay. (* $P < 0.05$), (** $P < 0.01$) versus compound untreated BMDCs in the presence of LPS.

Table 2. Anti-inflammatory effects of isolated compounds from *P. tenuifolia* roots on LPS-stimulated bone marrow-derived dendritic cells.

Compounds	IC$_{50}$ (µM) [a]		
	IL-12 p40	IL-6	TNF-α
3	9.25 ± 0.06	7.54 ± 0.08	5.77 ± 0.12
4	6.21 ± 0.27	6.44 ± 0.32	1.50 ± 0.25
5	14.34 ± 0.03	2.36 ± 0.08	1.04 ± 0.12
6	0.55 ± 0.03	0.35 ± 0.03	1.97 ± 0.03
7	2.99 ± 0.60	3.19 ± 0.03	6.08 ± 0.10
8	6.65 ± 0.03	5.82 ± 0.03	3.51 ± 0.03
9	5.89 ± 0.08	4.64 ± 0.08	5.08 ± 0.10
10	9.78 ± 0.09	9.04 ± 0.05	2.99 ± 0.29
11	21.05 ± 0.40	2.42 ± 0.03	6.34 ± 0.12
12	0.43 ± 0.05	0.83 ± 0.09	1.75 ± 0.02
14	10.52 ± 0.10	6.05 ± 0.03	2.05 ± 0.02
15	0.08 ± 0.01	0.24 ± 0.06	1.17 ± 0.13
SB203580 [b]	5.00 ± 0.01	3.50 ± 0.02	7.20 ± 0.02

[a] IC$_{50}$ values < 50 µM are considered to be active. [b] Positive control.

Natural products and related compounds are powerful alternatives for drug development. Phenolic and saponin constituents are important secondary metabolites in both medicinal plants and marine organisms that exhibit diverse pharmacological effects, such as anti-inflammatory [8], anti-cancer [19], anti-diabetic [20], anti-cardiovascular [21], and anti-oxidant activities [22]. Although there have been a few studies on the anti-inflammatory effects of *P. tenuifolia* [7], to the best of our knowledge, this is the first report on the inhibitory effects of individual constituents of *P. tenuifolia* roots on pro-inflammatory cytokine production in LPS-stimulated BMDCs.

In summary, 15 compounds (**1–15**) from *P. tenuifolia* roots were isolated via chromatographic separation techniques (silica gel CC with RP-C18 CC and Sephadex LH-20 columns). Their structures were unambiguously established by spectroscopic methods (1D-2D NMR and LC-MS), and their inhibitory effects on pro-inflammatory cytokine (IL-12 p40, IL-6, and TNF-α) production were characterized. The potential anti-inflammatory effects of the isolated compounds (**1–15**) increase our understanding of the chemotaxonomic properties of the Polygalaceae family and the mechanisms underlying the anti-inflammatory properties of *P. tenuifolia*. This work is the first to report the inhibitory effects of extracts and isolated constituents of *P. tenuifolia* roots on the pro-inflammatory cytokines IL-12 p40, IL-6, and TNF-α. Aqueous fractionation of *P. tenuifolia* extracts is a promising area for further anti-inflammation research. Additional in vivo mechanistic studies will help determine the potential of phenolic glycosides and triterpenoid saponins for use as anti-inflammatory drugs to suppress inflammatory and related diseases.

3. Materials and Methods

3.1. General Experimental Procedures

The optical rotation values were confirmed using a JASCO DIP-370 digital polarimeter (Hachioji, Tokyo, Japan). Electrospray ionization (ESI) mass spectra were obtained using an Agilent 1200 LC-MSD Trap spectrometer. LC-MS/MS analyses were performed using a Shimadzu LCMS-8040 system (Kyoto, Japan) in positive and negative mode. NMR spectra were analyzed on a JEOL ECA 400 and 600 spectrometer (JEOL Ltd, Tokyo, Japan) with TMS used as an internal standard. Sephadex LH-20 (GE Healthcare Bio-Science AB, Uppsala, Sweden) and Diaion HP-20 (Supelco, Bellefonte, PA, USA) resins were used. Thin layer chromatography (TLC) using YMC RP-18 resins was performed using pre-coated silica gel 60 F_{254} and RP-18 F_{254S} plates (both 0.25 mm, Merck, Darmstadt, Germany), and the spots were detected under UV light at 254 and 365 nm wavelengths and using 10% H_2SO_4, followed by heating for 3–5 min. The chemicals used were purchased from commercial suppliers and used as received. All chemical reagents were purchased from Sigma-Aldrich (St. Louis, MO, USA).

3.2. Plant Material

The root of *P. tenuifolia* was obtained from a herbal company. Plant identification was verified by an expert botanist (Y.H.K.). A representative specimen of *P. tenuifolia* (CNU PT 16005) was conserved in the Herbarium of the Natural Product Laboratory, Chungnam National University, Daejeon, Korea.

3.3. Extraction and Isolation

Dried roots of *P. tenuifolia* (2.5 kg) were extracted three times with methanol under reflux. The methanol extract (666.0 g) was suspended in H_2O (3.0 L) and partitioned with dichloromethane and ethyl acetate to afford a dichloromethane fraction (D, 100.0 g), ethyl acetate fraction (E, 40.0 g), and water layer (W), respectively. The water layer (W) was passed through a Diaion HP-20 column and eluted with increasing concentrations of MeOH in water (0%, 25%, 50%, and 100%) to obtain three fractions (W1–W3) after removing the fraction that was eluted with water. Fraction W3 (250.0 g) was separated by medium-pressure liquid chromatography (MPLC) on a silica gel column using a mobile phase of CH_2Cl_2-MeOH-H_2O (7:1:0.05, *v/v*) to obtain seven fractions (W3A–W3G). Fraction W3A (330.0 mg) was purified by YMC RP-18 CC using MeOH-H_2O (2:1, *v/v*) as the eluent to furnish compounds **6** (7.5 mg) and **8** (10.3 mg). Fraction W3B (10.0 g) was rechromatographed by Sephadex LH-20 and RP-C18 CC with MeOH–H_2O (1:1, *v/v*) to afford compounds **9** (6.8 mg), **1** (3.0 mg), **3** (5.0 mg), and **4** (8.9 mg). Fraction W3C (19.0 g) was separated on RP-C18 silica gel and Sephadex LH-20 to afford compounds **7** (25.0 mg), **5** (18.7 mg), and **2** (7.8 mg). Fraction W3G (28 g) was separated by YMC RP-18 and Sephadex LH-20 CC using solvent acetone–H_2O (1:1, 2:1, and 3:1 *v/v*) and further purified by silica gel CC with CH_2Cl_2–MeOH–H_2O (2.5:1:0.1, *v/v*) to afford compounds **10** (156.0 mg), **11** (220.3 mg) and **12** (188.9 mg). Repeating the same steps as for subfraction W3G, compounds **14** (266.6 mg) and **15** (15.3 mg) were obtained from subfraction W3F (16.0 g).

Physical Properties and Key Spectroscopic Data of Isolated Compounds:

Compound **1**. Yellow powder, ^{1}H (400 MHz in MeOD-d_4) δ_H 3.63 (d, J = 10.0 Hz), 3.67 (d, J = 9.0 Hz) and ^{13}C-NMR (100 MHz in MeOD-d_4): δ_C 65.6 (C-1), 104.9 (C-2), 79.8 (C-3), 74.1 (C-4), 84.2 (C-5), 63.4 (C-6), 93.2 (C-1′), 72.5 (C-2′), 74.9 (C-3′), 71.6 (C-4′), 72.5 (C-5′), 65.6 (C-6′), 131.6 (C-1″), 106.9 (C-2″), 154.9 (C-3″), 141.4 (C-4″), 154.9 (C-5″), 106.9 (C-6″), 147.4 (C-7″), 117.8 (C-8″), 168.2 (C-9″), 56.8 (3″, 5″-OMe), 61.2 (4″-OMe), 131.6 (C-1‴), 131.3 (C-2‴), 129.7 (C-3‴), 134.4 (C-4‴), 129.7 (C-5‴), 130.8 (C-6‴), 168.2 (C-7‴).

Compound **2**. Yellow powder, ^{1}H (400 MHz in MeOD-d_4) δ_H 3.63 (d, J = 10.0 Hz), 3.67 (d, J = 9.0 Hz) and ^{13}C-NMR (100 MHz in MeOD-d_4): δ_C 63.3 (C-1), 104.8 (C-2), 79.7 (C-3), 74.9 (C-4), 84.1 (C-5), 63.3 (C-6), 93.3 (C-1′), 73.1 (C-2′), 74.9 (C-3′), 71.2 (C-4′), 73.8 (C-5′), 65.3 (C-6′), 127.8 (C-1″), 112.1 (C-2″), 149.5 (C-3″), 150.8 (C-4″), 116.5 (C-5″), 124.2 (C-6″), 115.1 (C-7″), 147.8 (C-8″), 168.5 (C-9″), 56.9 (OMe).

Compound **3**. Yellow powder, ^{1}H (400 MHz in MeOD-d_4) δ_H 3.63 (d, J = 10.0 Hz), 3.67 (d, J = 9.0 Hz) and ^{13}C-NMR (100 MHz in MeOD-d_4): δ_C 65.3 (C-1), 104.8 (C-2), 79.7 (C-3), 74.6 (C-4), 84.1 (C-5), 63.3 (C-6), 93.3 (C-1′), 73.1 (C-2′), 74.9 (C-3′), 71.2 (C-4′), 73.8 (C-5′), 65.3 (C-6′), 127.8 (C-1″), 112.1 (C-2″), 149.5 (C-3″), 150.8 (C-4″), 116.5 (C-5″), 124.3 (C-6″), 115.1 (C-7″), 147.8 (C-8″), 168.5 (C-9″), 56.5 (OMe).

Compound **4**. White amorphous powder, ^{1}H (400 MHz in MeOD-d_4) δ_H 3.63 (d, J = 10.0 Hz), 3.67 (d, J = 9.0 Hz) and ^{13}C-NMR (100 MHz in MeOD-d_4): δ_C 65.6, 104.9, 79.7, 74.0, 84.1, 63.4, 93.1, 73.1, 74.9, 71.6, 72.5, 65.3, 131.5, 106.9, 154.9, 141.4, 154.9, 106.9, 147.4, 117.8, 168.9, 56.7 (3″, 5″-OMe), 61.2 (4″-OMe)

Compound **5**. White amorphous powder, ^{1}H (400 MHz in MeOD-d_4) δ_H 3.63 (d, J = 10.0 Hz), 3.67 (d, J = 9.0 Hz) and ^{13}C-NMR (100 MHz in MeOD-d_4): δ_C 65.6 (C-1), 104.8 (C-2), 79.3 (C-3), 74.1 (C-4), 84.2 (C-5), 63.7 (C-6), 92.6 (C-1′), 73.0 (C-2′), 75.0 (C-3′), 71.4 (C-4′), 72.4 (C-5′), 65.6 (C-6′), 126.5 (C-1″), 106.9 (C-2″), 154.3 (C-3″), 139.5 (C-4″), 149.3 (C-5″), 106.9 (C-6″), 147.4 (C-7″), 115.7 (C-8″), 168.3 (C-9″), 56.7 (3″-OMe), 126.5 (C-1‴), 149.3 (C-2‴), 149.3 (C-3‴), 139.5 (C-4‴), 149.3 (C-5‴), 106.9 (C-6‴), 147.4 (C-7‴), 115.7 (C-8‴), 169.2 (C-9‴), 56.5 (OMe), 56.8 (OMe).

Compound **6**. Off-white solid, ^{1}H (400 MHz in MeOD-d_4) δ_H 3.86 (6H, s), 3.79(3H, s), and ^{13}C-NMR (100 MHz in MeOD-d_4): δ_C 65.3 (C-1), 104.8 (C-2), 79.8 (C-3), 74.6 (C-4), 84.2 (C-5), 63.7 (C-6), 92.6 (C-1′), 73.0 (C-2′), 75.0 (C-3′), 71.8 (C-4′), 72.4 (C-5′), 65.7 (C-6′), 131.3 (C-1″), 106.9 (C-2″), 154.6 (C-3″), 141.2 (C-4″), 154.6 (C-5″), 106.8 (C-6″), 147.3 (C-7″), 117.6 (C-8″), 167.9 (C-9″), 56.7 (3″, 5″-OMe), 61.1 (4″,-OMe), 123.4 (C-1‴), 133.5(C-2‴), 114.9 (C-3‴), 165.3 (C-4‴), 114.9 (C-5‴), 132.9 (C-6‴), 167.8 (C-7‴), 56.0 (OMe).

Compound **7**. White amorphous powder, ^{1}H (400 MHz in MeOD-d_4) δ_H 3.63 (d, J = 10.0 Hz), 3.67 (d, J = 9.0 Hz) and ^{13}C-NMR (100 MHz in MeOD-d_4): δ_C 65.6 (C-1), 104.9 (C-2), 79.3 (C-3), 74.2 (C-4), 84.3 (C-5), 63.9 (C-6), 92.7 (C-1′), 73.1 (C-2′), 75.0 (C-3′), 71.9 (C-4′), 72.4 (C-5′), 65.6 (C-6′), 127.7 (C-1″), 112.1 (C-2″), 149.5 (C-3″), 150.8 (C-4″), 116.5 (C-5″), 124.3 (C-6″), 147.4 (C-7″), 115.0 (C-8″), 168.5 (C-9″), 56.5 (3″-OMe), 126.6 (C-1‴), 106.9 (C-2‴), 149.5 (C-3‴), 139.6 (C-4‴), 149.5 (C-5‴), 106.9 (C-6‴), 115.8 (C-7‴), 147.4 (C-8‴), 169.2 (C-9‴), 56.8 (OMe).

Compound **8**. Yellow powder, ^{1}H (400 MHz in MeOD-d_4) δ_H 3.63 (d, J = 10.0 Hz), 3.67 (d, J = 9.0 Hz) and ^{13}C-NMR (100 MHz in MeOD-d_4): δ_C 65.5 (C-1), 104.8 (C-2), 79.3 (C-3), 74.1 (C-4), 84.2 (C-5), 63.7 (C-6), 92.6 (C-1′), 72.9 (C-2′), 75.0 (C-3′), 71.8 (C-4′), 72.4 (C-5′), 65.6 (C-6′), 131.3 (C-1″), 106.8 (C-2″),

154.6 (C-3″), 141.2 (C-4″), 154.6 (C-5″), 106.8 (C-6″), 147.4 (C-7″), 117.6 (C-8″), 167.9 (C-9″), 56.7 (3″, 5″-OMe), 61.2 (4″-OMe), 126.5 (C-1‴), 106.8 (C-2‴), 149.3 (C-3‴), 139.4 (C-4‴), 149.3 (C-5‴), 106.8 (C-6‴), 169.2 (C-7‴), 115.7 (C-8‴), 147.3 (C-9‴), 56.5 (OMe), 56.8 (OMe).

Compound **9**. Yellow powder, ^{1}H (400 MHz in MeOD-d_4) δ_H 3.63 (d, J = 10.0 Hz), 3.67 (d, J = 9.0 Hz) and ^{13}C-NMR (100 MHz in MeOD-d_4): δ_C 65.5 (C-1), 104.8 (C-2), 79.3 (C-3), 74.1 (C-4), 84.2 (C-5), 63.7 (C-6), 92.6 (C-1′), 72.9 (C-2′), 75.0 (C-3′), 71.8 (C-4′), 72.4 (C-5′), 65.6 (C-6′), 131.3 (C-1″), 106.8 (C-2″), 154.6 (C-3″), 141.2 (C-4″), 154.6 (C-5″), 106.8 (C-6″), 147.4 (C-7″), 117.6 (C-8″), 167.9 (C-9″), 56.7 (3″, 5″-OMe), 61.2 (4″-OMe), 126.5 (C-1‴), 106.8 (C-2‴), 149.3 (C-3‴), 139.4 (C-4‴), 149.3 (C-5‴), 106.8 (C-6‴), 169.2 (C-7‴), 115.7 (C-8‴), 147.3 (C-9‴), 56.5 (OMe), 56.8 (OMe).

Compound **10**. White amorphous powder, ^{1}H (600 MHz in MeOD-d_4) δ_H 5.04 (d, J = 8.0 Hz), 6.08 (d, J = 8.0 Hz), 5.81 (br, s), 5.56 (br, s), 5.26 (d, J = 7.5 Hz), 6.09 (d, J = 3.0 Hz), 5.16 (d, J = 7.0 Hz), and ^{13}C-NMR (150 MHz in MeOD-d_4): δ_C 105.4 (C-1′, Glc), 95.0 (C-1″, Fuc), 102.2 (C-1‴, Rha-1), 104.9 (C-1‴′, Rha-2), 104.8 (C-1‴″, Xyl), 111.9 (C-1‴‴, Api), 105.5 (C-1‴‴′, Ara); LC-MS m/z 1697.8 [M + Na]$^+$ (calcd for C$_{79}$H$_{118}$NaO$_{38}{}^+$, 1697.7), and m/z 1673.9 [M − H]$^-$ (calcd for C$_{79}$H$_{117}$O$_{38}{}^-$, 1673.7).

Compound **11**. White amorphous powder, ^{1}H (600 MHz in MeOD-d_4) δ_H 4.93 (d, J = 7.8 Hz), 5.78 (d, J = 8.0 Hz), 6.23 (br, s), 4.77 (d, J = 7.5 Hz), 4.83 (d, J = 7.0 Hz), and ^{13}C-NMR (150 MHz in MeOD-d_4): δ_C 103.9 (C-1′, Glc), 94.2 (C-1″, Fuc), 100.9 (C-1‴, Rha-1), 106.3 (C-1‴″, Xyl), 103.3 (C-1‴‴, Gal); LC-MS m/z 1289.5 [M + Na]$^+$ (calcd for C$_{59}$H$_{94}$NaO$_{29}{}^+$, 1289.5), and m/z 1265.8 [M − H]$^-$ (calcd for C$_{59}$H$_{93}$O$_{29}{}^-$, 1265.5).

Compound **12**. White amorphous powder, ^{1}H (600 MHz in MeOD-d$_4$) δ_H 6.64 (d, J = 16.0 Hz), 7.13 (d, J = 9.0 Hz), 7.74 (d, J = 8.9 Hz), 7.88 (d, J = 16.0 Hz), and ^{13}C-NMR (150 MHz in MeOD-d$_4$): δ_C 105.0 (C-1′, Glc), 94.9 (C-1″, Fuc), 101.8 (C-1‴, Rha-1), 106.6 (C-1‴″, Xyl), 104.2 (C-1‴‴, Gal) 104.5 (C-1‴‴′, Rha-2); LC-MS m/z 1595.7 [M + Na]$^+$ (calcd for C$_{75}$H$_{112}$NaO$_{35}{}^+$, 1595.6), and m/z 1571.8 [M − H]$^-$ (calcd for C$_{75}$H$_{111}$O$_{35}{}^-$, 1571.6).

Compound **13**. White amorphous powder, ^{1}H (600 MHz in MeOD-d_4) δ_H 5.01 (d, J = 8.0 Hz), 5.86 (d, J = 8.0 Hz), 6.32 (br, s), 4.81 (d, J = 7.5 Hz), 4.87 (d, J = 8.0 Hz), and ^{13}C-NMR (150 MHz in MeOD-d_4): δ_C 104.1 (C-1′, Glc), 94.0 (C-1″, Fuc), 100.9 (C-1‴, Rha), 106.3 (C-1‴″, Xyl), 106.3 (C-1‴‴, Gal); LC-MS m/z 1268.7 [M+NH$_4$]$^+$ (calcd for C$_{59}$H$_{98}$NO$_{28}{}^+$, 1268.6), and m/z 1249.9 [M − H]$^-$ (calcd for C$_{57}$H$_{91}$O$_{28}{}^-$, 1249.5).

Compound **14**. White amorphous powder, ^{1}H (600 MHz in MeOD-d_4) δ_H 5.06 (d, J = 8.0 Hz), 6.06 (d, J = 8.0 Hz), δ_H 6.40 (br s), 5.40 (d, J = 7.0 Hz), and ^{13}C-NMR (150 MHz in MeOD-d4): δ_C 105.4 (C-1′, Glc), 94.8 (C-1″, Fuc), 101.2 (C-1‴, Rha), 107.4 (C-1‴″, Xyl); LC-MS m/z 1127.5 [M + Na]$^+$ (calcd for C$_{53}$H$_{84}$NaO$_{24}{}^+$, 1127.5), and m/z 1103.7 [M − H]$^-$ (calcd for C$_{53}$H$_{83}$O$_{24}{}^-$, 1103.7).

Compound **15**. White amorphous powder, ^{1}H (600 MHz in C$_5$D$_5$N-d_5) δ_H 5.09 (d, J = 7.5 Hz), 5.67 (d, J = 3.7 Hz), δ_H 5.84 (d, J = 1.6 Hz), 6.26 (d, J = 2.4 Hz), δ_H 6.45 (d, J = 3.3 Hz), and ^{13}C-NMR (150 MHz in C$_5$D$_5$N-d_5): δ_C 105.9 (C-1′, Glc), 93.7 (C-1″, Ara), 101.0 (C-1‴, Rha), 106.6 (C-1‴″, Xyl), Api (C-1‴‴); LC-MS m/z 1247.6 [M + Na]$^+$ (calcd for C$_{57}$H$_{92}$NaO$_{28}{}^+$, 1247.5), and m/z 1223.7 [M − H]$^-$ (calcd for C$_{57}$H$_{91}$O$_{28}{}^-$, 1223.5).

3.4. Cell Culture and Reagents

BMDCs were grown from wild-type C57BL/6 mice (Orient Bio Inc., Seoul, Korea) as previously described [18,23]. All animal procedures were approved by and performed according to the guidelines of the Institutional Animal Care and Use Committee of Jeju National University (#2016-0059). Briefly, bone marrow from the tibia and femur was obtained by flushing with Dulbecco's Modified Eagle

Medium (DMEM; Welgene, Gyeongsan, Korea) and bone marrow cells were cultured in RPMI 1640 medium containing 10% heat-inactivated fetal bovine serum (FBS; Gibco, New York, NY, USA), 50 µM of 2-ME, and 2 mM of glutamine, supplemented with 3% J558L hybridoma cell culture supernatant containing granulocyte-macrophage colony-stimulating factor (GM-CSF). The culture medium containing GM-CSF was replaced every other day. At day 6 of culture, non-adherent cells and loosely adherent DC aggregates were harvested, washed, and resuspended in RPMI 1640, supplemented with 5% FBS. DCs were incubated in 48-well plates at a density of 1×10^5 cells/0.5 mL and then treated with the isolated compounds at the indicated concentration for 1 h before stimulation with 10 ng/mL of LPS from *Salmonella minnesota* (Alexis, New York, USA). Supernatants were harvested 18 h after stimulation. Concentrations of murine IL-12 p40, IL-6, and TNF-α in the culture supernatants were determined by ELISA (BD PharMingen, San Diego, CA, USA) according to the manufacturer's protocols. All experiments were performed at least three times. Data are presented as the mean and the standard deviation (SD) of three independent experiments.

3.5. Cytokine Production Measurements

The BMDCs were incubated in 48-well plates in 0.5 mL containing 1×10^5 cells per well, and then treated with isolated compounds **1–15** at the indicated concentration for 1 h before stimulation with 10 ng/mL LPS from *Salmonella minnesota* (Alexis, NY, USA) Supernatants were collected 18 h after stimulation. Concentrations of murine IL-12 p40, IL-6, and TNF-α in the culture supernatants were identified by ELISA (BD PharMingen, CA, USA) according to the manufacturer's instructions.

The inhibitory activity (I) was expressed as the inhibition rate (%), which was calculated using the following formula:

$$ I = \frac{Cdvc - Cdcc}{Cdvc} \times 100 $$

where Cdvc is the cytokine level (ng/mL) in vehicle-treated DC, and Cdcc is the cytokine level (ng/mL) in compound-treated DC. The data were obtained by at least three independent experiments performed in triplicate.

3.6. Cell Viability Assay

To evaluate the effects of isolated compounds on cell viability, we conducted an MTT assay [18,24]. BMDCs were incubated with 1 to 50 µM of isolated compounds for 18 h. The results demonstrate that compounds **1**, **2**, and **13** showed strong cytotoxicity toward BMDCs. Other compounds displayed no notable cytotoxicity against BMDCs.

3.7. Statistical Analysis

All results are presented as means ± SD. Data were analyzed by one-factor analysis of variance (ANOVA). * P value < 0.05, and ** P value < 0.01 were considered statistically significant. All experiments were repeated at least three times independently.

Supplementary Materials: The following are available online at http://www.mdpi.com/2223-7747/9/9/1240/s1, Figures S1–S54: 1D (^{1}H, ^{13}C-NMR) and 2D (COSY, HMQC, and HMBC) spectra, as well as LC-MS data of all isolated compounds.

Author Contributions: L.B.V., M.H., and N.V.P. performed the isolation and structure elucidation of the constituents. I.A., and Y.S.K. carried out the bioassay experiments. Y.H.K. contributed to the revision of this manuscript. The research and the writing of the manuscript were carried out based on the planning of S.Y.Y. All authors have read and agreed to the published version of the manuscript.

Funding: This research was supported by Basic Science Research Program through the National Research Foundation of Korea (NRF) funded by the Ministry of Education, Science and Technology (NRF-2018R1A6A3A11047338).

Conflicts of Interest: The authors declare no conflict of interest.

References

1. Zhang, J.-M.; An, J. Cytokines, inflammation and pain. *Int. Anesthesiol. Clin.* **2007**, *45*, 27. [CrossRef]
2. Chen, L.; Deng, H.; Cui, H.; Fang, J.; Zuo, Z.; Deng, J.; Li, Y.; Wang, X.; Zhao, L. Inflammatory responses and inflammation-associated diseases in organs. *Oncotarget* **2018**, *9*, 7204. [CrossRef]
3. Qu, X.; Tang, Y.; Hua, S. Immunological approaches towards cancer and inflammation: A cross talk. *Front. Immunol.* **2018**, *9*, 563. [CrossRef]
4. Vinh, L.B.; Lee, Y.; Han, Y.K.; Kang, J.S.; Park, J.U.; Kim, Y.R.; Yang, S.Y.; Kim, Y.H. Two new dammarane-type triterpene saponins from Korean red ginseng and their anti-inflammatory effects. *Bioorg. Med. Chem. Lett.* **2017**, *27*, 5149–5153. [CrossRef] [PubMed]
5. Vinh, L.B.; Jo, S.J.; Nguyen Viet, P.; Gao, D.; Cho, K.W.; Koh, E.-J.; Park, S.S.; Kim, Y.H.; Yang, S.Y. The chemical constituents of ethanolic extract from *Stauntonia hexaphylla* leaves and their anti-inflammatory effects. *Nat. Prod. Res.* **2019**. [CrossRef]
6. Vinh, L.B.; Jang, H.-J.; Phong, N.V.; Cho, K.W.; Park, S.S.; Kang, J.S.; Kim, Y.H.; Yang, S.Y. Isolation, structural elucidation, and insights into the anti-inflammatory effects of triterpene saponins from the leaves of *Stauntonia hexaphylla*. *Bioorg. Med. Chem. Lett.* **2019**, *29*, 965–969. [CrossRef] [PubMed]
7. Cheong, M.-H.; Lee, S.-R.; Yoo, H.-S.; Jeong, J.-W.; Kim, G.-Y.; Kim, W.-J.; Jung, I.-C.; Choi, Y.H. Anti-inflammatory effects of *Polygala tenuifolia* root through inhibition of NF-κB activation in lipopolysaccharide-induced BV2 microglial cells. *J. Ethnopharmacol.* **2011**, *137*, 1402–1408. [CrossRef] [PubMed]
8. Li, C.; Yang, J.; Yu, S.; Chen, N.; Xue, W.; Hu, J.; Zhang, D. Triterpenoid saponins with neuroprotective effects from the roots of *Polygala tenuifolia*. *Planta Med.* **2008**, *74*, 133–141. [CrossRef] [PubMed]
9. Jiang, Y.; Tu, P.-F. Xanthone O-glycosides from *Polygala tenuifolia*. *Phytochemistry* **2002**, *60*, 813–816. [CrossRef]
10. Shi, T.-X.; Wang, S.; Zeng, K.-W.; Tu, P.-F.; Jiang, Y. Inhibitory constituents from the aerial parts of *Polygala tenuifolia* on LPS-induced NO production in BV2 microglia cells. *Bioorg. Med. Chem. Lett.* **2013**, *23*, 5904–5908. [CrossRef]
11. Dong, X.-Z.; Huang, C.-L.; Yu, B.-Y.; Hu, Y.; Mu, L.-H.; Liu, P. Effect of Tenuifoliside A isolated from *Polygala tenuifolia* on the ERK and PI3K pathways in C6 glioma cells. *Phytomedicine* **2014**, *21*, 1178–1188. [CrossRef] [PubMed]
12. Dao, T.T.; Dang, T.T.; Nguyen, P.H.; Kim, E.; Thuong, P.T.; Oh, W.K. Xanthones from *Polygala karensium* inhibit neuraminidases from influenza A viruses. *Bioorg. Med. Chem. Lett.* **2012**, *22*, 3688–3692. [CrossRef] [PubMed]
13. Vinh, L.B.; Jang, H.-J.; Phong, N.V.; Dan, G.; Cho, K.W.; Kim, Y.H.; Yang, S.Y. Bioactive triterpene glycosides from the fruit of *Stauntonia hexaphylla* and insights into the molecular mechanism of its inflammatory effects. *Bioorg. Med. Chem. Lett.* **2019**, *29*, 2085–2089. [CrossRef] [PubMed]
14. Vinh, L.B.; Nguyet, N.T.M.; Ye, L.; Dan, G.; Phong, N.V.; Anh, H.L.T.; Kim, Y.H.; Kang, J.S.; Yang, S.Y.; Hwang, I. Enhancement of an in vivo anti-inflammatory activity of oleanolic acid through glycosylation occurring naturally in *Stauntonia hexaphylla*. *Molecules* **2020**, *25*, 3699. [CrossRef] [PubMed]
15. Nguyen, D.H.; Doan, H.T.; Vu, T.V.; Pham, Q.T.; Khoi, N.M.; Huu, T.N.; Thuong, P.T. Oligosaccharide and glucose esters from the roots of *Polygala arillata*. *Nat. Prod. Res.* **2019**. [CrossRef]
16. Song, Y.-L.; Zhou, G.-S.; Zhou, S.-X.; Jiang, Y.; Tu, P.-F. Polygalins D–G, four new flavonol glycosides from the aerial parts of *Polygala sibirica* L.(Polygalaceae). *Nat. Prod. Res.* **2013**, *27*, 1220–1227. [CrossRef]
17. Quang, T.H.; Yen, D.T.H.; Nhiem, N.X.; Tai, B.H.; Ngan, N.T.T.; Anh, H.L.T.; Oh, H.; Van Kiem, P.; Van Minh, C. Oleanane-type triterpenoid saponins from the roots of *Polygala aureocauda* Dunn. *Phytochem. Let.* **2019**, *34*, 59–64. [CrossRef]
18. Ali, I.; Manzoor, Z.; Koo, J.-E.; Kim, J.-E.; Byeon, S.-H.; Yoo, E.-S.; Kang, H.-K.; Hyun, J.-W.; Lee, N.-H.; Koh, Y.-S. 3-Hydroxy-4, 7-megastigmadien-9-one, isolated from *Ulva pertusa*, attenuates TLR9-mediated inflammatory response by down-regulating mitogen-activated protein kinase and NF-κB pathways. *Pharm. Biol.* **2017**, *55*, 435–440. [CrossRef]
19. Vinh, L.B.; Nguyet, N.T.M.; Yang, S.Y.; Kim, J.H.; Thanh, N.V.; Cuong, N.X.; Nam, N.H.; Minh, C.V.; Hwang, I.; Kim, Y.H. Cytotoxic triterpene saponins from the mangrove *Aegiceras corniculatum*. *Nat. Prod. Res.* **2017**, *33*, 628–634. [CrossRef]

20. Kim, J.H.; Kim, H.Y.; Yang, S.Y.; Kim, J.-B.; Jin, C.H.; Kim, Y.H. Inhibitory activity of (−)-epicatechin-3, 5-O-digallate on α-glucosidase and in silico analysis. *Int. J. Biol. Macromol.* **2018**, *107*, 1162–1167. [CrossRef]

21. Sun, Y.N.; Kim, J.H.; Li, W.; Jo, A.R.; Yan, X.T.; Yang, S.Y.; Kim, Y.H. Soluble epoxide hydrolase inhibitory activity of anthraquinone components from Aloe. *Bioorg. Med. Chem.* **2015**, *23*, 6659–6665. [CrossRef] [PubMed]

22. Tung, N.H.; Song, G.Y.; Nhiem, N.X.; Ding, Y.; Tai, B.H.; Jin, L.G.; Lim, C.-M.; Hyun, J.W.; Park, C.J.; Kang, H.K. Dammarane-type saponins from the flower buds of *Panax ginseng* and their intracellular radical scavenging capacity. *J. Agric. Food Chem.* **2009**, *58*, 868–874. [CrossRef] [PubMed]

23. Thao, N.; Cuong, N.; Luyen, B.; Quang, T.; Hanh, T.; Kim, S.; Koh, Y.-S.; Nam, N.; Van Kiem, P.; Van Minh, C. Anti-inflammatory components of the starfish *Astropecten polyacanthus*. *Mar. Drugs* **2013**, *11*, 2917–2926. [CrossRef] [PubMed]

24. Ba Vinh, L.; Thi Minh Nguyet, N.; Young Yang, S.; Hoon Kim, J.; Thi Vien, L.; Thi Thanh Huong, P.; Van Thanh, N.; Xuan Cuong, N.; Hoai Nam, N.; Van Minh, C. A new rearranged abietane diterpene from Clerodendrum inerme with antioxidant and cytotoxic activities. *Nat. Prod. Res.* **2018**, *32*, 2001–2007. [CrossRef]

Review

Systematics, Phytochemistry, Biological Activities and Health Promoting Effects of the Plants from the Subfamily Bombacoideae (Family Malvaceae)

Gitishree Das [1], **Han-Seung Shin** [2], **Sanjoy Singh Ningthoujam** [3], **Anupam Das Talukdar** [4], **Hrishikesh Upadhyaya** [5], **Rosa Tundis** [6], **Swagat Kumar Das** [7] and **Jayanta Kumar Patra** [1,*]

[1] Research Institute of Biotechnology & Medical Converged Science, Dongguk University-Seoul, Goyangsi 10326, Korea; gdas@dongguk.edu
[2] Department of Food Science & Biotechnology, Dongguk University-Seoul, Goyangsi 10326, Korea; spartan@dongguk.edu
[3] Department of Botany, Ghanapriya Women's College, Dhanamanjuri University, Imphal 795001, India; ningthouja@hotmail.com
[4] Department of Life Science and Bioinformatics, Assam University, Silchar, Assam 788011, India; anupam@bioinfoaus.ac.in
[5] Department of Botany, Cotton University, Guwahati, Assam 781001, India; hkupbl_au@rediffmail.com
[6] Department of Pharmacy, Health and Nutritional Sciences, University of Calabria, Via P. Bucci, 87036 Rende, Italy; rosa.tundis@unical.it
[7] Department of Biotechnology, College of Engineering and Technology, Biju Patnaik University of Technology, Bhubaneswar, Odisha 751003, India; das.swagat@gmail.com
* Correspondence: jkpatra@dongguk.edu

Citation: Das, G.; Shin, H.-S.; Ningthoujam, S.S.; Talukdar, A.D.; Upadhyaya, H.; Tundis, R.; Das, S.K.; Patra, J.K. Systematics, Phytochemistry, Biological Activities and Health Promoting Effects of the Plants from the Subfamily Bombacoideae (Family Malvaceae). *Plants* **2021**, *10*, 651. https://doi.org/10.3390/plants10040651

Academic Editors: Jong Seong Kang and Narendra Singh Yadav

Received: 16 February 2021
Accepted: 17 March 2021
Published: 29 March 2021

Publisher's Note: MDPI stays neutral with regard to jurisdictional claims in published maps and institutional affiliations.

Abstract: Plants belonging to the subfamily Bombacoideae (family Malvaceae) consist of about 304 species, many of them having high economical and medicinal properties. In the past, this plant group was put under Bombacaceae; however, modern molecular and phytochemical findings supported the group as a subfamily of Malvaceae. A detailed search on the number of publications related to the Bombacoideae subfamily was carried out in databases like PubMed and Science Direct using various keywords. Most of the plants in the group are perennial tall trees usually with swollen tree trunks, brightly colored flowers, and large branches. Various plant parts ranging from leaves to seeds to stems of several species are also used as food and fibers in many countries. Members of Bombacoides are used as ornamentals and economic utilities, various plants are used in traditional medication systems for their anti-inflammatory, astringent, stimulant, antipyretic, microbial, analgesic, and diuretic effects. Several phytochemicals, both polar and non-polar compounds, have been detected in this plant group supporting evidence of their medicinal and nutritional uses. The present review provides comprehensive taxonomic, ethno-pharmacological, economic, food and phytochemical properties of the subfamily Bombacoideae.

Keywords: bombacoideae; pharmacology; phytochemical ingredients; bioactive compounds; medicine

1. Introduction

The plant group Bombacoideae is a subfamily of Malvaceae (kapok, cotton family). The subfamily contains about 304 species, most of them with high economical and medicinal values. Considering their importance, some of the plants are given special cultural status. For instance, the *Ceiba pentandra* tree is the national tree of Guatemala. Among the Mayan and Aztec civilizations in the Meso-America, the *Ceiba* species is considered as a sacred "World Tree". The Indian kapok tree, *Bombax ceiba,* is worshipped by the Hindu community in North India as a nakshatra tree and home of the female spirits Yakshi [1]. There is West African belief that the first human was born from the trunk of a baobab tree (*Adansonia* spp.) and these plants are regarded as the "Tree of Life". Many plants of the Bombacoideae are valued as ornamentals in various parts of the world because of their large branches and brightly colored flowers [2]. Moreover, many genera of this subfamily are known for producing fibers, timber, fruits, and vegetables, thereby, regarded as one of the important economic and commercial plant groups.

This group was previously recognized as a distinct family, Bombacaceae, based on the type genus *Bombax* by some traditional taxonomists. From the days of the natural system to the present days of phyletic classification, the status of this plant group is continuously debated. Apart from that, the number of genera under this family varied from one classification system to another. There are various arguments in favor of a distinct family or whether to subsume under a subfamily or tribe. The study of palyno-morphological characteristics supported the justification of separating Bombacaceae from Sterculiaceae, Malvaceae, and Tiliaceae [3]. Most of the traditional methodical educations related to the subfamily Bombacoideae are on the basis of the characteristics of the flower, especially the androecium [4]. Recently, morphological, anatomical, palynological, phytochemical, and molecular phylogenetic analyses have shown that separation of Bombacaceae from its related groups viz. Malvaceae, Tiliaceae, and Sterculiaceae is inconsistent [5]. This plant group includes several plants, which are used for medicinal and economic utilities. A detailed search on the number of publications related to the Bombacoideae subfamily was carried out in databases like PubMed and Science Direct and it was found that, as per the PubMed database, around 20 articles have been published during the years 1999–2020 and among them, 12 articles are full texts (https://pubmed.ncbi.nlm.nih.gov/?term=Bombacoideae; accessed on 12 October 2020). Interestingly, from a total of 20 articles, 16 were published during the last 10 years (2010–2020). Similarly, the Science Direct databases show a total of 53 articles were published during the years 1999–2020, of which, 42 are research articles, 2 are review articles, 4 book chapters, 1 short communication, 2 encyclopedia, and 2 others (https://www.sciencedirect.com/search?qs=Bombacoideae&show=100; accessed on 12 October 2020). Considering the importance, the authors attempted to extensively review the taxonomic, phytochemical, and medicinal utilities of the members of the subfamily Bombacoideae.

2. Taxonomy of the Subfamily Bombacoideae

The advent of new taxonomical tools has revolutionized taxonomical circumscriptions. Morphological and molecular analyses revealed that Bombacaceae is not a monophyletic group. Furthermore, families such as Tiliaceae, Sterculiaceae, and Malvaceae are largely nonmonophyletic. Singh [6] considered that traditional distinctions amongst these four families are random and unpredictable and fusion of four would form a monophyletic group. Bayer et al. [7], grouped these four families together into Malvaceae considering their common characteristics and assumed them to be monophyletic. The Malvaceae sensu lato is characterized by apomorphic inflorescence, presence of bicolor unit, 3-bracted cyme, and trimerous epicalyx. Bombacaceae was distributed into two subfamilies, Bombacoideae and Helicteroideae, within the family Malvaceae. The confinement of Bombacoideae and Malvoideae is still under controversy, as the former appears to be paraphyletic without the latter [8]. Most of the plants are included in the subfamily Bombacoideae. At present, Bombacoideae is one of the clades in the family Malvaceae (Figure 1). The taxonomic location of Bombacoideae as per different systems of classification is shown in Figure 2 [4].

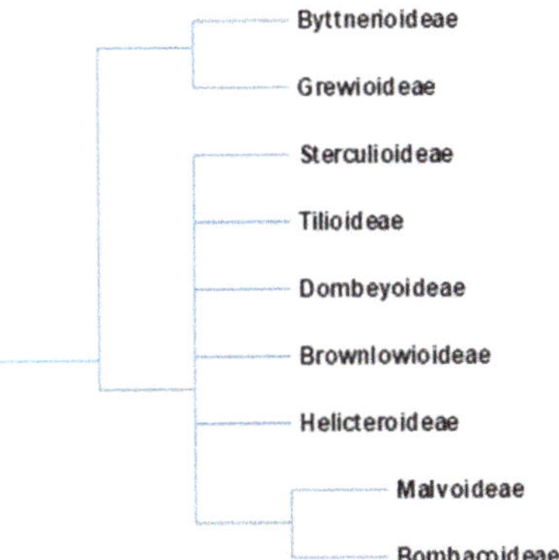

Figure 1. In the Angiosperm Phylogeny Group (APG) classification, erstwhile family Bombacaceae is allocated as subfamily Bombacoideae of the family Malvaceae. Cladogram of the Malvaceae is after Bayer et.al. 1999 and online version of APG (http://www.mobot.org/MOBOT/research/APweb; accessed on 10 January 2021).

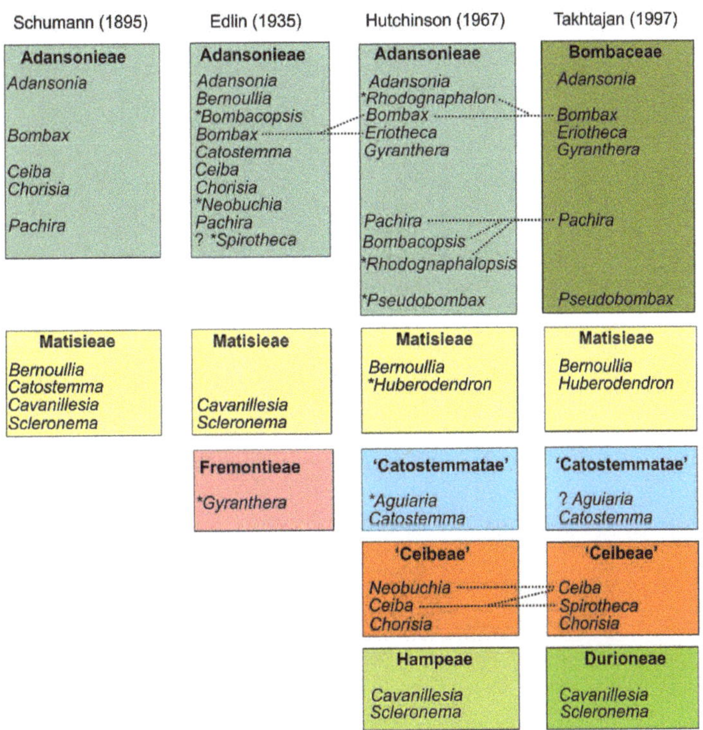

Figure 2. Taxonomic location of Bombacoideae as per different systems of classification. Dotted lines specify the changes in the genus limitation and the genera, which is described after the previous action, and are specified by a symbol (*), while the citation marks represent the tribes which are not validly published. Reproduced with permission from Carvalho-Sobrinho et al. [4] (originally Figure 1).

As a consequence of changes in circumscription and status of Bombacoideae, has led to the inclusion of 22 genera comprised of 120 species under this subfamily mainly distributed in the tropical regions. The Angiosperm Phylogeny Group (APG) IV Classification listed

24 genera in this subfamily. However, the classification by Maarten et.al. listed 27 genera under Bombacoideae [9]. This classification includes genera like *Camptostemon, Lagunaria,* and *Uladendron* in the Bombacoideae sub family. Molecular phylogenetic analysis established on nuclear (ETS, ITS) and plastid genes (matK, trnL-trnF, trnS-trnG) revealed that there are three key lineages noticeable by the kapok clade, seed, or fruit traits—the winged seed clade, and the spongy endocarp clade [4]. Such studies established the monophyly of the core Bombacoideae subfamily and the entire genera without *Pachira*. The monospecific *Septotheca* falls outside the core Bombacoideae in many studies [4,10].

3. Habitat, Distribution, and Characteristics of the Subfamily Bombacoideae

Bombacoideae occupies different habitats in various parts of the world (Figure 3) [11]. *Adansonia digitata* is confined to semi-arid, stony, hot, dry, and woodland areas, with low rainfall. This plant favors well-drained soils ranging from clays to sandy soils [12]. However, some other plants favor wet and humid habitats. For instance, *Bombax ceiba* favors humid lowland deciduous forest and is sometimes found near stream banks [13,14]. Species belonging to *Spirotheca* are epiphytic stranglers. Some species are part of mangrove vegetation in the tropical regions, for example, *Pachira aquatica, Camptostemon philippinense,* et cetera. The majority of the species in Bombacoideae prefer rain forest biome and seasonally dry biomes [11]. Several representative plant species belonging to the subfamily Bombacoideae grow in different habitats. *Adansonia digitata* L. grows in the hot, semi-arid region with poorly drained soil [15]. Plants such as *Bombax ceiba* L., *Ceiba pentandra*; and *Gyranthera caribensis* Pittier grow in wet habitats [16]. *Pachira aquatica* Aubl. and *Camptostemon philippinense* (S.Vidal) Becc. grow in mangrove habitats [17,18]. *Spirotheca rivieri* (Decne.) Ulbr. grows in the epiphyte environment [19] and *Ceiba pentandra* grows in the savannah habitat [15].

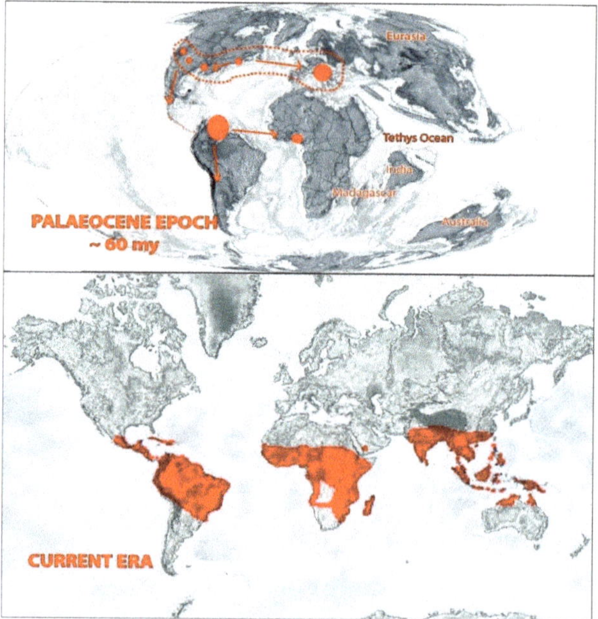

Figure 3. Distribution of Bombacoideae in past and present eras. Adapted from Krutzsch, [20], Zizka et.al. [11], and Angiosperm Phylogeny website version 14 (www.mobot.org/MOBOT/research/APweb; accessed on 10 January 2021).

The early distribution of this plant group can be ascertained from fossil records. There are various arguments for the distribution of Bombacoideae. Croizat (1952) favored the knowledge of an African entrance with Bombacoideae transferring northwards from Antarctica by Madagascar, across into Africa and through the East Indies to Australia [21]. However, the concept cannot be supported by floral evolution and geological shreds of evidence [22]. According to another view based on palyno-morphological characteristics exhibited by the members of this plant group, the subfamily is assumed to have a triphyletic origin—with southern Central America, East Africa, Madagascar, and Southeast Asia as centers of origin [3]. Fossil records of this subfamily mainly belonged to microfossils (classified as belonging to the pollen genus *Bombacacidites*) and some macrofossils [23]. This plant group occurred in the North Tethyan flora and reached tropical regions of South America through Central America during the transition phase between the Cretaceous and Tertiary periods. Then they moved to Central Africa in the Paleocene epoch. During the Pliocene and Pleistocene periods, this group extended its distribution to the Caribbean and Central America. When the tropical flora reduced along with the North Tethys during the Upper Paleogene, the Bombacoideae retreated to North India and reached South East Asia during the Miocene epoch. From there, they expanded to New Guinea and North Australia [20,23]. In the present era, the distribution of the extant species mainly falls in the tropical regions, particularly in Africa, America, and Australia. More than 80% of the species' richness of this subfamily lies in the Neotropical region [8,11].

There are many reports of native species in Asian countries. Various native species are introduced to other parts of the world through human activities and other influences. The center of origin of the species of this plant group differs according to the genus. Species of this plant group can be categorized into two groups—plants endemic to a certain area and plants widely distributed through introduction. Of the endemic species, *Adansonia suarezensis* H.Perrier and *Adansonia gregorii* F.Muell. are restricted to Madagascar and NW Australia, respectively [24]. Madagascar has many endemic species of *Adansonia* such as *Adansonia fony* Baill., *Adansonia madagascariensis* Baill., *Adansonia za* Baill., and *Adansonia perrieri* Capuron [25–27]. Wild regions of the endemic species are as follows: *Adansonia suarezensis, Adansonia fony* Baill., *Adansonia madagascariensis* Baill., *Adansonia za* Baill., *Adansonia perrieri* Capuron, and *Adansonia grandidieri* Baill. are the endemic species in the Madagascar region [24]. Similarly, *Adansonia gregorii* F.Muell., *Aguiaria excelsa* Ducke, *Uladendron codesuri* Marc.-Berti, *Gyranthera darienensis* Pittier, *Cavanillesia chicamochae* Fern. Alonso, *Gyranthera caribensis* Pittier, and *Neobuchia paulinae* Urb. are endemic to Australia, Brazil, Venezuela, Panama, Colombia, Venezuela, and Haiti, respectively [4,8,24,28–30].

Native regions of various species in this group fall within the tropical region of Africa, America, and Asia. From their native regions, many species have been introduced to other parts of the world. *Adansonia digitata* is amongst the most widely distributed ones covering Asia, Australia, Northern America, and some oceanic islands. Distribution of this plant in the Caribbean and parts of America is through human agencies, where people from West Africa were transported between the sixteenth and nineteenth centuries for sugarcane plantations in the New World countries. In the Indian subcontinent, Arab traders or medieval Muslim rulers who maintained African slave armies mainly introduced this species. However, genetic analyses conducted in Indian populations revealed that the introduction occurred through multiple phases [31]. Most of the species have neotropical distribution, with some species having native ranges in Asia. *Bombax ceiba* has wild distribution in South East Asia and India. The place of origin of some plants is uncertain. The origins of wild areas of *Ceiba pentandra* (L.) Gaertn. are uncertain but now it is distributed throughout tropical regions including Asia [32].

4. Characteristics

Plants of Bombacoideae are usually perennial tall trees usually with swollen tree trunks. Trees of wet forests are usually evergreen while those of dry forests are deciduous [33]. Tree trunks may contain parenchymatous water storage tissue or mucilage cells.

Pneumatophores are present in the *Camptostemon*, a mangrove genus [8]. Barks are usually thin, often green. Most of the plants of Bombacoideae are characterized by their large size gigantic flowers with brush types [8]. Plants in these groups have a terminal flower and three bracts that exhibit a "bicolor unit". The first, lowermost bracts remain sterile, however, other bracts subtend cymose partial inflorescences. Flowers are usually subtended by an involucre of bracts. Sepals are usually large and fused and petals are usually fused to the stamen tube [34]. The fruit capsule has a hairy endocarp. Leaves are usually peltately-palmate. Petioles are pulvinate, k connate with or without lobes. Monothecal anthers are present. These characteristics are assumed to have resulted from the splitting of whole stamens. Transitional forms are observed in some plants [8]. Anther walls have 5–7 cells across. Staminodes are usually absent. Pollen may be flattened, triangular in polar view. Seeds are usually large and usually more than two cm long.

Most of the members of this subfamily are trees, especially shrubs, with characteristic two to five carpels, fruit capsules, rarely indehiscent, endocarp usually pubescent, pollen usually without spines, seeds usually glabrous, and exceptionally spinulose [8]. Some plants have a ploidy level other than diploidy. The lowermost chromosome numbers in this group were witnessed in *Bombax insigne* (2n = 18) from India and *Pachira macrocarpa* (2x = 26) from China, while uppermost numbers were documented in *Eriotheca* species (6x = 276) in Brazil [35]. Distinguishing characteristucs of the genera in this subfamily is provided in Table 1.

Table 1. General synopsis of the genera under Bombacoideae.

Genus	Morphological Character	Number of Species	Distribution
Adansonia L.	Trunks swollen; leaves simple sometimes lobed; ovary 5–10 locular; fruits indehiscent; 2n = 72, 88, 144, 160. *Adansonia digitata* (2n = 144, 160)	8 species	Mainland Africa, Madagascar introduced to many countries
Aguiaria Ducke	Lepidote hairs; leaves simple; staminal tube short with various elongated free filaments; fruits dehiscent; seed ellipsoid	1 species	Amazon region of Brazil
Bernoullia Oliv.	Trees leave digitate, staminal tube long, stamens 15–20, fruits dehiscent; seeds numerous, winged	3 species	Mexico to Colombia
Bombax L.	Deciduous tree, trunk spiny; leaves digitate; deciduous sepals, fruits dehiscent, seeds winged, determined columella; 2n = 72, 92, 96.	9 species	Tropical Africa, Asia, and Australia
Camptostemon Mast. *	Mangrove tree or shrubs; epicalyx fused, enclosing flower; calyx fused; ovary bilocular; fruits dehiscent	3 species	Australia, New Guinea, Borneo, Phillippines
Catostemma Benth.	Trees, leaves simple, calyx campanulate; ovary trilocular, fruits dehiscent; cotyledons folded, unequal	15 species	The northern part of South America
Cavanillesia Ruiz. & Pav.	Trunks swollen sometimes, leaves simple or palmately lobed, ovary 3–5 locular; fruits winged, indehiscent; 2n = 72, 86, 88	5 species	Panama to Brazil and Peru
Ceiba Mill.	Trunks spiny, sometimes swollen; leaves digitate; staminal tube sometimes thickened, stamens 5–15, fruits dehiscent, seeds winged; 2n = 72, 74, 75, 76, 80, 84, 86, 88, 92	21 species	Tropical America, now introduced into the Old World
Chiranthodendron Sesse ex Larreat.	Leaves simple to lobed; flowers leaf-opposed; sepals dark red, petals absent, fruits dehiscent	1 species	Mexico, Guatemala

Table 1. *Cont.*

Genus	Morphological Character	Number of Species	Distribution
Eriotheca Schott & Endl.	Trees unarmed, leaves digitate, staminal tube without phalanges; fruits dehiscent; seeds small, winged; 2n = 92, 210, 270, 6n = 276	23 species	Tropical South America
Fremontodendron Coville	Shrubs; leaves simple or lobed, sepals yellow-orange, petals absent, fruits dehiscent	2 species	The southern part of North America
Gyranthera Pittier	Tall deciduous tree, leaves digitate, anthers spirally twisted, fruits dehiscent, seeds winged; 2n = 96	2 species	Panama, Venezuela
Huberodendron Ducke	Tall trees, hairs stellate, leaves simple; calyx campanulate; fruits dehiscent, seeds winged	3 species	Costa Rica to Brazil, Bolivia, and Peru.
Lagunaria (DC.) Rchb. *	Leaves simple, hairs lepidote, epicalyx fused, filaments diverging at different levels; fruits stinging, dehiscent	1 species	Norfolk and Howe Islands, Australia
Matisia Humb & Bonpl.	Leaves simple, inflorescences cauliflowers, flowers zygomorphic, fruit drupe	26 species	Tropical America
Neobuchia Urb.	Trunk spiny, leaves digitate, stamens 5, anthers twisted; stigmatic branches short; seeds exalbuminous	1 species	Haiti
Ochroma Sw.	Tree, leaves simple to lobed, venation palmate; stigma spirally grooved; fruits dehiscent; 2n = 78, 88, 90	1 species	Tropical America
Pachira Aubl.	Trunk spiny sometimes; leaves digitate; stamens 90–1000; fruits large, dehiscent, 2n = 72, 82, 88, 92 (neotropical species), 144, 150 (palaeotropical species)	47 species	Tropical Africa, neotropical regions
Patinoa Cuatrec.	Trees with verticillate branches, leaves simple, sessile anthers, ovules many, fruits indehiscent	4 species	Colombia to Brazil and Peru;
Pentaplaris L.O. Williams & Standl. *	Leaves simple, stipules fused; epicalyx fused, ovary bilocular, fruits indehiscent; cotyledons foliose	3 species	Costa Rica, Ecuador, Bolivia, and Peru
Phragmotheca Cuatrec.	Trees, lepidote hairs rare, leaves simple, flowers leaf-opposed; fruit a drupe; cotyledons flat or folded	5 species	Panama to Peru
Pseudobombax Dugand	Trunks swollen sometimes, leaves usually digitate, ovary 5 to 8 locular, fruits dehiscent, seeds winged; 2n = 72, 84, 88	22 species	Mexico, Tropical South America
Quararibea Aubl	Trees; lepidote hairs sometimes, calyx usually ridged; ovary 2 to 4 locular, fruit a drupe; n = 72(?)	88 species	Neotropical regions
Scleronema Benth.	Tall tree leaves simple, staminal tube short, ovary 2 to 4 locular, fruits dehiscent or indehiscent	3 species	Venezuela, Guyana and Brazil
Septotheca Ulbr.	Tall tree, lepidote hairs, simple leaves, cordate, anthers sessile; fruits dehiscent, seeds winged	1 species	Peru, Colombia, and Brazil.
Spirotheca Ulbr.	Epiphytic stranglers to the tree, trunk spiny sometimes, leaves digitate, stamens 5, anthers spirally twisted, fruits dehiscent; 2n = 88, 92	5 species	Panama to Peru and Brazil
Uladendron Marc.-Berti *	Leaves simple, slightly lobed, fruits dehiscent, seeds winged; cotyledon distorted	1 species	Venezuela

(www.theplantlist.org; accessed on 12 October 2020); * Genera incertae sedis (uncertain placement). Source: (Byng [34]; Fay [36]; Kubitzki and Bayer [8]; Lim [15]; Marinho et al. [35]).

The status of genera under Bombacoideae might be subjected to change in future revisions. The single species *Chiranthodendron pentadactylon* can be crossed with *Fremontodendron* sp. [8] exhibiting compatible genotypes. *Neobuchia paulinae* is an imperfectly known species that may be included in *Ceiba* [8].

5. Phytochemical Configuration of Bombacoideae Subfamily

Phytochemical investigations of Bombacoideae plant species resulted in the extraction and isolation of several classes of secondary metabolites. Among the most studied genera, there are *Adansonia*, *Bombax*, and *Chorisia* [2,37,38]. *Bombax ceiba* (syn. *Bombax malabaricum*, *Bombax malabarica*, *Salmalia malabaricum*, *Gossampinus malabarica*), *Adansonia digitata*, and *Chorisia speciosa* are the most chemically and biologically investigated species.

A wide spectrum of phytochemicals has been identified and has confirmed that this family is a rich source of phytochemicals. Table 2 lists the main alkaloids, anthocyanins, coumarins, flavonoids, lignans and neolignans, sesquiterpenes and sesquiterpene lactones, sterols, tannins, and triterpenes isolated from the Bombacoideae subfamily. Volatiles and fatty acids were also reported (Table 2).

Table 2. The main phytochemicals identified in plant species from the Bombacoideae subfamily.

Compound	Plant	Part	Reference
Alkaloids			
Adansonin	*A. digitata*		[39]
Funebral			[40]
Funebradiol	*Quararibea funebris*		[41]
Funebrine		Seeds and pulp	[42]
Anthocyanins		Flowers	
Cyanidin-3-glucoside	*Ceiba acuminata*		
	Chorisia speciose (*Ceiba speciosa* (A.St.-Hil., A.Juss. & Cambess.) Ravenna)		
	Ochroma lagopus (*Ochroma pyramidale* (Cav. ex Lam.) Urb.)	Calyx	[43]
Cyanidin-3,5-diglucoside	*Pachira aquatica*		
	Bombax ceiba		
	C. speciosa		
	Pseudobombax ellipticum		
	P. grandiflorum	Flowers	
Cyanidin-3-rutinoside	*Pachira aquatica*		
Cyanidin-7-methyl ether-3-β-d-glucoside	*B. ceiba*		[44]
Pelargonidin-5-β-d-glucoside	*B. ceiba*		
Pelargonidin-3,5-diglucoside	*B. ceiba*		[45]
Coumarins			
Cleomiscosine A	*Ochroma lagopus*	Heartwood	[46]
Esculetin			
Fraxetin	*B. ceiba*	Flowers	[47]
Scopoletin			
Scopolin			
Scopoletin	*P. aquatica*	Stems	[48]

Table 2. *Cont.*

Compound	Plant	Part	Reference
	Flavonoids		
Apigenin	*B. ceiba*	Flowers	[49]
Apigenin *O*-pentoside	*A. digitata*	Fruits	[50]
Apigenin-7-*O*-β-d-rutinoside	*Chorisia insignis* (*Ceiba insignis* (Kunth) P.E.Gibbs & Semir)	Leaves	[51]
Catechin	*A. digitata*	Fruits	[50]
	Ceiba pentandra	Stem bark	[52]
	Ochroma pyramidale	Leaves	[53]
Cosmetin	*B. ceiba*	Flowers	[49]
5,4′-Dihydroxy-3,6,7,8-tetramethoxyflavone	*P. aquatica*	Stems	[48]
5,4′-Dihydroxy-3,7-dimethoxyflavone			
Epicatechin	*A. digitata*	Fruits	[50,54]
Epicatechin	*O. pyramidale*	Leaves	[53]
3,5,6,7,8,3′,4′-Heptamethoxyflavone	*P. aquatica*	Stems	[48]
Hesperidin (5,3′-dihydroxy-4′-methoxy-flavan-7-*O*-α-Lrhamnopyranosyl-(1→ 6)-β-d-lucopyranoside	*B. ceiba*	Roots	[55]
5-Hydroxy-7,4′,5′-trimethoxy isoflavone-3′-*O*-α-l-arabinofuranosyl(1→6)-β-d-glucopyranoside	*Ceiba pentandra*	Stem bark	[56]
5-Hydroxyauranetin	*P. aquatica*	Stems	[48]
5-Hydroxy-7,4′-dimethoxy-flavone	*Bombax anceps*	Roots	[57]
5-Hydroxy-3,7,4′-trimethoxy- flavone			[58]
5-Hydroxy-3,6,7,4′-tetra-methoxyflavone	*Bombacopsis glabra* (*Pachira glabra* Pasq.)	Stem bark, root bark	
5-Hydroxy-3,6,7,8,4′-penta-methoxyflavone			[58]
Isovitexin	*B. ceiba*	Flowers	[49]
Linarin			
3,7-Dihydroxy-flavan-4-one-5-*O*-β-d-galactopyranosyl-(1→ 4)-β-d-glucopyranoside	*A. digitata*	Roots	[59]
5,7-Dimethoxy-flavone	*Bombax anceps*	Roots	[57]
3,5,7-Trimethoxy-flavone	*B. anceps*	Roots	
Luteolin-7-*O*-β-d-rutinoside	*C. insignis*	Leaves	[51]
Kaempferol	*A. digitata*	Fruits	[50]
	B. ceiba	Flowers	[60]
	C. pentandra	·	[61]
Kaempferol 3-*O*-galactoside	*A. digitata*	Fruits	[50]
Kaempferol 3-*O*-glucoside	*A. digitata*	Fruits	
Kaempferol 3,7,4′-trimethyl ether	*P. aquatica*	Stems	[48]
Pentandrin	*C. pentandra*	Stem bark	[62]
Pentandrin glucoside	*C. pentandra*	Stem bark	
Quercetin	*B. ceiba*	Flowers	[60]
	A. digitata	Fruits	[50]

215

Table 2. *Cont.*

Compound	Plant	Part	Reference
	C. pentandra	-	[61]
Quercetin-3-*O*-glucoside	*A. digitata*	Fruits	[50,54]
Quercetin-7-*O*-xylopyranoside	*A. digitata*	Stem	[63]
Retusin	*P. aquatica*	Stems	[48]
Rhoifolin	*Chorisia crispiflora*	Leaves	[64]
	Chorisia pubiflora	Leaves	
	C. speciosa	Leaves	[64,65]
Rutin	*A. digitata*	Leaves	[50]
	C. insignis	Leaves	[51]
Saponarin	*B. ceiba*	Flowers	[49]
Santin-7-methyl ether	*P. aquatica*	Stems	[48]
Shamimin	*B. ceiba*	Leaves	[66]
Shamimicin	*B. ceiba*	Stem bark	[67]
Tiliroside	*C. speciosa*	Leaves	[65]
Tiliroside isomer Tiliroside I Tiliroside II	*A. digitata*	Fruits, leaves	[50]
3,3′,4′-Trihydroxy flavan-4-one-7-*O*-α-L-rhamnopyranoside	*A. digitata*	Roots	[54,68]
Vicenin 2 Vitexin	*B. ceiba* *O. pyramidale*	Flowers Leaves	[49] [53]
Xanthomicrol	*B. ceiba*	Flowers	[49]
Lignans and neolignans			
Boehmenan	*Ochroma lagopus*	Heart wood	[46]
Boehmenan B			
Boehmenan C	*O. lagopus*	Heart wood	[69]
Boehmenan D			
Bombasin			
Bombasin-4-*O*-glucoside	*B. ceiba*	Flowers	[70]
Bombasinol A			[71]
Carolignan A			
Carolignan B			
Carolignan C	*O. lagopus*	Heart wood	[69]
Carolignan D			
Carolignan E			
Carolignan F			

Table 2. Cont.

Compound	Plant	Part	Reference
Dihydro-dehydro-diconiferyl alcohol- 4-*O*-glucopyranoside	*B. ceiba*	Flowers	[70]
5,6-Dihydroxymatairesinol			
Matairesinol	*B. ceiba*	Flowers	[71]
(+)-Pinoresinol			
Secoisolariciresinol diferulate	*O. lagopus*	Heart wood	[46]
Sesquiterpenes and *sesquiterpene lactones*			
Aquatidial	*Pachira aquatica*	Root bark	[72]
Bombamalabin		Root bark	[73]
Bombamalone A			
Bombamalone B			
Bombamalone C	*B. malabaricum*	Roots	[74]
Bombamalone D			
Bombamaloside			
7-Hydroxy-cadalene		Roots	[75]
Isohemigossypol-1-methyl ether	*B. anceps*	Roots	[57]
	B. ceiba	Root bark	[73,75]
Isohemigossypol-2-methyl ether	*B. anceps*	Roots	[57]
Isohemigossypol-1,2-dimethyl ether	*B. ceiba*	Roots, root bark	[75,76]
Isohemigossypol-2,7-dimethyl ether			
Lacinilene C	*B. ceiba*	Roots	[74,75]
Hemigossylic acid lactone-2-hydroxy-7-methyl ether			
Hemigossylic acid lactone-2-hydroxy-7-methyl ether	*C. pentandra*	Root bark	[77]
6-Hydroxy-5-isopropyl-3-methyl-7-methoxy-8,1-naphthalene carbolactone	*B. ceiba*	Roots	[78]
Isohemigossylic acid lactone-2-methyl ether	*B. ceiba*	Roots	[74,76]
Isohemigossylic acid lactone-2-methyl ether	*C. pentandra*	Root bark	[77]
5-Isopropyl-3-methyl-2,7-dimethoxy-8,1-naphthalene carbolactone	*B. ceiba*	Roots	[75]
5-Isopropyl-3-methyl-2,7-dimethoxy-8,1-naphthalene carbolactone	*C. pentandra*	Root bark	[77]

Table 2. *Cont.*

Compound	Plant	Part	Reference
	Sterols		
Campesterol	*A. digitata*	Seeds	[79]
	B. ceiba	Flowers	[49]
	A. fony		
	A. za		
	A. suarezensis	Seeds	[79]
	A. grandidieri		
	A. madagascariensis		
β-Sitosterol	*B. ceiba*	Stem bark	[37]
	B. ceiba	Root bark	[73]
	B. ceiba	Flowers	[80]
	A. digitata	Seeds	[79]
	C. pentandra	Stem bark	[62]
Stigmasterol	*A. digitata*	Seed	[79]
	B. ceiba	Flowers	[80]
	A. grandidieri		
	A. madagascariensis		
	A. fony	Seeds	[79]
	A. za		
	A. suarezensis		
	Tannins		
Epicatechin-(4β→8)-epicatechin			
Epicatechin-(4β→6)-epicatechin			
Epicatechin-(2β→O→7, 4β→8)-epicatechin	*A. digitata*	Fruits	[54]
Epicatechin-(4→β8)-epicatechin-(4→β8)-epicatechin			
Ethyl gallate		Seeds	[81]
Gallic acid		Stem bark	[37]
	B. ceiba		
1-Galloyl-β-d-glucose		Seeds	[81]
Tannic acid			

Table 2. Cont.

Compound	Plant	Part	Reference
	Triterpenes		
β-Amyrin	*C. speciosa*	Leaves	[65]
Lupeol	*B. glabra*	Stem bark, root bark	[58]
	B. ceiba	Stem bark	[37]
	B. malabarica	Root bark	[73]
	B. anceps	Roots	[57]
	Cavanillesia hylogeiton	Stem bark	[61]
	O. pyramidale	Leaves	[53]
	P. aquatica	Root bark	[72]
Oleanolic acid	*B. ceiba*	Roots	[55]
	O. pyramidale	Leaves	[53]
Ursolic acid	*A. digitata*	Fruits	[82]
	Other compounds		
Argentilactone I	*Chorisia crispiflora*	-	[83]
Argentilactone II	*C. crispiflora*	-	[83]
Bombalin	*B. ceiba*	Flowers	[70]
Bombaxquinone B	*B. anceps*	Roots	[57]
	B. ceiba	Roots	[74]
	C. pentandra	Root bark	[77]
	B. ceiba	Root bark	[84]
(R)-6-[(Z)-1-Heptenyl)]-5,6-dihydro-2H-pyran-2-one	*C. crispiflora*	-	[83]
Hemigossypolon-6-methyl ether	*B. ceiba*	Root bark	[85]
Isohemigossypolone	*B. glabra*	Stem bark, root bark	[58]
	B. ceiba	Root bark	[85]
	C. pentandra	Heart wood	[86]
	P. aquatica	Root bark	[58,87]
Isohemigossypolone-2-methyl ether	*P. aquatica*	Root bark	[87]
Neochlorogenic acid *trans*-3-(*p*-Coumaroyl)-quinic acid 3-Methyl-2(3H)-benzofuranone	*B. ceiba*	Flowers	[70]

The fruit pulp of *A. digitata* from Mali is characterized by flavonol glycosides and procyanidins as dominant classes of compounds [50]. Tiliroside was identified as a major constituent. *A. digitata* fruits from Nigeria showed hydroxycinnamic acid glycosides, iridoid glycosides, and phenylethanoid glycosides, secondary metabolites not detected in the fruits from Mali [88]. More recently, procyanidins, phenolic acids, and flavonol glycosides were identified in *A. digitata* fruits from Cameroon [89]. In particular, fruit pulp was characterized by the presence of non-flavonoid compounds such as hydroxycinnamic derivatives and flavonoids, mainly flavones, flavanols, proanthocyanidins, and flavonols.

Furthermore, polar compounds identified in leaf extracts consisted of several classes of flavonoids and hydroxycinnamic acids. Leaves from Cameroon [89] exhibited a very similar profile compared to the leaves from Mali [50].

Previously, Tembo et al. [90], quantifying several compounds in fresh *A. digitata* pulp and investigating quantitatively variations of some of these molecules induced by pasteurization and thermal preservation, described a high content of epicatechin, gallic acid, and procyanidin B2 in Malawi *A. digitata* fruits. Nasr et al. [65] isolated two flavonoid glycosides, namely, rhoifolin and tiliroside, in the alcoholic extract of *C. speciosa* leaves from Egypt, together with some sterols and triterpenes. The sesquiterpenes, bombamalin and isohemigossypol-1-methyl ether, and the phenols, 4-hydroxy-3,5-dimethoxybenzoic acid, 3,4,5-trimethoxyphenol-1-(β-xylopyranosyl-(1→2))-β-glucopyranoside, shorealactone, (−)-epicatechin 5-*O*-β-D-xylopyranoside, and 2-C-(β-D-apiofuranosyl-(1→6))-β-D-glucopyranosyl-1,3,6-trihydroxy-7-methoxyxanthone have been isolated from the ethanol extract of *B. malabarica* root bark [73].

Five new compounds, namely, bombamaloside and bombamalones A–D (Figure 4), were obtained by Zhang et al. [74] from the H$_2$O/acetone (3:7) extract of *B. malabaricum* roots, along with other known constituents such as bombaxquinone B, lacinilene C, isohemigossypol-1-methyl ester, and 2-*O*-methylisohemigossylic acid lactone.

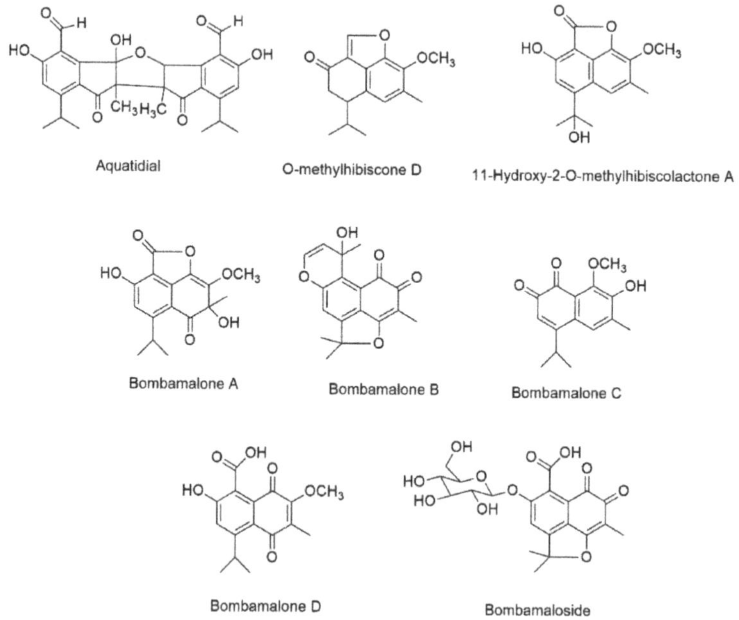

Aquatidial

O-methylhibiscone D

11-Hydroxy-2-O-methylhibiscolactone A

Bombamalone A

Bombamalone B

Bombamalone C

Bombamalone D

Bombamaloside

Figure 4. The chemical structures of new isolated compounds from Bombacoideae species.

Aquatidial (Figure 4) was previously isolated from a chloroform extracts of *P. aquatica* roots together with the known compounds lupeol, triacontyl *p*-coumarate, and isohemigossypolone [72]. Aquatidial is a new bis-norsesquiterpene with an uncommon skeleton, putatively derived from isohemigossypolone. Two new naphthofuranones, 11-hydroxy-2-*O*-methylhibiscolactone A and *O*-methylhibiscone D (Figure 4), have been extracted from the *P. aquatica* stems [48].

Several volatiles have also been described from some Bombacoideae species. Sulfur compounds (15.3%), benzenoids (7.8%), monoterpene hydrocarbons (0.6%), and oxygenated monoterpenes (0.2%) were identified in the flowers of *A. digitata* [91]. The oil obtained from the flowers of *C. pentandra* showed monoterpene hydrocarbons (34%), sesquiterpene hydrocarbons (26.9%), oxygenated monoterpenes (8.4%), benzenoids (7.8%), and miscellaneous compounds (2%) [91].

The most common fatty acids in the Bombacoideae subfamily are oleic, linoleic, linolenic, stearic, and palmitic acids. The cyclopropenoid fatty acids, malvalic acid and sterculic acid, have been identified in *A. digitata* [92–94], *A. fony* [94], and *Bombax oleagineum*, *C. acuminata*, and *C. pentandra* [61]. Recently, the seeds' *n*-hexane extract of *C. speciosa* from

Italy showed linoleic acid (28.22%) and palmitic acid (19.56%) as the most abundant fatty acids [95]. Percentages of 16.15 and 11.11% were found for malvalic acid and sterculic acid, respectively.

Linoleic acid (38.8%), palmitic acid (24.3%), and oleic acid (21.9%) were identified as the dominant fatty acids of *C. pentandra* seed oil from Malaysia [96]. Malvalic and sterculic acids were also identified. A lower percentage of linoleic acid was found in the seed oil of *C. pentandra* from India [97]. Saturated fatty acids and monounsaturated fatty acids were obtained from the seeds of *P. aquatica* by using the Soxhlet apparatus and *n*-hexane as solvent. Palmitic acid and oleic acid were the most abundant with percentages of 49.0 and 18.2%, respectively [98]. Linoleic acid (11.2%) is the only polyunsaturated fatty acid identified.

6. Details of the Extraction and Isolation Procedure of Major Compounds from Bombacoideae for Industrial Applications

Different bioactive constituents, mainly terpenes, flavonoids, alkaloids, steroids, and fatty acids, have been isolated from the Bombacoideae subfamily. The extraction technique is the first pivotal step to obtaining active phytochemicals from plants. The choice of extraction procedure would depend mainly on the advantages and disadvantages of the process, including yield, biological activity, environmental friendliness, and safety. The fruit pulp of *A. digitata* revealed the presence of iridoids and phenols by using 70% ethanol as solvent [88]. Proanthocyanidins were obtained as major constituents from the pericarp of *A. digitata* fruits [54] by using a hydroalcoholic solution (methanol/H_2O 80:20 *v/v*). Maceration with 95% ethanol of *B. malabarica* root bark led to the isolation of several sesquiterpenes, triterpenes, phenols, and sterols [73]. Conversely, cadinene sesquiterpenes were extracted from the roots of *B. malabaricum* by using H_2O/acetone (3:7 *v/v*) [74]. *B. malabaricum* flowers extracted by 70% (*v/v*) aq. ethanol is characterized by different lignans.

Until now, the most applied extraction technique to isolate phytochemicals from the Bombacoideae subfamily is maceration. Researchers are exploring other extraction procedures using less energy and less solvent while producing higher yields and that are more environmentally friendly. Some advanced methods (i.e., pressurized and accelerated fluid extraction, supercritical extraction) have demonstrated to be useful in mediating related extraction difficulties along with increased extraction yields. Two of the most commonly employed extraction techniques of flavonoids are microwave- (MAE) and ultrasound-assisted extraction (UAE). High extraction efficiency and less destruction of the active constituents are the many advantages of UAE [99–101]. Nevertheless, MAE is preferred over UAE because MAE has been shown to increase the mass transfusion through the solid matrix, faster mixing of the extraction solvent thus preserving the highest possible driving forces, and ensuring the highest quality and quantity of the extracted constituents. Indeed, several works have proven that MAE allows for great extraction yields, a reduction of the volumes of solvents used, and a reduction of the extraction times [99,100]. MAE has been applied to extract flavonoids, tanshinones, coumarins, and terpenes [101]. These characteristics along with the simplicity of operation would position MAE as a valuable and suitable technology for industries with the growing demand for increased productivity and efficiency. However, until now little progress has been described for the MAE application to Bombacoideae species. Surely, taking into account all the MAE features, in the future, it will be possible to optimize the process by exploiting the opportunity to apply this innovative extraction method to the study of species belonging to the Bombacoideae family.

7. Application in Food/Use as Food

From ancient periods until today, many plants of the Bombacoideae have been used as food in various corners of the world. Parts used may range from leaves, seeds, tuberous roots to stem, flowers, et cetera. There are various variations in the use of food according to genera and cultures associated. Native African populations commonly use fruits of *Adansonia digitata* as famine food to make sauces, decoctions, and refreshing beverages [102].

The leaves, seeds, and pulp of the fruit of this plant are all edible. Lim (2012) reported the use of young leaves, seeds, fruit pulp, and tuberous roots of *Adansonia gregorii* F. Muell as food. Along with *Adansonia* spp, *Ceiba pentandra* is another one of the plant foods common to West Africa. Leaves of this plant are cooked in the form of slurry sauce [103]. The utilization as food for this plant group is not restricted to Africa but observed in other parts of tropical countries. In Central and South America, flowers and tender leaves of *Pachira aquatica*, a wetland tree, are cooked and used as vegetables [15]. Young roots of *Bombax ceiba* are eaten raw or roasted in Cambodia. The cuipo tree (*Cavanillesia platanifolia*), growing in Central America, is used by the natives for getting water. To collect water, a piece of the root is cut and the bark is removed on one end after keeping the root horizontal. When the clean end of the root is lowered, the water drains out through the cut end [104].

The use and preparation of food from the members of Bombacoideae dates back to time immemorial. For instance, in South America, from the ancient pre-Colombian period [105], flowers of *Quararibea funebris* were used as an additive to chocolate drinks. Ancient Mayans used the sap from *Pseudobombax ellipticum* to make an intoxicating drink by fermentation. This drink was likely used in religious ceremonies such as sacrifice and self-mutilation [33]. The use of various members of Bombacoides as fruits, vegetables, and other forms are highlighted in Table 3.

Table 3. Plant and parts used as a food.

Name of the Species	Parts Used	Mode of Usage	Country	Reference
Adansonia digitata L.	Leaves and seeds	Soup, sauce, fermentation, gruel	Southern Africa, Italy	[15,106]
Adansonia gregorii F.Muell.	roots, fruit pulp, seeds, tuberous, young leaves,	Food	Aborigines in Australia	[15]
Bombax ceiba L.	Dry cores of the flower	soup	Shan State (Myanmar) and Northern Thailand	[107]
	Flower buds	Vegetable	South India	[108]
	Seeds	Roasted and eaten		[32]
Bombax costatum Pellegr. & Vuillet	Unripe fruits and flowers	Soup	Burkina Faso	[109]
Catostemma fragrans Benth.	Aril	Fresh	Guianas	[110]
Cavanillesia platanifolia (Humb. & Bonpl.) Kunth	Seed, Root	Sweet, water source	Peru	[104,111]
Ceiba pentandra (L.) Gaertn.	Young leaves, petals, capsules	Vegetable	Tropical countries of Asia and America, Thailand	[15,32]
Ceiba aesculifolia (Kunth) Britten & Baker f.	Young leaves, ripe fruits	Vegetable, Stew	Mexico	[32]
Pachira glabra Pasq.	Young Leaves	Vegetable	Equatorial Africa	[32]
Pachira insignis (Sw.) Savigny	Seeds, young leaves, flowers	Vegetable	South America	[15]
Patinoa almirajo Cuatrec.	Fruit	Edible fruit	Brazil, Colombia	[32]
Pseudobombax ellipticum (Kunth) Dugand		Beverage	South America	[33]
Quararibea cordata (Bonpl.) Vischer	Fruit	Juice, drinks	South America	[15]
Quararibea funebris (La Llave) Vischer	Flowers	Chocolate Drinks, desserts	South America	[33]
	Flowers	Spice	South America	[112]
Quararibea obliquifolia (Standl.) Standl.	Fruit	Fresh	Ecuador	[113]

8. Traditional and Economic Uses

Various members of Bombacoideae are used as fiber and other utilities and some are also used as ornamental plants. *Adansonia digitata* is a multipurpose plant with various economic and social values [106]. In African countries, *Adansonia digitata* is very popular and reported to have more than three hundred traditional uses [102]. *Ceiba* Mill. is now popular throughout the tropical regions for ornamental landscaping [114]. Many species of the genus *Ceiba* were sacred to the Mayan civilization as depicted in ancient ceramics because of their cultural importance [33].

Many Bombacoideae species are economically important. Some species are collected for their wood that is soft and can easily be carved into canoes and other useful products. One popular wood is balsa wood obtained from the *Ochroma pyramidale* [16] and other species were widely used for making dugout canoes in ancient South America. Ancient Peruvians are believed to have used legendary Kon-Tiki rafts made from balsa wood to navigate across the Pacific Ocean and settle in the Polynesian islands [115]. The silky cotton-like fluff (kapok) present in the seed pods of *Ceiba pentandra* is used for stuffing pillows, bedding, and soft toys in various parts [116]. Silk hair present in seeds of *Bombax ceiba* are used in India from time immemorial for stuffing cushions, mattresses, pillows, and making clothes [117]. Various traditional and economic uses of the members of this subfamily are summarized in Table 4.

Table 4. Economic and traditional uses of Bombacoideae members.

Name of the Species	Parts Used	Purpose	Country	Reference
Adansonia digitata L.	Fruit shell	Fuel	Tanzania	[118]
	Leaves	Fodder	The Sahelian region, Africa	[118]
	Fiber from bark	Ropes, textile, basketry, fishing lines	Africa	[118]
	Tree trunk	Reservoir of water	Sudan	[15]
	Roots	Red dye	East Africa	[118]
Aguiaria excelsa Ducke	Wood	Boat, construction	Brazil	[28]
Bombax ceiba L.	Fiber	Mattress, pillows, cloth	Asia	[117]
Bombax insigne Wall.	Wood	Timber, boat construction, matches, plywood	India, Sri Lanka, Nepal	[119,120]
Bombax costatum Pellegr. & Vuillet	Wood	Drum, xylophone, match stick, home appliances, door frame, fuelwood	Africa	[109]
	Tannin	Dye	Africa	[109]
	Fruits	Mattress, cushion, pillow	Africa	[109,121]
Bombax rhodognaphalon K. Schum.	Leaves, roots	Witchcraft	Africa	[122]
Catostemma commune Sandwith	Wood	Timber	Central and Latin America	[123]
Cavanillesia umbellata Ruiz & Pav.	Bark	Drum hoops	Peru	[111]
	Wood	Door fillings, light boxes, toothpicks, paper pulps	Peru	[111]
Ceiba aesculifolia (Kunth) Britten & Baker f.	Fiber	Fiber	Mexico, Guatemala	[32]

Table 4. *Cont.*

Name of the Species	Parts Used	Purpose	Country	Reference
Ceiba pentandra (L.) Gaertn.	Fiber, wood	Paper, fiber, insulation material, pillows, toys	Tropical countries	[32,116]
Ceiba samauma (Mart. & Zucc.) K.Schum.	Seed	Thermal insulation	Ecuador	[124]
Ceiba trischistandra (A.Gray) Bakh.	Fruit wall	Fiber	Java, Peru, and Brazil	[32]
Huberodendron patinoi Cuatrec.	Wood	Timber	Colombia	[125]
Ochroma pyramidale (Cav. ex Lam.) Urb.	Wood	Bowls, rafts, canoes, toys, carvings (Balsa)	Venezuela	[16]
Pachira aquatica Aubl.	Whole tree	Ornamental, fortune tree	East Asia, South East Asia	[15]
Pachira insignis (Sw.) Savigny	Wood	Paper	South America	[32]
Pentaplaris davidsmithii Dorr & C. Bayer	Wood	Firewood	Bolivia	[126]
Quararibea funebris (La Llave) Vischer	Flowers	Perfume	South America	[33]
Quararibea malacocalyx A.Robyns & S.Nilsson	Seed fiber	Thermal and acoustic insulation	Ecuador	[124]
Scleronema micranthum (Ducke) Ducke	Wood	Construction, joinery, flooring, furniture	Brazil	[127]
Spirotheca rivieri (Decne.) Ulbr.	Wood	Box, Linings	Brazil	[128]

9. Ethnopharmacology

In various tropical countries, plants of Bombacoideae are used in traditional medicine mainly for pharmacological properties like anti-inflammatory, astringent, antimicrobial, stimulant, antipyretic, analgesic, and diuretic [2]. For instance, various parts of *Bombax ceiba* such as the stem bark, flowers, fruits, seeds, leaves, and root of young plants, are traditionally used as remedy in South India [108]. Its main therapeutic applications include diabetes, urinogenital disorders, gastrointestinal and skin diseases, gynecological, and general debility [129]. Another important plant from this subfamily in the Indian ayurvedic system is *Ceiba pentandra* known as Sweta Salmali for its acrid, bitter, thermogenic, diuretic, and purgative properties. The known pharmacological activities of *Ceiba pentandra* include hepatoprotective, antidiabetic, antipyretic, laxative, and anti-inflammatory [130]. *Adansonia digitata* is one of the most studied species for its therapeutic properties against antipyretic, diarrhea, dysentery, and as a substitute for cinchona in traditional medicinal preparations [105]. Different species under Bombacoideae having reported ethnopharmacological uses are summarized in Table 5.

Table 5. Plants belonging to Bombacoideae with ethnopharmacological uses.

Species	Country	Parts Used	Disease	Mode of Usage	Reference
Adansonia digitata L.	India	Pulp	Diarrhea and dysentery	External application	[118]
	India	Leaves	Swellings	Crushed and applied	[118]
	South and East Africa	Leaves	Malaria and fever	Mixed with water	[131]
	Cameroon, Central Africa	Fruits, seeds	Dysentery, fever	Decoction	[131]
	South Africa	Leaves	Diarrhea, fever, kidney and liver diseases, inflammation, asthma	Infusion	[132]
	Nigeria	Bark	Sickle-cell anemia	Aqueous extract	[15,133]
	Burkina Faso	Leaves	Toothache, gingivitis		[134]
Bernoullia flammea Oliv.	Guatemala	Seeds	Intoxication	Smoke	[135]
Bombax ceiba L.	India	Root	A nocturnal emission, cold, and cough, dysentery, diarrhea, snake bite, gonorrhea, leucorrhea	Drink the powdered solution; applied the paste	[1,136]
	India, Nepal	Bark	Wounds, diarrhea, digestive disorder, heartburn, kidney stone	Paste, Juice	[1,136,137]
	India, Pakistan	Stem, root	Acne, skin blemishes, pimples	Powder	[136,137]
	India, Pakistan	Root	Diabetes		[129,137]
	China	Bark, root	Muscular injury		[137]
	Bangladesh	Seeds, roots	Leprosy		[137]
	India	Fruits	Urolithiasis	Oral administration	[129]
	India	Gum	Asthma, piles, diarrhea and dysentery, dental caries, scabies		[1]
	India	Flower	Hematuria, anemia, leucorrhea, hydrocoele, gonorrhea, menstrual disorders, boils and sores		[1]
Bombax insigne Wall.	India	Bark	Dysentery	Tea	[122]
Bombax buonopozense P.Beauv.	Africa	Leaves	Venereal disease, constipation, infections		[122]
Bombax costatum Pellegr. & Vuillet	Senegal, Sierra Leone, Burkina Faso	Bark	Diuretic properties, dysentery, epilepsy		[109,122]
	Senegal	Leaves	Oedema, snake bite, convulsions, measles	Extract, decoction, paste	[109,121]

Table 5. *Cont.*

Species	Country	Parts Used	Disease	Mode of Usage	Reference
Bombax rhodognaphalon K. Schum.	Tanzania, Mozambique	Bark	Diarrhea		[122]
Catostemma fragrans Benth.	Guianas	Bark	Fever	Decoction	[138]
Catostemma commune Sandwith	Guianas	Seed	Snoring		[138]
Cavanillesia platanifolia (Humb. & Bonpl.) Kunth	South America, Peru	Bark, oil	Underweight	Infusion	[111,122]
Ceiba pentandra (L.) Gaertn.	South America	Immature fruits, roots, leaves barks	Cough, hair shampoo; component of ayahuasca, psychoactive drugs		[15,32]
	Java	Leaves	Intestinal catarrh and urethritis, gonorrhea	Infusion	[15]
	Congo	Bark	Management of sickle cell anemia	Aqueous extracts	[139]
	Philippines	Bark	Vomitive and aphrodiastic	Decoction	[15]
Ceiba ventricosa (Nees & Mart.) Ravenna	Brazil		Skin disease, inflammation		[122]
Chiranthodendron pentadactylon Larreat.	Mexico	Flowers	Gastrointestinal disorder, diarrhea, dysentery, blood pressure	Infusion	[122]
Eriotheca globosa (Aubl.) A.Robyns	South America	Ripe fruits	Cuts, wounds	Application	[122]
Fremontodendron californicum (Torr.) Coult.	North America	Bark	Throat irritation	Infusion	[122]
Huberodendron patinoi Cuatrec.	Colombia	Bark	Leishmania		[140]
Huberodendron swietenioides (Gleason) Ducke	Ecuador	Leaves	Diabetes	Aqueous infusion	[124]
Matisia glandifera Planch. & Triana	Colombia	Bark, leaves	Malaria		[141]
Ochroma pyramidale (Cav. ex Lam.) Urb.	Brazil	Root bark	Emetic		[142]
Pachira aquatica Aubl.	Nicaragua	Bark	Stomach complaint, headache		[15]
Pachira glabra Pasq.	India	Leave	Blood pressure, Anemia		[122]
Pseudobombax ellipticum (Kunth) Dugand	Guatemala	Bark	Cough and catarrh	Decoction	[143]
Pseudobombax grandiflorum (Cav.) A.Robyns	Brazil	Bark	Wound healing	Decoction	[144]
Quararibea cordata (Bonpl.) Vischer	South America		Astringent, tonic, antiseptic, for skin infections		[122]
Quararibea funebris (La Llave) Vischer	South America	Flowers	Hallucinogenic, psychopathic fears		[40]
Scleronema micranthum (Ducke) Ducke	Brazil	Leaf	Toothache		[145]

10. Pharmacological Potential of Bombacoideae

The different species, viz. *Adansonia digitata, Bombax ceiba, B. malabaricum,* and *Ceiba pentandra* of the Bombacoideae family [136,146], were reported for their various pharmacological potentials, which are summarized in the following section (Table 6; Figure 5).

Table 6. Pharmacological studies on some of the plant species of Bombacoideae subfamily.

Plant Species and Part	Part (s) and Solvent	Assay	Results	References
		Antioxidant activity		
Adansonia digitata L.	Methanolic leaf extracts; ethanolic leaf	*In vitro* DPPH, ABTS, FRAP, β-carotene bleaching test, superoxide scavenging assay; CAT and SOD, and GSH assay	The DPPH scavenging activity recorded highest in seed extract (27.69%) and lowest in fruit wall (20.69%) extract. The antioxidant status of the STZ induced diabetic rats are normalized by reducing the elevated levels of reduced glutathione (GSH) superoxide dismutase (SOD), and catalase (CAT)	[50,147–150]
	Methanolic fruit extracts;	DPPH, ABTS, FRAP assay, β-carotene bleaching test, superoxide scavenging assay	Scavenge the DPPH free radicals with the percentage of inhibition of 13.4, 29.23, and 39.21%, respectively	[50,149]
Bombax malabaricum DC.	*n*-hexane and methanol extracts of flower	DPPH radical scavenging, lipid peroxidation, myeloperoxidase activity	Scavenged DPPH radicals over a concentration range of 0.55–0.0343 mg/mL and 0.5–0.0312 mg/mL, respectively	[49,151]
Bombax ceiba L.	Methanolic root; aqueous soluble partitioned of the methanolic root; methanol, dichloromethane, and petroleum ether extracts of roots	The extract exhibited dose-dependent DPPH and reducing power assay. Phenolic constituents donate· OH leading to resonance stabilization	Methanolic root extract could scavenge DPPH radicals, lipid peroxidation, and ascorbyl radicals with an EC_{50} value of 87 µg/mL	[152–155]
	Aqueous and ethanolic bark Methanolic stem bark	DPPH, ABTS, nitric oxide and superoxide radical scavenging activity, lipid peroxidation, metal chelating, and total antioxidant capacity	Inhibited lipid peroxidation in rat liver microsome induced by ascorbyl and peroxynitrite radicals with an IC_{50} value of 141 µg/mL and 115 µg/mL, respectively	[156,157]
	Methanolic extract of the whole plant	DPPH scavenging assay	IC_{50} values of aqueous extracts of *B. ceiba* varied between 85.71 and 102.45 µg/mL and for ethanolic extract, it varied between 85.48 and 103.4 µg/mL	[158]
	Diethyl ether and light petroleum ether extracts of flowers; Aqueous flower extracts; Methanolic flower extracts	DPPH, metal chelating and beta carotene bleaching test, hydroxyl radical, hydrogen peroxide radical, FRAP assay, reducing power assay	Petroleum ether of *B. ceiba* flowers exhibited DPPH and Fe-chelating activities with IC_{50} values of 37.6 and 33.5 µg/mL and diethyl ether extracts exhibited beta-carotene bleaching test with an IC_{50} value of 58.3 µg/mL.	[80,159–162]
	the aqueous methanol extract of the calyx	Methylglyoxal induced oxidative stress in HEK-293 cells	Reduced the level of reactive oxygen species (ROS), NADPH oxidase (NOX), and thereby lowered the mitochondrial dysfunction in methylglyoxal induced protein glycation	[163]

Table 6. Cont.

Plant Species and Part	Part (s) and Solvent	Assay	Results	References
		Antioxidant activity		
Ceiba pentandra (L.) Gaertn	seed extracts	DPPH, FRAP, reducing assay, and hydroxyl radical scavenging assay	Decoction, maceration, and methanol scavenged DPPH radical with IC_{50} values of 87.84, 54.77, and 6.15 µg/mL, respectively.	[164]
	Methanol extracts of stem bark; ethyl acetate fraction of stem bark	hydroxyl radical, against lipid peroxidation; DPPH radical scavenging	Scavenge DPPH, nitric oxide, and hydroxyl radicals with IC_{50} values of 27.4, 24.45, and 51.65 µg/mL	[165,166]
	ethanol leaf extract; aqueous and methanol extracts of stem bark	DPPH, nitric oxide, and hydroxyl radical scavenging	The aqueous and methanol stem bark extracts inhibited superoxide (IC_{50} values of 51.81 and 34.26 µg/mL), hydrogen peroxide (44.84 and 1.78 µg/mL) and protein oxidation induced by H_2O_2 (120.60 and 140.40 µg/mL).	[167,168]
		Antimicrobial activity		
Adansonia digitata L.	Methanolic, ethanolic leaf, and stem bark extracts	agar well diffusion method		[169]
Bombax ceiba L.	Methanolic stem bark	Agar well diffusion method	The order of sensitivity from highest to least was *Staphylococcus aureus* > *Escherechia coli* > *Pseudomonas aeruginosa* > *Bacillus subtilis* > *Salmonella typhi*	[157]
	Methanolic flower extracts	Agar disc diffusion assay and MIC study.	Exhibited antibacterial activity against *Klebsiella pneumonia, E. coli, P. aeruginosa* (Gram-negative), and *S. aureus, B. subtilis* (Gram-positive) bacteria with the MIC value ranging between 3.125 and 12.500 µg/mL	[38,161]
	methanol, dichloromethane, and petroleum ether extracts of roots	Agar disc diffusion assay	The methanol, dichloromethane, and PE extracts exhibited mild to moderate antibacterial activity against different bacterial strains including *Sarcina lutea, Bacillus megaterium, B. subtilis, S. aureus, B. cereus, P. aeruginosa, Salmonella typhi, E. coli, Vibrio mimicus, Shigella boydii,* and *Shigella dysenteriae* with 7–13 mm zone of inhibition	[155]
Bombax malabaricum DC.	*n*-hexane and methanol extracts of flower	Agar disc diffusion method	n-hexane and methanol extracts (at 100 µg/mL) of demonstrated antimicrobial activities	[49]
Ceiba pentandra (L.) Gaertn	Ethyl acetate fraction of leaf and bark; ethanol leaf extract	Agar dilution method	ethyl acetate fraction of leaf and bark of *C. pentandra* showed antimicrobial activity against *E. coli, Salmonella typhi, B. subtilis, Kleibsiella pneumonia,* and *S. aureus*	[167,170]
	aqueous, methanol, ethanol, and acetone extract of seed	Disc diffusion method	dose-dependently inhibits antibacterial activity against *E. coli* and *S. aureus*	[171]

Table 6. *Cont.*

Plant Species and Part	Part (s) and Solvent	Assay	Results	References
		Anticancer activity		
Adansonia digitata L.	seed and pulp extracts	MTT assay	At 10, 100, and 500 μg/mL dose, the inhibition ranges between 22.57 and 29.96% for MCF-7 cell line; 25.85 and 37.81% for Hep-G2 cell line and 20.75 and 27.34% for COLO-205 cell line. Dichloromethane and methanolic extract demonstrated cytotoxic activity against human bBreast development cell lines BT474 with IC_{50} value of 15.3 $\pm$ 0.4 μg/mL	[172]
Bombax ceiba L.	diethyl ether and light petroleum ether extracts of flowers	sulforhodamine B (SRB) assay brine shrimp lethality bioassay	Antiproliferative activity against human renal adenocarcinoma cell (ACHN) with respective IC_{50} values of 53.2 and 45.5 μg/mL The petroleum ether, dichloromethane, and methanol extracts of *B. ceiba* roots exhibited cytotoxic effect with LC_{50} values of 22.58, 37.72, and 70.72 μg/mL, respectively	[80]
Ceiba pentandra (L.) Gaertn	petroleum and acetone stem bark extracts	Dalton's lymphoma ascites (DLA or solid tumor) model	At 15 and 30 mg/kg doses could reduce tumor weight by >50% and tumor volume on the 30th day in Dalton's lymphoma ascites. The petroleum ether, benzene, chloroform, acetone, and ethanolic extract of this plant demonstrated cytotoxicity in a concentration dependent manner after 3 h of incubation with EAC cells with EC_{50} values of 53.30, 70.58, 250.48, 67.30, and 56.11 μg/mL, respectively	[50,148,150,173]
Bombax ceiba L.	dichloromethane, ethanol, and aqueous extracts of thalamus and flower; *n*-hexane fraction of sepal	Alpha-amylase and alpha-glucosidase inhibition assay	The IC_{50} values for alpha amylase inhibition for water extract of thalamus, ethanolic extract of thalamus, ethanolic extract of flower, dichloromethane extract of thalamus, water extract of flower, and dichloromethane extract of flower were 32.95, 33.45, 33.85, 34.95, 35.15, and 35.65 μg/mL, respectively.	[174,175]
	ethanolic root extracts	Alloxan induced diabetic rat	At 400 mg/kg decreased the blood glucose level in diabetic mice	[176]
	Ethanolic leaf extracts	STZ- induced diabetic mice	At 70, 140, and 280 mg/kg doses it decreased fasting blood glucose, glycosylated hemoglobin in diabetic rats	[177]
	Bark extracts	STZ- induced diabetic rats	At 600 mg/kg dose the extract could significantly decrease elevated levels of blood glucose in diabetic rats.	[178]

Table 6. Cont.

Plant Species and Part	Part (s) and Solvent	Assay	Results	References
Anticancer activity				
Ceiba pentandra (L.) Gaertn	Aqueous stem bark extracts; aqueous (AE) and methanol (ME) extracts of bark	Dexamethasone-induced insulin resistant rats; STZ- induced diabetic rats; Alpha-amylase and alpha-glucosidase assay	At 75 or 150 mg/kg doses could decrease the level of glycemia in insulin resistant rats. Aqueous stem bark extracts of inhibited alpha-amylase and glucosidase with IC_{50} values of 6.15 and 76.61 µg/mL, respectively, whereas the methanol extract inhibited alpha-amylase and glucosidase with IC_{50} values of 54.52 and 86.49 µg/mL, respectively	[165,168,179,180]
Anti-inflammatory activity				
Adansonia digitata L.	methanol leaf extracts, aqueous leaf extract	iNOS and NF-*k*B expression in LPS-stimulated RAW264.7 cell	Inhibit NO production with an IC_{50} value of 28.6 µg/mL.	[181]
	fruit pulp extract	inhibition of proinflammatory cytokine IL-8 expression	Leaf extract (70 µg/mL) exhibited better anti-inflammatory activity when compared to pulp extract (247 µg/mL).	[182]
Bombax ceiba L.	Petroleum ether, ethanol, and aqueous extracts	HRBC membrane stabilization method.	At 1000 µg/mL concentration exhibited anti-inflammatory potential by stabilizing the HRBC membrane	[183]
Ceiba pentandra (L.) Gaertn	ethyl acetate extract of aerial part	MTX-induced nephrotoxic rats	At 400 mg/kg dose could inhibit methotrexate (MTX)-initiated apoptotic and inflammatory cascades	[184]
Hepatoprotective activity				
Adansonia digitata L.	aqueous extract of fruit; methanolic extract of the fruit	CCL_4 induced hepatotoxic rats; paracetamol-induced hepatotoxicity in rats	Reduction in serum AST, ALT, ALP, bilirubin levels were observed in carbon tetrachloride (CCL_4) induced hepatotoxic rat. Level of ALT, AST, ALP, total bilirubin, and total protein measurements were normalized in paracetamol-induced hepatotoxic rats.	[185–189]
Bombax ceiba L.	Aqueous flower extracts; Methanolic flower extracts	Histological studies; enzyme assay alkaline phosphates, alanine transaminases, aspartate transaminases, and total bilirubin assay	Decreased elevated levels of glutamic-oxaloacetic transaminase (SGOT), glutamic pyruvic transaminase (SGPT), alkaline phosphatize (ALP), bilirubin, and triglycerides, total protein.	[159,190]
	ethanolic root extracts	Enzyme assay in alloxan induced diabetic mice	At 400 mg/kg decreased the hepatotoxicity in diabetic mice by reducing the elevated levels of SGOT and SGPT	[176]
Ceiba pentandra (L.) Gaertn	the methanol extract of stem bark	Enzyme assay paracetamol-induced liver damage in rats	Reduces levels of SGOT, SGPT, ALP, and total bilirubin content.	[191]

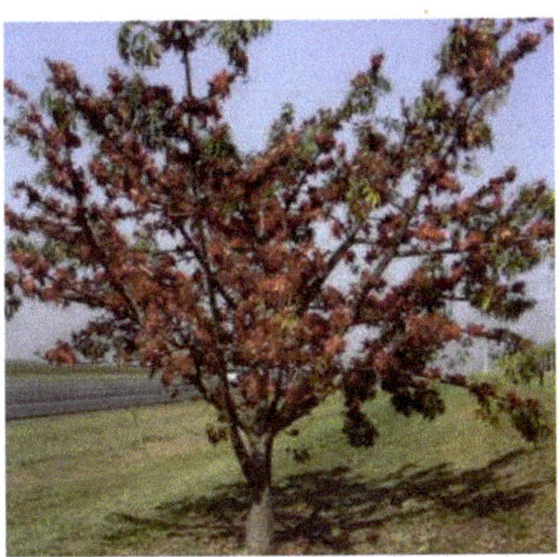

Figure 5. Photo of representative plant species from the Bombacoideae subfamily. Reproduced under Creative Commons Attribution-Non-Commercial 4.0 International License (https://creativecommons.org/licenses/by-nc/4.0/; accessed on 27 February 2021) from Rameshwar et al. [136] (originally Figure 1).

10.1. Antioxidant Properties

Adansonia digitata L.

The methanolic fruit pulp and leaf extracts of *A. digitata* exhibited in vitro antioxidant activities as studied by 2,2-diphenyl-1-picryl-hydrazyl-hydrate (DPPH), 2,2-azinobis—(3-ethylbenzothiazoline-6-sulfonate) (ABTS), ferric reducing antioxidant power (FRAP), β-carotene bleaching test, superoxide-scavenging assays [50,150]. The methanol extracts of leaf, seed, bark, fruit wall, and floral extracts of *A. digitata* were reported for their DPPH scavenging potential [147]. The DPPH scavenging activity was highest in seed extract (27.69%) and lowest in fruit wall (20.69%) extract. The methanolic leaf extract of *A. digitata* could maintain the antioxidant status of the streptozotocin (STZ) induced diabetic rats by normalizing the elevated levels of reduced glutathione (GSH) superoxide dismutase (SOD), and catalase (CAT) [148]. The ethanolic leaf, bark, and fruit extracts of *A. digitata* could scavenge the DPPH free radicals with percentages of inhibition of 13.4, 29.23, and 39.21%, respectively [149].

Bombax ceiba L.

The methanolic root extract of *Bombax ceiba* could scavenge DPPH radicals, lipid peroxidation, and ascorbyl radicals with an EC_{50} value of 87 µg/mL. The extract also inhibited lipid peroxidation in rat-liver microsome induced by ascorbyl and peroxynitrite radicals with IC_{50} values of 141 µg/mL and 115 µg/mL, respectively [192]. In another study, the methanol root extract of *B. ceiba* scavenged DPPH radical with an EC_{50} value of 15.07 µg. The extract also could reduce the Fe^{3+} to Fe^{2+} in a dose-dependent manner with the maximum activity at 500 µg. The study also demonstrated that the administration of 3 g root powder could raise the antioxidant status in the human volunteer. The antioxidant activity properties of the root extract are attributed to their high phenolic and tannin contents [152]. The aqueous soluble partition (AQSF) of the methanolic root extracts of *B. ceiba* scavenged DPPH radical with an IC_{50} value of 3.33 µg/mL [153]. Further, the methanol and petroleum ether root extract of *B. ceiba* was reported to scavenge DPPH radical with IC_{50} values of 144.77 and 214.83 µg/mL [155]. The methanolic stem bark extract of *B. ceiba* exhibited antiradical activity with EC_{50} values of 18.78, 23.62, and 139.4 µg/mL for nitric oxide, DPPH, and reducing power activity assay, respectively [193]. Similarly, Hossain et al. [154] reported the antioxidant activity of methanolic root extract of *B. ceiba* by DPPH scavenging assay (IC_{50} value of 58.6 µg/mL). Gandhare et al. [156] reported that aqueous and ethanolic extracts of the *B. ceiba* bark exhibited DPPH, ABTS, nitric oxide, and superoxide radical scavenging activity along with total antioxidant activity. Besides the extract also inhibited lipid peroxidation and reduced ferric ions. The IC_{50} values of aqueous extracts of *B. ceiba* varied between 85.71 and 102.45 µg/mL, and for ethanolic extract, it varied between 85.48 and 103.4 µg/mL. Komati et al. [163] reported that aqueous methanol extract of *B. ceiba* calyx reduced the level of reactive oxygen species (ROS), NADPH oxidase (NOX), and thereby lowered the mitochondrial dysfunction in methylglyoxal induced protein glycation. Further, in HEK-293 cells, Mn and Cu/Zn-superoxide dismutase and glutathione reductase antioxidant enzymes levels were improved. The whole plant methanolic extract of *B. ceiba* scavenged DPPH radical with an IC_{50} value of 68 µg/mL [158]. The petroleum ether (PE) of *B. ceiba* flowers exhibited DPPH and Fe-chelating activities with IC_{50} values of 37.6 and 33.5 µg/mL and diethyl ether extracts (DE) exhibited beta-carotene bleaching test with an IC_{50} value of 58.3 µg/mL. The antioxidant properties of *B. ceiba* flower extracts are attributed to the presence of beta-sitosterol and some fatty acids [80]. Similarly, another study reported that aqueous flower extracts of *B. ceiba* could scavenge DPPH radicals with an IC_{50} value of 50.21 µg/mL [159]. The aqueous flower extracts of *B. ceiba* exhibited antioxidant activities against DPPH, hydroxyl, hydrogen peroxide, and ferric ion reducing antioxidant power (FRAP) activity with IC_{50} values of 1.70 mg/mL, 4.20 mg/mL, 3.51 mg/mL, and 2.15 mg/mL, respectively [160]. The hexane, benzene, chloroform, ethyl acetate, acetone, methanol, and ethanol extracts prepared from methanolic flower extract of *B. ceiba* exhibited DPPH scavenging activity [161]. The hexane, chloroform, and methanolic extracts prepared from dried powder extracts of *B. ceiba* flower exhibited antioxidant activity in terms of FRAP, DPPH, and reducing power assay [162].

Bombax malabaricum DC.

The n-hexane and methanol flower extracts of *B. malabaricum* scavenged DPPH radicals over a concentration range of 0.55–0.0343 mg/mL and 0.5–0.0312 mg/mL, respectively. The maximum DPPH scavenging was observed in the range of 0.55–0.5 mg/mL for both extracts [49]. The antioxidant potential of flower extract was attributed to the presence of bioactive constituent, viz. apigenin, cosmetin, xanthomicrol, saponarin, vicenin 2, isovitexin, and linarin. Similarly, in another study, the aqueous, acetone, and ethanol flower extracts of *B. malabaricum* flowers showed DPPH radical-scavenging properties along with Oxygen radical absorbance capacity (ORAC), reducing power, and liposome peroxidation inhibition activities [151].

Ceiba pentandra L.

The different stem bark extracts of *C. pentandra* such as decoction, maceration, and methanol scavenged DPPH radical with IC_{50} values of 87.84, 54.77, and 6.15 µg/mL, respectively. The extracts also restrained the H_2O_2-induced hemolysis and lipid peroxidation [165]. The Soxhlet seed oil extracts at 100 mg/mL concentration of *C. pentandra* exhibited DPPH, and OH radical scavenging along with FRAP, reducing power activities by 47.65%, 39.69%, and 309 FRAP units, and 20.52 µg of ascorbic acid equivalent, respectively [193]. The in vitro antioxidant evaluation of *C. pentandra* ethanol leaf extract demonstrated that the extract could scavenge DPPH, nitric oxide, and hydroxyl radicals with IC_{50} values of 27.4, 24.45, and 51.65 µg/mL, respectively. The Gas chromatography-mass spectrometry (GC-MS) study revealed the presence of 9 compounds, amongst which, hexadecanoic acid was found to be the most prominent compound [167]. In another study, Fitria et al. [166] demonstrated that a compound vavain or 5, 3'-dihydroxy-7, 4', 5'-trimethoxyisoflavone isolated from the ethyl acetate fraction of stem bark of *C. pentandra* could scavenge DPPH radical with IC_{50} value of 81.66 µg/mL. However, the ethyl acetate extract of the aerial part of *C. pentandra* scavenged the DPPH radicals with an IC_{50} value of 0.0716 mg/mL [184]. The aqueous and methanol stem bark extracts of *C. pentandra* inhibited superoxide ($O_2^{\bullet-}$) (IC_{50} values of 51.81 and 34.26 µg/mL), hydrogen peroxide (44.84 and 1.78 µg/mL), and protein oxidation induced by H_2O_2 (120.60 and 140.40 µg/mL) [168].

10.2. Anti-Inflammatory Activity

Adansonia digitata L.

The methanol leaf extracts of *A. digitata* reduced iNOS and NF-*k*B expression in LPS-stimulated RAW264.7, thereby showing its anti-inflammatory potential [181]. The extract could inhibit NO production with an IC_{50} value of 28.6 µg/mL. Similarly, the Dimethyl sulfoxide (DMSO) fruit pulp and aqueous leaf extract of *A. digitata* inhibited expressions of proinflammatory cytokine IL-8 [182]. The leaf extract (70 µg/mL) exhibited better anti-inflammatory activity compared to pulp extract (247 µg/mL).

Bombax ceiba L.

The petroleum ether, ethanol, and aqueous bark extracts of *B. ceiba* at 1000 µg/mL concentration exhibited anti-inflammatory potential by stabilizing the Human red blood cell (HRBC) membrane. Amongst the different solvent extracts, better anti-inflammatory activity is shown by ethanol extract followed by aqueous and petroleum ether extract [183].

Ceiba pentandra (L.) Gaertn

The ethyl acetate extracts of aerial parts of *C. pentandra* upon oral administration at 400 mg/kg dose could inhibit methotrexate (MTX)-initiated apoptotic and inflammatory cascades. The extract could improve the architecture of histopathological changes observed in the renal tissue of MTX-induced nephrotoxic rats [184].

10.3. Antimicrobial Activity

Adansonia digitata L.

The methanolic and ethanolic leaf and stem bark extracts of *A. digitata* inhibited the growth of *S. aureus* and *E. coli* at different concentrations, viz, 100, 200, 500, and 1000 mg/mL with a minimum bactericidal concentration (MIC) at 100 mg/mL [169].

Bombax ceiba L.

The methanolic stem bark extract of *B. ceiba* could inhibit the growth of both Gram-negative (*Escherichia coli, Pseudomonas aeruginosa,* and *Salmonella typhi)* and Gram-positive bacteria (*Bacillus subtilis, Staphylococcus aureus)* dose-dependently. The order of sensitivity from highest to lowest was *S. aureus* > *E. coli* > *P. aeruginosa* > *B. subtilis* > *S. typhi* [157]. The methanolic flower extract of *B. ceiba* exhibited antibacterial activity against *Klebsiella pneumonia, E. coli, P. aeruginosa* (Gram-negative), and *S. aureus, B. subtilis* (Gram-positive) bacteria with the MIC value ranging between 3.125 and 12.500 µg/mL [161]. The methanol, dichloromethane, and PE extracts of *B. ceiba* roots exhibited mild to moderate antibacterial activity against different bacterial strains including *Sarcina lutea, Bacillus megaterium, B. subtilis, S. aureus, B. cereus, P. aeruginosa, Salmonella typhi, E. coli, Vibrio mimicus, Shigella boydii,* and *S. dysenteriae* with a 7–13 mm zone of inhibition [155].

Bombax malabaricum DC.

The n-hexane and methanol extracts (at 100 µg/mL) of *B. malabaricum* demonstrated antimicrobial activities against Gram-positive (*E. coli, Neisseria gonorrhoeae, P. aeruginosa),* Gram-negative (*S. aureus, B. subtilis, Streptococcus faecalis)* bacteria and fungi (*Aspergillus niger, A. flavus Candida albicans).* Of the two extracts, the methanol extract showed better activity against all the studied bacterial strains and *C. albicans.* Further, only the methanol extract exhibited moderate activities against *A. niger* and *A. flavus* [49].

Ceiba pentandra (L.) Gaertn

The ethyl acetate fraction of leaf and bark of *C. pentandra* showed antimicrobial activity against *E. coli, Salmonella typhi, B. subtilis, Kleibsiella pneumonia,* and *S. aureus* [170]. Similarly, aqueous, methanol, ethanol, and acetone seed extracts of *C. pentandra* exhibited antimicrobial activity against *E. coli, S. aureus, K. pneumonia, Enterobacter aerogenes, P. aeruginosa, Salmonella typhimurium, S.typhi, Staphylococcus epidermidis,* and *Proteus vulgaris* [171]. Another study revealed that ethanol leaf extract of *C. pentandra* dose-dependently inhibits antibacterial activity against *E. coli* and *S. aureus* [167].

10.4. Anticancer and Cytotoxicity Activity

Adansonia digitata L.

The seed and pulp extracts of *A. digitata* (at 10, 100, and 500 µg/mL) exhibited anti-cancer activity against MCF-7 (breast cancer cell), Hep-G2 (liver cancer cell), and COLO-205 (colon cancer cell) in a dose dependent manner [172]. The results of the MTT study revealed that the inhibition ranges between 22.57 and 29.96% for MCF-7 cell line; 25.85 and 37.81% for Hep-G2 cell line, and 20.75 and 27.34% for COLO-205 cell line. The dichloromethane and methanolic leaf extracts of *A. digitata* demonstrated cytotoxic activity against human breast development cell lines BT474 evaluated by MTT assay. The methanol leaves of the plant exhibited moderate cytotoxic activity (56%) against the BT474 cell line with IC_{50} values of 15.3 ± 0.4 µg/mL [185].

Bombax ceiba L.

The diethyl ether and light petroleum ether extracts of *B. ceiba* flowers exhibited antiproliferative activity against human renal adenocarcinoma cell (ACHN) with respective IC_{50} values of 53.2 and 45.5 µg/mL. The antiproliferative properties were attributed to the presence of beta-sitosterol and some fatty acids in *B. ceiba* flowers [80]. The brine shrimp lethality bioassay revealed that the petroleum ether, dichloromethane, and methanol extracts of *B. ceiba* roots exhibited cytotoxic effect with LC_{50} values of 22.58, 37.72, and 70.72 µg/mL, respectively [155].

Ceiba pentandra (L.) Gaertn

The petroleum and acetone stem bark extracts of *C. pentandra* at 15 and 30 mg/kg doses could reduce tumor weight by >50% and tumor volume on the 30th day in Dalton's lymphoma ascites (DLA) model [173]. Similarly, both these extracts of *C. pentandra* exhibited cytotoxic effects against Ehrlich ascites carcinoma (EAC) cells as evaluated by trypan blue assay [173]. At 15 mg/kg doses, both the extracts showed improvement in mean survival time and decline in tumor induced increase in body weight. Further, the petroleum ether, benzene, chloroform, acetone, and ethanolic extract of this plant demonstrated cytotoxicity in a concentration dependent manner after 3 h of incubation with EAC cells with EC_{50} values of 53.30, 70.58, 250.48, 67.30, and 56.11 µg/mL, respectively.

10.5. Hepatoprotective Activity

Adansonia digitata L.

The aqueous fruit pulp extract of *A. digitata* showed hepatoprotective potential in carbon-tetrachloride (CCL_4) -induced hepatotoxic rat models as significant reductions in serum aspartate aminotransferase (AST), alanine aminotransferase (ALT), alkaline phosphatize (ALP), and bilirubin levels were observed in extract-treated hepatotoxic rats [186,187]. The liver protection potential could be attributed to the presence of triterpenoids, β-sitosterol, β-amyrin palmitate, and α-amyrin or without, and ursolic acid in the fruit pulp [188]. The methanolic fruit pulp extract of *A. digitata* exhibited hepatoprotective potential in paracetamol-induced hepatotoxic rat models. The disturbances in the liver function such as ALT, AST, ALP, total bilirubin, and total protein measurements of the hepatotoxic rats were normalized due to the administration of paracetamol [189].

Bombax ceiba L.

The hepatoprotective property of aqueous [159] and methanolic [190] flower extracts of *B. ceiba* was studied in CCl_4-induced hepatotoxic rats. Treatment with extracts decreased elevated levels of glutamate oxaloacetate transaminase (SGOT), glutamic pyruvic transaminase (SGPT), alkaline phosphatize (ALP), bilirubin, triglycerides, and total protein. Treatment with the extract further attenuated the damage caused to the liver as seen by histological studies. The young roots of *B. ceiba* exhibited hepatoprotective activities in alloxan induced diabetic mice. Administration of ethanolic root extracts at 400 mg/kg decreased the hepatotoxicity in diabetic mice by reducing the elevated levels of SGOT and SGPT [176].

Ceiba pentandra (L.) Gaertn

The ethyl acetate fraction of methanolic stem bark extract of *C. pentandra* exhibited a hepatoprotective effect against paracetamol-induced hepatotoxic rats by reducing the serum enzyme levels of SGOT, SGPT, ALP, and total bilirubin content [191].

10.6. Antidiabetic Activity

Adansonia digitata L.

The methanolic fruit pulp and leaf extracts of *A. digitata* exhibited in vitro antidiabetic activities by inhibiting the digestive enzyme α-glucosidase dose-dependently [50]. The IC_{50} values of the fruit extracts ranged between 1.71 ± 0.23 and 2.39 ± 0.22 µg/mL while the leaf extract had an IC_{50} value of 1.71 ± 0.23 µg/mL. Similarly, the methanolic leaf extract of this plant inhibited α-amylase, α-glucosidase, and aldolase reductase [150]. The antidiabetic potency of the extracts may be attributed to the presence of catechin, epicatechin, rutin, quercitrin, quercetin, kaempferol, luteolin (flavonoids), gallic, chlorogenic, caffeic, and ellagic acids (phenolic acids). The methanolic leaf extract of *A. digitata* reduced the elevated blood glucose, glycosylated hemoglobin levels in streptozotocin (STZ) induced diabetic rats [148].

Bombax ceiba L.

The dichloromethane, ethanol, and aqueous thalamus and flower extracts of *B. ceiba* were reported for their antidiabetic properties in terms of their capacity to inhibit alpha-amylase and alpha-glucosidase enzymes under in vitro condition. The corresponding IC_{50} values for alpha-amylase inhibition activities for thalamus were 36.22 µg/mL (dichloromethane extract), 35.32 µg/mL (ethanolic extract), and 31.31 µg/mL (aqueous extract) and for flowers, 38.13 µg/mL (dichloromethane extract), 35.23 µg/mL (ethanolic extract), and 33.00 µg/mL (aqueous extract) [174]. The *n*-hexane fraction of sepals [175] and ethanolic leaf extracts [177] of *B. ceiba* exhibited antidiabetic activities in STZ-induced diabetic rats. The *n*-hexane fraction at 0.1 gm/kg bw, b.d. dose reduced the fasting blood sugar level and restored the levels of serum insulin, Hb, and glycated hemoglobin in diabetic rats. Histological studies of also showed marked improvement in diminution in the area of the islets of Langerhans of pancreases in diabetic rats treated with the plant extracts [175]. Similarly, the leaf extract of *B. ceiba* (at 70, 140, and 280 mg/kg doses) decreased the fasting blood glucose, glycosylated hemoglobin, and increased the oral glucose tolerance in the STZ-induced diabetic rats. The antidiabetic property may be attributed to the antioxidant activity and protecting pancreatic β-cells of the extract [177]. The young roots of *B. ceiba* exhibited antidiabetic activities in alloxan-induced diabetic mice. Administration of ethanolic root extracts at 400 mg/kg decreased blood glucose levels in diabetic mice as compared to untreated diabetic mice at different time points (0–24 h). [176]. However, at 600 mg/kg dose the extract could significantly decrease elevated levels of blood glucose in diabetic rats [178].

Ceiba pentandra (L.) Gaertn

The aqueous stem bark extracts of *C. pentandra* exhibited antihyperglycemic, insulin-sensitizing potential, and cardioprotective effects in dexamethasone-induced insulin-resistant rats. Extracts of both 75 or 150 mg/kg doses could decrease the level of glycemia [179]. The decoction extracts of stem bark of *C. pentandra* decreased glucose level by increasing glucose uptake in the liver and skeletal muscle cells by 56.57% and 94.19%, respectively. The extract also reduced the glucose release in liver cells by 33.94% in a hypoglycemic milieu [165]. The ethanolic bark extract of *C. pentandra* at 200 mg/kg dose exhibited antihyperglycemic activity in STZ-induced diabetic rats by decreasing the levels of blood glucose, total cholesterol, and triglycerides, preventing degeneration of liver and pancreas, and increasing serum insulin and liver glycogen content [180]. The aqueous stem bark extracts of *C. pentandra* inhibited alpha-amylase and glucosidase with IC_{50} values of 6.15 and 76.61 µg/mL, respectively, whereas the methanol extract inhibited alpha-amylase and glucosidase with IC_{50} values of 54.52 and 86.49 µg/mL, respectively [168].

10.7. Miscellaneous Activities

The petroleum ether and methanol extract from *B. ceiba* stem bark displayed increased osteogenic activity as demonstrated by Chauhan et al. [37] in UMR-106 cells and surgical ovariectomy models in female Wistar albino rats. It has been reported that the administration of the extracts for 28 days ameliorated the consequences of ovariectomy-induced bone porosity, restoring the normal architecture of bone in experimented rats. The in vitro osteogenic activity of the extracts could be attributed to the presence of lupeol, gallic acid, and β-sitosterol in *B. ceiba*.

Komati et al. [163] reported the antiglycation properties of aqueous methanolic calyx extract of *B. ceiba* in methylglyoxal-induced protein glycation and oxidative stress in HEK-293 cells. The extract could inhibit advanced glycation end products (AGEs) formation and restrained Receptor for advanced glycation end products (RAGE) up-regulation in HEK-293 cells.

The aqueous and crude ethanol fruit extracts of *B. ceiba* exhibited diuretic effects in rats. Both aqueous and ethanol extracts could increase the urine output in the rats. The aqueous extract increased the urinary Na+ and K+ levels demonstrating the diuretic effect of the extracts [194]. The ethanolic leaf, bark, and fruit extracts of *A. digitata* exhibited antipyretic activity in albino rats at 400 and 800 mg/kg doses [149].

11. Mechanism of Action of Extracts and Bioactive Compounds of the Plants' Species with Pharmacological Properties

The different plant species of Bombacoideae are well known for their medicinal properties and can act as a useful bio-resource for medicines, nutraceuticals, pharmaceuticals, and chemical analogs for synthetic drugs. Bombacoideae plant species contain several bioactive phytocompounds such as alkaloids, anthocyanins, coumarins, flavonoids, lignans and neolignans, sesquiterpenes, sesquiterpene lactones, sterols, tannins, triterpenes, et cetera, which may be responsible for their antimicrobial properties. The antimicrobial action of phytocompounds might be due to their capacity to disintegrate cytoplasmic membrane, destabilize proton motive force, electron flow, active transport, and coagulation of the cell content in microbes [195]. Silva and Fernandes [196] also reviewed the antimicrobial properties of plants and concluded that different chemical classes of phytochemicals including alkaloids, flavonoids, terpenoids, phenols, tannins, et cetera may be responsible for their antimicrobial potential.

Several phytochemicals of the different classes of compounds such as alkaloids, flavonoids, saponins, terpenoids, vitamins, glycosides, phenols, et cetera play significant roles in inhibiting or arresting cancer cell progression by different mechanisms such as (a) by inhibiting cancer cell-activating signaling pathways such as Cdc2, CDK2, and CDK4 kinases, topoisomerase enzyme, cyclooxygenase, and COX-2, Bcl-2, cytokines, PI3K, Akt, MAPK/ERK, MMP, and TNK; (b) activating mechanisms of DNA repairing, viz. p21, p27, p51, and p53 genes, and Bax, Bid, and Bak proteins; or (c) by stimulating the formation of protective enzymes, viz. Caspase-3, 7, 8, 9, 10, and 12 [197].

Plants enriched with phenolic acids, flavonoids, coumarins, lignans, terpenoids, et cetera can exert antioxidant action by scavenging radicals and chelating metal ions by acting as reducing agents, hydrogen donors, singlet oxygen quenchers, metal chelators, or reductants of ferryl hemoglobin [198]. Therefore, the antioxidant potential of different species of Bombacoideae may be due to the presence of several classes of phytoconstituents including vicenin 2, linarin, saponarin, cosmetin, isovitexin, xanthomicrol, vavain, apigenin, beta-sitosterol, et cetera [151,184].

The different bioactive phytocomponents could exhibit anti-inflammatory activities by down regulating of signaling pathways like NF-κB pathway. This is done by different mechanisms such as (a) inhibiting common mediators of inflammation like NO, iNOS, and pro-inflammatory cytokines like TNF-α, IL-1β, IL-6, and IL-12p40; (b) inhibition of chemokines such as RANTES and MCP-1; (c) downregulating mediators of inflammation such as cycloxygenase-2 (COX-2), prostaglandins, and leukotrienes; (d) reducing the production of ROS and lipid peroxidation; and (e) upregulating enzymatic (superoxide dismutase, catalase, etc.) and non-enzymatic (glutathione, etc.) defense systems [199]. The different species of Bombacoideae such as *A. digitata* and *C. pentandra* could inhibit inhibition against proinflammatory cytokine IL-8 expression or by reducing iNOS and NF-*k*B expression [181,182], and the activity is attributed to the presence of different phytoconstituents, viz. quercitrin, cinchonains 1a and 1b, cis-clovamide, trans-clovamide, and glochidioboside [184]. The phytoconstituents of different plants could show antidiabetic activities by inhibiting carbohydrate metabolizing enzymes like amylase and glucosidase enzymes, or by stimulating insulin release or by increasing glucose uptake by cells or by decreasing insulin resistance [200]. Several studies have rightly pointed out that different Bombacoideae plants could exhibit antidiabetic activities by inhibiting α-amylase and α-glucosidase enzymes [148].

12. Conclusions

Plants are considered important natural resources as food supplements and in traditional and modern medicine in different regions of the world. Bioactive phytochemicals are valued candidates for the discovery of new drugs. Detailed reporting of plants with food value, and therapeutic and economic importance, of subfamily Bombacoideae, was undertaken in this review. Isolated phytochemicals of diversified classes of secondary metabolites are reported to possess numerous therapeutic properties against different ailments. The bioactive phytochemicals from plants of this subfamily will play important roles in the development of new drug leads with less toxicity and side effects.

Author Contributions: G.D., H.-S.S., S.S.N., A.D.T., H.U., R.T., S.K.D. and J.K.P., writing—original draft preparation, investigation, resources, data curation; G.D., A.D.T., S.K.D., R.T. and J.K.P., writing—review and editing, methodology, formal analysis; J.K.P., conceptualization, supervision, project administration, funding acquisition, visualization. All authors have read and agreed to the published version of the manuscript.

Funding: This work is supported by the National Research Foundation of Korea (NRF) grant funded by the Korea government (MSIT) (No. 2020R1G1A1004667), Republic of Korea for support.

Institutional Review Board Statement: Not applicable.

Informed Consent Statement: Not applicable.

Data Availability Statement: All data presented in the manuscript are available in the form of tables and figures in the manuscript.

Acknowledgments: All authors are grateful to their respective institutions for support. J.K.P. acknowledges the National Research Foundation of Korea (NRF) grant funded by the Korea government (MSIT) (No. 2020R1G1A1004667), Republic of Korea for support.

Conflicts of Interest: The authors declare no conflict of interest.

References

1. Jain, V.; Verma, S.; Katewa, S. Myths, traditions and fate of multipurpose *Bombax ceiba* L.—An appraisal. *Indian J. Tradit. Knowl.* **2009**, *8*, 638–644.
2. Refaat, J.; Desoky, S.Y.; Ramadan, M.A.; Kamel, M.S. Bombacaceae: A phytochemical review. *Pharm. Biol.* **2013**, *51*, 100–130. [CrossRef] [PubMed]
3. Fuchs, H.P. Pollen morphology of the family Bombacaceae. *Rev. Palaeobot. Palynol.* **1967**, *3*, 119–132. [CrossRef]
4. Carvalho-Sobrinho, J.G.; Alverson, W.S.; Alcantara, S.; Queiroz, L.P.; Mota, A.C.; Baum, D.A. Revisiting the phylogeny of Bombacoideae (Malvaceae): Novel relationships, morphologically cohesive clades, and a new tribal classification based on multilocus phylogenetic analyses. *Mol. Phylogenetics Evol.* **2016**, *101*, 56–74. [CrossRef] [PubMed]
5. Lima, J.B.; Bovini, M.G.; da Souza Conceição, A. Bombacoideae, Byttnerioideae, Grewioideae and Helicterioideae (Malvaceae s.l.) in the Raso da Catarina Ecoregion, Bahia, Brazil. *Biota Neotrop.* **2019**, *19*, 1–21. [CrossRef]
6. Singh, G. *Plant Systematics: An Integrated Approach*, 4th ed.; CRC Press: Boca Raton, FL, USA, 2019.
7. Bayer, C.; Fay, M.; Bruijn, A.; Savolainen, V.; Morton, C.; Kubitzki, K.; Alverson, W.; Chase, M. Support for an expanded family concept of Malvaceae within a recircumscribed order Malvales: A combined analysis of plastid atpB and rbcL DNA sequences. *Bot. J. Linn. Soc.* **1999**, *129*, 267–303. [CrossRef]

8. Kubitzki, K.; Bayer, C. *The Families and Genera of Vascular Plants*; Springer: Berlin/Heidelberg, Germany, 2003; Volume 5.
9. Christenhusz, M.J.M.; Fay, M.F.; Chase, M.W. *Plants of the World: An Illustrated Encyclopedia of Vascular Plants*; University of Chicago Press: Chicago, IL, USA, 2017.
10. Duarte, M. Phylogenetic Analyses of *Eriotheca* and Related Genera (Bombacoideae, Malvaceae). *Syst. Bot.* **2011**, *36*, 690–701. [CrossRef]
11. Zizka, A.; Carvalho-Sobrinho, J.G.; Pennington, R.T.; Queiroz, L.P.; Alcantara, S.; Baum, D.A.; Bacon, C.D.; Antonelli, A. Transitions between biomes are common and directional in Bombacoideae (Malvaceae). *J. Biogeogr.* **2020**, *47*, 1–12. [CrossRef]
12. Kamatou, G.P.P.; Vermaak, I.; Viljoen, A.M. An updated review of *Adansonia digitata*: A commercially important African tree. *S. Afr. J. Bot.* **2011**, *77*, 908–919. [CrossRef]
13. Barwick, M. *Tropical & Subtropical Trees: A Worldwide Encyclopaedic Guide*; Thames & Hudson: London, UK, 2004.
14. Wickens, G.E. *Edible Nuts*; FAO: Rome, Italy, 1995; Volume 5.
15. Lim, T.K. *Edible Medicinal and Non-Medicinal Plants*; Springer: Berlin/Heidelberg, Germany, 2012; Volume 1.
16. Roth, I.; Lindorf, H. *South American Medicinal Plants: Botany, Remedial Properties and General Use*; Springer: Berlin/Heidelberg, Germany, 2013.
17. Middeljans, M. *The Species Composition of the Mangrove Forest along the Abatan River in Lincod, Maribojoc, Bohol, Philippines and the Mangrove Forest Structure and its Regeneration Status between Managed and Unmanaged Nipa palm (Nypa fruticans Wurmb)*; Wageningen University: Wageningen, The Netherlands, 2015. [CrossRef]
18. Infante-Mata, D.; Moreno-Casasola, P.; Madero-Vega, C. *Pachira aquatica*, un indicador del límite del manglar? *Rev. Mex. Biodivers.* **2014**, *85*, 143–160. [CrossRef]
19. Gibbs, P.E.; Alverson, W.S. How many species of *Spirotheca* (Malvaceae sl, Bombacoideae)? *Brittonia* **2006**, *58*, 245–258. [CrossRef]
20. Krutzsch, W. Paleogeography and historical phytogeography (paleochorology) in the Neophyticum. *Plant Syst. Evol.* **1989**, *162*, 5–61. [CrossRef]
21. Croizat, L. *Manual of Phytogeography: An Account of Plant-Dispersal Throughout the World*; Springer Science + Business: Dordrecht, The Netherlands, 1952.
22. Wickens, G.E.; Lowe, P. *The Baobabs: Pachycauls of Africa, Madagascar and Australia*; Springer: Dordrecht, The Netherlands, 2008.
23. Ehrendorfer, F. *Woody Plants—Evolution and Distribution Since the Tertiary: Proceedings of a Symposium Organized by Deutsche Akademie Der Naturforscher LEOPOLDINA in Halle/Saale, German Democratic Republic, 9–11 October 1986*; Springer: Vienna, Austria, 1989.
24. Odetokun, S. The nutritive value of Baobab fruit (*Adansonia digitata*). *Riv. Ital. Sostanze Grasse* **1996**, *73*, 371–373.
25. Chevalier, M.A. Les Baobabs (Adansonia) de l'Afrique continentale. *Bull. Société Bot. Fr.* **1906**, *53*, 480–496. [CrossRef]
26. Hostettmann, K.; Schaller, F. Antimicrobial Diterpenes. U.S. Patent US5929124A, 27 July 1999.
27. Inngjerdingen, K.; Nergård, C.; Diallo, D.; Mounkoro, P.; Paulsen, B. An Ethnopharmacological survey of plants used for wound healing in Dogonland, Mali, West Africa. *J. Ethnopharmacol.* **2004**, *92*, 233–244. [CrossRef] [PubMed]
28. Cardoso, D.; Carvalho-Sobrinho, J.; Zartman, C.; Komura, D.; Queiroz, L. Unexplored Amazonian diversity: Rare and phylogenetically enigmatic tree species are newly collected. *Neodiversity* **2015**, *8*, 55–73. [CrossRef]
29. Mitré, M. Gyranthera darienensis. *IUCN Red List Threat. Species 1998* **1998**, *e.T30572A9554110*. [CrossRef]
30. Díaz-Pérez, C.N.; Puerto-Hurtado, M.A.; Fernández-Alonso, J.L. Evaluación Del Hábitat, Las Poblaciones Y El Estatus De Conservación Del Barrigón (*Cavanillesia chicamochae*, Malvaceae–Bombacoideae). *Caldasia* **2011**, *33*, 105–119.
31. Bell, K.L.; Rangan, H.; Kull, C.A.; Murphy, D.J. The history of introduction of the African baobab (*Adansonia digitata*, Malvaceae: Bombacoideae) in the Indian subcontinent. *R. Soc. Open Sci.* **2015**, *2*, 150370. [CrossRef] [PubMed]
32. Mansfeld, R.; Hanelt, P. *Mansfeld's Encyclopedia of Agricultural and Horticultural Crops: (Except Ornamentals)*; Springer: Berlin/Heidelberg, Germany, 2001.
33. Zidar, C.; Elisens, W. Sacred Giants: Depiction of Bombacoideae on Maya Ceramics in Mexico, Guatemala, and Belize. *Econ. Bot.* **2009**, *63*, 119–129. [CrossRef]
34. Byng, J.W. *The Flowering Plants Handbook: A Practical Guide to Families and Genera of the World*; Plant Gateway Ltd.: Hertford, UK, 2014.
35. Marinho, R.C.; Mendes-Rodrigues, C.; Balao, F.; Ortiz, P.L.; Yamagishi-Costa, J.; Bonetti, A.M.; Oliveira, P.E. Do chromosome numbers reflect phylogeny? New counts for Bombacoideae and a review of Malvaceae s.l. *Am. J. Bot.* **2014**, *101*, 1456–1465. [CrossRef]
36. Fay, M. *Index to Plant Chromosome Numbers 2004–2006. Regnum Vegetabile 152*; Oxford University Press: Oxford, UK, 2011.
37. Chauhan, S.; Sharma, A.; Upadhyay, N.K.; Singh, G.; Lal, U.R.; Goyal, R. In-vitro osteoblast proliferation and in-vivo anti-osteoporotic activity of *Bombax ceiba* with quantification of Lupeol, gallic acid and β-sitosterol by HPTLC and HPLC. *BMC Complement. Altern. Med.* **2018**, *18*, 233. [CrossRef]
38. Deepshikha, R.; Geetanjali; Ram, S. Phytochemistry and Pharmacology of Genus *Bombax*. *Nat. Prod. J.* **2019**, *9*, 184–196. [CrossRef]
39. Osman, M.A. Chemical and nutrient analysis of baobab (*Adansonia digitata*) fruit and seed protein solubility. *Plant Foods Hum. Nutr.* **2004**, *59*, 29–33. [CrossRef] [PubMed]
40. Zennie, T.M.; Cassady, J.M.; Raffauf, R.F. Funebral, a new pyrrole lactone alkaloid from *Quararibea funebris*. *J. Nat. Prod.* **1986**, *49*, 695–698. [CrossRef] [PubMed]
41. Zennie, T.M.; Cassady, J.M. Funebradiol, a New Pyrrole Lactone Alkaloid from *Quararibea funebris* Flowers. *J. Nat. Prod.* **1990**, *53*, 1611–1614. [CrossRef] [PubMed]

42. Raffauf, R.F.; Zennie, T.M.; Onan, K.D.; Le Quesne, P.W. Funebrine, a structurally novel pyrrole alkaloid, and other. gamma.-hydroxyisoleucine-related metabolites of *Quararibea funebris* (Llave) Vischer (Bombacaceae). *J. Org. Chem.* **1984**, *49*, 2714–2718. [CrossRef]

43. Scogin, R. Reproductive phytochemistry of Bombacaceae: Floral anthocyanins and nectar constituents. *Aliso: A J. Syst. Evol. Bot.* **1986**, *11*, 377–385. [CrossRef]

44. Niranjan, G.; Gupta, P. Anthocyanins from the flowers of *Bombax malabaricum*. *Planta Med.* **1973**, *24*, 196–199. [CrossRef]

45. Rizk, A.; Al-Nowaihi, A. *The Phytochemistry of the Horticultural Plants of Qatar*; Alden Press: Oxford, UK, 1989.

46. Paula, V.F.; Barbosa, L.C.; Demuner, A.J.; Howarth, O.W.; Veloso, D.P. Chemical constituents of *Ochroma lagopus* Swartz. *Quim. Nova* **1996**, *19*, 225–229.

47. Joshi, K.R.; Devkota, H.P.; Yahara, S. Chemical analysis of flowers of *Bombax ceiba* from Nepal. *Nat. Prod. Commun.* **2013**, *8*, 1934578X1300800508. [CrossRef]

48. Cheng, L.-Y.; Liao, H.-R.; Chen, L.-C.; Wang, S.-W.; Kuo, Y.-H.; Chung, M.-I.; Chen, J.-J. Naphthofuranone derivatives and other constituents from *Pachira aquatica* with inhibitory activity on superoxide anion generation by neutrophils. *Fitoterapia* **2017**, *117*, 16–21. [CrossRef] [PubMed]

49. El-Hagrassi, A.M.; Ali, M.M.; Osman, A.F.; Shaaban, M. Phytochemical investigation and biological studies of *Bombax malabaricum* flowers. *Nat. Prod. Res.* **2011**, *25*, 141–151. [CrossRef] [PubMed]

50. Braca, A.; Sinisgalli, C.; De Leo, M.; Muscatello, B.; Cioni, P.L.; Milella, L.; Ostuni, A.; Giani, S.; Sanogo, R. Phytochemical Profile, Antioxidant and Antidiabetic Activities of *Adansonia digitata* L. (Baobab) from Mali, as a Source of Health-Promoting Compounds. *Molecules* **2018**, *23*, 3104. [CrossRef] [PubMed]

51. El-Alfy, T.; El-Sawi, S.; Sleem, A.; Moawad, D. Investigation of flavonoidal content and biological activities of *Chorisia insignis* Hbk. leaves. *Aust. J. Basic Appl. Sci.* **2010**, *4*, 1334–1348.

52. Noreen, Y.; El-Seedi, H.; Perera, P.; Bohlin, L. Two new isoflavones from *Ceiba pentandra* and their effect on cyclooxygenase-catalyzed prostaglandin biosynthesis. *J. Nat. Prod.* **1998**, *61*, 8–12. [CrossRef] [PubMed]

53. Vázquez, E.; Martínez, E.M.; Cogordán, J.A.; Delgado, G. Triterpenes, Phenols, and Other Constituents from the leaves of *Ochroma pyramidale* (Balsa Wood, Bombacaceae): Preferred Conformations of 8-C-β-D-Glucopyranosyl-apigenin (vitexin). *Rev. Soc. Química México* **2002**, *46*, 254–258.

54. Shahat, A.A. Procyanidins from *Adansonia digitata*. *Pharm. Biol.* **2006**, *44*, 445–450. [CrossRef]

55. Qi, Y.; Guo, S.; Xia, Z.; Xie, D. Chemical constituents of Gossampinus malabarica (L.) Merr.(II). *Zhongguo Zhong Yao Za Zhi Zhongguo Zhongyao Zazhi China J. Chin. Mater. Med.* **1996**, *21*, 234.

56. Ueda, H.; Kaneda, N.; Kawanishi, K.; Alves, S.M.; Moriyasu, M. A new isoflavone glycoside from *Ceiba pentandra* (L.) Gaertner. *Chem. Pharm. Bull.* **2002**, *50*, 403–404. [CrossRef] [PubMed]

57. Sichaem, J.; Siripong, P.; Khumkratok, S.; Tip-Pyang, S. Chemical constituents from the roots of *Bombax anceps*. *J. Chil. Chem. Soc.* **2010**, *55*, 325–327. [CrossRef]

58. Paula, V.F.; Cruz, M.P.; Barbosa, L.C.d.A. Chemical constituents of *Bombacopsis glabra* (Bombacaceae). *Química Nova* **2006**, *29*, 213–215. [CrossRef]

59. Chauhan, J.; Kumar, S.; Chaturvedi, R. A new flavanonol glycoside from *Adansonia digitata* roots. *Planta Med.* **1984**, *50*, 113. [CrossRef]

60. Gopal, H.; Gupta, R. Chemical constituents of *Salmalia malabarica* Schott and Endl. flowers. *J. Pharm. Sci.* **1972**, *61*, 807–808. [CrossRef] [PubMed]

61. Bravo, J.A.; Lavaud, C.; Bourdy, G.; Giménez, A.; Sauvainc, M. First bioguided phytochemical approach to *Cavanillesia* Aff. hylogeiton. *Rev. Boliv. Química* **2002**, *19*, 18–24.

62. Ngounou, F.; Meli, A.; Lontsi, D.; Sondengam, B.; Choudhary, M.I.; Malik, S.; Akhtar, F. New isoflavones from *Ceiba pentandra*. *Phytochemistry* **2000**, *54*, 107–110. [CrossRef]

63. Chauhan, J.; Chaturvedi, R.; Kumar, S. A new flavonol glycoside from the stem of Adansonia-Digitata. *Indian J. Chem. B Org. Incl. Med. Chem.* **1982**, *21*, 254–255.

64. Coussio, J. Isolation of rhoifolin from *Chorisia* species (Bombacaceae). *Experientia* **1964**, *20*, 562. [CrossRef] [PubMed]

65. Nasr, E.M.; Assaf, M.H.; Darwish, F.M.; Ramadan, M.A. Phytochemical and biological study of *Chorisia speciosa* A. St. Hil. cultivated in Egypt. *J. Pharmacogn.* **2018**, *7*, 649–656.

66. Faizi, S.; Ali, M. Shamimin: A new flavonol C-glycoside from leaves of *Bombax ceiba*. *Planta Med.* **1999**, *65*, 383–385. [CrossRef]

67. Saleem, R.; Ahmad, S.I.; Ahmed, M.; Faizi, Z.; Zikr-ur-Rehman, S.; Ali, M.; Faizi, S. Hypotensive activity and toxicology of constituents from *Bombax ceiba* stem bark. *Biol. Pharm. Bull.* **2003**, *26*, 41–46. [CrossRef]

68. Chauhan, J.; Kumar, S.; Chaturvedi, R. A new flavanone glycoside from the roots of *Adansonia digitata*. *Natl. Acad. Sci. Lett. India* **1987**, *10*, 177–179.

69. Paula, V.F.; Barbosa, L.C.; Howarth, O.W.; Demuner, A.J.; Cass, Q.B.; Vieira, I.J. Lignans from *Ochroma lagopus* Swartz. *Tetrahedron* **1995**, *51*, 12453–12462. [CrossRef]

70. Wu, J.; Zhang, X.-H.; Zhang, S.-W.; Xuan, L.-J. Three Novel Compounds from the Flowers of *Bombax malabaricum*. *Helv. Chim. Acta* **2008**, *91*, 136–143. [CrossRef]

71. Wang, G.K.; Lin, B.B.; Rao, R.; Zhu, K.; Qin, X.Y.; Xie, G.Y.; Qin, M.J. A new lignan with anti-HBV activity from the roots of *Bombax ceiba*. *Nat. Prod. Res.* **2013**, *27*, 1348–1352. [CrossRef] [PubMed]

72. Paula, V.F.; Rocha, M.E.; Barbosa, L.C.d.A.; Howarth, O.W. Aquatidial, a new bis-Norsesquiterpenoid from *Pachira aquatica* Aubl. *J. Braz. Chem. Soc.* **2006**, *17*, 1443–1446. [CrossRef]

73. Lam, S.-H.; Chen, J.-M.; Tsai, S.-F.; Lee, S.-S. Chemical investigation on the root bark of *Bombax malabarica*. *Fitoterapia* **2019**, *139*, 104376. [CrossRef]

74. Zhang, X.; Zhu, H.; Zhang, S.; Yu, Q.; Xuan, L. Sesquiterpenoids from *Bombax malabaricum*. *J. Nat. Prod.* **2007**, *70*, 1526–1528. [CrossRef]

75. Sankaram, A.V.B.; Reddy, N.S.; Shoolery, J.N. New sesquiterpenoids of *Bombax malabaricum*. *Phytochemistry* **1981**, *20*, 1877–1881. [CrossRef]

76. Puckhaber, L.; Stipanovic, R. Revised structure for a sesquiterpene lactone from *Bombax malbaricum*. *J. Nat. Prod.* **2001**, *64*, 260–261. [CrossRef]

77. Rao, K.V.; Sreeramulu, K.; Gunasekar, D.; Ramesh, D. Two new sesquiterpene lactones from *Ceiba pentandra*. *J. Nat. Prod.* **1993**, *56*, 2041–2045. [CrossRef]

78. Sood, R.P.; Suri, K.A.; Suri, O.P.; Dhar, K.L.; Atal, C.K. Sesquiterpene lactone from *Salmalia malbarica*. *Phytochemistry* **1982**, *21*, 2125–2126. [CrossRef]

79. Bianchini, J.-P.; Ralaimanarivo, A.; Gaydou, E.M.; Waegell, B. Hydrocarbons, sterols and tocopherols in the seeds of six *Adansonia* species. *Phytochemistry* **1982**, *21*, 1981–1987. [CrossRef]

80. Tundis, R.; Rashed, K.; Said, A.; Menichini, F.; Loizzo, M.R. In vitro cancer cell growth inhibition and antioxidant activity of *Bombax ceiba* (Bombacaceae) flower extracts. *Nat. Prod. Commun.* **2014**, *9*, 691–694. [CrossRef] [PubMed]

81. Dhar, D.; Munjal, R. Chemical examination of the seeds of *Bombax malabaricum*. *Planta Med.* **1976**, *29*, 148–150. [CrossRef] [PubMed]

82. Sipra, D.; Dan, S.S. Phytochemical study of Adansonia digitata, Coccoloba excoriata, Psychotria adenophylla and Schleichera oleosa. *Fitoterapia* **1986**, *57*, 445–446.

83. Matsuda, M.; Endo, Y.; Fushiya, S.; Endo, T.; Nozoe, S. Cytotoxic 6-substituted 5, 6-dihydro-2H-pyran-2-ones from a Brazilian medicinal plant, *Chorisia crispiflora*. *Heterocycles* **1994**, *38*, 1229–1232. [CrossRef]

84. Vijaya Bhaskar Reddy, M.; Kesava Reddy, M.; Gunasekar, D.; Marthanda Murthy, M.; Caux, C.; Bodo, B. A new sesquiterpene lactone from *Bombax malabaricum*. *Chem. Pharm. Bull.* **2003**, *51*, 458–459. [CrossRef] [PubMed]

85. Bell, A.A.; Stipanovic, R.D.; O'Brien, D.H.; Fryxell, P.A. Sesquiterpenoid aldehyde quinones and derivatives in pigment glands of *Gossypium*. *Phytochemistry* **1978**, *17*, 1297–1305. [CrossRef]

86. Kishore, P.H.; Reddy, M.V.B.; Gunasekar, D.; Caux, C.; Bodo, B. A new naphthoquinone from *Ceiba pentandra*. *J. Asian Nat. Prod. Res.* **2003**, *5*, 227–230. [CrossRef] [PubMed]

87. Shibatani, M.; Hashidoko, Y.; Tahara, S. A Major fungitoxin from *Pachira aquatica* and its accumulation in outer bark. *J. Chem. Ecol.* **1999**, *25*, 347–353. [CrossRef]

88. Li, X.-N.; Sun, J.; Shi, H.; Yu, L.L.; Ridge, C.D.; Mazzola, E.P.; Okunji, C.; Iwu, M.M.; Michel, T.K.; Chen, P. Profiling hydroxycinnamic acid glycosides, iridoid glycosides, and phenylethanoid glycosides in baobab fruit pulp (*Adansonia digitata*). *Food Res. Int.* **2017**, *99*, 755–761. [CrossRef]

89. Tsetegho Sokeng, A.J.; Sobolev, A.P.; Di Lorenzo, A.; Xiao, J.; Mannina, L.; Capitani, D.; Daglia, M. Metabolite characterization of powdered fruits and leaves from *Adansonia digitata* L. (baobab): A multi-methodological approach. *Food Chem.* **2019**, *272*, 93–108. [CrossRef]

90. Tembo, D.T.; Holmes, M.J.; Marshall, L.J. Effect of thermal treatment and storage on bioactive compounds, organic acids and antioxidant activity of baobab fruit (*Adansonia digitata*) pulp from Malawi. *J. Food Compos. Anal.* **2017**, *58*, 40–51. [CrossRef]

91. Pettersson, S.; Ervik, F.; Knudsen, J.T. Floral scent of bat-pollinated species: West Africa vs. the New World. *Biol. J. Linn. Soc.* **2004**, *82*, 161–168. [CrossRef]

92. Gaydou, E.; Bianchini, D.J.P.; Ralaimanarivo, A. Cyclopropenoid fatty acids in Malagasy baobab: *Adansonia grandidieri* (Bombacaceae) seed oil. *Fette Seifen Anstrichm.* **1982**, *84*, 468–472. [CrossRef]

93. Ralaimanarivo, A.; Gaydou, E.M.; Bianchini, J.P. Fatty acid composition of seed oils from six *Adansonia* species with particular reference to cyclopropane and cyclopropene acids. *Lipids* **1982**, *17*, 1–10. [CrossRef] [PubMed]

94. Ramesh, D.; Dennis, T.; Shingare, M. Constituents of *Adansonia digitata* root bark. *Fitoterapia* **1992**, *63*, 278–279.

95. Rosselli, S.; Tundis, R.; Bruno, M.; Leporini, M.; Falco, T.; Gagliano Candela, R.; Badalamenti, N.; Loizzo, M.R. *Ceiba speciosa* (A. St.-Hil.) Seeds Oil: Fatty Acids Profiling by GC-MS and NMR and Bioactivity. *Molecules* **2020**, *25*, 1037. [CrossRef] [PubMed]

96. Berry, S.K. The characteristics of the kapok (*Ceiba pentandra*, Gaertn.) seed oil. *Pertanika* **1979**, *2*, 1–4.

97. Ravi Kiran, C.; Raghava Rao, T. Lipid Profiling by GC-MS and Anti-inflammatory Activities of *Ceiba pentandra* Seed Oil. *J. Biol. Act. Prod. Nat.* **2014**, *4*, 62–70.

98. Sunday, A.S.; Gillian, I.O.; John, I.O. Fatty acid composition of seed oil from *Pachira aquatica* Grown in Nigeria. *J. Agric. Ecol. Res. Int.* **2019**, *18*, 1–9. [CrossRef]

99. Krishnan, R.Y.; Chandran, M.N.; Vadivel, V.; Rajan, K. Insights on the influence of microwave irradiation on the extraction of flavonoids from *Terminalia chebula*. *Sep. Purif. Technol.* **2016**, *170*, 224–233. [CrossRef]

100. Wang, X.-H.; Wang, J.-P. Effective extraction with deep eutectic solvents and enrichment by macroporous adsorption resin of flavonoids from *Carthamus tinctorius* L. *J. Pharm. Biomed. Anal.* **2019**, *176*, 112804. [CrossRef] [PubMed]

101. Pérez-Serradilla, J.; Japon-Lujan, R.; de Castro, M.L. Simultaneous microwave-assisted solid–liquid extraction of polar and nonpolar compounds from *Alperujo*. *Anal. Chim. Acta* **2007**, *602*, 82–88. [CrossRef]

102. Buchmann, C.; Prehsler, S.; Hartl, A.; Vogl, C.R. The Importance of Baobab (*Adansonia digitata* L.) in Rural West African Subsistence—Suggestion of a Cautionary Approach to International Market Export of Baobab Fruits. *Ecol. Food Nutr.* **2010**, *49*, 145–172. [CrossRef]

103. Friday, E.T.; James, O.; Olusegun, O.; Gabriel, A. Investigations on the nutritional and medicinal potentials of *Ceiba pentandra* leaf: A common vegetable in Nigeria. *Int. J. Plant Physiol. Biochem.* **2011**, *3*, 95–101.

104. Brown, T. *Tom Brown's Guide to Wild Edible and Medicinal Plants*; Berkley Books: New York, NY, USA, 1986.

105. Paula, V.F.; Barbosa, L.C.A.; Demuner, A.J.; Piló-Veloso, D. A química da família bombacaceae. *Química Nova* **1997**, *20*, 627–630. [CrossRef]

106. Chadare, F.J.; Linnemann, A.; Hounhouigan, D.; Nout, M.J.; Boekel, M. Baobab Food Products: A Review on their Composition and Nutritional Value. *Crit. Rev. Food Sci. Nutr.* **2009**, *49*, 254–274. [CrossRef] [PubMed]

107. Divakar, T.; Rao, Y.H. Biological synthesis of silver nano particles by using *Bombax ceiba* plant. *J. Chem. Pharm. Sci.* **2017**, *10*, 574–576.

108. R Nayagam, D.J. IBA Induced Rooting Characteristics in Marathi Moggu Stem Cuttings: Evaluation Using SVI Concept. *Int. J. Curr. Microbiol. Appl. Sci.* **2016**, *5*, 734–739. [CrossRef]

109. Assogba, G.A.; Fandohan, A.B.; Gandji, K.; Salako, V.K.; Assogbadjo, A.E. *Bombax costatum* (Malvaceae): State of knowns, unknowns and prospects in West Africa. *BASE* **2018**, *22*, 267–275.

110. Van Roosmalen, M.G. *Fruits of the Guianan Flora*; Institute of Systematic Botany, Utrecht University: Utrecht, The Netherlands, 1985.

111. Macbride, J.F. *Flora of Peru*; Field Museum of Natural History: Chicago, IL, USA, 1936; Volume 13.

112. Rosengarten, F. An unusual spice from Oaxaca: The flowers of *Quararibea funebris*. *Bot. Mus. Leafl. Harv. Univ.* **1977**, *25*, 183–202.

113. Vickers, W.T.; Plowman, T. *Useful Plants of the Siona and Secoya Indians of Eastern Ecuador*; Field Museum of Natural History: Chicago, IL, USA, 1984.

114. Tripathi, S.; Farooqui, A.; Singh, V.; Singh, S.; Kumar Roy, R. Morphometrical analysis of *Ceiba* Mill. (Bombacoideae, Malvaceae) pollen: A sacred plant of the Mayan (Mesoamerican) civilisation. *Palynology* **2019**, *43*, 551–573. [CrossRef]

115. Easterling, K.; Harrysson, R.; Gibson, L.; Ashby, M.F. On the mechanics of balsa and other woods. *Proc. R. Soc. Lond. A Math. Phys. Sci.* **1982**, *383*, 31–41.

116. Zheng, Y.; Wang, A. Kapok Fiber: Applications. In *Biomass and Bioenergy: Applications*; Hakeem, K.R., Jawaid, M., Rashid, U., Eds.; Springer: Cham, Switzerland, 2014; pp. 251–266. [CrossRef]

117. Bhardwaj, A.; Rani, S.; Rana, J. Traditionally used common fibre plants in outer siraj area, Himachal Pradesh. *Indian J. Nat. Prod. Resour.* **2014**, *5*, 190–194.

118. Sidibe, M.; Williams, J.T. *Baobab: Adansonia digitata L.*; International Centre for Underutilised Crops: Southampton, UK, 2002; Volume 4.

119. Watt, G. *A Dictionary of the Economic Products of India*; Cambridge University Press: Cambridge, UK, 2014.

120. Sint, K.M.; Adamopoulos, S.; Koch, G.; Hapla, F.; Militz, H. Wood Anatomy and Topochemistry of *Bombax ceiba* L. and *Bombax insigne* Wall. *BioResources* **2013**, *8*, 530–544. [CrossRef]

121. Oyen, L.P.A. Bombax costatum Pellegr. & Vuillet. In *Record from PROTA4U*; PROTA: Wageningen, The Netherlands, 2011.

122. Quattrocchi, U. *CRC World Dictionary of Medicinal and Poisonous Plants: Common Names, Scientific Names, Eponyms, Synonyms, and Etymology*; CRC Press: Boca Raton, FL, USA, 2016.

123. Günter, S.; Weber, M.; Stimm, B.; Mosandl, R. *Silviculture in the Tropics*; Springer: Berlin/Heidelberg, Germany, 2011.

124. Ballesteros, J.L.; Bracco, F.; Cerna, M.; Vita Finzi, P.; Vidari, G. Ethnobotanical Research at the Kutukú Scientific Station, Morona-Santiago, Ecuador. *BioMed Res. Int.* **2016**, *2016*, 9105746. [CrossRef]

125. Weniger, B.; Estrada, A.; Aragon, R.; Deharo, E.; Arango, G.; Lobstein, A.; Anton, R. Ethnopharmacology in the search for new leishmanicidal drugs. *Etudes Chim. Pharmacol.* **2002**, *1*, 391–394.

126. DeWalt, S.J.; Geneviève, B.; Lia, R.C.d.M.; Celin, Q. Ethnobotany of the Tacana: Quantitative Inventories of Two Permanent Plots of Northwestern Bolivia. *Econ. Bot.* **1999**, *53*, 237–260. [CrossRef]

127. Ruffinatto, F.; Crivellaro, A. *Atlas of Macroscopic Wood Identification: With a Special Focus on Timbers Used in Europe and CITES-Listed Species*; Springer International Publishing: Cham, Switzerland, 2019.

128. Lorenzi, H. *Instituto Plantarum de Estudos da Flora. Brazilian Trees: A Guide to the Identification and Cultivation of Brazilian Native Trees*; Instituto Plantarum de Estudos da Flora: Nova Odessa, Brazil, 2002.

129. Jain, V.; Verma, S. Assessment of credibility of some folk medicinal claims on *Bombax ceiba* L. *Indian J. Tradit. Knowl.* **2014**, *13*, 87–94.

130. Elumalai, A.; Mathangi, N.; Didala, A.; Kasarla, R.; Venkatesh, Y. A Review on *Ceiba pentandra* and its medicinal features. *Asian J. Pharma. Technol.* **2012**, *2*, 83–86.

131. Watt, J.M.; Breyer-Brandwijk, M.G. *The Medicinal and Poisonous Plants of Southern and Eastern Africa Being an Account of Their Medicinal and other Uses, Chemical Composition, Pharmacological Effects and Toxicology in Man and Animal*; E & S Livingstone Ltd.: Edinburgh, UK, 1962.

132. Van Wyk, B.E.; Gericke, N. *People's Plants: A Guide to Useful Plants of Southern Africa*; Briza Publications: Pretoria, South Africa, 2000.

133. Adesanya, S.A.; Idowu, T.B.; Elujoba, A.A. Antisickling activity of *Adansonia digitata*. *Planta Med.* **1988**, *54*, 374. [CrossRef]

134. Tapsoba, H.; Deschamps, J.P. Use of medicinal plants for the treatment of oral diseases in Burkina Faso. *J. Ethnopharmacol.* **2006**, *104*, 68–78. [CrossRef]

135. Rosado, F.L. *Manual de los Árboles de Tikal*; Agencia Española de Cooperación Internacional, Programa de Preservación del Patrimonio Cultural en Iberoamérica: Barcelona, Spain, 1996.

136. Rameshwar, V.; Kishor, D.; Tushar, G.; Siddharth, G.; Sudarshan, G. A Pharmacognostic and pharmacological overview on *Bombax ceiba*. *Sch. Acad. J. Pharm.* **2014**, *3*, 100–107.

137. Chaudhary, P.H.; Khadabadi, S.S. *Bombax ceiba* Linn.: Pharmacognosy, ethnobotany and phyto-pharmacology. *Pharmacogn. Commun.* **2012**, *2*, 2–9. [CrossRef]

138. DeFilipps, R.; Maina, S.; Crepin, J. *Medicinal Plants of the Guianas (Guyana, Surinam, French Guiana)*; DC Department of Botany: Washington, DC, USA, 2004.

139. Nsimba, M.M.; Lami, J.N.; Hayakawa, Y.; Yamamoto, T.K. Decreased thrombin activity by a Congolese herbal medicine used in sickle cell anemia. *J. Ethnopharmacol.* **2013**, *148*, 895–900. [CrossRef]

140. Weniger, B.; Robledo, S.; Arango, G.J.; Deharo, E.; Aragón, R.; Muñoz, V.; Callapa, J.; Lobstein, A.; Anton, R. Antiprotozoal activities of Colombian plants. *J. Ethnopharmacol.* **2001**, *78*, 193–200. [CrossRef]

141. Pabon, A.; TV, V.M.; Cardona, F.; Blair, S.; Ramírez, O. Ethnobotany of medicinal plants used to treat malaria by traditional healers from ten indigenous Colombian communities located in Waupes Medio. *Biodivers. Int. J.* **2017**, *1*, 151–167. [CrossRef]

142. Mors, W.B.; Rizzini, C.T.; Pereira, N.A. *Medicinal Plants of Brazil*; Reference Publications Inc.: Algonac, MI, USA, 2000.

143. Standley, P.C.; Steyermark, J.A. *Flora of Guatemala*; Natural History Museum: Chicago, IL, USA, 1946.

144. Ribeiro, R.V.; Bieski, I.G.C.; Balogun, S.O.; Martins, D.T.d.O. Ethnobotanical study of medicinal plants used by Ribeirinhos in the North Araguaia microregion, Mato Grosso, Brazil. *J. Ethnopharmacol.* **2017**, *205*, 69–102. [CrossRef]

145. Moraes, L.L.C.; Freitas, J.L.; Matos Filho, J.R.; Silva, R.B.L.; Borges, C.H.A.; Santos, A.C. Ethno-knowledge of medicinal plants in a community in the eastern Amazon. *Revista Ciências Agrárias* **2019**, *42*, 291–300.

146. Rahul, J.; Jain, M.K.; Singh, S.P.; Kamal, R.K.; Anuradha; Naz, A.; Gupta, A.K.; Mrityunjay, S.K. *Adansonia digitata* L. (baobab): A review of traditional information and taxonomic description. *Asian Pac. J. Trop. Biomed.* **2015**, *5*, 79–84. [CrossRef]

147. Talari, S.; Gundu, C.; Koila, T.; Nanna, R.S. In vitro free radical scavenging activity of different extracts of *Adansonia digitata* L. *Int. J. Environ. Agric. Biotechnol.* **2017**, *2*, 238781. [CrossRef]

148. Ebaid, H.; Bashandy, S.A.; Alhazza, I.M.; Hassan, I.; Al-Tamimi, J. Efficacy of a methanolic extract of *Adansonia digitata* leaf in alleviating hyperglycemia, hyperlipidemia, and oxidative stress of diabetic rats. *BioMed Res. Int.* **2019**, *2019*, 2835152. [CrossRef]

149. Kiyawa, S.A.; Mukhtar, Z.G. Antioxidant and Antipyretic Activities of *Adansonia Digitata* (African Baobab) Fruit, Leaf And Bark Extracts. *Chem. J.* **2019**, *3*, 1–8.

150. Irondi, E.A.; Akintunde, J.K.; Agboola, S.O.; Boligon, A.A.; Athayde, M.L. Blanching influences the phenolics composition, antioxidant activity, and inhibitory effect of *Adansonia digitata* leaves extract on α-amylase, α-glucosidase, and aldose reductase. *Food Sci. Nutr.* **2017**, *5*, 233–242. [CrossRef]

151. Yu, Y.-G.; He, Q.-T.; Yuan, K.; Xiao, X.-L.; Li, X.-F.; Liu, D.-M.; Wu, H. In vitro antioxidant activity of *Bombax malabaricum* flower extracts. *Pharm. Biol.* **2011**, *49*, 569–576. [CrossRef] [PubMed]

152. Jain, V.; Verma, S.; Katewa, S.; Anandjiwala, S.; Singh, B. Free radical scavenging property of *Bombax ceiba* Linn. root. *Res. J. Med. Plant* **2011**, *5*, 462–470. [CrossRef]

153. Rahman, M.H.; Rashid, M.A.; Chowdhury, T.A. Studies of Biological Activities of the Roots of *Bombax ceiba* L. *Bangladesh Pharm. J.* **2019**, *22*, 219–223. [CrossRef]

154. Hossain, S.; Islam, J.; Ahmed, F.; Hossain, M.; Siddiki, M.; Hossen, S. Free radical scavenging activity of six medicinal plants of Bangladesh: A potential source of natural antioxidant. *J. Appl. Pharm.* **2015**, *7*, 96–104. [CrossRef]

155. Hoque, N.; Rahman, S.; Jahan, I.; Shanta, M.A.; Tithi, N.S.; Nasrin, N. A Comparative Phytochemical and Biological Study between Different Solvent Extracts of *Bombax ceiba* Roots Available in Bangladesh. *Pharmacol. Pharm.* **2018**, *9*, 53–66. [CrossRef]

156. Gandhare, B.; Soni, N.; Dhongade, H.J. In vitro antioxidant activity of *Bombax ceiba*. *Int. J. Biomed. Res* **2010**, *1*, 31–36. [CrossRef]

157. Masood Ur, R.; Akhtar, N.; Mustafa, R. Antibacterial And Antioxidant Potential Of Stem Bark Extract Of *Bombax ceiba* Collected Locally From South Punjab Area Of Pakistan. *Afr. J. Tradit. Complementary Altern. Med.* **2017**, *14*, 9–15. [CrossRef]

158. Shyur, L.-F.; Tsung, J.-H.; Chen, J.-H.; Chiu, C.-Y.; Lo, C.-P. Antioxidant properties of extracts from medicinal plants popularly used in Taiwan. *Int. J. Appl. Sci. Eng.* **2005**, *3*, 195–202.

159. Wanjari, M.M.; Gangoria, R.; Dey, Y.N.; Gaidhani, S.N.; Pandey, N.K.; Jadhav, A.D. Hepatoprotective and antioxidant activity of *Bombax ceiba* flowers against carbon tetrachloride-induced hepatotoxicity in rats. *Hepatoma Res.* **2016**, *2*, 144–150. [CrossRef]

160. Sharma, A.; Batish, D.R.; Singh, H.P. In vitro antioxidant potential of the edible flowers of *Bombax ceiba*—An underutilized tropical tree. In Proceedings of the 6th International Conference on Recent Trends in Engineering, Science & Management, Punjab, India, 8 January 2017; pp. 364–374.

161. Rathore, D.; Srivastava, R.; Geetanjali, G.; Singh, R. Phytochemical screening, antioxidant and antibacterial activities of Bombax ceiba flower. *Alger. J. Nat. Prod.* **2019**, *7*, 651–656.

162. Donipati, R.S.; Subhasini, P. Determining the Antioxidant Activity of *Bombax ceiba* Flower Extracts. *Int. J. Pharm. Res. Sch.* **2016**, *5*, 146–150.

163. Komati, A.; Anand, A.; Shaik, H.; Mudiam, M.K.R.; Babu, K.S.; Tiwari, A.K. *Bombax ceiba* (Linn.) calyxes ameliorate methylglyoxal-induced oxidative stress via modulation of RAGE expression: Identification of active phytometabolites by GC-MS analysis. *Food Funct.* **2020**, *11*, 5486–5497. [CrossRef]

164. Kiran, C.R.; Rao, D.B.; Sirisha, N.; Rao, T.R. Assessment of phytochemicals and antioxidant activities of raw and germinating *Ceiba pentandra* (kapok) seeds. *J. Biomed. Res.* **2015**, *29*, 414.

165. Fofie, C.K.; Wansi, S.L.; Nguelefack-Mbuyo, E.P.; Atsamo, A.D.; Watcho, P.; Kamanyi, A.; Nole, T.; Nguelefack, T.B. In vitro anti-hyperglycemic and antioxidant properties of extracts from the stem bark of *Ceiba pentandra*. *J. Complementary Integr. Med.* **2014**, *11*, 185–193. [CrossRef] [PubMed]

166. Fitria, Z.; Efdi, A.; Efdi, M. Isolation and characterization of antioxidative constituent from stem bark extract of *Ceiba pentandra* L. *J. Chem. Pharm. Res.* **2015**, *7*, 257–260.

167. Bhavani, R.; Bhuvaneswari, E.; Rajeshkumar, S. Antibacterial and Antioxidant activity of Ethanolic extract of *Ceiba pentandra* leaves and its Phytochemicals Analysis using GC-MS. *Res. J. Pharm. Technol.* **2016**, *9*, 1922–1926. [CrossRef]

168. Nguelefack, T.B.; Fofie, C.K.; Nguelefack-Mbuyo, E.P.; Wuyt, A.K. Multimodal α-Glucosidase and α-Amylase Inhibition and Antioxidant Effect of the Aqueous and Methanol Extracts from the Trunk Bark of *Ceiba pentandra*. *BioMed Res. Int.* **2020**, *2020*, 3063674. [CrossRef]

169. Datsugwai, M.S.S.; Yusuf, A.S. Phytochemical analysis and antimicrobial activity of baobab (*Adansonia digitata*) leaves and stem bark extracts on *Staphylococcus aureus* and *Escherichia coli*. *J. Biosci. Biotechnol.* **2017**, *6*, 9–16.

170. Osuntokun, O.; AjayiAyodele, A.M. Assessment of Antimicrobial and Phytochemical Properties of Crude Leaf and Bark Extracts of *Ceiba Pentandra* on Selected Clinical Isolates. *JBMOA* **2017**, *4*, 17–23.

171. Parulekar, G. Antibacterial and phytochemical analysis of *Ceiba pentandra* (L.) seed extracts. *J. Pharmacogn. Phytochem.* **2017**, *6*, 586–589.

172. Kadam, S.D.; Kondawar, M. Evaluation of In-Vitro Anticancer Activity and Quantitation of Active Ingredient of *Adansonia digitata* L. Fruit. *Int. J. Pharm. Sci. Drug Res.* **2019**, *11*, 358–369.

173. Kumar, R.; Kumar, N.; Ramalingayya, G.V.; Setty, M.M.; Pai, K.S. Evaluation of *Ceiba pentandra* (L.) Gaertner bark extracts for in vitro cytotoxicity on cancer cells and in vivo antitumor activity in solid and liquid tumor models. *Cytotechnology* **2016**, *68*, 1909–1923. [CrossRef] [PubMed]

174. Mir, M.A.; Mir, B.A.; Bisht, A.; Rao, Z.; Singh, D. Antidiabetic properties and metal analysis of *Bombax ceiba* flower extracts. *Glob. J. Addict. Rehabil. Med.* **2017**, *1*, 1–7. [CrossRef]

175. De, D.; Ali, K.M.; Chatterjee, K.; Bera, T.K.; Ghosh, D. Antihyperglycemic and antihyperlipidemic effects of n-hexane fraction from the hydro-methanolic extract of sepals of *Salmalia malabarica* in streptozotocin-induced diabetic rats. *J. Complement. Integr. Med.* **2012**, *9*, 12. [CrossRef]

176. Sharmin, R.; Joarder, H.; Alamgir, M.; Mostofa, G.; Islam, M. Antidiabetic and hepatoprotective activities of *Bombax ceiba* young roots in alloxan-induced diabetic mice. *J. Nutr. Health Food Sci.* **2018**, *6*, 1–7.

177. Guang-Kai, X.; Xiao-Ying, Q.; Guo-Kai, W.; Guo-Yong, X.; Xu-Sen, L.; Chen-Yu, S.; Bao-Lin, L.; Min-Jian, Q. Antihyperglycemic, antihyperlipidemic and antioxidant effects of standard ethanol extract of *Bombax ceiba* leaves in high-fat-diet-and streptozotocin-induced Type 2 diabetic rats. *Chin. J. Nat. Med.* **2017**, *15*, 168–177.

178. Bhavsar, C.; Talele, G.S. Potential anti-diabetic activity of *Bombax ceiba*. *Bangladesh J. Pharmacol.* **2013**, *8*, 102–106. [CrossRef]

179. Fofié, C.K.; Nguelefack-Mbuyo, E.P.; Tsabang, N.; Kamanyi, A.; Nguelefack, T.B. Hypoglycemic properties of the aqueous extract from the stem bark of *Ceiba pentandra* in dexamethasone-induced insulin resistant rats. *Evid. Based Complement. Altern. Med.* **2018**, *2018*, 4234981. [CrossRef]

180. Satyaprakash, R.; Rajesh, M.; Bhanumathy, M.; Harish, M.; Shivananda, T.; Shivaprasad, H.; Sushma, G. Hypoglycemic and antihyperglycemic effect of *Ceiba pentandra* L. gaertn in normal and streptozotocininduced diabetic rats. *Ghana Med. J.* **2013**, *47*, 121–127.

181. Ayele, Y.; Kim, J.-A.; Park, E.; Kim, Y.-J.; Retta, N.; Dessie, G.; Rhee, S.-K.; Koh, K.; Nam, K.-W.; Kim, H.S. A methanol extract of *Adansonia digitata* L. leaves inhibits pro-inflammatory iNOS possibly via the inhibition of NF-κB activation. *Biomol. Ther.* **2013**, *21*, 146. [CrossRef]

182. Selvarani, V. Multiple inflammatory and antiviral activities in *Adansonia digitata* (Baobab) leaves, fruits and seeds. *J. Med. Plants Res.* **2009**, *3*, 576–582.

183. Anandarajagopal, K.; Sunilson, J.A.J.; Ajaykumar, T.; Ananth, R.; Kamal, S. In-vitro anti-inflammatory evaluation of crude *Bombax ceiba* extracts. *Eur. J. Med. Plants* **2013**, *3*, 99–104. [CrossRef]

184. Abouelela, M.E.; Orabi, M.A.; Abdelhamid, R.A.; Abdelkader, M.S.; Madkor, H.R.; Darwish, F.M.; Hatano, T.; Elsadek, B.E. Ethyl acetate extract of *Ceiba pentandra* (L.) Gaertn. reduces methotrexate-induced renal damage in rats via antioxidant, anti-inflammatory, and antiapoptotic actions. *J. Tradit. Complement. Med.* **2019**, *10*, 478–486. [CrossRef]

185. Basipogu, D.; Basha Syed, N.; Chitta, S.K. Cytotoxic effect of *Adansonia digitata* L. on breast cancer cell lines. *Eur. J. Biomed. Pharm. Sci.* **2018**, *5*, 956–959.

186. Sa'id, A.; Musa, A.; Mashi, J.; Maigari, F.; Nuhu, M. Phytochemical Screening and Hepatoprotective Potential of Aqueous Fruit Pulp Extract of *Adansonia digitata* against CCL4 Induced Liver Damage in Rats. *Asian J. Biochem. Genet. Mol. Biol.* **2020**, 12–21. [CrossRef]

187. Mohamed, M.A.; Mohamed, I.; El-Din, T.T.; Hassan, A.; Hassan, M. Hepatoprotective effect of *Adansonia digitata* L.(Baobab) fruits pulp extract on CCl4-induced hepatotoxicity in rats. *World J. Pharm. Res.* **2015**, *4*, 368–377.

188. Al-Qarawi, A.; Al-Damegh, M.; El-Mougy, S. Hepatoprotective influence of *Adansonia digitata* pulp. *J. Herbs Spices Med. Plants* **2003**, *10*, 1–6. [CrossRef]

189. Badr-Eldin, S.M.; Aldawsari, H.M.; Hanafy, A.; Abdlmoneim, M.S.; Abdel-Hady, S.E.-S.; Hasan, A. Enhanced Hepatoprotective Effect of *Adansonia Digitata* Extract on Paracetamol-Induced Hepatotoxicity. *Res. J. Pharm. Biol. Chem. Sci.* **2017**, *8*, 787–794.

190. Lin, C.-c.; Chen, S.-y.; Lin, J.-m.; Chiu, H.-f. The Pharmacological and Pathological Studies on Taiwan Folk Medicine (Vlll): The Anti-inflammatory and Liver Protective Effects of Mu-mien. *Am. J. Chin. Med.* **1992**, *20*, 135–146. [CrossRef]
191. Bairwa, N.K.; Sethiya, N.K.; Mishra, S. Protective effect of stem bark of *Ceiba pentandra* linn. against paracetamol-induced hepatotoxicity in rats. *Pharmacogn. Res.* **2010**, *2*, 26–30. [CrossRef]
192. Vieira, T.O.; Said, A.; Aboutabl, E.; Azzam, M.; Creczynski-Pasa, T.B. Antioxidant activity of methanolic extract of *Bombax ceiba*. *Redox Rep.* **2009**, *14*, 41–46. [CrossRef] [PubMed]
193. Ch, R.K.; Madhavi, Y.; Raghava Rao, T. Evaluation of phytochemicals and antioxidant activities of *Ceiba pentandra* (kapok) seed oil. *J. Bioanal. Biomed.* **2012**, *4*, 068–073.
194. Jalalpure, S.; Gadge, N. Diuretic effects of young fruit extracts of *Bombax ceiba* L. in rats. *Indian J. Pharm. Sci.* **2011**, *73*, 306. [PubMed]
195. Burt, S. Essential oils: Their antibacterial properties and potential applications in foods—A review. *Int. J. Food Microbiol.* **2004**, *94*, 223–253. [CrossRef]
196. Silva, N.C.C.; Fernandes Júnior, A. Biological properties of medicinal plants: A review of their antimicrobial activity. *J. Venom. Anim. Toxins Incl. Trop. Dis.* **2010**, *16*, 402–413. [CrossRef]
197. Iqbal, J.; Abbasi, B.A.; Mahmood, T.; Kanwal, S.; Ali, B.; Shah, S.A.; Khalil, A.T. Plant-derived anticancer agents: A green anticancer approach. *Asian Pac. J. Trop. Biomed.* **2017**, *7*, 1129–1150. [CrossRef]
198. Kumaran, A. Antioxidant and free radical scavenging activity of an aqueous extract of *Coleus aromaticus*. *Food Chem.* **2006**, *97*, 109–114. [CrossRef]
199. Sarkar, D. The anti-inflammatory activity of plant derived ingredients: An analytical review. *Int. J. Pharm. Sci. Res.* **2020**, *11*, 496–506. [CrossRef]
200. Das, S.K.; Samantaray, D.; Patra, J.K.; Samanta, L.; Thatoi, H. Antidiabetic potential of mangrove plants: A review. *Front. Life Sci.* **2016**, *9*, 75–88. [CrossRef]

MDPI
St. Alban-Anlage 66
4052 Basel
Switzerland
Tel. +41 61 683 77 34
Fax +41 61 302 89 18
www.mdpi.com

Plants Editorial Office
E-mail: plants@mdpi.com
www.mdpi.com/journal/plants